Advances in Computer Vision and Pattern Recognition

For further volumes:
http://www.springer.com/series/4205

Shaogang Gong · Marco Cristani
Shuicheng Yan · Chen Change Loy
Editors

Person Re-Identification

Springer

Editors
Shaogang Gong
Queen Mary University
London
UK

Marco Cristani
University of Verona
Verona
Italy

Shuicheng Yan
National University of Singapore
Singapore

Chen Change Loy
The Chinese University of Hong Kong
Shatin
Hong Kong SAR

Series editors
Sameer Singh
Rail Vision Europe Ltd.
Castle Donington
Leicestershire, UK

Sing Bing Kang
Interactive Visual Media Group
Microsoft Research
Redmond, WA, USA

ISSN 2191-6586 ISSN 2191-6594 (electronic)
Advances in Computer Vision and Pattern Recognition
ISBN 978-1-4471-6295-7 ISBN 978-1-4471-6296-4 (eBook)
DOI 10.1007/978-1-4471-6296-4
Springer London Heidelberg New York Dordrecht

Library of Congress Control Number: 2013957125

Printed on acid-free paper

Springer is part of Springer Science+Business Media (www.springer.com)

Preface

Person re-identification is the problem of recognising and associating a person at different physical locations over time after the person had been previously observed visually elsewhere. Solving the re-identification problem has gained a rapid increase in attention in both academic research communities and industrial laboratories in recent years. The problem has many manifestations from different application domains. For instance, the problem is known as "re-acquisition" when the aim is to associate a target (person) when it is temporarily occluded during the tracking in a single camera view. On the other hand, in domotics applications or personalised healthcare environments, the primary aim is to retain the identity of a person whilst one is moving about in a private home of distributed spaces, e.g. crossing multiple rooms. Re-identification can provide a useful tool for validating the identity of impaired or elderly people in a seamless way without the need for more invasive biometric verification procedures, e.g. controlled face or fingerprint recognition. Moreover, in a human–robot interaction scenario, solving the re-identification problem can be considered as "non-cooperative target recognition", where the identity of the interlocutor is maintained, allowing the robot to be continuously aware of the surrounding people. In larger distributed spaces such as airport terminals and shopping malls, re-identification is mostly considered as the task of "object association" in a distributed multi-camera network, where the goal is to keep track of an individual across different cameras with non-overlapping field of views. For instance, in a multi-camera surveillance system, re-identification is needed to trace the inter-camera whereabouts of individuals of interest (a watch-list), or simply to understand how people move in complex environments such as an airport and a train station for better crowd traffic management and crowding control. In a retail environment, re-identification can provide useful information for improving customer service and shopping space management. In a more general setting of online shopping, re-identification of visual objects of different categories, e.g. clothing, can help in tagging automatically huge volumes of visual samples of consumer goods in Internet image indexing, search and retrieval.

Solving the person re-identification problem poses a considerable challenge that requires visually detecting and recognising a person (subject) at different space time locations observed under substantially different, and often unknown, viewing conditions without subject collaboration. Early published work on re-identification can date back a decade ago to 2003, but most contemporary techniques have been developed since 2008, and in particular in the last 2–3 years. In the past 5 years, there has been a tremendous increase in computer vision research on solving the re-identification problem, evident from a large number of academic papers published in all the major conferences (ICCV, CVPR, ECCV, BMVC, ICIP) and journals (TPAMI, IJCV, Pattern Recognition). This trend will increase further in the coming years, given that many open problems remain unsolved.

Inspired by the First International Workshop on Re-Identification held at Florence in Italy in October 2012, this book is a collection of invited chapters from some of the world's leading researchers working on solving the re-identification problem. It aims to provide a comprehensive and in-depth presentation of recent progress and the current state-of-the-art approaches to solving some of the fundamental challenges in person re-identification, benefiting from wider research in the computer vision, pattern recognition and machine learning communities, and drawing insights from video analytics system design considerations for engineering practical solutions. Due to its diverse nature, the development of person re-identification methods by visual matching has been reported in a wide range of fields, from multimedia to robotics, from domotics to visual surveillance, but all with an underlying computer vision theme. Re-identification exploits extensively many core computer vision techniques that aim at extracting and representing an individual's visual appearance in a scene, e.g. pedestrian detection and tracking, and object representation; and machine learning techniques for discriminative matching, e.g. distance metric learning and transfer learning. Moreover, solving the person re-identification problem can benefit from exploiting heterogeneous information by learning more effective semantic attributes, exploiting spatio-temporal statistics, estimating feature transformation across different cameras, taking into account soft-biometric cues (e.g. height, gender) and considering contextual cues (e.g. baggage, other people nearby).

This book is the first dedicated treatment on the subject of *Person Re-Identification* that aims to address a highly focused problem with a strong multidisciplinary appeal to practitioners in both fundamental research and practical applications. In the context of video content analysis, visual surveillance and human recognition, there are a number of other books published recently that aim to address a wider range of topics, e.g. *Video Analytics for Business Intelligence,* by Caifeng Shan, Fatih Porikli, Tao Xiang and Shaogang Gong (2012); *Visual Analysis of Behaviour: From Pixels to Semantics,* by Shaogang Gong and Tao Xiang (2011); and *Visual Analysis of Humans: Looking at People*, by Thomas Moeslund, Adrian Hilton, Volker Kruger and Leonid Sigal (2011). In contrast to those other books, this book provides a more in-depth analysis and a more comprehensive presentation of techniques required specifically for solving the problem of person re-identification. Despite aiming to address a highly focused problem, the techniques presented in this book, e.g. feature representation, attribute learning, ranking, active learning and transfer learning, are highly applicable to other more general problems in computer vision, pattern recognition and machine learning. Therefore, the book should also be of considerable interest to a wider audience.

We anticipate that this book will be of special interest to academics, postgraduates and industrial researchers specialised in computer vision and machine learning, database (including internet) image retrieval, big data mining and search engines. It should also be of interest to commercial developers and managers keen to exploit this emerging technology for a host of applications including security and surveillance, personalised healthcare, commercial information profiling, business intelligence gathering, smart city, public space infrastructure management, consumer electronics and retails. Finally, this book will also be of use to postgraduate students of computer science, engineering, applied mathematics and statistics, cognitive and social studies.

London, October 2013 — Shaogang Gong
Verona — Marco Cristani
Singapore — Shuicheng Yan
Hong Kong — Chen Change Loy

Acknowledgments

The preparation of this book has required the dedication of many people. First of all, we thank all the contributing authors for their extraordinary effort and dedication in preparing the book chapters within a very tight time frame. Second, we express our gratitude to all the reviewers. Their critical and constructive feedback helped in improving the quality of the book. Finally, we thank Simon Rees and Wayne Wheeler at Springer for their support throughout the preparation of this book. The book was typeset using LaTeX.

This book was inspired by the First International Workshop on Re-Identification (Re-Id 2012), in conjunction with the European Conference on Computer Vision, held at Florence in Italy in October 2012. To that end, we thank the workshop programme committee and the authors who made the workshop a huge success. We also thank the workshop industrial sponsors Bosch, KAI Square, Vision Semantics and Embedded Vision Systems who sponsored the Best Paper Award prize and made the workshop a more rewarding experience.

Contents

Contributors

Tamar Avraham Technion Israel Institute of Technology, Haifa, Israel, e-mail: tammya@cs.technion.ac.il

Andrew D. Bagdanov University of Florence, Florence, Italy, e-mail: bagdanov@dsi.unifi.it

Sławomir Bąk INRIA, Sophia Antipolis, France, e-mail: slawomir.bak@inria.fr

Alberto Basso University of Padua, Padua, Italy, e-mail: bassoal1@dei.unipd.it

Loris Bazzani Istituto Italiano di Tecnologia, Genova, Italy, e-mail: loris.bazzani@iit.it

Csaba Beleznai Austrian Institute of Technology, Vienna, Austria, e-mail: Csaba.Beleznai@ait.ac.at

Alberto Del Bimbo University of Florence, Florence, Italy, e-mail: delbimbo@dsi.unifi.it

Horst Bischof Graz University of Technology, Graz, Austria, e-mail: bischof@icg.tugraz.at

François Brémond INRIA, Sophia Antipolis, France, e-mail: francois.bremond@inria.fr

Rama Chellappa University of Maryland, College Park, USA, e-mail: rama@umiacs.umd.edu

Dong Seon Cheng Hankuk University of Foreign Studies, Seoul, Korea, e-mail: cheng_ds@hufs.ac.kr

Marco Cristani University of Verona, Verona, Italy, e-mail: marco.cristani@univr.it

Rita Cucchiara University of Modena and Reggio Emilia, Modena, Italy, e-mail: rita.cucchiara@unimore.it

Henry Detmold Snap Network Surveillance, Adelaide, Australia, e-mail: henry@snapcontrol.com

Anthony Dick University of Adelaide, Adelaide, Australia, e-mail: anthony.dick@adelaide.edu.au

François Fleuret IDIAP, Martigny, Switzerland, e-mail: francois.fleuret@idiap.ch

Andrea Fossati ETH Zurich, Zurich, Switzerland, e-mail: fossati@vision.ee.ethz.ch

Pascal Fua EPFL, Lausanne, Switzerland, e-mail: pascal.fua@epfl.ch

Giorgio Fumera University of Cagliari, Cagliari, Italy, e-mail: fumera@diee.unica.it

Shaogang Gong Queen Mary University of London, London, UK, e-mail: sgg@eecs.qmul.ac.uk

Martin Hirzer Graz University of Technology, Graz, Austria, e-mail: hirzer@icg.tugraz.at

Timothy M. Hospedales Queen Mary University of London, London, UK, e-mail: tmh@eecs.qmul.ac.uk

Frédéric Jurie University of Caen Basse-Normandie, Caen, France, e-mail: frederic.jurie@unicaen.fr

Svebor Karaman University of Florence, Florence, Italy, e-mail: svebor.karaman@unifi.it

Martin Köstinger Graz University of Technology, Graz, Austria, e-mail: koestinger@icg.tugraz.at

Ryan Layne Queen Mary University of London, London, UK, e-mail: rlayne@eecs.qmul.ac.uk

Annan Li National University of Singapore, Singapore, Singapore, e-mail: elela@nus.edu.sg

Xinggang Lin Tsinghua University, Beijing, China, e-mail: xglin@mail.tsinghua.edu.cn

Michael Lindenbaum Technion Israel Institute of Technology, Haifa, Israel, e-mail: mic@cs.technion.ac.il

Giuseppe Lisanti University of Florence, Florence, Italy, e-mail: lisanti@dsi.unifi.it

Chunxiao Liu Tsinghua University, Beijing, China, e-mail: lcx08@mails.tsinghua.edu.cn

Luoqi Liu National University of Singapore, Singapore, Singapore, e-mail: liuluoqi@nus.edu.sg

Chen Change Loy The Chinese University of Hong Kong, Shatin, Hong Kong, e-mail: ccloy@ie.cuhk.edu.hk

Bingpeng Ma University of Chinese Academy of Sciences, Beijing, China, e-mail: bpma@ucas.ac.cn

Emanuele Menegatti University of Padua, Padua, Italy, e-mail: emg@dei.unipd.it

Matteo Munaro University of Padua, Padua, Italy, e-mail: munaro@dei.unipd.it

Vittorio Murino Istituto Italiano di Tecnologia, Genova, Italy, e-mail: vittorio.murino@iit.it

Jie Ni University of Maryland, College Park, USA, e-mail: jni@umiacs.umd.edu

Federico Pala University of Cagliari, Cagliari, Italy, e-mail: federico.pala@diee.unica.it

Qiang Qiu Duke University, Durham, USA, e-mail: qiang.qiu@duke.edu

Yogesh Raja Vision Semantics Ltd, London, UK, e-mail: yraja@visionsemantics.com

Fabio Roli University of Cagliari, Cagliari, Italy, e-mail: roli@diee.unica.it

Peter M. Roth Graz University of Technology, Graz, Austria, e-mail: pmroth@icg.tugraz.at

Riccardo Satta European Commission JRC Institute for the Protection and Security of the Citizen, Ispra, Italy, e-mail: Riccardo.Satta@jrc.ec.europa.eu

Horesh Ben Shitrit EPFL, Lausanne, Switzerland, e-mail: horesh.benshitrit@epfl.ch

Yu Su University of Caen Basse-Normandie, Caen, France, e-mail: yu.su@unicaen.fr

Luc Van Gool ETH Zurich, Zurich, Switzerland, e-mail: vangool@vision.ee.ethz.ch

Anton van den Hengel University of Adelaide, Adelaide, Australia, e-mail: anton.vandenhengel@adelaide.edu.au

Roberto Vezzani University of Modena and Reggio Emilia, Modena, Italy, e-mail: roberto.vezzani@unimore.it

Xiaogang Wang The Chinese University of Hong Kong, Shatin, Hong Kong, e-mail: xgwang@ee.cuhk.edu.hk

Tao Xiang Queen Mary University of London, London, UK, e-mail: txiang@eecs.qmul.ac.uk

Shuicheng Yan National University of Singapore, Singapore, Singapore, e-mail: eleyans@nus.edu.sg

Rui Zhao The Chinese University of Hong Kong, Shatin, Hong Kong, e-mail: rzhao@ee.cuhk.edu.hk

Wei-Shi Zheng Sun Yat-sen University, Guangzhou, China, e-mail: wszheng@ieee.org

Chapter 1
The Re-identification Challenge

Shaogang Gong, Marco Cristani, Chen Change Loy and Timothy M. Hospedales

Abstract For making sense of the vast quantity of visual data generated by the rapid expansion of large-scale distributed multi-camera systems, automated person re-identification is essential. However, it poses a significant challenge to computer vision systems. Fundamentally, person re-identification requires to solve two difficult problems of '*finding needles in haystacks*' and '*connecting the dots*' by identifying instances and associating the whereabouts of targeted people travelling across large distributed space–time locations in often crowded environments. This capability would enable the discovery of, and reasoning about, individual-specific long-term structured activities and behaviours. Whilst solving the person re-identification problem is inherently challenging, it also promises enormous potential for a wide range of practical applications, ranging from security and surveillance to retail and health care. As a result, the field has drawn growing and wide interest from academic researchers and industrial developers. This chapter introduces the re-identification problem, highlights the difficulties in building person re-identification systems, and presents an overview of recent progress and the state-of-the-art approaches to solving some of the fundamental challenges in person re-identification, benefiting from research in computer vision, pattern recognition and machine learning, and drawing insights from video analytics system design considerations for engineering practical solutions. It also provides an introduction of the contributing chapters of this book.

S. Gong (✉) · T. M. Hospedales
Queen Mary University of London, London, UK
e-mail: sgg@eecs.qmul.ac.uk

M. Cristani
University of Verona and Istituto Italiano di Tecnologia, Verona, Italy
e-mail: marco.cristani@univr.it

C. C. Loy
The Chinese University of Hong Kong, Shatin, Hong Kong
e-mail: ccloy@ie.cuhk.edu.hk

T. M. Hospedales
e-mail: tmh@eecs.qmul.ac.uk

S. Gong et al. (eds.), *Person Re-Identification*,
Advances in Computer Vision and Pattern Recognition,
DOI: 10.1007/978-1-4471-6296-4_1, © Springer-Verlag London 2014

The chapter ends by posing some open questions for the re-identification challenge arising from emerging and future applications.

1.1 Introduction

A fundamental task for a distributed multi-camera surveillance system is to associate people across camera views at different locations and time. This is known as the person re-identification (re-id) problem, and it underpins many crucial applications such as long-term multi-camera tracking and forensic search. More specifically, re-identification of an individual or a group of people collectively is the task of visually matching a single person or a group in diverse scenes, obtained from different cameras distributed over non-overlapping scenes (physical locations) of potentially substantial distances and time differences. In particular, for surveillance applications performed over space and time, an individual disappearing from one view would need to be matched in one or more other views at different physical locations over a period of time, and be differentiated from numerous visually similar but different candidates in those views. Potentially, each view may be taken from a different angle and distance, featuring different static and dynamic backgrounds under different lighting conditions, degrees of occlusion and other view-specific variables. A re-identification computer system aims to automatically match and track individuals either retrospectively or on-the-fly when they move across different locations.

Relying on human operator manual re-identification in large camera networks is prohibitively costly and inaccurate. Operators are often assigned more cameras than they can feasibly monitor simultaneously, and even within a single camera, manual matching is vulnerable to inevitable attentional gaps [1]. Moreover, baseline human performance is determined by the individual operator's experience amongst other factors. It is difficult to transfer this expertise directly between operators, and it is difficult to obtain consistent performance due to operator bias [2]. As public space camera networks have grown quickly in recent years, it is becoming increasingly clear that manual re-identification is not scalable. There is therefore a growing interest within the computer vision community in developing automated re-identification solutions.

In a crowded and uncontrolled environment observed by cameras from an unknown distance, person re-identification relying upon conventional biometrics such as face recognition is neither feasible nor reliable due to insufficiently constrained conditions and insufficient image detail for extracting robust biometrics. Instead, visual features based on the appearance of people, determined by their clothing and objects carried or associated with them, can be exploited more reliably for re-identification. However, visual appearance is intrinsically weak for matching people. For instance, most people in public spaces wear dark clothes in winter, so most colour pixels are not informative about identity in a unique way. To further compound the problem, a person's appearance can change significantly between different camera views if large changes occur in view angle, lighting, background clutter and occlusion. This results in different people often appearing more alike than the same person across different camera views. That is, intra-class variability can be, and is often, significantly larger than inter-class variability when camera view changes are involved. Current research efforts for solving the re-identification problem have primarily focused on two aspects:

1. Developing feature representations which are discriminative for identity, yet invariant to view angle and lighting [3–5];
2. Developing machine learning methods to discriminatively optimise parameters of a re-identification model [6]; and with some studies further attempting to bridge the gap by learning an effective class of features from data [7, 8].

Nevertheless, achieving automated re-identification remains a significant challenge due to the inherent limitation that most visual features generated from people's visual appearance are either insufficiently discriminative for cross-view matching, especially with low resolution images, or insufficiently robust to viewing condition changes, and under extreme circumstances, totally unreliable if clothing is changed substantially.

Sustained research on addressing the re-identification challenge benefits other computer vision domains beyond visual surveillance. For instance, feature descriptor design in re-identification can be exploited to enhance tracking [9] and identification of people (e.g. players in sport videos) from medium to far distance; the metric learning and ranking approaches developed for re-identification can be adapted for face verification and content-based image analysis in general. Research efforts in re-identification also contribute to the development of various machine learning topics, e.g. similarity and distance metric learning, ranking and preference learning, sparsity and feature selection, and transfer learning.

This chapter is organised as follows. We introduce the typical processing steps of re-id in Sect. 1.2. In Sect. 1.3, we highlight the challenges commonly encountered in formulating a person re-identification framework. In particular, we discuss challenges related to feature construction, model design, evaluation and system implementation. In Sect. 1.4, we review the most recent developments in person re-identification, introduce the contributing chapters of this book and place them in context. Finally in Sect. 1.5, we discuss a few possible new directions and open questions to be solved in order to meet the re-identification challenge in emerging and future real-world applications.

1.2 Re-identification Pipeline

Human investigators tasked with the forensic analysis of video from multi-camera CCTV networks face many challenges, including data overload from large numbers of cameras, limited attention span leading to important events and targets being missed, a lack of contextual knowledge indicating what to look for, and limited ability or inability to utilise complementary non-visual sources of knowledge to assist the search process. Consequently, there is a distinct need for a technology to alleviate the burden placed on limited human resources and augment human capabilities.

An automated re-identification mechanism takes as input either tracks or bounding-boxes containing segmented images of individual persons, as generated by a localised tracking or detection process of a visual surveillance system. To automatically match people at different locations over time captured by different camera views, a re-identification process typically takes the following steps:

1. Extracting imagery features that are more reliable, robust and concise than raw pixel data;
2. Constructing a descriptor or representation, e.g. a histogram of features, capable of both describing and discriminating individuals; and
3. Matching specified probe images or tracks against a gallery of persons in another camera view by measuring the similarity between the images, or using some model-based matching procedure. A training stage to optimise the matching parameters may or may not be required depending on the matching strategy.

Such processing steps raise certain demands on algorithm and system design. This has led to both the development of new and the exploitation of existing computer vision techniques for addressing the problems of feature representation, model matching and inference in context.

Representation: Contemporary approaches to re-identification typically exploit low-level features such as colour [10], texture, spatial structure [5] or combinations thereof [4, 11, 12]. This is because these features can be relatively easily and reliably measured, and provide a reasonable level of inter-person discrimination together with inter-camera invariance. Such features are further encoded into fixed-length person descriptors, e.g. in the form of histograms [4], covariances [13] or fisher vectors [14].

Matching: Once a suitable representation has been obtained, nearest-neighbour [5] or model-based matching algorithms such as support-vector ranking [4] may be used for re-identification. In each case, a distance metric (e.g. Euclidean or Bhattacharyya) must be chosen to measure the similarity between two samples. Model-based matching approaches [15, 16] and nearest-neighbor distance metrics [6, 17] can both be discriminatively optimised to maximise re-identification performance given annotated training data of person images. Bridging these two stages, some studies [7, 8, 18] have also attempted to learn discriminative low-level features directly from data.

Context: Other complementary aspects of the re-identification problem have also been pursued to improve performance, such as improving robustness by combining multiple frames worth of features along a trajectory tracklet [9, 12], set-based analysis [19, 20], considering external context such as groups of persons [21], and learning the topology of camera networks [22, 23] in order to reduce the matching search space and hence reduce false-positives.

1.2.1 A Taxonomy of Methods

Different approaches (as illustrated in different chapters of this book) use slightly different taxonomies in categorising existing person re-identification methods.

In general, when only an image pair is matched, the method is considered as a *single-shot recognition* method. If matching is conducted between two sets of images, e.g. frames obtained from two separate trajectories, the method is known as a *multi-shot recognition* approach. An approach is categorised as a *supervised* method if prior to application, and it exploits labelled samples for tuning model parameters such as distance metrics, feature weight or decision boundaries. Otherwise a method is regarded as an *unsupervised* approach if it concerns the extraction of robust visual features and does not rely on training data. Blurring these boundaries somewhat are methods which do learn from training data prior to deployment, but do not rely on annotation for these data.

1.3 The Challenge

1.3.1 Feature Representation

Designing suitable feature representation for person re-identification is a critical and challenging problem. Ideally, the features extracted should be robust to changes in illumination, viewpoint, background clutter, occlusion and image quality/resolution. In the context of re-id, however, it is unclear whether there exists universally important and salient features that can be applied readily to different camera views and for all individuals. The discriminative power, reliability and computability of features are largely governed by the camera-pair viewing conditions and unique appearance characteristics of different persons captured in the given views. Moreover, the difficulty in obtaining an aligned bounding box, and accurately segmenting a person from cluttered background makes extracting pure and reliable features depicting the person of interest even harder.

1.3.2 Model and System Design

There are a variety of challenges that arise during model and system design:

1. *Inter- and Intra-class variations*: A fundamental challenge in constructing a re-id model is to overcome the inter-class confusion, i.e. different persons can look alike across camera views; and intra-class variation, i.e. the same individual may look different when observed under different camera views. Such variations between camera view pairs are in general complex and multi-modal, and therefore are necessarily non-trivial for a model to learn.
2. *Small sample size*: In general a re-id module may be required to match single probe images to single gallery images. This means from a conventional classification perspective, there is likely to be insufficient data to learn a good model of each person's intra-class variability. 'One-shot' learning may be required under

which only a single pair of examples is available for model learning. For this reason, many frameworks treat re-id as a pairwise binary classification (same vs. different) problem [4, 16] instead of a conventional multi-class classification problem.

3. *Data labelling requirement*: For exploiting a supervised learning strategy to train a good model robust to cross-camera view variations, persons from each view annotated with identity or binary labels depicting same versus different are required. Consequently, models which can be learned with less training data are preferred since for a large camera network, collecting extensive labelled data from every camera would be prohibitively expensive.
4. *Generalisation capability*: This is the flip side of training data scalability. Once trained for a specific pair of cameras, most models do not generalise well to another pair of cameras with different viewing conditions [24]. In general, one seeks for a model with good generalisation ability that can be trained once and then applied to a variety of different camera configurations from different locations. This would sidestep the issue of training data scalability.
5. *Scalability*: Given a topologically complex and large camera network, the search space for person matching can be extremely large with numerous potential of candidates to be discriminated. Thus test-time (probe-time) scalability is crucial, as well as real-time low latency implementation for processing numerous input video streams, and returning query results promptly for on-the-fly real-time response.
6. *Long-term re-identification*: The longer the time and space separation between views is, the greater the chance will be that people may appear with some changes of clothes or carried objects in different camera views. Ideally a re-identification system should have some robustness to such changes.

1.3.3 Data and Evaluation

Many standard benchmark datasets reflect a 'closed-world' scenario, e.g. exactly two camera views with exactly one instance of each person per camera and 1:1 exact identity correspondence between the cameras. This is in contrast to a more realistic 'open-world' scenario, where persons in each camera may be only partially overlapping and the number of cameras, spatial size of the environment and number of people may be unknown and at a significantly larger scale. Thus the search space is of unknown size and contains a potentially unlimited number of candidate matches for a target. Re-identification of targets in such open environments can potentially scale to arbitrary levels, covering huge spatial areas spanning not just different buildings but different cities, countries or even continents, leading to an overwhelming quantity of 'big data'.

There are a variety of metrics that are useful for quantifying the effectiveness of a re-identification system. The two most common metrics are 'Rank-1 accuracy', and the 'CMC curve'. Rank-1 accuracy refers to the conventional notion of classification accuracy: the percentage of probe images which are perfectly matched to their corresponding gallery image. High Rank-1 accuracy is notoriously hard to obtain on challenging re-id problems. More realistically a model is expected to report a ranked list of matches which the operator can inspect manually to confirm the true match. The question is how high true matches typically appear on the ranked list. The CMC (Cumulative Match Characteristic) curve summarises this: the chance of the true match appearing in the top 1, 2, ..., N of the ranked list (the first point on the CMC curve being Rank-1 accuracy). Other metrics which can be derived from the CMC curve include the scalar area under the curve, and expected rank (on average how far down the list is the true match). Which of these two metrics is the most relevant arguably depends on the specific application scenario: Whether a (probably low in absolute terms) chance of perfect match or a good average ranking is preferred. This dichotomy raises the further interesting question of which evaluation criterion is the relevant one to optimise when designing discriminatively trained re-identification models.

1.4 Perspectives and Progress

1.4.1 On Feature Representation

Seeking Robust Features

A large number of feature types have been proposed for re-identification, e.g. colour, textures, edges, shape, global features, regional features, and patch-based features. In order to cope with sparsity of data and the challenging view conditions, most person re-identification methods benefit from integrating several types of features with complementary nature [4–6, 9, 11, 12, 25–29]. Often, each type of visual feature is represented by a bag-of-words scheme in the form of a histogram. Feature histograms are then concatenated with some weighting between different types of features in accordance to their perceived importance, i.e. based on some empirical or assumed discriminative power of certain type of features in distinguishing visual appearance of individuals. Spatial information about the layout of these features is also an important cue. However, there is a tradeoff between more granular spatial decomposition providing a more detailed cue and increasing risk of mis-alignment between regions in image pairs, and thus brittleness of the match. To integrate spatial information into the feature representation, images are typically partitioned into different segments or regions, from which features are extracted. Existing partitioning schemes include horizontal stripes [4, 6, 18, 29], triangulated graphs [30], concentric rings [21], and localised patches [8, 13]. Chapters 2, 3 and 4 introduce some examples of robust feature representations for re-identification, such as fisher vectors and covariance descriptors. Chapters 5 and 6 take a different view of learning mid-level semantic attribute features reflecting a low-dimensional human-interpretable description of each person's appearance. Chapter 17 provides a detailed analysis and comparison of the different feature types used in re-identification.

Exploiting Shape and Structural Constraints

Re-identification requires first detecting a person prior to feature extraction. Performance of existing pedestrian detection techniques is still far from accurate for the re-identification purpose. Without a tight detection bounding box, the features extracted are likely to be affected by background clutter. Many approaches start with attempting to segment the pixels of a person in the bounding box (foreground) from background also included in the bounding box. This increases the purity of extracted features by eliminating contamination by background information.

If different body parts can be detected with pose estimation (parts configuration rather than 3D orientation) and human parsing systems, the symmetry and shape of a person can be exploited to extract more robust and relevant imagery features from different body parts. In particular, natural objects reveal symmetry in some form and background clutter rarely exhibits a coherent and symmetric pattern. One can exploit these symmetric and asymmetric principles to segregate meaningful body parts as the foreground, while discard distracting background clutter. Chapter 3 presents a robust symmetry-based descriptor for modelling the human appearance, which localises perceptually relevant body parts driven by asymmetry and/or symmetry principles. Specifically, the descriptor imposes higher weights to features located near to the vertical symmetry axis than those that are far from it. This permits higher preference to internal body foreground, rather than peripheral background portions in the image. The descriptor, when enriched with chromatic and texture information, shows exceptional robustness to low resolution, pose, viewpoint and illumination variations.

Another way of reducing the influence of background clutter is by decomposing a full pedestrian image into articulated body parts, e.g. head, torso, arms and legs. In this way, one wishes to focus selectively on similarities between the appearance of body parts whilst filtering out as much of the background pixels in proximity to the foreground as possible. Naturally, a part-based re-identification representation exhibits better robustness to partial (self) occlusion and changes in local appearances. Chapters 6 and 7 describe methods for representing the pedestrian body parts as 'Pictorial Structures'. Chapter 7 further demonstrates an approach to obtaining robust signatures from the segmented parts not only for 'single-shot' but also 'multi-shot' recognition.

Beyond 2D Appearance Features

Re-identification methods based on entirely 2D visual appearance-based features would fail when individuals change their clothing completely. To address this problem, one can attempt to measure soft biometric cues that are less sensitive to clothing appearance, such as the height of a person, the length of his arms and legs and the ratios between different body parts. However, soft biometrics are exceptionally difficult to measure reliably in typical impoverished surveillance video at 'stand off' distances and unconstrained viewing angles. Chapter 8 describes an approach to recover skeleton lengths and global body shape from calibrated 3D depth images obtained from depth-sensing cameras. It shows that using such non-2D appearance features as a form of soft biometrics promises more robust re-identification for long-term video surveillance.

Exploiting Local Contextual Constraints

In crowded public spaces such as transport hubs, achieving accurate pedestrian detection is hard, let alone extracting robust features for re-identification purpose. The problem is further compounded by the fact that many people are wearing clothing with similar colour and style, increasing the ambiguity and uncertainty in the matching process. Where possible, one aims to seek more holistic contextual constraints in addition to localised visual appearance of isolated (segmented) individuals.

In public scenes people often walk in groups, either with people they know or strangers. The availability of more and richer visual content in a group of people over space and time could provide vital contextual constraints for more accurate matching of individuals within the group. Chapter 9 goes beyond the conventional individual person re-identification by casting the re-identification problem in the context of associating groups of people in proximity over different camera views [21]. It aims to address the problem of associating groups of people over large space and time gaps. Solving the group association problem is challenging in that a group of people can be highly non-rigid with changing relative position of people within the group, as well as individuals being subject to severe self-occlusions.

Not All Are Equal: Salient Feature Selection

Two questions arise: (1) Are all features equal? (2) Does the usefulness of a feature (type) universally hold? Unfortunately, not all features are equally important or useful for re-identification. Some features are more discriminative for identity, whilst others more tolerant or invariant to camera view changes. It is important to determine both the circumstances and the extent of the usefulness of each feature. This is considered as the problem of feature weighting or feature selection. Existing re-identification techniques [4, 6, 11, 31] mostly assume implicitly a feature weighting or selection mechanism that is *global*, i.e. a set of generic weights on feature types invariant to a population. That is, to assume a single weight vector or distance metric (e.g. mahalanobis distance metric) that is globally optimal for all people. For instance, one often assumes colour is the most important (intuitively so) and universally a good feature for matching all individuals. Besides heuristic or empirical tuning, such weightings can be learned through boosting [11], ranking [4], or distance metric learning [6] (see Sect. Learning Distance Metric).

Humans often rely on *salient* features for distinguishing one from the other. Such feature saliency is valuable for person re-identification but is often too subtle to be captured when computing generic feature weights using existing techniques. Chapter 10 considers an alternative perspective that different appearance features are more important or salient than others in describing each particular individual and distinguishing him/her from other people. Specifically, it provides empirical evidence to demonstrate that some re-identification advantages can be gained from unsupervised feature importance mining guided by a person's appearance attribute classification. Chapter 17 considers a similar concept in designing a patch-based re-identification system, which aims to discover salient patches of each individual in an unsupervised manner in order to achieve more robust re-identification [8].

Exploiting Semantic Attributes

When performing person re-identification, human experts rely upon matching appearance or functional attributes that are discrete and unambiguous in interpretation, such as hairstyle, shoe-type or clothing-style [32]. This is in contrast to the continuous and more ambiguous 'bottom-up' imagery features used by contemporary computer vision based re-identification approaches, such as colour and texture [3–5]. This 'semantic attribute' centric representation is similar to a description provided verbally to a human operator, e.g. by an eyewitness.

Attribute representations may start with the same low-level feature representation that conventional re-identification models use. However, they use these to generate a low-dimensional attribute description of an individual. In contrast to standard unsupervised dimensionality reduction methods such as Principal Component Analysis (PCA), attribute learning focuses on representing persons by projecting them onto a basis set defined by axes of appearance which are semantically meaningful to humans.

Semantic attribute representations have various benefits: (1) In re-identification, a single pair of images may be available for each target. This exhibits the challenging case of 'one-shot' learning. Attributes can be more powerful than low-level features [33–35], as pre-trained attribute classifiers learn implicitly the variance in appearance of each particular attribute and invariances to the appearance of that attribute across cameras. (2) Attributes can be used synergistically in conjunction with raw data for greater effectiveness [7, 35]. (3) Attributes are a suitable representation for direct human interaction, therefore allowing searches to be specified, initialised or constrained using human-labelled attribute-profiles [33, 34, 36], i.e. enabling forensic person search. Chapter 5 defines 21 binary attributes regarding clothing-style, hairstyle, carried objects and gender to be learned with Support Vector Machines (SVMs). It evaluates the theoretical discriminative potential of the attributes, how reliably they can be detected in practice, how their weighting can be discriminatively learned and how they can be used in synergy with low-level features to re-identify accurately. Finally, it is shown that attributes are also useful for zero-shot identification, i.e. replacing the probe image with a specified attribute semantic description without visual probe. Chapter 6 embeds middle-level cloth attributes via a latent SVM framework for more robust person re-identification. The pairwise potentials in the latent SVM allow attribute correlation to be considered. Chapter 10 takes a different approach to discover a set of prototypes in an unsupervised manner. Each prototype reveals a mixture set of attributes to describe a specific population of people with similar appearance characteristics. This alleviates the labelling effort for training attribute classifiers.

1.4.2 On Model Learning

Learning Feature Transforms

If camera pair correspondences are known, one can learn a feature transfer function for modelling camera-dependent photometric or geometric transformations. In particular, a photometric function captures the changes of colour distribution of objects transiting from one camera view to another. The changes are mainly caused by different lighting and viewing conditions. Geometric transfer functions can also be learned from the correspondences of interest points. Following the work of Porikli [37], a number of studies have proposed different ways for estimating the Brightness Transfer Function (BTF) [4, 38–42]. The BTF can be learned either separately on different colour channels, or taking into account the dependencies between channels [41]. Some BTFs are defined on each individual, whilst other studies learn a cumulative function on the full available training set [4]. A detailed review of different BTF approaches can be found in Chap. 11. Most BTF approaches assume the availability of perfect foreground segments, from which robust colour features can be extracted. This assumption is often invalid in real-world scenarios. Chapter 11 relaxes this assumption through performing automatic feature selection with the aim to discard background clutter irrelevant to re-identification. It further demonstrates an approach to estimate a robust transfer function given only limited training pairs from two camera views.

In many cases the transfer functions between camera view pairs are complex and multi-modal. Specifically, the cross-views transfer functions can be different under the influence of multiple factors such as lighting, poses, camera calibration parameters and the background of a scene. Therefore, it is necessary to capture these different configurations during the learning stage. Chapter 17 provides a solution to this problem and demonstrates that the learned model is capable of generalising better to a novel view pair.

Learning Distance Metric

A popular alternative to colour transformation learning is distance metric learning. The idea of distance metric learning is to search for the optimal metric under which instances belonging to the same person are more similar, and instances belonging to different people are more different. It can be considered as a data-driven feature importance mining technique [18] to suppress cross-view variations.

Existing distance metric learning methods for re-identification include Large Margin Nearest Neighbour (LMNN) [43], Information Theoretic Metric Learning (ITML) [44], Logistic Discriminant Metric Learning (LDML) [45], KISSME [46], RankSVM [4], and Probabilistic Relative Distance Comparison (PRDC) [6]. Chapter 8 provides an introduction to using RankSVM for re-identification. In particular, it details how the re-identification task can be converted from a matching problem into a pairwise binary classification problem (correct match vs. incorrect match), and aims to find a linear function to weigh the absolute difference of samples via optimisation given pairwise relevance constraints.

In contrast to RankSVM which solely learns an independent weight for each feature, full Mahalanobis matrix metric learners optimise a full distance matrix, which is potentially significantly more powerful. Early metric learning methods [43, 44] are relatively slow and data hungry. More recently, re-identification research has driven the development of faster and lighter methods [46, 47]. Chapter 12 presents a metric learner for single-shot person re-identification and provides extensive comparisons on some of the widely used metric learning approaches. It has been shown that in general, metric learning is capable of boosting the re-identification performance without complicated and handcrafted feature representations. All the aforementioned methods learn a single metric space for matching. Chapter 17 suggests that different groups of people may be better distinguished by different types of features (a similar concept is also presented in Chap. 10). It proposes a candidate-set-specific metric for more discriminative matching given a specific group with small number of subjects.

Reduce the Need for Exhaustive Data Labelling

A major weakness of pairwise metric learning and other discriminative methods is the construction of a training set. This process requires manually annotating pairs of individuals across each camera pair. Such a requirement is reasonable for training and testing splits on controlled benchmark datasets, but limits their scalability to more realistic open-world problems, where there may be very many pairs of cameras, making this 'calibration' requirement impossible or prohibitively expensive. One possible solution has been presented in [48], where a per-patch representation of the human body is adopted, and each patch of the images of the original training dataset has been sampled many times in order to simulate diverse illumination conditions. Alternatively, other techniques have also been proposed [29, 49] that aim to exploit the structure of unlabelled samples in a semi-supervised multi-feature learning framework given very sparse labelled samples. Chapter 13 attempts to resolve this problem by dictionary-based domain adaptation, focusing on face re-identification. In particular, it assumes that the source domain (early location) has plenty of labelled data (subjects with known identities), whilst the target domain (different location) has limited labelled images. The approach learns a domain invariant sparse representation as a shared dictionary for cross-domain (cross-camera) re-identification. In this way, the quantity of pairwise correspondence annotations may be reduced.

Another perspective on this data scalability problem is that of transfer learning. Ideally one wishes to construct a re-identification system between a pair of cameras with minimal calibration/training annotation. To achieve this, re-identification models learned from an initial set of annotated camera-pairs should be able to be exploited and/or adapted to a new target camera pair (possibly located at a different site) without exhaustive annotation in the new camera pair. Adapting and transferring re-id models is a challenging problem which despite some initial work [24, 50] is still an open problem.

Re-identification as an Inference Problem

In many cases one would like to infer the identity of past and unlabelled observations on the basis of very few labelled examples of each person. In practice, the number of labelled images available is significantly fewer than the quantity of images for which one wants to identify. Chapter 14 introduces formally the problem of identity inference as a generalisation of the person re-identification problem. Identity inference addresses the situation of using few labelled images to label many unknown images without explicit knowledge that groups of images represent the same individual. The standard single- and multi-shot recognition problem commonly known in the literature can then be regarded as special cases of this formulation. This chapter discusses how such an identity inference task can be effectively solved through using a CRF (Conditional Random Field) model. Chapter 15 discusses a different facet of the re-identification problem. Instead of matching people across different camera views, the chapter explores identify inference within the same camera view. This problem is essentially a multi-object tracking problem, of which the aim is to mitigate the identity switching problem with the use of appearance cues. The study formulates a minimum-cost maximum-flow linear program to achieve robust multi target tracking.

1.4.3 From Closed- to Open-World Re-identification

Limitations of Existing Datasets

Much effort has been expended on developing methods for automatic person re-identification, with particular attention devoted to the problems of learning discriminative features and formulating robust discriminative distance metrics. Nevertheless, existing work is generally conditioned towards maximising ranking performance on small, carefully constructed *closed-world* benchmark datasets largely unrepresentative of the scale and complexity of more realistic *open-world* scenarios.

To bring re-identification from closed- to open-world deployment required by real-world applications, it is important to first understand the characteristics and limitations of existing benchmark datasets. Chapter 16 provides a comprehensive list of established re-identification benchmark datasets with highlights on their specific challenges and limitations. The chapter also discusses evaluation metrics such as Cumulative Match Characteristic (CMC) curve, which are commonly adopted by re-identification benchmarking methods. Chapter 17 provides an overview of various person re-identification systems and their evaluation on closed-world benchmark datasets. In addition, the chapter highlights a number of general limitations inherent to current re-identification databases, e.g. non-realistic assumption of perfectly aligned images, and limited number of camera views and test images for evaluation.

Exploiting Environmental Contextual Knowledge

Person re-identification cannot be achieved ultimately by matching imagery information alone. In particular, given a large camera network, the search space for re-identification can be enormous, leading to huge number of false matches. To reduce the very large number of possible candidates for matching, it is essential to discover and model the knowledge about inter-camera relationships as environmental contextual constraints to assist re-identification over different camera views.

The problem of inferring the spatial and temporal relationships among cameras is often known as *camera topology inference* [22, 23, 51–53], which involves the estimation of camera transition probabilities, i.e. (1) how likely people detected in one view are to appear in other views; and (2) an inter-camera transition time distribution, i.e. how much travel time is needed to cross a blind area [54]. State-of-the-art methods infer topology through searching for consistent spatiotemporal relationships from population activity patterns (rather than individual whereabouts) across views. For instance, methods presented in [51, 52] accumulate a large set of cross-camera entrance and exit events to establish a transition time distribution. Anton van den Hengel et al. [53] accumulate occupancy statistics in different regions of an overlapping camera network for scalable topology mapping. Loy et al. [22, 23] present a tracking-free method to infer camera transition probability and the associated time delay through correlating activity patterns in segmented regions across non-overlapping camera views over time. Chapter 19 describes a scalable approach based on [53] to automatically derive overlap topology for camera networks and evaluate its use for large-scale re-identification. Chapter 20 presents a re-identification prototype system that employs the global space–time profiling method proposed in [22] for real-world re-identification in disjoint cameras with non-overlapping fields of views.

Improving Post-Rank Search Efficiency

In open-world re-identification one may need to deal with an arbitrarily large number of individuals in multiple camera views during the query stage. After the ranking process, a ranked list of possibly hundreds of likely match images are returned by an appearance-based matching method. The final judgement is left to a human operator, who needs to inspect the list and manually localise the correct match against the query (probe) image. Existing re-identification methods generally assume the ranked list is good enough for decision making. In reality, such a ranking list is far from good and necessarily suboptimal, due to (1) visual ambiguities and disparities, and (2) lack of sufficient labelled pairs of training samples to cover diverse appearance variations from unknown changes in viewing conditions. Often, an operator needs to scroll down hundreds of images to find the true re-identification. For viable open-world re-identification, this *post-rank searching* problem needs be resolved.

Zheng et al. [19] takes a *set-based verification* perspective. More precisely, the study re-defines the re-identification problem as a verification problem of a small set of target people (which they call a watch list) from a large group of irrelevant individuals. The post-rank search thus becomes more realistic and relevant, as one only needs to verify a query against a watch-list, rather than matching the query against everyone in the scene exhaustively. Liu et al. [55] further present a *man-in-the-loop* method to make the post-rank search much more efficient. Specifically, they propose a manifold-based re-ranking method that allows a user to quickly refine their search by either 'one-shot' or a couple of sparse negative selections. Their study shows that the method allows correct re-identification converges three times faster than ordinary exhaustive search.

Chapter 18 proposes an attribute-centric alternative to improve target search by using textual description such as 'white upper garment and blue trousers'. Such a complex description can be conveniently obtained through combining a set of 'atomic' or basic attribute descriptions using Boolean operators. The resulting description is subsequently matched against the attribute profile of every image in the gallery to locate the target. Chapter 5 also explores a similar idea, which they call as 'zero-shot' re-identification. In a more practical sense, rather than using textual description solely for target search, Chap. 20 exploits the description to complement the ranking of candidate matches. In particular, a user may select multiple attributes describing the target to re-rank the initial list so as to promote target with similar attributes to a higher rank, leading to much faster target search in the rank list.

System Design and Implementation Considerations

To date, very little work has focused on addressing the practical question of how to best leverage the current state-of-the-art in re-identification techniques whilst tolerating their limitations in *engineering* practical systems that are *scalable* to typical real-world operational scenarios. Chapter 20 describes design rationale and implementational considerations of building a practical re-identification system that scales to arbitrarily large, busy, and visually complex spaces. The chapter defines three scalability requirements, i.e. associativity, capacity and accessibility. Associativity underpins the system's capability of accurate target extraction from a large search space. Several computer vision techniques such as tracklet association and global space–time profiling are implemented to achieve robust associativity. In terms of capacity requirement, the chapter also examines the system's computational speed in processing multiple video streams. The analysis concludes that person detection and feature extraction are among the most computationally expensive components in the re-identification pipeline. To accelerate the computations, it is crucial and necessary to exploit Graphics Processing Unit (GPU) and multi-threading. In the discussion of accessibility requirement, the chapter provides detailed comparative analysis concerning the effect of user query time versus database size, and the efficiency difference when a database is accessed locally or remotely.

1.5 Discussions

This chapter has provided a wide panorama of the re-identification challenge, together with an extensive overview of current approaches to addressing this challenge. The rest of the book will present more detailed techniques and methods for solving different aspects of the re-identification problem, reflecting the current state-of-the-art on person re-identification. Nevertheless, these techniques and approaches by no means have covered exhaustively all the opening problems associated with solving the re-identification challenge. There remain other open problems to be addressed in addition to the need for improving existing techniques based on existing concepts. We consider a few as follows.

1.5.1 Multi-spectral and Multimodal Analysis

Current re-identification methods mostly rely upon only visual information. However, there are other sensory technologies that can reinforce and enrich the detection and description of the presence of human subjects in a scene. For example, infrared signals are often used together with visual sensory input to capture people under extremely limited lighting conditions [56, 57]. An obviously interesting approach to be exploited is to utilise thermal images. This can also extend to the case of exploiting the energy signature of a person, including movement and consumption of energy unique to each individual, e.g. whilst walking and running. Such information may provide unique characteristics of a person in crowds. Equally interesting is to exploit audio/vocal signatures of individual human subjects including but not limited to vocal outburst or gait sound, similar to how such techniques are utilised in human–robot interaction system designs [58, 59].

1.5.2 PTZ Cameras and Embedded Sensors

Human operators are mostly trained to perform person re-identification by focusing on particular small parts (unique attributes) of a person of interest [32]. To exploit such a cognitive process is not realistic without active camera pan-tilt-zoom control in order to provide selective focus on body parts from a distance. Pan-Tilt-Zoom (PTZ) cameras are widespread and many tracking algorithms have been developed to automatically zoom on particular areas of interest [60]. Embedding such a technology in the re-identification system can be exploited, with some early simulated experiments giving encouraging results [61], in which saliency detection is utilised to drive automatically the PTZ camera to focus on certain parts of a human body, to learn the most discriminative attribute which characterises a particular individual. A similar approach can be extended to wearable sensors.

1.5.3 Re-identification of Crowds

This considers an extension of the group re-identification concept described early in this book. Instead of re-identification of small groups of people, one may consider the task of re-identifying masses of people (or other visual objects such as vehicles) in highly crowded scenes, e.g. in a public rally or a traffic jam. Adopting local static features together with elastic/dynamical crowd properties may permit the modelling of extreme variability of single individuals in fluid dynamics of crowds.

1.5.4 Re-identification on the Internet ('Internetification')

This is a further extension of re-identification from multi-camera networks to distributed Internet spaces, necessarily across multi-sources over the Internet taking images from, for instance, the Facebook profile, Flickr and other social media. Such a functionality may create a virtual avatar, composed of multiple and heterogeneous shots, as an intermediate representation, which can then be projected in diverse scenarios and deployed to discover likely matches of a gallery subject across the Internet. In such a way, re-identification can be highly pervasive with a wide spectra of potential applications in the near and far future.

1.6 Further Reading

Interested readers may wish to refer to the following material:

- [62] for a review of re-identification methods in surveillance and forensic scenarios.

- [54] for a general introduction to a variety of applications and emerging techniques in surveillance.
- [63] for a review on video analysis in multiple camera network.

References

1. Keval, H.: CCTV control room collaboration and communication: does it work? In: Human Centred Technology Workshop (2006)
2. Williams, D.: Effective CCTV and the challenge of constructing legitimate suspicion using remote visual images. J. Invest. Psychol. Offender Profiling **4**(2), 97–107 (2007)
3. Gray, D., Brennan, S., Tao, H.: Evaluating appearance models for recognition, reacquisition, and tracking. In: IEEE International Workshop on Performance Evaluation of Tracking and Surveillance (2007)
4. Prosser, B., Zheng, W., Gong, S., Xiang, T.: Person re-identification by support vector ranking. In: British Machine Vision Conference, pp. 21.1–21.11 (2010)
5. Farenzena, M., Bazzani, L., Perina, A., Murino, V., Cristani, M.: Person re-identification by symmetry-driven accumulation of local features. In: IEEE Conference Computer Vision and Pattern Recognition, pp. 2360–2367 (2010)
6. Zheng, W., Gong, S., Xiang, T.: Re-identification by relative distance comparison. IEEE Trans. Pattern Anal. Mach. Intell. **35**(3), 653–668 (2013)
7. Layne, R., Hospedales, T.M., Gong, S.: Person re-identification by attributes. In: British Machine Vision Conference (2012)
8. Zhao, R., Ouyang, W., Wang, X.: Unsupervised salience learning for person re-identification. In: IEEE Conference on Computer Vision and Pattern Recognition (2013)
9. Bazzani, L., Cristani, M., Murino, V.: Symmetry-driven accumulation of local features for human characterization and re-identification. Comput. Vis. Image Underst. **117**(2), 130–144 (2013)
10. Madden, C., Cheng, E.D., Piccardi, M.: Tracking people across disjoint camera views by an illumination-tolerant appearance representation. Mach. Vis. Appl. **18**(3), 233–247 (2007)
11. Gray, D., Tao, H.: Viewpoint invariant pedestrian recognition with an ensemble of localized features. In: European Conference on Computer Vision, pp. 262–275 (2008)
12. Bazzani, L., Cristani, M., Perina, A., Murino, V.: Multiple-shot person re-identification by chromatic and epitomic analyses. Pattern Recogn. Lett. **33**(7), 898–903 (2012)
13. Bak, S., Corvee, E., Brémond, F., Thonnat, M.: Person re-identification using spatial covariance regions of human body parts. In: IEEE International Conference on Advanced Video and Signal Based Surveillance, pp. 435–440 (2010)
14. Ma, B., Su, Y., Jurie, F.: Local descriptors encoded by fisher vectors for person re-identification. In: European Conference on Computer Vision, First International Workshop on Re-Identification, pp. 413–422 (2012)
15. Zheng, W., Gong, S., Xiang, T.: Person re-identification by probabilistic relative distance comparison. In: IEEE Conference Computer Vision and Pattern Recognition, pp. 649–656 (2011)
16. Avraham, T., Gurvich, I., Lindenbaum, M., Markovitch, S.: Learning implicit transfer for person re-identification. In: European Conference on Computer Vision, First International Workshop on Re-Identification, pp. 381–390 (2012)
17. Hirzer, M., Beleznai, C., Roth, P., Bischof, H.: Person re-identification by descriptive and discriminative classification. In: Heyden, A., Kahl, F. (eds.) Image Analysis, pp. 91–102. Springer, New York (2011)
18. Liu, C., Gong, S., Loy, C.C., Lin, X.: Person re-identification: what features are important? In: European Conference on Computer Vision, First International Workshop on Re-identification, pp. 391–401 (2012)

19. Zheng, W., Gong, S., Xiang, T.: Transfer re-identification: from person to set-based verification. In: IEEE Conference on Computer Vision and Pattern Recognition, pp. 2650–2657 (2012)
20. Li, W., Zhao, R., Wang, X.: Human reidentification with transferred metric learning. In: Asian Conference on Computer Vision, pp. 31–44 (2012)
21. Zheng, W., Gong, S., Xiang, T.: Associating groups of people. In: British Machine Vision Conference, pp. 23.1–23.11 (2009)
22. Loy, C.C., Xiang, T., Gong, S.: Time-delayed correlation analysis for multi-camera activity understanding. Int. J. Comput. Vision **90**(1), 106–129 (2010)
23. Loy, C.C., Xiang, T., Gong, S.: Incremental activity modelling in multiple disjoint cameras. IEEE Trans. Pattern Anal. Mach. Intell. **34**(9), 1799–1813 (2012)
24. Layne, R., Hospedales, T.M., Gong, S.: Domain transfer for person re-identification. In: ACM Multimedia International Workshop on Analysis and Retrieval of Tracked Events and Motion in Imagery Streams, pp. 25–32. http://dl.acm.org/citation.cfm?id=2510658. (2013)
25. Wang, X.G., Doretto, G., Sebastian, T., Rittscher, J., Tu, P.: Shape and appearance context modeling. In: International Conference on Computer Vision, pp. 1–8 (2007)
26. Alahi, A., Vandergheynst, P., Bierlaire, M., Kunt, M.: Cascade of descriptors to detect and track objects across any network of cameras. Comput. Vis. Image Underst. **114**(6), 624–640 (2010)
27. Schwartz, W.R., Davis, L.S.: Learning discriminative appearance-based models using partial least squares. In: Brazilian Symposium on Computer Graphics and Image Processing, pp. 322–329 (2009)
28. Cheng, D.S., Cristani, M., Stoppa, M., Bazzani, L., Murino, V.: Custom pictorial structures for re-identification. In: British Machine Vision Conference, pp. 68.1–68.11 (2011)
29. Loy, C.C., Liu, C., Gong, S.: Person re-identification by manifold ranking. In: IEEE International Conference on Image Processing (2013)
30. Gheissari, N., Sebastian, T., Hartley, R.: Person reidentification using spatiotemporal appearance. In: IEEE Conference Computer Vision and Pattern Recognition, pp. 1528–1535 (2006)
31. Mignon, A., Jurie, F.: PCCA: A new approach for distance learning from sparse pairwise constraints. In: IEEE Conference Computer Vision and Pattern Recognition, pp. 2666–2672 (2012)
32. Nortcliffe, T.: People Analysis CCTV Investigator Handbook. Home Office Centre of Applied Science and Technology, Holland (2011)
33. Lampert, C.H., Nickisch, H., Harmeling, S.: Learning to detect unseen object classes by between-class attribute transfer. In: IEEE Conference on Computer Vision and Pattern Recognition, pp. 951–958 (2009)
34. Siddiquie, B., Feris, R.S., Davis, L.S.: Image ranking and retrieval based on multi-attribute queries. In: IEEE Conference on Computer Vision and Pattern Recognition, pp. 801–808 (2011)
35. Liu, J., Kuipers, B.: Recognizing human actions by attributes. In: IEEE Conference on Computer Vision and Pattern Recognition, pp. 3337–3344 (2011)
36. Kumar, N., Berg, A., Belhumeur, P.: Describable visual attributes for face verification and image search. In: IEEE Trans. Pattern Anal. Mach. Intell. **33**(10), 1962–1977 (2011)
37. Porikli, F.: Inter-camera color calibration by correlation model function. In: IEEE International Conference on Image Processing (2003)
38. Chen, K.W., Lai, C.C., Hung, Y.P., Chen, C.S.: An adaptive learning method for target tracking across multiple cameras. In: IEEE Conference on Computer Vision and Pattern Recognition, pp. 1–8 (2008)
39. D'Orazio, T., Mazzeo, P.L., Spagnolo, P.: Color brightness transfer function evaluation for non overlapping multi camera tracking. In: ACM/IEEE International Conference on Distributed Smart Cameras, pp. 1–6 (2009)
40. Javed, O., Shafique, K., Rasheed, Z., Shah, M.: Modeling inter-camera space-time and appearance relationships for tracking across non-overlapping views. Comput. Vis. Image Underst. **109**(2), 146–162 (2008)
41. Jeong, K., Jaynes, C.: Object matching in disjoint cameras using a color transfer approach. Mach. Vis. Appl. **19**(5–6), 443–455 (2008)

42. Lian, G., Lai, J.H., Suen, C.Y., Chen, P.: Matching of tracked pedestrians across disjoint camera views using CI-DLBP. IEEE Trans. Circuits Syst. Video Technol. **22**(7), 1087–1099 (2012)
43. Weinberger, K.Q., Saul, L.K.: Distance metric learning for large margin nearest neighbor classification. J. Mach. Learn. Res. **10**, 207–244 (2009)
44. Davis, J.V., Kulis, B., Jain, P., Sra, S., Dhillon, I.S.: Information-theoretic metric learning. In: International Conference on Machine learning, pp. 209–216 (2007)
45. Guillaumin, M., Verbeek, J., Schmid, C.: Is that you? metric learning approaches for face identification. In: International Conference on Computer Vision, pp. 498–505 (2009)
46. Kostinger, M., Hirzer, M., Wohlhart, P., Roth, P.M., Bischof, H.: Large scale metric learning from equivalence constraints. In: IEEE Conference on Computer Vision and Pattern Recognition, pp. 2288–2295 (2012)
47. Hirzer, M., Roth, P., Köstinger, M., Bischof, H.: Relaxed pairwise learned metric for person re-identification. In: European Conference on Computer Vision, pp. 780–793 (2012)
48. Satta, R., Fumera, G., Roli, F., Cristani, M., Murino, V.: A multiple component matching framework for person re-identification. In: International Conference on Image analysis and Processing, pp. 140–149 (2011)
49. Figueira, D., Bazzani, L., Quang, M.H., Cristani, M., Bernardino, A., Murino, V.: Semi-supervised multi-feature learning for person re-identification. In: IEEE International Conference on Advanced Video and Signal-Based Surveillance (2013)
50. Wu, Y., Li, W., Minoh, M., Mukunoki, M.: Can feature-based inductive transfer learning help person re-identification? In: IEEE International Conference on Image Processing (2013)
51. Makris, D., Ellis, T., Black, J.: Bridging the gaps between cameras. In: IEEE Conference on Computer Vision and Pattern Recognition, pp. 205–210 (2004)
52. Tieu, K., Dalley, G., Grimson, W.E.L.: Inference of non-overlapping camera network topology by measuring statistical dependence. In: International Conference on Computer Vision, vol. 2, pp. 1842–1849 (2005)
53. van den Hengel, A., Dick, A., Hill, R.: Activity topology estimation for large networks of cameras. In: IEEE International Conference on Video and Signal Based Surveillance (2006)
54. Gong, S., Loy, C.C., Xiang, T.: Security and surveillance. In: Visual Analysis of Humans, pp. 455–472. Springer, New York (2011)
55. Liu, C., Loy, C.C., Gong, S., Wang, G.: POP: Person re-identification post-rank optimisation. In: International Conference on Computer Vision (2013)
56. Han, J., Bhanu, B.: Fusion of color and infrared video for moving human detection. Pattern Recogn. **40**(6), 1771–1784 (2007)
57. Correa, M., Hermosilla, G., Verschae, R., Ruiz-del Solar, J.: Human detection and identification by robots using thermal and visual information in domestic environments. J. Intell. Rob. Syst. **66**(1–2), 223–243 (2012)
58. Hofmann, M., Geiger, J., Bachmann, S., Schuller, B., Rigoll, G.: The TUM gait from audio, image and depth (GAID) database: multimodal recognition of subjects and traits. J. Vis. Commun. Image Represent. (2013)
59. Choudhury, T., Clarkson, B., Jebara, T., Pentland, A.: Multimodal person recognition using unconstrained audio and video. In: International Conference on Audio- and Video-Based Person Authentication, pp. 176–181 (1999)
60. Choi, H., Park, U., Jain, A.: PTZ camera assisted face acquisition, tracking and recognition. In: IEEE International Conference on Biometrics: Theory, Applications and Systems (2010)
61. Salvagnini, P., Bazzani, L., Cristani, M., Murino, V.: Person re-identification with a PTZ camera: an introductory study. In: IEEE International Conference on Image Processing (2013)
62. Vezzani, R., Baltieri, D., Cucchiara, R.: People re-identification in surveillance and forensics: a survey. ACM Comput. Surv. **46**(2), 1–36 (2014)
63. Wang, X.: Intelligent multi-camera video surveillance: a review. Pattern Recogn. Lett. **34**(1), 3–19 (2012)

Part I
Features and Representations

Chapter 2
Discriminative Image Descriptors for Person Re-identification

Bingpeng Ma, Yu Su and Frédéric Jurie

Abstract This chapter looks at *person re-identification* from a *computer vision* point of view, by proposing two new image descriptors designed for matching people bounding boxes in images. Indeed, one key issue of person re-identification is the ability to measure the similarity between two person-centered image regions, allowing to predict if these regions represent the same person despite changes in illumination, viewpoint, background clutter, occlusion, and image quality/resolution. They hence heavily rely on the signatures or descriptors used for representing and comparing the regions. The first proposed descriptor is a combination of Biologically Inspired Features (BIF) and covariance descriptors, while the second builds on the recent advances of Fisher Vectors. These two image descriptors are validated through experiments on two different person re-identification benchmarks (VIPeR and ETHZ), achieving state-of-the-art performance on both datasets.

2.1 Introduction

In recent years, person re-identification in unconstrained videos (i.e. without subjects' knowledge and in uncontrolled scenarios) has attracted more and more research interest. Generally speaking, person re-identification consists of recognizing an individual through different images (e.g., coming from cameras in a distributed network

B. Ma (✉)
School of Computer and Control Engineering, University of Chinese Academy of Sciences, Beijing, China
e-mail: bpma@ucas.ac.cn

Y. Su · F. Jurie
GREYC—CNRS UMR 6072, University of Caen Basse-Normandie, Caen, France
e-mail: yu.su@unicaen.fr

F. Jurie
e-mail: frederic.jurie@unicaen.fr

S. Gong et al. (eds.), *Person Re-Identification*,
Advances in Computer Vision and Pattern Recognition,
DOI: 10.1007/978-1-4471-6296-4_2,

or from the same camera at different time). It is done by measuring the similarity between two person-centered bounding boxes and predicting—based on this similarity—if they represent the same person. This is challenging in unconstrained scenarios because of illumination, viewpoint, and background changes, as well as occlusions or low resolution.

In order to tackle this problem, researchers have concentrated their effort on either (1) the design of visual features to describe individual images or (2) the use of adapted distance measures (e.g., obtained by metric learning). This chapter focuses on the former by proposing two novel image representations. The proposed image representations can be used to measure effectively the similarity between two persons, without requiring any preprocessing step (e.g., background subtraction or body part segmentation).

The first representation is based on Biologically Inspired Features (BIF) [30] extracted through the use of Gabor filters (S1 layer) and MAX operator (C1 layer). They are encoded by the covariance descriptor of [37], used to compute the similarity of BIF features at neighboring scales. The Gabor filters and the covariance descriptor improve the robustness to the illumination variation, while the MAX operator increases the tolerance to scale changes and image shifts. Furthermore, we argue that measuring the similarity of neighboring scales limits the influence of the background (see Sect. 2.3.3 for details). By overcoming illumination, scale, and background changes, the performance of person re-identification is widely improved.

The second one builds on the recently proposed Fisher Vectors for image classification [26] which encodes higher order statistics of local features, and gives excellent performance for several object recognition and image retrieval tasks [27, 28]. Motivated by the success of Fisher Vector, we combine Fisher Vectors with a novel and very simple seven-dimensional local descriptor adapted to the representation of person images, and use the resultant representation (*Local Descriptors encoded by Fisher Vector* or LDFV) as a person descriptor.

These two representations have been experimentally validated on two person re-identification databases (namely the VIPeR and ETHZ datasets), which are challenging since they contain pose changes, viewpoint and lighting variations, and occlusions. Furthermore, as they are commonly used in the recent literature, they allow comparisons with state-of-the-art approaches.

The remainder of this chapter is organized as follows: Sect. 2.2 reviews the related works on image representation for person re-identification in videos. Section 2.3 describes the first proposed descriptor in detail, analyzes its advantages, and then shows its effectiveness on the VIPeR and ETHZ datasets. The second person descriptor and its experimental validation are given Sect. 2.4. Finally, Sect. 2.5 concludes the chapter.

2.2 Related Work

Person re-identification in the literature has been considered either as a *on the fly* [21] or as an *offline* [33] problem. More formally, person re-identification can be defined as finding the correspondences between the images of a *probe set* representing a single person and the corresponding images in a *gallery set*. Depending on the number of available images per individual (i.e., the size of the probe set), different scenarios have been addressed: (a) Single versus Single (S vs. S) if only one exemplar per individual is available both in probe and in gallery sets [17]; (b) Multiple versus Single (M vs. S) if multiple exemplars per individual are available in the gallery set [12]; (c) Multiple versus Multiple (M vs. M) if multiple exemplars per individual are available both in the probe and gallery sets [33].

As explained before, the image descriptors used for comparing persons are important as they strongly impact the overall performance. The recent literature abounds with such image descriptors. They can be based on (1) color—widely used since the color of clothing constitutes a simple but efficient visual signature—usually encoded within histograms of RGB or HSV values [6], (2) shape, e.g., HOG-based signatures [25, 33], (3) texture, often represented by Gabor filters [18, 29, 40], differential filters [18, 29], Haar-like representations [4] and Co-occurrence Matrices [33], (4) interest points, e.g., SURF [15] and SIFT [21, 41] and (5) image regions [6, 25].

Region-based methods usually split the human body into different parts and extract features for each part. In [6, 9], Maximally Stable Color Regions (MSCR) are extracted, by grouping pixels of similar color into small stable clusters. Then, the regions are described by their area, centroid, second moment matrix, and average color. The Region Covariance Descriptor (RCD) [1, 5, 40] has also been widely used for representing regions. In RCD, the pixels of a region are first represented by a feature vector which captures their intensity, texture, and shape statistics. The so-obtained feature vectors are then encoded by a covariance matrix.

Besides these generic representations, there are some more specialized representations. For example, Epitomic Analysis [7], Spin Images [2, 3], Bag-of-Words based descriptors [41], Implicit Shape Models (ISM) [21], or Panoramic Maps [14] have also been applied to person re-identification.

Since the elementary features (color, shape, texture, etc.) capture different aspects of the information contained in images, they are often combined to give a richer signature. For example, [29] combined 8 color features with 21 texture filters (Gabor and differential filters). Bazzani et al. [6] and Cheng et al. [9] combined MSCR descriptors with weighted Color Histograms, achieving state-of-the-art results on several widely used person re-identification datasets. Interestingly, RCD can be generalized to any type of images such as one-dimensional intensity images, three channel color images, or even other types of images (e.g., infrared). For example, in [40], Gabor features and Local Binary Patterns (LBP) are combined to form a Covariance descriptor which handles the difficulties of varying illumination, viewpoint changes, and nonrigid body deformations.

Different representations need different similarity functions. For example, representations based on histograms can be compared with Bhattacharyya distance [6, 7, 9] or Earth Mover's Distance (EMD) [2, 3]. When the dimensionalities of the representations to be compared are different, EMD can also be used as it allows many-to-many association [25]. Feature selection has been used to improve the discriminative power of the distance function, *e.g.* with boosting. In [18], the authors select the most relevant features (color and texture) by a weighted ensemble of likelihood ratio tests, obtained with AdaBoost. Similarly, in [4] Haar-like features are extracted from the whole body and the most discriminative ones are selected by AdaBoost.

Metric learning has also been used to provide a metric adapted to person re-identification (e.g. [17, 29, 41]). Most distance metric learning approaches learn a Mahalanobis-like distance such as Large Margin Nearest Neighbors (LMNN) [38], Information Theoretic Metric Learning (ITML) [10], Logistic Discriminant Metric Learning (LDML) [19], or PCCA [23]. LMMN minimizes the distance between each training point and its K nearest similarly labeled neighbors, while maximizing the distance between all differently labeled points which are closer than the aforementioned neighbors' distances plus a constant margin. In [11], the authors improved the LMNN with rejection and successfully applied their method to person re-identification. Besides Adaboost and metric learning, RVM [29], Partial Least Squares (PLS) and multiple instance learning [31, 32] have also been applied to person re-identification, with the same idea of improving the performance.

Our approach builds on these recent works, and shows that carefully designed visual features can provide us with state-of-the art results, without the need for any complex distance functions.

2.3 Bio-inspired Covariance Descriptor for Person Re-identification

Our first descriptor is a covariance descriptor using bio-inspired features, BiCov for short. It is a two-stage representation (see Fig. 2.1) in which biologically inspired features are encoded by computing the difference of covariance descriptors at different scales. In the following, the two stages are presented and motivated.

2.3.1 Low-Level Biologically Inspired Features (BIF)

Based on the study of the human visual system, bio-inspired features [30] have obtained excellent performances on several computer vision tasks such as object category recognition [34], face recognition [22], age estimation [20], and scene classification [36].

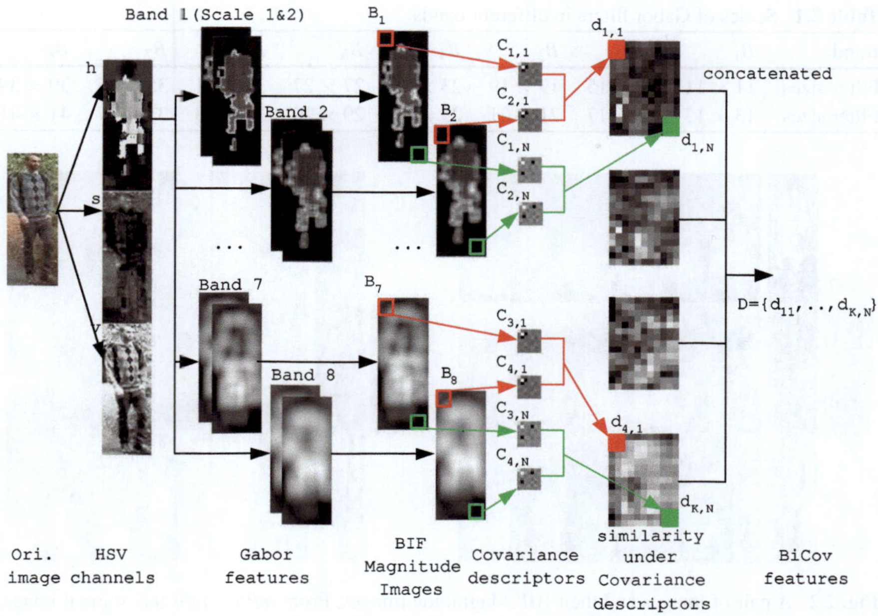

Fig. 2.1 Flowchart of the proposed approach: (1) color images are split into three color channels (HSV), (2) for each channel, Gabor filters are computed at different scales, (3) pairs of neighboring scales are grouped to form one band, (4) magnitude images are produced by applying the MAX operator within the same band, (5) magnitude images are divided into small bins and each bin is represented by a covariance descriptor, and (6) the difference of covariance descriptors between two consecutive bands is computed for each bin and concatenated to form the image representation

Considering the great success of these BIFs, the first step consists of extracting such features to model the low-level properties of images. For an image $I(x, y)$, we compute its convolution with Gabor filters according to the following equations [39]:

$$G(\mu, \nu) = I(x, y) * \psi_{\mu,\nu}(z) \tag{2.1}$$

with:

$$\psi_{\mu,\nu}(z) = \frac{\left\| k_{\mu,\nu} \right\|^2}{\sigma^2} \mathrm{e}^{\left(\frac{-\|k_{\mu,\nu}\|^2 \|z\|^2}{2\sigma^2}\right)} \left[e^{ik_{\mu,\nu}z} - \mathrm{e}^{\frac{-\sigma^2}{2}} \right] \tag{2.2}$$

$$k_{\mu,\nu} = k_\nu e^{i\phi_\mu}, k_\nu = 2^{-\frac{\nu+2}{2}}\pi, \phi_\mu = \mu\frac{\pi}{8} \tag{2.3}$$

where μ and ν are scale and orientation parameters, respectively. In our work, μ is quantized into 16 scales while the ν is quantized into eight orientations.

In practice, we have observed that for person re-identification, the image representations $G(\mu, \nu)$ for different orientations can be averaged without significant loss of performance. Thus, in this case, we replace $\psi_{\mu,\nu}(z)$ in Eq. 2.1 by

Table 2.1 Scales of Gabor filters in different bands

Band	B_1	B_2	B_3	B_4	B_5	B_6	B_7	B_8
Filter sizes	11 × 11	15 × 15	19 × 19	23 × 23	27 × 27	31 × 31	35 × 35	39 × 39
Filter sizes	13 × 13	17 × 17	21 × 21	25 × 25	29 × 29	33 × 33	37 × 37	41 × 41

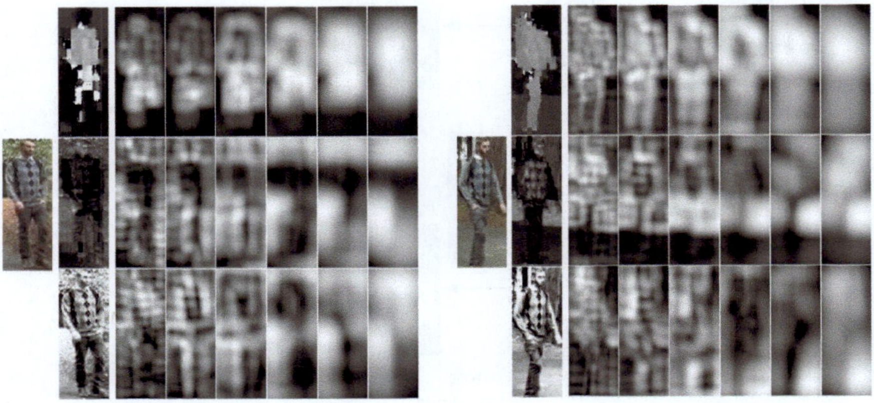

Fig. 2.2 A pair of images and their BIF Magnitude Images. From *left* to *right* the original image, its three HSV channels, six BIF Magnitude Images for different bands

$\psi_\mu(z) = \frac{1}{8}\sum_{\nu=1}^{8} \psi_{\mu,\nu}(z)$. This simplification makes the computations of $G(\mu)$—which is the average of $G(\mu,\nu)$ over all orientations—more efficient.

In all our experiments, the number of scales is fixed to 16 and two neighborhood scales are grouped into one band (we therefore have eight different bands). The scales of Gabor filters in different bands are shown in Table 2.1. We then apply the *MAX pooling* over two consecutive scales (within the same orientation if the orientations are not merged):

$$B_i = \max(G(2i-1), G(2i)) \tag{2.4}$$

The MAX pooling operation increases the tolerance to small-scale changes which often occur, even for the same person, since images are only roughly aligned. We refer to B_i $i \in [1,\ldots,8]$ as the *BIF Magnitude Images*. Figure 2.2 shows a pair of images of one person and its respective BIF Magnitude Images. The image in the first column is the input image, while the ones in the second column are three HSV channels. The images from the third to the eigth column are the BIF Magnitude Images for six different bands.

2.3.2 BiCov Descriptor

In the second stage, BIF Magnitude Images are divided into small overlapping rectangular regions, allowing the preservation of some spatial information. Then, each region is represented by a covariance descriptor [37]. Covariance descriptors can capture shape, location, and color information, and their performances have been shown to be better than other methods in many situations, as rotation and illumination changes are absorbed, to some extent, by the covariance matrix [37].

In order to do this, each pixel of the BIF Magnitude Image B_i is encoded into a seven-dimensional feature vector which captures the intensity, texture, and shape statistics:

$$f_i(x, y) = [x, y, B_i(x, y), B_{i_x}(x, y), B_{i_y}(x, y), B_{i_{xx}}(x, y), B_{i_{yy}}(x, y)] \tag{2.5}$$

where x and y are the pixel coordinates, $B_i(x, y)$ is the raw pixel intensity at position (x, y), $B_{i_x}(x, y)$ and $B_{i_y}(x, y)$ are the derivatives of image B_i with respect to x and y, and $B_{i_{xx}}(x, y)$ and $B_{i_{yy}}(x, y)$ are the second-order derivatives.

Finally, the covariance descriptor is computed for each region of the image:

$$C_{i,r} = \frac{1}{n-1} \sum_{(x,y) \in region\ r} (f_i(x, y) - \bar{f}_i)(f_i(x, y) - \bar{f}_i)^T \tag{2.6}$$

where $\bar{f}_i$ is the mean of $f_i(x, y)$ over the region r and n is the size of region r (in pixels).

Usually, the covariance matrices computed by Eq. 2.6 are considered as the image representation. Covariance matrices are positive definite symmetric matrices lying on a manifold of the Euclidean space. Hence, many usual operations (like the l2 distance) cannot be used directly.

In this chapter, differently from past approaches using covariance descriptors, we compute (for each region separately) the difference of covariance descriptors between two consecutive bands:

$$d_{i,b} = d(C_{2i-1,r}, C_{2i,r}) = \sqrt{\sum_{p=1}^{P} \ln^2 \lambda_p(C_{2i-1,r}, C_{2i,r})} \tag{2.7}$$

where $\lambda_p(C_{2i-1,r}, C_{2i,r})$ is the p-th generalized eigenvalues of $C_{2i-1,r}$ and $C_{2i,r}$, $i = 1, 2, 3, 4$. Finally, the differences are concatenated to form the image representation:

$$D = (d_{1,1}, \cdots, d_{1,R}, \cdots, d_{K,1}, \cdots, d_{K,R}) \tag{2.8}$$

where R is the number of regions and K is the number of band pairs (four in our case). The distance between two images I_i and I_j is obtained by computing the Euclidian distance between their representations D_i and D_j:

$$d(I_i, I_j) = ||D_i - D_j|| \tag{2.9}$$

It is worth pointing out that color images are processed by splitting the image into three color channels (HSV), extracting the proposed descriptor on each channel separately, and finally concatenating the three descriptors into a single signature.

As mentioned in Sect. 2.2, it is usually better to combine several image descriptors. In this chapter, we combine the BiCov descriptor with two other ones, namely the (a) Weighted Color Histogram (wHSV) and (b) the MSCR, such as that defined in [6]. For simplicity, we denote this combination as eBiCov (enriched BiCov). The difference between two eBicov signatures $D_1 = (HA_1, MSCR_1, BiCov_1)$ and $D_2 = (HA_2, MSCR_2, BiCov_2)$ is computed as:

$$d_{eBiCov}(D_1, D_2) = \frac{1}{3} d_{wHSV}(HA_1, HA_2) + \frac{1}{3} d_{MSCR}(MSCR_1, MSCR_2) + \frac{1}{3} d(BiCov_1, BiCov_2) \quad (2.10)$$

Obviously, further improvements could be obtained by optimizing the weights (i.e., *using a supervised approach*), but as we are looking for an unsupervised method, we fix them once for all. Regarding the definition of d_{wHSV} and d_{MSCR}, we use the ones given in [6].

2.3.3 BiCoV Analysis

By combing Gabor filters and covariance descriptors—which are both known to be tolerant to illuminations changes [37]—the BiCov representation is robust to illumination variations.

In addition, BiCov is also robust to background variations. Roughly speaking, background regions are not as contrasted as foreground ones, making their Gabor features (and therefore their covariance descriptors) at different neighboring scales very similar. Since the BiCov descriptor is based on the difference of covariance descriptors, background regions are, to some extent, filtered out.

Finally, it is worth pointing out that our approach makes a very different use of the covariance descriptor. In the literature, covariance-based similarity is defined by the difference between covariance descriptors computed on two different images. Knowing how time-consuming it is to compute eigenvalues, the standard approach which requires to evaluate Eq. 2.7 for computing the distance between the query and each image of the gallery can hardly be used with large galleries. In contrast, BiCov computes the similarity of covariance descriptors *within* the same image, between two consecutive scales, once for all. These similarities are then concatenated to obtain the image signature, and the difference of probe and gallery images is obtained by simply computing the $l2$ distance between their signatures.

Fig. 2.3 VIPeR dataset: Sample images showing same subjects from different viewpoints

2.3.4 Experiments

The proposed representation has been experimentally validated on two datasets for person re-identification (VIPeR [17] and ETHZ [12]).

Person Re-identification on the VIPeR Dataset

VIPeR is specifically made for viewpoint-invariant pedestrian re-identification. It contains 1,264 images of 632 pedestrians. There are exactly two views per pedestrian, taken from two nonoverlapping viewpoints. All images are normalized to 128×48 pixels. The VIPeR dataset contains a high degree of viewpoint and illumination variations: most of the examples contain a viewpoint change of 90 degrees, as can be seen in Fig. 2.3. This dataset has been widely used and is considered to be one of the benchmarks of reference for person re-identification. All the experiments on this dataset address the unsupervised setting, i.e., without using training data, and therefore not involving any metric leaning.

We use the Cumulative Matching Characteristic (CMC) curve [24] and Synthetic Reacquisition Rate (SRR) curve [17], which are the two standard performance measurements for this task. CMC measures the expectation of the correct match at rank r while SRR measures the probability that any of the m best matches is correct.

Figure 2.4 shows the performance of the eBicov representation, and gives comparisons with SDALF [6] which is the state-of-the-art approach for this dataset. We follow the same experimental protocol as [6] and report the average performance over 10 different random sets of 316 pedestrians. We can see that eBiCov

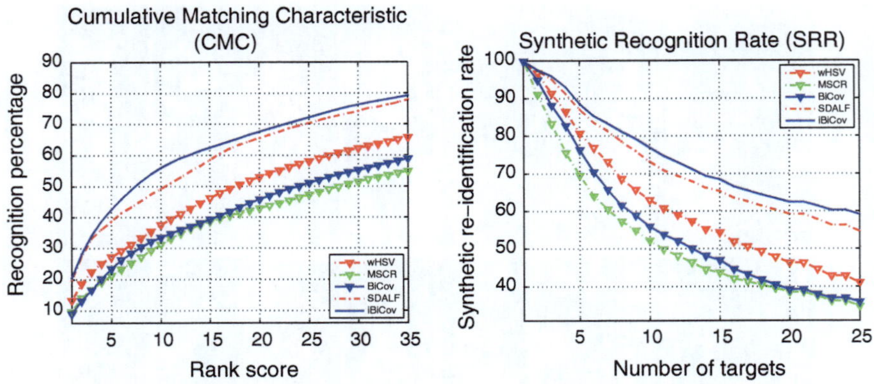

Fig. 2.4 VIPeR dataset: CMC and SRR curves

consistently outperforms SDALF: matching rate at rank 1 for eBiCov is 20.66 % while that of SDALF is 19.84 %. The matching rate at rank 10 for eBiCov is 56.18 while that of SDALF is 49.37. This improvement can be explained in two ways: on one hand, most of the false positives are due to severe lighting changes, which the combination of Gabor filters and covariance descriptors can handle efficiently. On the other hand, since many people tend to dress in very similar ways, it is important to capture as fine image details as possible. This is what BIF does. In addition, it is worth noting that for these experiments the orientation of Gabor filters is not used, allowing to reduce the computational cost. We have indeed experimentally observed that the performance is almost as good as that with orientations.

Finally, Fig. 2.4 also reports the performance of the three components of the eBicov components (i.e., BiCov, wHSV, and MSCR) when used alone.

Person Re-identification on the ETHZ Dataset

The ETHZ dataset contains three video sequences of crowded street scenes captured by two moving cameras mounted on a chariot. SEQ. #1 includes 4,857 images of 83 pedestrians, SEQ. #2 1,961 images of 35 pedestrians, and SEQ. #3 1,762 images of 28 pedestrians. The most challenging aspects of ETHZ are illumination changes and occlusions. We follow the evaluation framework proposed by [6] to perform these experiments.

Figure 2.5 shows the CMC curves for the three different sequences, for both single ($N = 1$) and multiple shots ($N = 2, 5, 10$) cases. In the single-shot case, we can see that the performance of BiCov alone is already much better than that of SDALF, on all of the three sequences. The performance of eBiCov[1] is greatly improved on SEQ. 1 and 2. In particular, on SEQ. 1, eBiCov is 7 % better than SDALF at ranks between

[1] Remember that eBiCov is the combination of BiCov, MSCR, and wHSV.

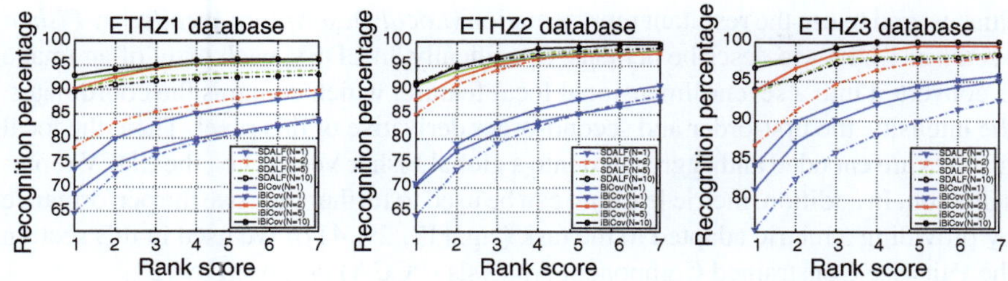

Fig. 2.5 The CMC *curves* on the ETHZ dataset

1 and 7. In SEQ. 2, matching rate at rank 1 is around 71 % for eBiCov and 64 % for SDALF. Compared with the improvements observed on VIPeR, improvements on ETHZ are even more obvious. As the images come from a few video sequences, they are rather similar and the performance is more heavily dependent on the quality of the descriptor.

Besides the single-shot setting, we also tested our method in the multishot case. As in [6], N is set to 2, 5, 10. The results are given in Fig. 2.5. It can be seen that on SEQ. 1 and 3, the proposed eBiCoV gives much better results than SDALF. It is even more obvious on SEQ. 3 for which our method's CMC is equal to 100 % for $N = 5, 10$, which experimentally validates our descriptor.

2.4 Fisher Vector Encoded Local Descriptors for Person Re-identification

This section presents our second descriptor and experimentally demonstrates its effectiveness on the two previously mentioned benchmarks.

As explained in the Introduction, this descriptor is based on local features embedding. The most common approach for combining local features into a global signature is the Bag-of-Words (BoW) model [35], in which local features extracted from an image are mapped to a set of pre-learned visual words, the image being represented as a histogram of visual word occurrences. The BoW model has been used for person re-identification in [41], where the authors built groups of descriptors by embedding the visual words into concentric spatial structures and by enriching the BoW description of a person by the contextual information coming from the surrounding people. Recently, the BoW model has been greatly enhanced by the Fisher Vector [26] which encodes higher order statistics of local features. Compared with BoW, Fisher Vectors encode how the parameters of the model should be changed to optimally represent the image, rather than only the number of visual words occurrences. It has been shown that the resultant Fisher Vector gives excellent performance for several challenging object recognition and image retrieval tasks [27, 28]. Motivated by these recent advances, we propose to combine Fisher Vectors with a novel and very simple seven-dimensional local descriptor adapted to the representation of persons images, and to use the resultant representation (*Local Descriptors encoded by Fisher Vector* or LDFV) to describe persons. Specifically, in LDFV, each pixel of an image is converted into a seven-dimensional local feature, which contains the coordinates, the intensity, the first-order and second-order derivative of this pixel. Then, the local features are encoded and aggregated into a global Fisher Vector, i.e., the LDFV representation. In addition, metric learning can be used to further improve the performance by providing a metric adapted to the task (e.g. [17, 29, 41]). We used in this section the Pairwise Constrained Component Analysis (PCCA) proposed by [23].

2.4.1 Local Image Descriptor

In order to capture the local properties of images, we have designed a very simple seven-dimensional descriptor inspired by [37] as well as by the method proposed in the first section of this chapter:

$$f(x, y, I) = (x, y, I(x, y), I_x(x, y), I_y(x, y), I_{xx}(x, y), I_{yy}(x, y)) \tag{2.11}$$

where x and y are the pixel coordinates, $I(x, y)$ is the raw pixel intensity at position (x, y), I_x and I_y are the first-order derivatives of image I with respect to x and y, and I_{xx} and I_{yy} are the second-order derivatives.

Let $M = \{m_t, t = 1, \dots, T\}$ be the set of the T local descriptors extracted from an image. The key idea of Fisher Vectors [26] is to model the data with a generative model and compute the gradient of the likelihood of the data with respect to the parameters of the model, i.e., $\nabla_\lambda \log p(M|\lambda)$. We model M with a Gaussian mixture model (GMM) using Maximum Likelihood (ML) estimation. Let $\hat{u}_\lambda$ be the GMM model: $\hat{u}_\lambda(m) = \sum_{i=1}^{K} w_i u_i(\mu_i, \sigma_i)$, where K is the number of Gaussian components. The parameters of the models are $\lambda = \{w_i, \mu_i, \sigma_i, i = 1, \dots, K\}$, where w_i denotes the weight of the i-th component, while μ_i and σ_i are its mean and its standard deviations. We assume the covariance matrices are diagonal and σ_i represents the vector of standard deviations of the i-th component of the model. It is worth pointing out that, considering the computational efficiency for each image in the training set, only a randomly selected subset of local features is used to train the GMM model.

After getting the GMM, image representations are computed using Fisher Vector, which is a powerful method for aggregating local descriptors and has been demonstrated to outperform the BoW model by a large margin [8].

Let $\gamma_t(i)$ be the soft assignment of the descriptor m_t to the component i:

$$\gamma_t(i) = \frac{w_i u_i(m_t)}{\sum_{j=1}^{K} w_j u_j(m_t)} \tag{2.12}$$

$G_{\mu,i}^M$ and $G_{\sigma,i}^M$ are the 7-dimensional gradients with respect to μ_i and σ_i of the component i. They can be computed using the following derivations:

$$G^M_{\mu,i} = \frac{1}{T\sqrt{w_i}} \sum_{t=1}^{T} \gamma_t(i) \left(\frac{m_t - \mu_i}{\sigma_i} \right) \tag{2.13}$$

$$G^M_{\sigma,i} = \frac{1}{T\sqrt{2w_i}} \sum_{t=1}^{T} \gamma_t(i) \left[\frac{(m_t - \mu_i)^2}{\sigma_i^2} - 1 \right] \tag{2.14}$$

where the division between vectors is performed as a term-by-term operation. The final gradient vector G is the concatenation of the $G^M_{\mu,i}$ and $G^M_{\sigma,i}$ vectors for $i = 1, \ldots, K$ and is therefore $2 \times 7 \times K$-dimensional.

LDFV on color images. Previous works have shown that using color is a useful cue for person re-identification. We use the color information by splitting the image into three color channels (HSV), extract the proposed descriptor on each channel separately, and finally concatenate the three descriptors into a single signature.

Similarity between LDFV representations. Finally, the distance between two images I_i and I_j can be obtained by computing the Euclidean distance between their representations :

$$d(I_i, I_j) = ||LDFV_i - LDFV_j||. \tag{2.15}$$

2.4.2 Extending the Descriptor

Adding spatial Information. To provide a rough approximation of the spatial information, we divide the image into many rectangular bins and compute one LDFV descriptor per bin. Please note that for doing this we compute one GMM per bin. Then, the descriptors of the different bins are concatenated to form the final representation. It is denoted by bLDFV, for bin-based LDFV.

It must be pointed out that our method does not use any body part segmentation. However, adapting the bins to body parts would be possible and could make the results even better.

Combining LDFV with other features. As mentioned in the Introduction, combining different types of image descriptors is generally useful. In this chapter, we combine our bLDFV descriptor with two other descriptors: the Weighted Color Histogram (wHSV) and the MSCR, shown to be efficient for this task [6]. We denote this combination as eLDFV (enriched LDFV). In eLDFV, the difference between two image signatures $eD_1 = (HA_1, MSCR_1, bLDFV_1)$ and $eD_2 = (HA_2, MSCR_2, bLDFV_2)$ is computed as:

$$d_{eLDFV}(eD_1, eD_2) = \frac{1}{6} d_{wHSV}(HA_1, HA_2) + \frac{1}{6} d_{MSCR}(MSCR_1, MSCR_2) + \frac{2}{3} d_{bLDFV}(bLDFV_1, bLDFV_2). \tag{2.16}$$

Regarding the definition of d_{wHSV} and d_{MSCR}, we use those given in [6]. For simplicity reasons and because it is not the central part of the chapter, we have set the mixing weights by hand, giving more importance to the proposed descriptor. Learning them could certainly improve the results further.

Using metric learning. In addition to the unsupervised similarity function (Eq. 2.15), we have also evaluated a supervised similarity function in which we use PCCA [23] to learn the metric. This variant is denoted sLDFV for supervised bLDFV. Any metric learning could have been done but we chose PCCA because of its success in person re-identification [23]. PCCA learns a projection into a low-dimensional space where the distance between pairs of data points respects the desired constraints, exhibiting good generalization properties in the presence of high-dimensional data. Please note that the bLDFV descriptors are preprocessed by applying a whitened PCA before PCCA, to make the computation faster. In sLDFV, PCCA is used with a linear kernel.

2.4.3 Experiments

The proposed approach has been experimentally validated on the two previously introduced person re-identification datasets (VIPeR [17] and ETHZ [12, 33]). We present in this section several experiments showing the efficiency of our simple LDFV descriptor and its extensions.

Evaluation of the Image Descriptor

In this section, our motivation is to evaluate the intrinsic properties of the descriptor. For this reason we do not use any metric learning but simply measure the similarity between two persons using the Euclidean distance between their representations.

Evaluation of the simple feature vector. The core of our descriptor is the seven-dimensional simple feature vector given by Eq. 2.11. This first set of experiments aims at validating this feature vector by comparing it with several alternatives, the rest of the framework being exactly the same. We performed experiments with (1) SIFT features (reduced to 64 dimensions by PCA) and (2) Gabor features [13] (with eight scales and eight orientations). For these experiments, we divide the bounding box into 12 bins (3×4) and the number of GMM components is set to 16. For each bin and each one of the three color channels (HSV), we compute the FV model and concatenate the 12 descriptors for obtaining the final representation. The size of the final descriptor is therefore $7 \times 16 \times 12 \times 2 \times 3$ for our 7-d descriptor, $64 \times 16 \times 12 \times 2 \times 3$ for both the SIFT and Gabor descriptor based FV. We then

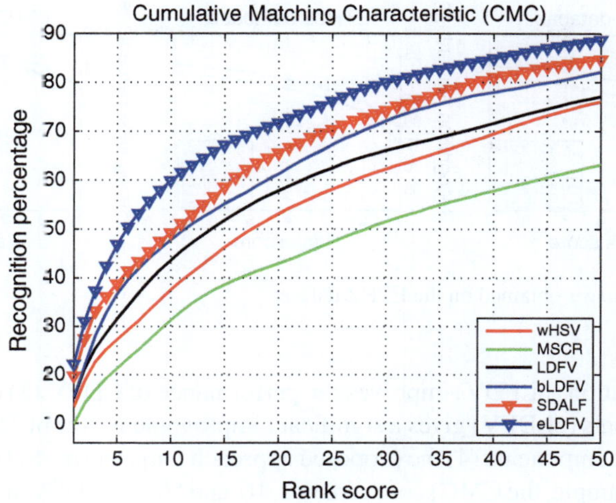

Fig. 2.6 VIPeR dataset: CMC *curves* obtained with LDFV, bLDFV, eLDFV and SDALF

compute CMC normalized Area under Curve (nAUC) on VIPeR and get 83.17, 86.37, and 91.60 %, respectively, for SIFT, Gabor and bLDFV using our seven-dimensional feature vector. Consequently, the proposed descriptor, in addition to being compact and very simple to compute, gives much better results than SIFT and Gabor filters for this task.

We have evaluated the performance of our descriptor for different number of GMM components (16, 32, 50, and 100), and have observed that the performance is not very sensitive to this parameter. Consequently, we use 16 components in all of our experiments, which is a good tradeoff between performance and efficiency.

A set of representative images is required to learn the GMM. We conducted a set of experiments in order to evaluate how critical the choice for these images is. Our experiments have shown that using the whole dataset or only a smaller training set independent from the test set makes almost no difference, showing that, in practice, a small set of representative images is more than enough for learning the GMM.

Single-shot experiments. Single-shot means that a single image is used as the query. We first present some experiments on the VIPeR dataset, showing the relative importance of the different components of our descriptor. The full descriptor (eLDFV) is based on a basic Fisher encoding of the simple seven-dimensional feature vector (LFDV) computed on the three color channels (HSV). The two extensions are (1) bLFDV which embeds spatial encoding and (2) the combination with two other features (namely wHSV and MSCR).

Figure 2.6 shows the performance of eLDFV as well as the performance of wHSV, MSCR, and bLDFV alone. We follow the same experimental protocol as that of [6], and report the average performance over 10 random splits of 316 persons. The figure also gives the performance of the state-of-the-art SDALF [6]. We can draw several conclusions: (1) LDFV alone performs much better than MSCR and wHSV (2) using

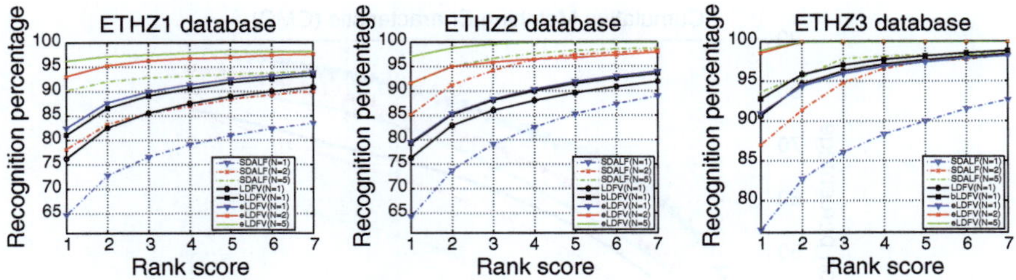

Fig. 2.7 CMC *curves* obtained on the ETHZ dataset

spatial information (bLFDV) improves the performance of LDFV (3) combining the three components (eLDFV) gives a significant improvement over bLDFV and any of the individual components (4) the proposed approach outperforms SDALF by a large margin. For example, the CMC score at rank 1, 10, and 50 for eLDFV are 22.34, 60.04, and 88.82 %, respectively, while those of SDALF are 19.84, 49.37, and 84.84 %.

We have also tested the proposed descriptor on the ETHZ database, in the single-shot scenario ($N = 1$). Here again we follow the evaluation protocol proposed by [6]. Figure 2.7 shows the CMC curves for the three different sequences. In the figure, dashed results come from [6]. The solid line is given by the proposed method. We can see that the performances of LDFV, bLDFV, and eLDFV are all much better than that of SDALF, on all the three sequences, and improvements are even more visible than on VIPeR. Especially, on SEQ. 1 and 3, the performances of eLDFV are much worse than those of bLDFV though eLDFV is the combination of bLDFV, wHSV, and MSCR. We attribute this to the low accuracy of wHSV and MSCR. In particular, on SEQ. 1, the minimum and maximum of the matching rate between the eLDFV and SDALF is about 10 and 18 %, respectively. In SEQ. 2, the matching rate at rank 1 is around 80 % for eLDFV and 64 % for SDALF. The average difference of the matching rate between eLDFV and SDALF, at rank 7, is about 10 % in SEQ. 3.

Multishot experiments on ETHZ. Besides the single-shot case, we also test our descriptors in the multishot case. In this case $N \geq 2$ images are used as queries. We again follow the evaluation framework proposed by [6], the number of query images N being set to 2 and 5. Results are also shown in Fig. 2.7. We can see that on SEQ. 1 and 3, eLDFV gives almost perfect results. Especially, on SEQ. 3, the performance of eLDFV is 100 % with $N \geq 2$, for ranks greater than 2.

Comparison with Recent Approaches

In this section we compare our framework with recent approaches. For making comparison fair, we use here the metric learning algorithm described in Sect. 2.4.2.

We first present some experiments done on the VIPeR dataset. Following the standard protocol for this dataset, the dataset is split into a train and a test set by randomly selecting 316 persons out of the 632 for the test set, the remaining persons

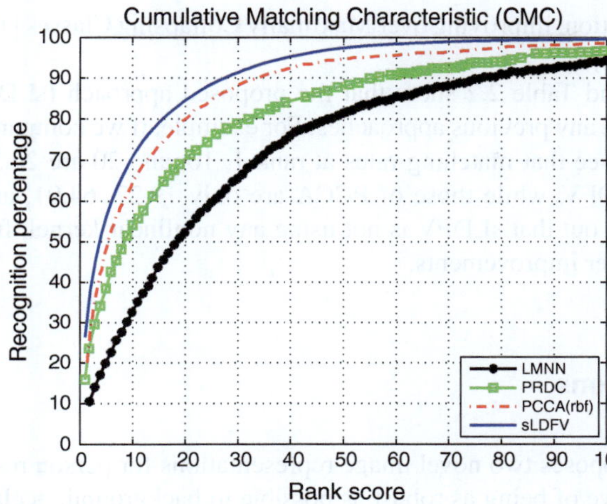

Fig. 2.8 VIPeR dataset: CMC *curves* with 316 persons

Table 2.2 VIPeR dataset: matching rates (%) at rank *r* with 316 persons

Method	$r = 1$	$r = 5$	$r = 10$	$r = 20$
PRDC [42]	15.66	38.42	53.86	70.09
MCC [42]	15.19	41.77	57.59	73.39
ITML [42]	11.61	31.39	45.76	63.86
LMNN [42]	6.23	19.65	32.63	52.25
CPS [9]	21.00	45.00	57.00	71.00
PRSVM [29]	13.00	37.00	51.00	68.00
ELF [18]	12.00	31.00	41.00	58.00
PCCA-sqrt $n^- = 10$ [23]	17.28	42.41	56.68	74.53
PCCA-rbf $n^- = 10$ [23]	19.27	48.89	64.91	80.28
sLDFV $n^- = 10$	**26.53**	**56.38**	**70.88**	**84.63**

The values of bold are the best performance of different methods at the specific rank.

being in the training set. As in [23], one negative pair is produced for each person, by randomly selecting one image of another person. We produce 10 times more negative pairs than positive ones. The process is repeated 100 times and the results are reported as the mean/std values over the 100 runs.

Figure 2.8 and Table 2.2 compare our approach (sLDFV) with three different approaches using metric learning: PRDC [42], LMNN [38] and PCCA [23]. The results of PRDC and LMNN are taken from [42] while the ones of PCCA come from [23]. For PRDC and LMNN, the image representation is the combination of RGB, YCbCr, and HSV color features and two texture features extracted by local derivatives and Gabor filters on six horizontal strips. For PCCA, the feature descriptor is a 16 bin color histogram in three color spaces (RGB, HSV, and YCrCb) as well as texture histograms based on Local Binary Patterns (LBP) computed on six nonoverlapping horizontal strips. PCCA [23] reports state-of-the-art results for person re-identification, improving over Maximally Collapsing Classes [16], ITML [10] or LMNN-R [11].

Figure 2.8 and Table 2.2 show that the proposed approach (sLDFV) performs much better than any previous approaches. For example, if we compare sLDFV with PCCA, we can see that matching rates at rank 1, 10, and 20 are 26.53, 70.88, and 84.63 % for sLDFV, while those of PCCA are only 19.27, 64.91, and 80.28 %. It must be pointed out that sLDFV is not using any nonlinear kernel, from which we can expect further improvements.

2.5 Conclusions

This chapter proposes two novel image representations for person re-identification, with the objective of being as robust as possible to background, occlusions, illumination, or viewpoint changes. The first representation—so-called BiCov—combines Biologically Inspired Features (BIF) and covariance descriptors. BiCov is more robust to illumination, scale, and background variations than competing approaches which makes it suitable for person re-identification. The second representation—namely LDFV—is based on a simple seven-dimensional feature representation encoded by Fisher Vectors. We have validated these two descriptors on two challenging public datasets (VIPeR and ETHZ) for which they outperformed all current state-of-the-art methods.

Though both the two proposed representations outperform state-of-the-art approaches, they have their own characteristics. While BiCov is usually not as well performing as LDVF, it is worth pointing out that it does not need any training images, which is a huge advantage for real applications. In addition, it is very fast as the most computational demanding step is to extract the low-level features. On the other hand, LDFV requires to build a GMM model during the training stage, which is time-consuming. However, after getting the GMM model, the computation of the representation of the testing sample is very fast, which makes it usable in online systems.

Acknowledgments This work was partly realized as part of the Quaero Program funded by OSEO, French State agency for innovation and by the ANR, grant reference ANR-08-SECU-008-01/SCARFACE. The first author is partially supported by National Natural Science Foundation of China under contract No. 61003103.

References

1. Ayedi, W., Snoussi, H., Abid, M.: A fast multi-scale covariance descriptor for object re-identification. Pattern Recogn. Lett. (2011)
2. Aziz, K., Merad, D., Fertil, B.: People re-identification across multiple non-overlapping cameras system by appearance classification and silhouette part segmentation. In: Proceedings

of International Conference on Advanced Video and Signal-Based Surveillance, pp. 303–308 (2011)
3. Aziz, K., Merad, D., Fertil, B.: Person re-identification using appearance classification. In: International Conference on Image Analysis and Recognition, Burnaby (2011)
4. Bak, S., Corvee, E., Bremond, F., Thonnat, M.: Person re-identification using haar-based and DCD-based signature. In: Proceedings of International Workshop on Activity Monitoring by Multi-camera Surveillance Systems (2010)
5. Bak, S., Corvee, E., Bremond, F., Thonnat, M.: Multiple-shot human re-identification by mean Riemannian covariance grid. In: Proceedings of International Conference on Advanced Video and Signal-Based Surveillance (2011)
6. Bazzani, L., Cristani, M., Murino, V.: Symmetry-driven accumulation of local features for human characterization and re-identification. Comput. Vis. Image Underst. **117**(2), 130–144 (2013)
7. Bazzani, L., Cristani, M., Perina, A., Murino, V.: Multiple-shot person re-identification by chromatic and epitomic analyses. Pattern Recogn. Lett. **33**(7), 898–903 (2012). (Special Issue on Awards from ICPR 2010)
8. Chatfield, K., Lempitsky, V., Vedaldi, A., Zisserman, A.: The devil is in the details: an evaluation of recent feature encoding methods. In: Proceedings of British Machine Vision Conference (2011)
9. Cheng, D., Cristani, M., Stoppa, M., Bazzani, L., Murino, V.: Custom pictorial structures for re-identification. In: Proceedings of British Machine Vision Conference (2011)
10. Davis, J.V., Kulis, B., Jain, P., Sra, S., Dhillon, I.S.: Information-theoretic metric learning. In: Proceedings of International Conference on Machine Learning, pp. 209–216 (2007)
11. Dikmen, M., Akbas, E., Huang, T., Ahuja, N.: Pedestrian recognition with a learned metric. Proc. Asian Conf. Comput. Vis. **4**, 501–512 (2010)
12. Ess, A., Leibe, B., Schindler, K., van Gool, L.: A mobile vision system for robust multi-person tracking. In: Proceedings of the IEEE Conference on Computer Vision and Pattern Recognition (2008)
13. Fisher, R.A.: The use of multiple measures in taxonomic problems. Ann. Eugenics **7**, 179–188 (1936)
14. Gandhi, T., Trivedi, M.: Person tracking and re-identification: introducing panoramic appearance map (PAM) for feature representation. Mach. Vis. Appl. **18**(3–4), 207–220 (2007)
15. Gheissari, N., Sebastian, T., Tu, P., Rittscher, J., Hartley, R.: Person reidentification using spatiotemporal appearance. In: Proceedings of the IEEE Conference on Computer Vision and Pattern Recognition, vol. 2, pp. 1528–1535 (2006)
16. Globerson, A., Roweis, S.: Metric learning by collapsing classes. In: Advances in Neural Information Processing Systems (2006)
17. Gray, D., Brennan, S., Tao, H.: Evaluating appearance models for recognition, reacquisition, and tracking. In: IEEE International Workshop on Performance Evaluation of Tracking and Surveillance (2007)
18. Gray, D., Tao, H.: Viewpoint invariant pedestrian recognition with an ensemble of localized features. In: Proceedings of the European Conference on Computer Vision, pp. 262–275 (2008)
19. Guillaumin, M., Verbeek, J., Schmid, C.: Is that you? metric learning approaches for face identification. In: Proceedings of the IEEE International Conference on Computer Vision(2009)
20. Guo, G., Mu, G., Fu, Y., T.S. Huang: Human age estimation using bio-inspired features. In: Proceedings of the IEEE Conference on Computer Vision and Pattern Recognition, pp. 112–119 (2009)
21. Kai, J., Bodensteiner, C., Arens, M.: Person re-identification in multi-camera networks. In: Proceedings of the IEEE Conference on Computer Vision and Pattern Recognitio Workshops, pp. 55–61 (2011)
22. Meyers, E., Wolf, L.: Using biologically inspired features for face processing. Int. J. Comput. Vis. **76**(1), 93–104 (2008)
23. Mignon, A., Jurie, F.: PCCA: a new approach for distance learning from sparse pairwise constraints. In: Proceedings of the IEEE Conference on Computer Vision and Pattern Recognition (2012)

24. Moon, H., Phillips, P.: Computational and performance aspects of PCA-based face-recognition algorithms. Perception **30**(3), 303–321 (2001)
25. Oreifej, O., Mehran, R., Shah, M.: Human identity recognition in aerial images. In: Proceedings of the IEEE Conference on Computer Vision and Pattern Recognition (2010)
26. Perronnin, F., Dance, C.: Fisher kernels on visual vocabularies for image categorization. In: Proceedings of the IEEE Conference on Computer Vision and Pattern Recognition, pp. 1–8 (2007)
27. Perronnin, F., Liu, Y., Sánchez, J., Poirier, H.: Large-scale image retrieval with compressed Fisher vectors. In: Proceedigs of the IEEE Conference on Computer Vision and Pattern Recognition (2010)
28. Perronnin, F., Sánchez, J., Mensink, T.: Improving the Fisher kernel for large-scale image classification. In: Proceedings of the European Conference on Computer Vision, pp. 143–156 (2010)
29. Prosser, B., Zheng, W., Gong, S., Xiang, T.: Person re-identification by support vector ranking. In: Proceedings of the British Machine Vision Conference (2010)
30. Riesenhuber, M., Poggio, T.: Hierarchical models of object recognition in cortex. Nat. Neurosci. **2**(11), 1019–1025 (1999)
31. Satta, R., Fumera, G., Roli, F.: Exploiting dissimilarity representations for person re-identification. In: Proceedings of the International Workshop on Similarity-Based Pattern Analysis and Recognition (2011)
32. Satta, R., Fumera, G., Roli, F., Cristani, M., Murino, V.: A multiple component matching framework for person re-identification. In: International Conference on Image Analysis and Processing (2011)
33. Schwartz, W., Davis, L.: Learning discriminative appearance based models using partial least squares. In: Brazilian Symposium on Computer Graphics and Image Processing (2009)
34. Serre, T., Wolf, L., Poggio, T.: Object recognition with features inspired by visual cortex. In: Proceedings of IEEE Conference on Computer Vision and Pattern Recognition, vol. 2, pp. 994–1000 (2005)
35. Sivic, J., Zisserman, A.: Video google: a text retrieval approach to object matching in videos. In: Proceedings of IEEE International Conference on Computer Vision (2003)
36. Song, D., Tao, D.: Biologically inspired feature manifold for scene classification. IEEE Trans. Image Process. **19**, 174–184 (2010)
37. Tuzel, O., Porikli, F., Meer, P.: Pedestrian detection via classification on riemannian manifolds. IEEE Trans. Pattern Anal. Mach. Intell. **30**(10), 1713–1727 (2008)
38. Weinberger, K., Saul, L.: Distance metric learning for large margin nearest neighbor classification. J. Mach. Learn. Res. **10**, 207–244 (2009)
39. Wiskott, L., Fellous, J.M., Krüger, N., Malsburg, C.V.D.: Face recognition by elastic bunch graph matching. IEEE Trans. Pattern Anal. Mach. Intell. **19**(7), 775–779 (1997)
40. Zhang, Y., Li, S.: Gabor-LBP based region covariance descriptor for person re-identification. In: International Conference on Image and Graphics, pp. 368–371 (2011)
41. Zheng, W., Gong, S., Xiang, T.: Associating groups of people. In: Proceedings of British Machine Vision Conference (2009)
42. Zheng, W., Gong, S., Xiang, T.: Re-identification by relative distance comparison. IEEE Trans. Pattern Anal. Mach. Intell. **35**(3), 653–668 (2013)

Chapter 3
SDALF: Modeling Human Appearance with Symmetry-Driven Accumulation of Local Features

Loris Bazzani, Marco Cristani and Vittorio Murino

Abstract In video surveillance, person re-identification (re-id) is probably *the* open challenge, when dealing with a camera network with non-overlapped fields of view. Re-id allows the association of different instances of the same person across different locations and time. A large number of approaches have emerged in the last 5 years, often proposing novel visual features specifically designed to highlight the most discriminant aspects of people, which are invariant to pose, scale and illumination. In this chapter, we follow this line, presenting a strategy with three important key-characteristics that differentiate it with respect to the state of the art: (1) a symmetry-driven method to automatically segment salient body parts, (2) an accumulation of features making the descriptor more robust to appearance variations, and (3) a person re-identification procedure casted as an image retrieval problem, which can be easily embedded into a multi-person tracking scenario, as the observation model.

3.1 Introduction

Modeling the human appearance in surveillance scenarios is challenging because people are often monitored at low resolution, under occlusions, bad illumination conditions, and in different poses. Robust modeling of the body appearance of a person becomes mandatory for re-identification and tracking, especially when other

L. Bazzani (✉) · M. Cristani · V. Murino
Pattern Analysis & Computer Vision, Istituto Italiano di Tecnologia, Genova, Italy
e-mail: loris.bazzani@iit.it

M. Cristani · V. Murino
University of Verona, Verona, Italy
e-mail: marco.cristani@univr.it

V. Murino
e-mail: vittorio.murino@iit.it

S. Gong et al. (eds.), *Person Re-Identification*,
Advances in Computer Vision and Pattern Recognition,
DOI: 10.1007/978-1-4471-6296-4_3,

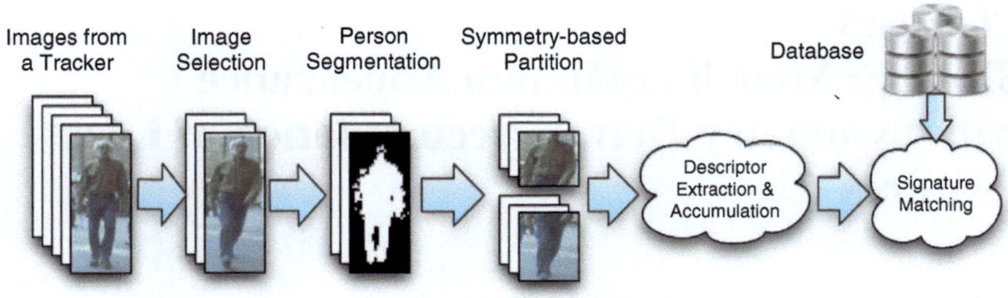

Fig. 3.1 Person re-id and re-acquisition pipeline. See the text for details

classical biometric cues (e.g., face, gait, or fingerprint) are not available or difficult to acquire.

Appearance-based re-id can be considered as a general *image retrieval* problem, where the goal is to find the images from a database that are more similar to the query. The only constraint is that there is an assumption of the presence of a person in the query image and the images in the database. On the other hand, person re-identification is also seen as a fundamental module for *cross-camera tracking* to keep unique identifiers in a camera network. In this setup, temporal and geometric constraints can be added to make easier re-id. In general, we define re-identification as matching the signature of each *probe* individual to a *gallery* database composed by hundreds or thousands of candidates which have been captured in various locations by different cameras and in different instants. Similarly to re-identification, multi-person tracking is another problem where the description of individuals plays an important role to ensure consistent tracks across time. In this context, the problem can be seen as matching across time the signature (also called template) of the tracked person with the set of detected individuals. Both re-identification and people tracking share the problem of modeling the human appearance in a way that is robust to occlusion, low resolution, illumination and other issues.

In this chapter, we describe the pipeline for re-identification that has become a standard in the last few years [14]. The pipeline and the descriptor used for characterizing the human appearance are called *Symmetry-Driven Accumulation of Local Features* (SDALF). The re-id pipeline is defined in six steps (Fig. 3.1): (1) *image gathering* collects images from a tracker; (2) *image selection* discards redundant information; (3) *person segmentation* discards the noisy background information; (4) *symmetry-based silhouette partition* discovers parts from the foreground exploiting symmetric and asymmetric principles; (5) *descriptor extraction and accumulation* over time, using different frames in a multi-shot modality; (6) *signature matching* between the probe signature and the gallery database.

SDALF is composed by a symmetry-based description of the human body, and it is inspired by the well-known principle that the natural objects reveal symmetry in some form. For this reason, detecting and characterizing symmetries is useful to understand the structure of objects. This claim is strongly supported by the Gestalt psychology school [30] that considers symmetry as a fundamental principle of perception: symmetrical elements are more likely integrated into one coherent object than asymmetric regions. The principles of the Gestaltism have been largely exploited in computer vision for characterizing salient parts of structured objects [10, 31, 43, 44]. In SDALF, asymmetry principles allow to segregate meaningful body parts (head, upper body, lower body). Symmetries help to extract features from the actual human body, pruning out distracting background clutter. The idea is that features near the vertical symmetry axis are weighted more than those that are far from it, in order to obtain information from the internal part of the body, trusting less the peripheral portions more prone to noise.

Once parts have been localized, complementary aspects of the human body appearance are extracted in SDALF, highlighting: (i) the global chromatic content, by the color histogram (see Fig. 3.4c); (ii) the per-region color displacement, employing Maximally Stable Colour Regions (MSCR) [18] (see Fig. 3.4d); (iii) the presence of *Recurrent Highly Structured Patches* (RHSP) [14] (see Fig. 3.4e).

Different feature accumulation strategies can be considered for re-id, and in this regard the literature is divided in single-shot and multi-shot modes, reflecting the way the descriptors are designed (see Sect. 3.2 for more details). In the former case, the signature is built using only one image for each individual, whereas in the latter multiple images are utilized. The multi-shot mode is strongly motivated by the fact that in several surveillance scenarios it is really easy to extract multiple images of the same individual from consecutive frames. For example, if an automatic tracking system is available, consecutive shots of a tracked individual can be used in refining the object model against appearance changes. SDALF takes into account these situations: it accumulates the descriptors from all the available images of an individual, increasing the robustness and the expressiveness of its description. After the signature is built, the matching phase consists of a distance minimization strategy to search for a probe signature across the gallery set, in a similar spirit of image retrieval algorithms.

In this chapter, we discuss how SDALF can be easily adapted to deal with the multi-person tracking problem, in the same spirit of [5]. The idea is to build a signature for each tracked target (the template). Then, the signature is matched against a gallery set: this set is composed by diverse hypotheses that come from a detection module or from the tracking dynamics. The idea is to employ the matching scores as probabilistic evaluations of the hypotheses. The template is then updated with SDALF, as multiple images are gathered over time in the multi-shot mode.

The proposed method is tested on challenging benchmarks: VIPeR [20], iLIDS for re-id [54], ETHZ [47], and CAVIAR4REID [9], giving convincing performance. These benchmarks represent different challenges for the re-id problem: pose, viewpoint and lighting variations, and occlusions. We test the limit of SDALF by sub-sampling these datasets up to dramatic resolutions (11×22 pixels). Moreover, the multi-person tracker based on SDALF was tested on CAVIAR, which represents a challenging real tracking scenario, due to pose, resolution and illumination changes, and severe occlusions.

The rest of the chapter is organized as follows. In Sect. 3.2, the state of the art of re-id is described, highlighting our peculiarities with respect to other approaches.

Table 3.1 Taxonomy of the existing appearance-based re-identification methods

	Single-shot	Multiple-shot
Learning-based	[21, 32, 36, 41, 46, 47] [1, 16, 23, 54, 55]	[48, 53]
Direct methods	[2] **SDALF**	[7, 19, 22, 45, 52] **SDALF**

Section 3.3 details the re-id pipeline and the SDALF descriptor. Section 3.4 describes how the signature matching is performed. Section 3.5 describes how SDALF can be embedded into a particle filtering-based tracker. Several results and comparative analyses are reported in Sect. 3.6, and, finally, conclusions and future perspectives are discussed in Sect. 3.7.

3.2 Related Work

Re-id methods that rely only on visual information are addressed as *appearance-based* techniques. Other approaches assume less general operative conditions: *geometry-based techniques* exploit geometrical constrains in a scenario with overlapped camera views [39, 50]. *Temporal methods* deal with non-overlapped views adding a temporal reasoning on the spatial layout of the monitored environment, in order to prune the candidate set to be matched [25, 33, 42]. The assumption is that people usually enter in a few locations, spend a fixed period (learned beforehand) in the blind spots, and re-appear somewhere else in the field of view of a pre-selected set of cameras. *Depth-based approaches* consider other sensors (such as RGB-D cameras) to extract 3D soft-biometric cues from depth images in order to be robust to the change of clothes [3].

Appearance-based methods can be divided into two groups (see Table 3.1): the *learning-based* methods and the *direct* methods. Learning-based techniques are characterized by the use of a training dataset of *different individuals* where the features and/or the policy for combining them are utilized. The common assumption is that the knowledge extracted from the training set could be generalized to unseen examples. In [36], local and global features are accumulated over time for each subject, and fed into a multi-class SVM for recognition and pose estimation, employing different learning schemes. Viewpoint invariance is instead the main issue addressed by [21]: spatial and color information are here combined using an ensemble of discriminant localized features and classifiers selected by boosting. In [32], pairwise dissimilarity profiles between individuals are learned and adapted for a nearest neighbor classification. Similarly, in [47], a high-dimensional signature composed by texture, gradient and color information is projected into a low-dimensional discriminant latent space by Partial Least Squares (PLS) reduction. Multiple Component Learning is casted into the re-id scenario, dubbing it a Multiple Component Matching and exploiting SDALF as a descriptor, in [46]. The descriptor proposed in [54] uses contextual visual knowledge coming from the surrounding people that form a group, assuming that groups can be detected. Re-id is casted as a binary classification problem (one vs. all) by [1] using Haar-like features and a part-based MPEG7 dominant color descriptor. In [41, 53, 55], the authors formulate re-id as a ranking problem and an informative subspace is learned where the potential true match corresponds to the highest ranking. Metric learning methods, which learn a distance metric from pairs of samples from different cameras, are becoming popular, see [23, 34]. In [16], re-id is defined as a semi-supervised single-shot recognition problem where multiple features are fused at the classification output level using the recent multi-view learning framework in [35].

The main disadvantage of learning-based methods is the need of retraining for environment covariates, e.g., night–day, indoor–outdoor. In addition, some learning–based approaches also depend on the cardinality and the kind of training set: once a new individual is added to the gallery set, the classifier should be retrained from scratch.

The other class of approaches, the direct method class, does not consider training datasets of multiple people and works on each person independently, usually focusing on the design of features that capture the most distinguishing aspects of an individual. In [7], the bounding box of a pedestrian is equally subdivided into ten horizontal stripes, and the median HSL value is extracted in order to manage x-axis pose variations. These values, accumulated over different frames, generate a multiple signature. A spatio-temporal local feature grouping and matching is proposed by [19], considering ten consecutive frames for each person, and estimating a region-based segmented image. The same authors present a more expressive model, building a decomposable triangulated graph that captures the spatial distribution of the local descriptions over time, so as to allow a more accurate matching. In [52], the method consists in segmenting a pedestrian image into regions, and registering their color spatial relationship into a co-occurrence matrix. This technique proved to work well when pedestrians are seen under small variations of the point of view. In [22], the person re-id scheme is based on the matching of SURF interest points [4] collected in several images, during short video sequences. Covariance features, originally employed for pedestrian detection, are extracted from coarsely located body parts and tailored for re-id purposes [2].

Considering the features employed for re-id, in addition to color information, which is universally adopted, several other cues are textures [21, 41, 47], edges [47], Haar-like features [1], interest points [19], image patches [21], and segmented regions [52]. These features, when not collected densely, can be extracted from horizontal stripes [7], triangulated graphs [19], concentric rings [54], and localized patches [2].

Besides, the taxonomy (Table 3.1) for the re-identification algorithms distinguishes the class of the *single-shot* approaches, focusing on associating pairs of images, each containing one instance of an individual, from the class of *multiple-shot* methods. The latter employs multiple images of the same person as probe or gallery elements. The assumption of the multi-shot methods is that individuals are tracked, so that it is possible to gather lots of images. The hope is that the system will obtain a set of images that vary in terms of resolution, partial occlusions, illumination, poses, etc. In this way, we can build a significant signature of each individual.

Looking at Table 3.1, which reports all these four paradigms of re-id, it is worth noting that direct single-shot approaches represent the case where the least information is employed. For each individual, we have a single image, whose features are independently extracted and matched against hundreds of candidates. The learning-based multi-shot approaches, instead, are in the opposite situation. The proposed method lies in the class of the direct strategies and works both in the single and in the multi-shot modality.

3.3 Symmetry-Driven Accumulation of Local Features (SDALF)

As discussed in the previous section, we assume to have a set of trackers that estimate the trajectories of each person in the several (non-)overlapped camera views. For each individual, a set of bounding boxes can be obtained (from one or more consecutive frames), and SDALF analyzes these images to build a signature while performing matching for recognizing individuals in a database of pre-stored individuals. The proposed re-id pipeline of SDALF consists of six phases as depicted in Fig. 3.1:

1. *Image Gathering* aggregates images given by the trajectories of the individuals and their bounding boxes.
2. *Image Selection* selects a small set of representative images, when the number of images is very big (e.g., in tracking) in order to discard redundant information. [Sect. 3.3.1]
3. *Person Segmentation* separates the pixels of the individual (foreground) from the rest of the image (background) that usually "distracts" the re-id. [Sect. 3.3.2]
4. *Symmetry-based Silhouette Partition* detects perceptually salient body regions exploiting symmetry and asymmetry principles. [Sect. 3.3.3]
5. *Descriptor Extraction and Accumulation* composes the signature as an ensemble of global or local features extracted from each body part and from different frames. [Sect. 3.3.4]
6. *Signature Matching* minimizes a certain similarity score between the probe signature and a set of signatures collected in a database (gallery set). [Sect. 3.4]

The nature of this process is slightly different (steps 5 and 6) depending on if-whether we have one or more images, that is, single- or multiple-shot case, respectively.

3.3.1 Image Gathering and Selection

The first step consists in gathering images of the tracked people. Since there is a temporal correlation between images of each tracked individual, redundancy is expected. Redundancy is therefore eliminated by applying the unsupervised Gaussian clustering method [17] that is able to automatically select the number of clusters. Hue Saturation Value (HSV) histogram of the cropped image of the individual is used as the feature for clustering, in order to capture appearance similarities across different frames. HSV histograms are invariant to small changes in illumination, scale and pose, so different clusters will be obtained. The output of the algorithm is a set of N_k clusters for each person (k stays for the k-th person). Then, we build the set $\mathbf{X}^k = \{X_r^k\}_{n=1}^{N_k}$ by randomly selecting an image of the k-th person for each cluster. Experimentally, we found that clusters with a small number of elements (=3 in our experiments) usually contain outliers, such as occlusions or partial views of the person, thus these clusters are discarded. It is worth noting that the selected clusters can still contain occlusions and bad images, hard for the re-id task.

3.3.2 Person Segmentation

Person segmentation allows the descriptor to focus on the individual foreground, avoiding being distracted from the noisy background. When videos are available (e.g., a video-surveillance scenario), foreground extraction can be performed with standard motion-based background subtraction strategies such as [11, 13, 40, 49]. In this work, the standard re-id datasets, which contain only still images, constrained us to use the Stel component analysis (SCA) [27]. However, we claim that any other person segmentation method can be used as a component of SDALF.

SCA lies on the notion of "structure element" (stel), which can be intended as an image portion whose topology is consistent over an image class. In a set of given objects, a stel is able to localize common parts over all the instances (e.g., the body in a set of images of pedestrians). SCA extends the stel concept as it captures the common structure of an image class by blending together multiple stels. SCA has been learned beforehand on a person database not considering the experimental data, and the segmentation over new samples consists in a fast inference (see [5, 27] for further details).

3.3.3 Symmetry-Based Silhouette Partition

The goal of this phase is to partition the human body into salient parts, exploiting asymmetry and symmetry principles. Considering a pedestrian acquired at very low resolution (see some examples in decreasing resolutions in Fig. 3.2), it is easy to note that the most distinguishable parts are three: head, torso and legs. We present a method that is able to work at very low resolution, where more accurate part detectors, such as the pictorial structures [9], fail.

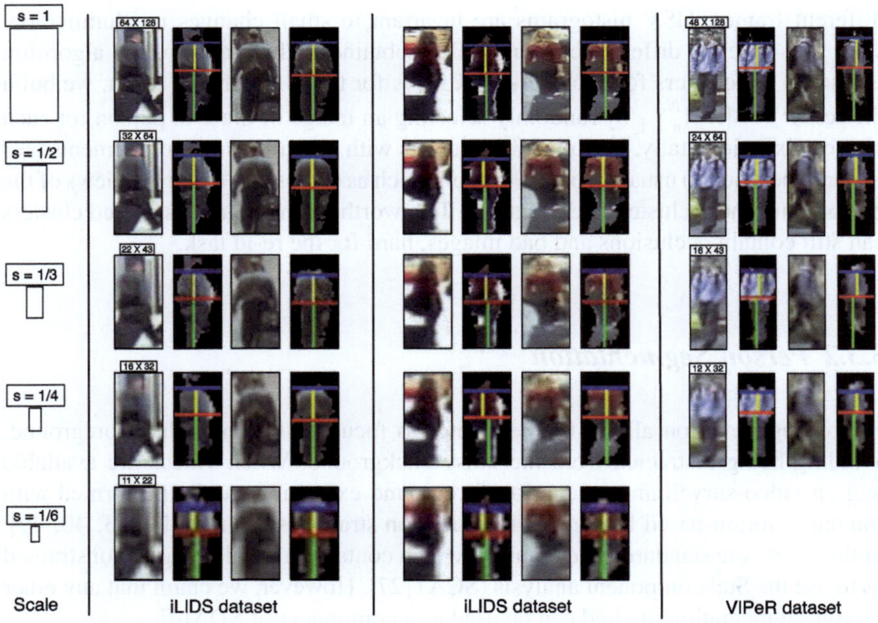

Fig. 3.2 Images of individuals at different resolutions (from 64 × 128 to 11 × 22) and examples of foreground segmentation and symmetry-based partitions

Let us first introduce the *chromatic bilateral operator* defined as:

$$C(i,\delta) \propto \sum_{B_{[i-\delta,i+\delta]}} d^2\left(p_i,\hat{p}_i\right) \tag{3.1}$$

where $d(\cdot,\cdot)$ is the Euclidean distance, evaluated between HSV pixel values p_i, $\hat{p}_i$, located symmetrically with respect to the horizontal axis at height i. This distance is summed up over $B_{[i-\delta,i+\delta]}$, i.e., the foreground region (as estimated by the object segmentation phase) lying in the box of width J and vertical extension $2\delta+1$ around i (see Fig. 3.3). We fix $\delta = I/4$, proportional to the image height, so that scale independency can be achieved.

The second operator is the *spatial covering operator*, which calculates the difference of foreground areas for two regions:

$$S(i,\delta) = \frac{1}{J\delta}\left|A\left(B_{[i-\delta,i]}\right) - A\left(B_{[i,i+\delta]}\right)\right|, \tag{3.2}$$

where $A\left(B_{[i-\delta,i]}\right)$, similarly as above, is the foreground area in the box of width J and vertical extension $[i-\delta, i]$.

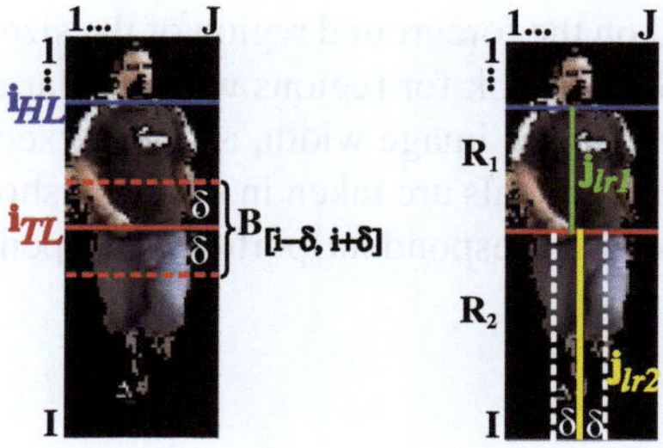

Fig. 3.3 Symmetry-based silhouette partition. First the asymmetrical axis i_{TL} is extracted, then i_{HT}; afterwards, for each $R_k, k = \{1, 2\}$ region the symmetrical axes j_{LRk} are computed

Combining opportunely C and S gives the axes of symmetry and asymmetry. The main x-axis of asymmetry is located at height i_{TL}:

$$i_{TL} = \underset{i}{\operatorname{argmin}}\ (1 - C(i, \delta)) + S(i, \delta), \tag{3.3}$$

i.e., we look for the x-axis that separates regions with strongly different appearance and similar area. The values of C are normalized by the numbers of pixels in the region $B_{[i-\delta,i+\delta]}$. The search for i_{TL} holds in the interval $[\delta,\ I - \delta]$: i_{TL} usually separates the two biggest body portions characterized by different colors (corresponding to t-shirt/pants or suit/legs, for example).

The other x-axis of asymmetry is positioned at height i_{HT}, obtained as:

$$i_{HT} = \underset{i}{\operatorname{argmin}}\ (-S(i, \delta))\,. \tag{3.4}$$

This asymmetry axis separates regions that strongly differ in area and places i_{HT} between head and shoulders. The search for i_{HT} is limited in the interval $[\delta, i_{TL} - \delta]$.

The values i_{HT} and i_{TL} isolate three regions $R_k,\ k = \{0, 1, 2\}$, approximately corresponding to head, body and legs, respectively (see Fig. 3.3). The head part R_0 is discarded, because it often consists in few pixels, carrying very low informative content.

At this point, for each part $R_k,\ k = \{1, 2\}$, a (vertical) symmetry axis is estimated, in order to individuate the areas that most probably belong to the human body, i.e., pixels near the symmetry axis. In this way, the risk of considering background clutter is minimized.

On both R_1 and R_2, the y-axis of symmetry is estimated in $j_{LRk},\ (k = 1, 2)$, obtained using the following operator:

$$j_{LRk} = \underset{j}{\operatorname{argmin}}\ C(j, \delta) + S(j, \delta). \tag{3.5}$$

This time, C is evaluated on the foreground region of the size of the height R_k timing the width δ (see Fig. 3.3). We look for regions with similar appearance and area. In this case, δ is proportional to the image width, and it is fixed to $J/4$.

In Fig. 3.2, different individuals are taken in different shots. As one can observe, our subdivision segregates correspondent portions independently on the assumed pose and the adopted resolution.

3.3.4 Accumulation of Local Features

Different features are extracted from the detected parts R_1 and R_2 (torso and legs, respectively). The goal is to extract as much complementary information as possible in order to encode heterogeneous information of the individuals. Each feature is extracted by considering its distance with respect to the vertical axes. The basic idea is that locations far from the symmetry axis belong to the background with higher probability. Therefore, features coming from those areas have to be either a) weighted accordingly or b) discarded.

Considering the literature in human appearance modeling, features may be grouped by considering the kind of information to focus on, that is, chromatic (histograms), region-based (blobs), and edge-based (contours, textures) information. Here, we consider a feature for each aspect, showing later their importance (see Fig. 3.4c–e for a qualitative analysis of the feature for the SDALF descriptor).

Weighted Color Histograms (WCH)

The chromatic content of each part of the pedestrian is encoded by color histograms. We evaluate different color spaces, namely, HSV, RGB, normalized RGB (where each channel is normalized by the sum of all the channels), per-channel normalized RGB [2], and CIELAB. Among these, HSV has been shown to be superior and also allows an intuitive quantization against different environmental illumination conditions and camera acquisition settings.

We define *WCH* of the foreground regions that take into consideration the distance to the vertical axes. In particular, each pixel is weighted by a 1-dimensional Gaussian kernel $\mathcal{N}(\mu, \sigma)$, where μ is the y-coordinate of j_{LRk}, and σ is a priori set to $J/4$. The nearer a pixel is to j_{LRk}, the more important it will be. In the single-shot case, a single histogram for each part is built. Instead, in the multiple-shot case, as M instances, all the M histograms for each part are considered during matching (see Sect. 3.4).

The advantage of using the weighted histogram is that in practice the person segmentation algorithm is prone to error especially in the contour of the silhouette. The weighted histogram is able to reduce the noise of the masks that contain background pixels wrongly detected as foreground.

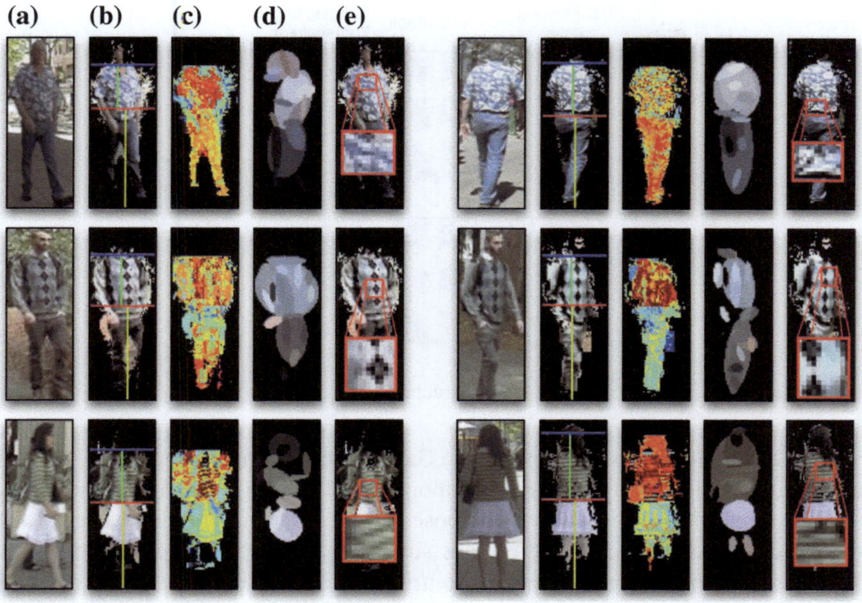

Fig. 3.4 Sketch of the SDALF descriptor for single-shot modality. **a** given an image or a set of images, **b** SDALF localizes meaningful body parts. Then, complementary aspects of the human body appearance are extracted: **c** weighted color histogram, the values accumulated in the histogram are back-projected into the image to show which colors of the image are more important. **d** Maximally stable color regions [18] and **e** recurrent highly structured patches. The objective is to correctly match SDALF descriptors of the same person (first column vs. sixth column)

Maximally Stable Color Regions (MSCR)

The MSCR operator[1] [18] detects a set of blob regions by looking at successive steps of an agglomerative clustering of image pixels. Each step clusters neighboring pixels with similar color, considering a threshold that represents the maximal chromatic distance between colors. These maximal regions that are stable over a range of steps represent the maximally stable color regions of the image. The detected regions are then described by their area, centroid, second moment matrix and average RGB color, forming 9-dimensional patterns. These features exhibit desirable properties for matching: covariance to adjacency, preserving transformations and invariance to scale changes, and affine transformations of image color intensities. Moreover, they show high repeatability, i.e., given two views of an object, MSCRs are likely to occur in the same correspondent locations.

In the single-shot case, we extract MSCRs separately from each part of the pedestrian. In order to discard outliers, we select only MSCRs that lie inside the foreground regions. In the multiple-shot case, we opportunely accumulate the MSCRs coming

[1] Code available at http://www2.cvl.isy.liu.se/~perfo/software/.

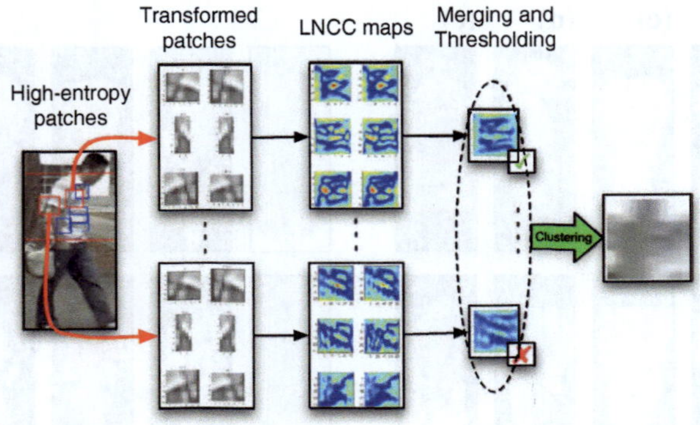

Fig. 3.5 Recurrent high-structured patches extraction

from the different images by employing a Gaussian clustering procedure [17], which automatically selects the number of components. Clustering is carried out using the 5-dimensional MSCR sub-pattern composed by the centroid and the average RGB color of each blob. We cluster the blobs similar in appearance and position, since they yield redundant information. The contribution of the clustering is twofold: (i) it captures only the relevant information, and (ii) it keeps low the computational cost of the matching process, when the clustering results are used. The final descriptor is built by a set of 4-dimensional MSCR sub-pattern composed by the y coordinate and the average RGB color of each blob. Please note that x coordinates are discarded because they are strongly dependent on the pose and viewpoint variation.

Recurrent High-Structured Patches (RHSP)

This feature was designed in [14], taking inspiration from the image epitome [26]. The idea is to extract image patches that are highly recurrent in the human body figure (see Fig. 3.5). Differently from the epitome, we want to take into account patches that are (1) informative, and (2) can be affected by rigid transformations. The first constraint selects only those patches that are informative in an information theoretic sense. Inspired by [51], RHSP uses entropy to select textural patches with strong edges. The higher the entropy is, the more likely it is to have a strong texture. The second requirement takes into account that the human body is a 3D entity whose parts may be captured with distortions, depending on the pose. For simplicity, we modeled the human body as a vertical cylinder. In these conditions, the RHSP generation consists in three phases.

The first step consists in the random extraction of patches p of size $J/6 \times I/6$, independently of each foreground body part of the pedestrian. In order to take the vertical symmetry into consideration, we mainly sample the patches around the j_{LRk} axes, exploiting the Gaussian kernel used for the color histograms computation. In order to focus on informative patches, we operate a thresholding on the entropy values of the patches, pruning away patches with low structural information (e.g., uniformly colored). This entropy is computed as the sum H_p of the pixel entropy of each RGB channel. We choose those patches with H_p higher than a fixed threshold τ_H (=13 in all our experiments). The second step applies a set of transformations $T_i, i = 1, 2, \ldots, N_T$ on the generic patch p, for all the sampled p's in order to check their invariance to (small) body rotations, i.e., considering that the camera may capture the one's front, back or side, and supposing the camera is at the face's height. We thus generate a set of N_T *simulated* patches p_i, gathering an enlarged set $\hat{p} = \{p_1, \ldots, p_{N_T}, p\}$.

In the third and final phase, we investigate how much recurrent a patch is. We evaluate the Local Normalized Cross-Correlation (LNCC) of each patch in $\hat{p}$ with respect to the original image. All the $N_T + 1$ LNCC maps are then summed together forming an average map. Averaging again over the elements of the map indicates how much a patch, and its transformed versions, are present in the image. Thresholding this value ($\tau_\mu = 0.4$) generates a set of candidates RHSP patches. The set of RHSPs is generated through clustering [17] of the LBP description [37] in order to capture patches with similar textural content. For each cluster, the patch closer to each centroid composes the RHSP.

Given a set of RHSPs for each region R_1 and R_2, the descriptor consists of an HSV histogram of these patches. We have tested experimentally the LBP descriptor, but it turned out to be less robust than color histograms. The single-shot and the multiple-shot methods are similar, with the only difference that in the multi-shot case the candidate RHSP descriptors are accumulated over different frames.

Please note that, even if we have several thresholds that regulate the feature extraction, they have been fixed once, and left unchanged in all the experiments. The best values have been empirically selected using the first 100 image pairs of the VIPeR dataset.

3.4 Signature Matching

In a general re-id problem two sets of signatures are available: a gallery set A and a probe set B. Re-id consists in associating the signature $\mathbf{P}^B$ of each person in B to the corresponding signature $\mathbf{P}^A$ of each person in A. The matching mechanism depends on how the two sets are organized, more specifically, on how many pictures are present for each individual. This gives rise to three matching philosophies: (1) *single-shot versus single-shot* (SvsS), if each image in a set represents a different individual; (2) *multiple-shot versus single-shot* (MvsS), if each image in B represents a different individual, while in A each person is portrayed in different images, or *instances*; (3) *multiple-shot versus multiple-shot* (MvsM), if both A and B contain multiple instances per individual.

In general, we can define re-id as a maximum log-likelihood estimation problem. More specifically, given a probe B matching is carried out by:

$$A^* = \arg\max_A \left(\log P(\mathbf{P}^A | \mathbf{P}^B)\right) = \arg\min_A \left(d(\mathbf{P}^A, \mathbf{P}^B)\right) \tag{3.6}$$

where the equality is valid because we define $P(\mathbf{P}^A|\mathbf{P}^B)$ in Gibbs form $P(\mathbf{P}^A|\mathbf{P}^B) = e^{-d(\mathbf{P}^A,\mathbf{P}^B)}$ and $d(\mathbf{P}^A, \mathbf{P}^B)$ measures the distance between two descriptors.

The *SDALF matching distance d* is defined as a convex combination of the local features:

$$d(\mathbf{P}^A, \mathbf{P}^B) = \sum_{f \in F} \beta_f \cdot d_f(f(\mathbf{P}^A), f(\mathbf{P}^B)) \tag{3.7}$$

where the $F = \{\text{WCH, MSCR, RHSP}\}$ is the set of the feature extractors, and βs are normalized weights.

The distance d_{WCH} considers the weighted color histograms. In the SvsS case, the HSV histograms of each part are concatenated channel by channel, then normalized, and finally compared via Bhattacharyya distance [28]. Under the MvsM and MvsS policies, we compare each possible pair of histograms contained in the different signatures, keeping the lowest distance.

For d_{MSCR}, in the SvsS case, we estimate the minimum distance of each MSCR element b in $\mathbf{P}^B$ to each element a in $\mathbf{P}^A$. This distance is defined by two components: d_y^{ab}, which compares the y component of the MSCR centroids; the x component is ignored, in order to be invariant with respect to body rotations. The second component is d_c^{ab}, which compares the MSCR color. In both cases, the comparison is carried out using the Euclidean distance.

The two components are combined as:

$$d_{\text{MSCR}} = \sum_{b \in \mathbf{P}^B} \min_{a \in \mathbf{P}^A} \gamma \cdot d_y^{ab} + (1 - \gamma) \cdot d_c^{ab} \tag{3.8}$$

where γ takes values between 0 and 1. In the multi-shot cases, the set $\mathbf{P}^A$ becomes a subset of blobs contained in the most similar cluster to the MSCR element b.

The distance d_{RHSP} is obtained by selecting the best pair of RHSPs, one in $\mathbf{P}^A$ and one in $\mathbf{P}^B$, and evaluating the minimum Bhattacharyya distance among the RHSP's HSV histograms. This is done independently for each body part (excluding the head), summing up all the distances achieved, and then normalizing for the number of pairs.

In our experiments, we fix the values of the parameters as follows: $\beta_{\text{WCH}} = 0.4$, $\beta_{\text{MSCR}} = 0.4$, $\beta_{\text{RHSP}} = 0.2$ and $\gamma = 0.4$. These values are estimated by cross validating over the first 100 image pairs of the VIPeR dataset, and left unchanged for all the experiments.

3.4.1 Analysis

The signature of SDALF and its characteristics for both the single-shot and multi-shot descriptors are summarized in Table 3.2. The second column reports which feature the basic descriptor is constructed from. The third and forth columns show the encoding used as description and the distance used in the matching module, respectively, in

Table 3.2 Summary of the characteristics of SDALF

		Single-shot		Multi-shot	
	Construction Cue	Encoding	Distance	Encoding	Distance
WCH	Color	HSV hist. per region	Bhattacharyya	Accumulate	Min over distance pairs
MSCR	Color	RGB color + y position per blob	(Eq. 3.8)	Clustering	(Eq. 3.8) using clusters
RHSP	Texture	HSV hist. per recurrent patch	Bhattacharyya	Accumulate	Min over distance pairs

the case of the single-shot version of SDALF. The last two columns report the same information for the multi-shot version.

Please note that even though the encoding of each descriptor is based on the color component, the way in which they are constructed is completely different. Therefore, the descriptors give a different mode/view of the same data. Color description has revealed one of the most useful features in appearance-based person re-id that usually gives the main contribution in terms of accuracy.

In terms of computational speed,[2] we evaluate how long the computation of the descriptor and the matching phase (Eq. 3.6) take in average on images of size 48 × 128. Partitioning of the silhouette in (a-)symmetric parts takes 56 ms per image. SDALF is then composed by three descriptors WCH, MSCR and RHSP that take 6, 31 and 4843 ms per image, respectively. It is easy to note that the actual bottleneck of the computation of SDALF is the RHSP. Matching is performed independently for each descriptor and it takes less than 1 ms per pair of images for WCH and RHSP and 4 ms per image for MSCR. In terms of computational complexity, the computation of the SDALF descriptor is linear in the number of images, while the matching phase is quadratic.

3.5 SDALF for Tracking

In tracking, a set of hypotheses of the object position on the image are analyzed at each frame, in order to find the one which best fits the target appearance, i.e., the *template*. The paradigm is different from the classical re-id: the gallery set is now the hypothesis set, which is different for each target. And the goal is to distinguish the target from the background and from the other visible targets. The problem of tracking shares some aspects with re-id: for example, the background can be hardly discernible from the background. Another example is when people are relatively close to each other in the video. In that case, hypotheses of a person position may

[2] The following values have been computed using our non-optimized MATLAB code on a quad-core Intel Xeon E5440, 2.83 GHz with 30 GB of RAM.

go to the background or the wrong person. A descriptor specifically created for re-id better handles these situations.

The goal of tracking is thus to perform a soft matching, i.e., compute the likelihood between the probe set (the target template) and the gallery set (the hypothesis set) without performing hard matching, like in re-id.

In this section, we briefly describe particle filtering for tracking (Sect. 3.5.1) and we exploit SDALF as appearance model (Sect. 3.5.2).

3.5.1 Particle Filter

Particle filter offers a probabilistic framework for recursive dynamic state estimation [12] that fits the tracking problem. The goal is to determine the posterior distribution $p(x_t|z_{1:t})$, where x_t is the current state, z_t is the current measurement, and $x_{1:t}$ and $z_{1:t}$ are the states and the measurements up to time t, respectively. The Bayesian formulation of $p(x_t|z_{1:t})$ enables us to rewrite the problem as:

$$p(x_t|z_{1:t}) \propto p(z_t|x_t) \int_{x_{t-1}} p(x_t|x_{t-1}) p(x_{t-1}|z_{1:t-1}) dx_{t-1}. \tag{3.9}$$

Particle filter is fully specified by an initial distribution $p(x_0)$, a dynamical model $p(x_t|x_{t-1})$, and an observation model $p(z_t|x_t)$. The posterior distribution at previous time $p(x_{t-1}|z_{1:t-1})$ is approximated by a set of S weighted particles, i.e., $\{(x_{t-1}^{(s)}, w_{t-1}^{(s)})\}_{s=1}^{S}$, because the integral in Eq. (3.9) is often analytically intractable. Equation (3.9) can be rewritten by its Monte Carlo approximation:

$$p(x_t|z_{1:t}) \approx \sum_{s=1}^{S} w_t^{(s)} \, \delta(x_t - x_t^{(s)}). \tag{3.10}$$

where

$$w_t^{(s)} \propto w_{t-1}^{(s)} \frac{p(z_t|x_t^{(s)}) \, p(x_t^{(s)}|x_{t-1}^{(s)})}{q(x_t^{(s)}|x_{t-1}^{(s)}, z_t)} \tag{3.11}$$

where q is called *proposal distribution*. The design of an optimal proposal distribution is a critical task. A common choice is $q(x_t^{(n)}|x_{t-1}^{(n)}, z_t) = p(x_t^{(n)}|x_{t-1}^{(n)})$ because it simplifies Eq. (3.11) in $w_t^{(n)} \propto w_{t-1}^{(n)} \, p(z_t|x_t^{(n)})$. However, this is not an optimal choice. We can make use of the observation z_t in order to propose particles in more interesting regions of the state space. As in [38], detections are used in the proposal distribution to guide tracking and make it more robust.

Given this framework, tracking consists of observing the image z_t at each time t and updating the distribution over the state x_t by propagating particles as in Eq. (3.11).

3.5.2 *SDALF as Observation Model*

The basic idea is to propose a new observation model $p(z_t|x_t^{(s)})$ so that the object representation is made up by the SDALF descriptor. We define the observation model considering the distance defined in (Eq. 3.6) $d(\mathbf{P}_A|\mathbf{P}_B) := d(x_t^{(s)}(z_t), \tau_t)$, where $\mathbf{P}_B$ becomes the object template τ_t made by SDALF descriptors, and $\mathbf{P}_A$ is the current hypothesis $x_t^{(s)}$. Minimization of Eq. (3.6) over the gallery set elements is not performed for tracking. Instead, the probability distribution over the hypotheses is kept in order to approximate Eq. (3.9).

Some simplifications are required when embedding SDALF into the proposed tracking framework. First of all, since the descriptor has to be extracted for each hypothesis $x_t^{(s)}$, it should be reasonably efficient to compute. In our current implementation, the computation of RHSP for each particle is not feasible as the transformations T_i performed on the original patches to make the descriptor invariant to rigid transformations constitute a too high burden. Therefore, the RHSP is not used in the descriptor.

The observation model becomes:

$$p(z_t|x_t^{(s)}) = e^{-D(x_t^{(s)}(z_t),\tau_t)}, \qquad D(x_t^{(s)}(z_t), \tau_t) = \sum_{f \in F_R} \beta_f \cdot d_f(f(x_t^{(s)}), f(\tau_t)) \tag{3.12}$$

where $x_t^{(s)}$ is the hypothesis extracted from the image z_t, and τ_t is the template of the object and $F_R = \{\text{WCH, MSCR}\}$. During tracking, the object template has to be updated in order to model the different aspects of the captured object (for example, due to different poses). Therefore, τ_t is composed by a set of images accumulated over time (previous L frames). Then, in order to balance the number of images employed for building the model and the computational effort required, $N = 3$ images are randomly selected at each time step to form $\mathbf{P}_A$.

3.6 Experiments

In this section, an exhaustive analysis of SDALF for re-identification and tracking is presented. SDALF is evaluated on the re-id task against the state-of-the-art methods in Sect. 3.6.1. Then, it is evaluated on a tracking scenario in Sect. 3.6.2.

3.6.1 Results: Re-identification

In literature, several different datasets are available: VIPeR[3] [20], iLIDS for re-id [54], ETHZ[4] 1, 2, and 3 [47], and the more recent CAVIAR4REID[5] [9]. These datasets cover challenging aspects of the person re-id problem, such as shape deformation, illumination changes, occlusions, image blurring, very low resolution images, etc.

Datasets The VIPeR dataset [20] contains image pairs of 632 pedestrians normalized to 48 $\times$ 128 pixels. It represents one of the most challenging single-shot datasets currently available for pedestrian re-id.

The ETHZ dataset [47] is captured from moving cameras in a crowded street and contains three sub-datasets: ETHZ1 with 83 people (4.857 images), ETHZ2 with 35 people (1.936 images), and ETHZ3 contains 28 with (1.762 images). ETHZ does not represent a genuine re-id scenario (no different cameras are employed), and it still carries important challenges not exhibited by other public datasets, as the big number of images per person.

The iLIDS for re-id [54] dataset is composed by 479 images of 119 people acquired from non-overlapping cameras. However, iLIDS does not fit well in a multi-shot scenario because the average number of images per person is four, and thus some individuals have only two images. For this reason, we also created a modified version of the dataset of 69 individuals, named iLIDS$_{\geq 4}$, where we selected the subset of individuals with at least four images.

The CAVIAR4REID dataset [9] contains images of pedestrians extracted from the shopping center scenario of the CAVIAR dataset.[6] The ground truth of the sequences was used to extract the bounding box of each pedestrian, resulting in a set of 10 images of 72 unique pedestrians: 50 with the two camera views and 22 with one camera view. The main differences of CAVIAR4REID with respect to the already-existing datasets for re-id are: (1) it has broad changes of resolution, and the minimum and maximum size of the images contained on CAVIAR4REID dataset are 17 $\times$ 39 and 72 $\times$ 144, respectively. (2) Unlike ETHZ, it is extracted from a real scenario where re-id is necessary due to the presence of multiple cameras and (3) pose variations are severe. (4) Unlike VIPeR, it contains more than one image for each view. (5) It contains all the images variations of the other datasets.

Evaluation Measures. State-of-the-art measurements are used in order to compare the proposed methods with the others: the Cumulative Matching Characteristic (CMC) curve represents the expectation of finding the correct match in the top n matches and the normalized Area Under the Curve (nAUC) is the area under the entire CMC curve normalized over the total area of the graph. We compare the proposed method with some of the best re-id methods on the available datasets: Ensemble of Localized Features (ELF) [21] and Primal-based Rank-SVM (PRSVM) [41]

---

[3] Available at http://users.soe.ucsc.edu/~dgray/VIPeR.v1.0.zip.

[4] Available at http://www.liv.ic.unicamp.br/~wschwartz/datasets.html.

[5] Available at http://www.lorisbazzani.info/code-datasets/caviar4reid/.

[6] Available at http://homepages.inf.ed.ac.uk/rbf/CAVIARDATA1/.

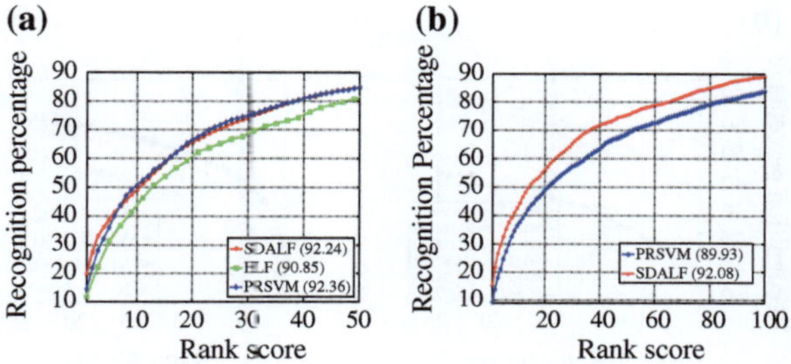

Fig. 3.6 Performances on the VIPeR dataset in terms of CMC and nAUC (*within brackets*). In **a** and **b**, comparative profiles of SDALF against ELF [21] and PRSVM [41] on the 316-pedestrian dataset and the 474-pedestrian dataset, respectively. In **c**, SDALF at different scales

in VIPeR, PLS by [47] in ETHZ, Context-based re-id [54] and Spatial Covariance Region (SCR) [2] in iLIDS.

Results. Considering first the VIPeR dataset, we define CAM B as the gallery set, and CAM A as the probe set; each image of the probe set is matched with the images of the gallery. This provides a ranking for every image in the gallery with respect to the probe. We followed the same experimental protocol of [21]. In this work, the dataset is split evenly into a training and a test set, and matching is performed. In both algorithms a set of few random permutations are performed (five runs for PRSVM, 10 runs for ELF), and the averaged score is kept. In order to fairly compare our results with theirs, we should know precisely the splitting assignment. Since this information is not provided we compare the existent results with the average of the results obtained by our method for 10 different random sets of 316 pedestrians and 474 pedestrians. In Fig. 3.6, we depict a comparison among ELF, PRSVM and SDALF in terms of CMC curves. We provided also the nAUC score for each method (within brackets in the legend of the plots of Fig. 3.6). Considering the experiment on 316 pedestrians (Fig. 3.6a), SDALF outperforms ELF in terms of nAUC, and we obtain comparable results with respect to PRSVM. Even if PRSVM is slightly superior to SDALF, one can note that the differences between it and SDALF are negligible (less than 0.12 %). This is further corroborated looking at the different philosophy underlying the PRSVM and our approach. In the former case, PRSVM uses the 316 pairs as training set, whereas in our case we act directly on the test images, operating on each single image as an independent entity. Thus, no learning phase is needed for our descriptor. In addition, it is worth noting that SDALF slightly outperforms PRSVM in the first positions of the CMC curve (rank 1–6). This means that in a real scenario where only the first ranks are considered, our method performs better.

Figure 3.6b shows a comparison between PRSVM and SDALF when dealing with a larger test dataset where a set of 474 individuals has been extracted, as done in the PRSVM paper. This is further evidence about how the performance of PRSVM depends on the training set, which is now composed by 158 individuals. In this case, our approach outperforms PRSVM showing an advantage in terms of nAUC of about 2.15 %.

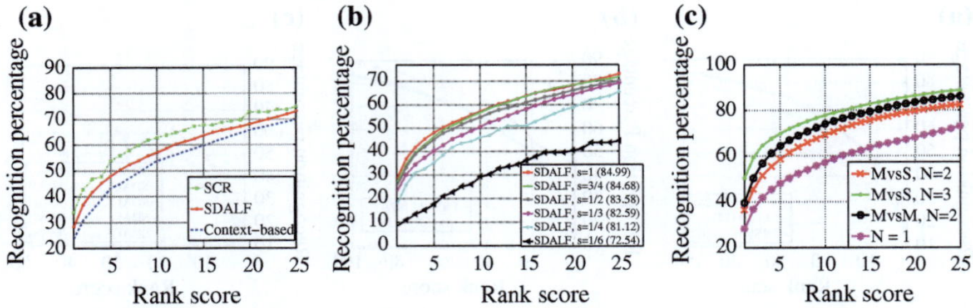

Fig. 3.7 Performances on iLIDS dataset. **a** CMC curves comparing context-based re-id [54], SCR [2] and single-shot SDALF. **b** analysis of SDALF performance at different resolutions. **c** CMC curves for MvsS and MvsM cases varying the average number of images N for each pedestrian. For reference, we put also the single-shot case ($N = 1$). In accordance with what reported by [54], only the first 25 ranking positions of the CMC curves are displayed

The last analysis of this dataset is conducted by testing the robustness of SDALF when the image resolution decreases. We scaled the original images of the VIPeR dataset by factors $s = \{1, 3/4, 1/2, 1/3, 1/4\}$ reaching a minimum resolution of 12×32 pixels (Fig. 3.2 on the right). The results, depicted in Fig. 3.6, show that the performance decreases, as expected, but not drastically. nAUC slowly drops down from 92.24 % at scale 1 to 86.78 % at scale 1/4.

Now let us analyze the results on iLIDS dataset. We reproduce the same experimental settings of [54] in order to make a fair comparison. We randomly select one image for each pedestrian to build the gallery set, while the others form the probe set. Then, the matching between probe and gallery set is estimated. For each image in the probe set the position of the correct match is obtained. The whole procedure is repeated 10 times, and the average CMC curves are displayed in Fig. 3.7.

SDALF outperforms the Context-based method [54] without using any additional information about the context (Fig. 3.7a) even using images at lower resolution (Fig. 3.7b). The experiments of Fig. 3.7b show SDALF when scaling factors are $s = \{1, 3/4, 1/2, 1/3, 1/4, 1/6\}$ with respect to the original size of the images, reaching a minimum resolution of 11×22 pixels. Fig. 3.7a shows that we get lower performance with respect to SCR [2]. Unfortunately, it has been impossible to test SCR on low resolution images (no public code available), but since it is based on covariance of features we expect that second order statistics on very few values may be uninformative and not significant.

Concerning the multiple-shot case, we run experiments on both MvsS and MvsM cases. In the former trial, we built a gallery set of multi-shot signatures and we matched it with a probe set of one-shot signatures. In the latter, both gallery and probe sets are made up of multi-shot signatures. In both cases, the multiple-shot signatures are built from N images of the same pedestrian randomly selected. Since the dataset contains an average of about four images per pedestrian, we tested our algorithm with $N = \{2, 3\}$ for MvsS, and just $N = 2$ for MvsM running 100 independent trials for each case. It is worth noting that some of the pedestrians have less than four images, and in this case, we simply build a multi-shot signature composed by less instances. In the MvsS strategy, this applies to the gallery signature only, and in the MvsM signature, we start by decreasing the number of instances that compose the probe signature, leaving unchanged the gallery signature; once we reach just one instance for the probe signature, we start decreasing the gallery signature too. The results, depicted in Fig. 3.7c, show that, in the MvsS case, just two images are enough to increment the performance by about 10 % and to outperform the Context-based method [54] and SCR [2]. Adding another image induces an increment of 20 % with respect to the single-shot case. It is interesting to note that the results for MvsM lie in between these two figures.

In ETHZ dataset, PLS [47] produces the best performance. In the single-shot case, the experiments are carried out exactly as for iLIDS. The multiple-shot case is carried out considering $N = 2, 5, 10$ for MvsS and MvsM, with 100 independent trials for each case. Since the images of the same pedestrian come from video sequences, many are very similar and picking them for building the multi-shot signature would not provide new useful information about the subject. Therefore, we apply the clustering procedure discussed in Sect. 3.3.1.

The results for both single and multiple-shot cases for SEQ.#1 are reported on Fig. 3.8, and we compare the results with those reported by [47]. In SEQ. #1 we do not obtain the best results in the single-shot case, but adding more information to the signature we can get up to 86 % rank 1 correct matches for MvsS and up to 90 % for MvsM. We think that the difference with PLS is due to the fact that PLS uses all foreground and background information, while we use only the foreground. Background information helps here because each pedestrian is framed and tracked in the same location, but it is not valid in general in a multi-camera setting.

In SEQ. #2 (Fig. 3.8) we have a similar behavior: rank 1 correct matches can be obtained in 91 % of the cases for MvsS, and in 92 % of the cases for MvsM. The results for SEQ. #3 show instead that SDALF outperforms PLS even in the single-shot case. The best performance as to rank 1 correct matches is 98 % for MvsS and 94 % for MvsM. It is interesting to note that there is a point after that adding more information does not enrich the descriptive power of the signature any more. $N = 5$ seems to be the correct number of images to use.

Results: AHPE. To prove that the ideas introduced by SDALF should be used in combination with other descriptors, we modified the Histogram Plus Epitome (HPE) descriptor of [6]. HPE is made by two parts: color histograms accumulated over time in the same spirit of SDALF, and the epitome to describe local recurrent motifs. We extended HPE to Asymmetry HPE (AHPE) [6], where HPE is extracted from (a-)symmetric parts in the same partition method used by SDALF. The quantitative evaluation of HPE and AHPE considers the six multi-shot datasets: ETHZ 1, 2, and 3, iLIDS for re-id, iLIDS$_{\geq 4}$, and CAVIAR4REID.

A comparison between different state-of-the-art methods in the multi-shot setup ($N = 5$), HPE and AHPE descriptor is shown in Fig. 3.9. On ETHZ, AHPE gives the best results, showing consistent improvements on ETHZ1 and ETHZ3. On ETHZ2, AHPE gives comparable results with SDALF, since the nAUC is 98.93 and 98.95 % for AHPE and SDALF, respectively. Note that if we remove the image selection step

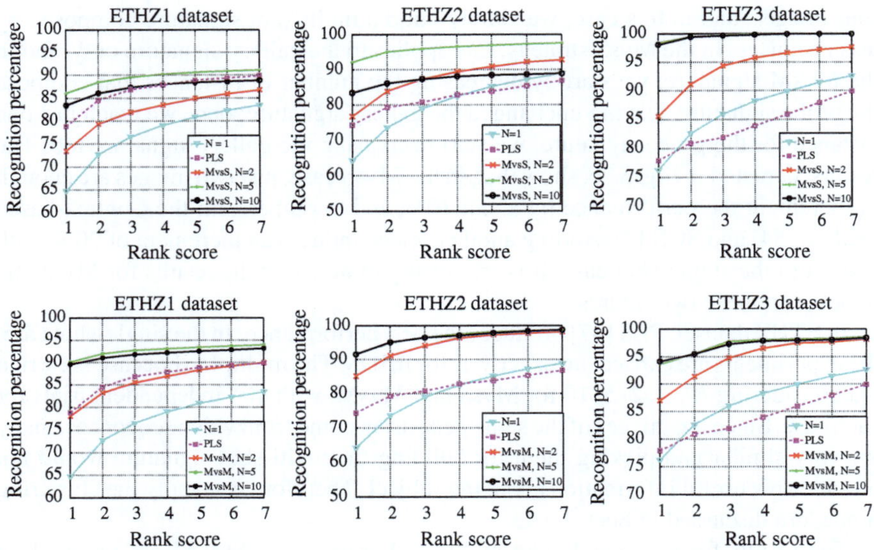

Fig. 3.8 Performances on ETHZ dataset. *Left* column, results on SEQ. #1; *middle* column, on SEQ. #2; *right* column, on SEQ. #3. We compare our method with the results of PLS [47]. On the *top* row, we report the results for single-shot SDALF ($N = 1$) and MvsS SDALF; on the *bottom* row, we report the results for MvsM SDALF. In accordance with [47], only the first 7 ranking positions are displayed

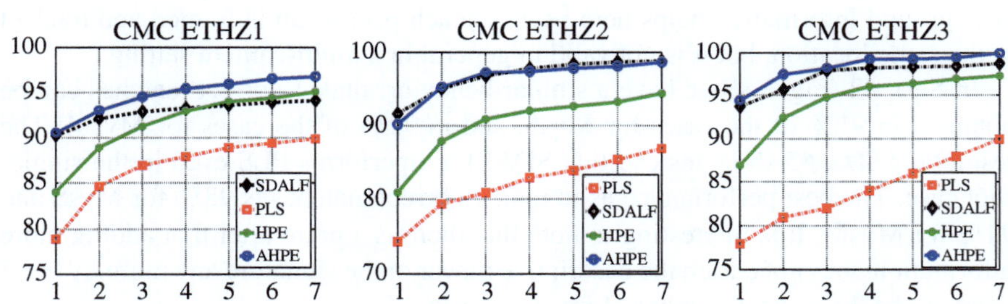

Fig. 3.9 Comparisons on ETHZ 1, 2, 3 between AHPE (*blue*), HPE (*green*), SDALF (*black*), PLS [47] (*red*). For the multi-shot case we set $N = 5$

(used for ETHZ), the performance decreases of 5 % in terms of CMC, because the intra-variance between images of the same individual is low, and thus the multi-shot mode does not gain new discriminative information.

On iLIDS (Fig. 3.10, left), AHPE is outperformed only by SDALF. This witnesses again the fact, explained in the previous experiment, that the epitomic analysis works very well when the number of instances is appropriate (say, at least $N = 5$). This statement is clearer by the experiments on iLIDS$_{\geq 4}$ and CAVIAR4REID (Fig. 3.10, last two columns). Especially, if we remove from iLIDS the instances with less than four images, then AHPE outperforms SDALF (Fig. 3.10, center). The evaluation on

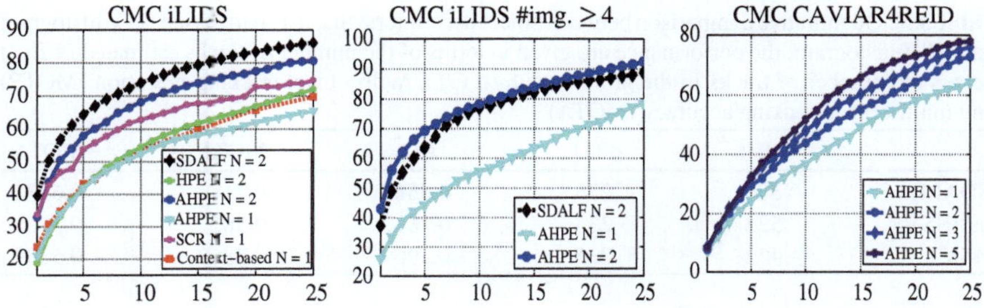

Fig. 3.10 Comparisons on iLIDS (*first column*), iLIDS$_{\geq 4}$ (*second column*) and CAVIAR4REID (*third column*) between AHPE (*blue*), HPE (*green, only iLIDS*), SDALF (*black*), SCR [2] (*magenta, only iLIDS*), and context-based [54] (*red, only iLIDS*). For iLIDS and iLIDS$_{\geq 4}$ we set $N = 2$. For CAVIAR4REID, we analyze different values for N. Best viewed in colors

CAVIAR4REID (Fig. 3.10, right) shows that: (1) the accuracy increases with N, and (2) the real, worst-case scenario of re-id is still a very challenging open problem.

3.6.2 Results: Tracking

As benchmark, we adopt CAVIAR, as it represents a challenging real tracking scenario, due to pose, resolution and illumination changes, and severe occlusions. The dataset consists of several sequences along with the ground truth captured in the entrance lobby of the INRIA Labs and in a shopping center in Lisbon. We select the shopping center scenario, because it mirrors a real situation where people move in the scene. The shopping center dataset is composed by 26 sequences recorded from two different points of view, at the resolution of 384 $\times$ 288 pixels. It includes individuals walking alone, meeting with others, window shopping, entering and exiting shops.

We aim to show the capabilities of SDALF as appearance descriptor in a multi-person tracking case. We use the particle filtering approach described in Sect. 3.5, since it represents a general tracking engine employed by many algorithms. As proposal distribution, we use the already-trained person detector [15] in the same way exploited by the boosted particle filter [38]. For generating new tracks, weak tracks (tracks initialized for each not associated detection) are kept in memory, and it is checked whether they are supported continuously by a certain number of detections. If this happens, the tracks are initialized [8].

The proposed SDALF-based observation model is compared against two classical appearance descriptors for tracking: joint HSV histogram and part-based HSV histogram (partHSV) [24] where each of three body parts (head, torso, legs) are described by a color histogram.

Table 3.3 Quantitative comparison between object descriptors: SDALF, part-based HSV histogram and HSV histogram; the performance are given in terms of the number of tracks estimated (# Est.) versus the number of tracks in the ground truth (# GT), multi-object tracking precision (MOTP) and multi-object tracking accuracy (MOTA)

	# Est.	# GT	ATA	MOTP	MOTA
SDALF	**300**	235	**0.4567**	**0.7182**	**0.6331**
partHSV	522	235	0.1812	0.5822	0.5585
HSV	462	235	0.1969	0.5862	0.5899

The quantitative evaluation of the method is provided by adopting the metrics presented in [29]:[7]

- Average Tracking Accuracy (*ATA*): measures penalizing fragmentation phenomena in both the temporal and spatial dimensions, while accounting for the number of objects detected and tracked, missed objects, and false positives;
- Multi-Object Tracking Precision (*MOTP*): considers the spatio-temporal overlap between the reference tracks and the tracks produced by the test method.
- Multi-Object Tracking Accuracy (*MOTA*): considers missed detections, false positives, and ID switches by analyzing consecutive frames.

For more details, please refer to the original paper [29]. In addition, we provide also an evaluation in terms of:

- the number of tracks estimated by our method (# **Est.**) versus the number of tracks in the ground truth (# *GT*): an estimate of how many tracks are wrongly generated (for example, because weak appearance models cause tracks drifting).

The overall tracking results averaged over all the sequences are reported in Table 3.3. The number of estimated tracks using SDALF is closer to the correct number than partHSV and HSV. Experimentally, we noted that HSV and partHSV fail very frequently in the case of illumination, pose, and resolution changes and partial occlusions. In addition, several tracks are frequently lost and then re-initialized.

Considering the temporal consistency of the tracks (ATA, MOTA, and MOTP), wa can notice that SDALF definitely outperforms HSV and partHSV. The values of ATA are not so high, because track fragmentation is frequent. This is due to the fact that the tracking algorithm does not explicitly cope with complete occlusions. ATA shows that SDALF gives the best results. This experiment promotes SDALF as an accurate person descriptor for tracking, able to manage the natural noisy evolution of the appearance of people.

[7] For the sake of fairness, we use the code provided by the authors. For the metric ATA, we use the association threshold suggested by the authors (0.5).

3.7 Conclusions

In this chapter, we presented a pipeline for re-identification and a robust symmetry-based descriptor for modeling the human appearance. SDALF relies on perceptually relevant parts localization driven by asymmetry/symmetry principles. It consists of three features that encode different information, namely, chromatic and structural information, as well as recurrent high-entropy textural characteristics. In this way, robustness to low resolution, pose, viewpoint and illumination variations is achieved. SDALF was shown to be versatile, being able to work using a single image of a person (single-shot modality), or several frames (multiple-shot modality). Moreover, SDALF was also showed to be robust to very low resolutions, maintaining high performance up to 11 $\times$ 22 windows size.

References

1. Bak, S., Corvee, E., Bremond, F., Thonnat, M.: Person re-identification using haar-based and DCD-based signature. In: 2nd Workshop on Activity Monitoring by Multi-camera Surveillance Systems (2010)
2. Bak, S., Corvee, E., Bremond, F., Thonnat, M.: Person re-identification using spatial covariance regions of human body parts. In: International Conference on Advanced Video and Signal-Based Surveillance (2010)
3. Barbosa, I.B., Cristani, M., Del Bue, A., Bazzani, L., Murino, V.: Re-identification with rgb-d sensors. In: European Conference on Computer Vision. Workshops and Demonstrations, Lecture Notes in Computer Science, vol. 7583, pp. 433–442 (2012)
4. Bay, H., Tuytelaars, T., Van Gool, L.: SURF: speeded up robust features. In: Proceedings of the European Conference on Computer Vision, pp. 404–417 (2006)
5. Bazzani, L., Cristani, M., Murino, V.: Symmetry-driven accumulation of local features for human characterization and re-identification. Comput. Vis. Image Underst. **117**(2), 130–144 (2013)
6. Bazzani, L., Cristani, M., Perina, A., Murino, V.: Multiple-shot person re-identification by chromatic and epitomic analyses. Pattern Recogn. Lett. **33**(7), 898–903 (2012)
7. Bird, N., Masoud, O., Papanikolopoulos, N., Isaacs, A.: Detection of loitering individuals in public transportation areas. IEEE Trans. Intell. Transp. Syst. **6**(2), 167–177 (2005)
8. Breitenstein, M.D., Reichlin, F., Leibe, B., Koller-Meier, E., Gool, L.V.: Robust tracking-by-detection using a detector confidence particle filter. In: IEEE International Conference on Computer Vision (2009)
9. Cheng, D.S., Cristani, M., Stoppa, M., Bazzani, L., Murino, V.: Custom pictorial structures for re-identification. In: British Machine Vision Conference (BMVC) (2011)
10. Cho, M., Lee, K.M.: Bilateral symmetry detection and segmentation via symmetry-growing. In: British Machine Vision Conference (2009)
11. Cristani, M., Bicego, M., Murino, V.: Multi-level background initialization using hidden markov models. In: First ACM SIGMM International Workshop on Video Surveillance, IWVS '03, pp. 11–20. ACM, New York (2003). http://doi.acm.org/10.1145/982452.982455
12. Doucet, A., Freitas, N.D., Gordon. N.: Sequential monte carlo methods in practice (2001)
13. Elgammal, A., Duraiswami, R., Harwood, D., Davis, L.S.: Background and foreground modeling using nonparametric kernel density estimation for visual surveillance. Proc. IEEE **90**(7), 1151–1163 (2002)

14. Farenzena, M., Bazzani, L., Perina, A., Murino, V., Cristani, M.: Person re-identification by symmetry-driven accumulation of local features. In: IEEE Conference on Computer Vision and Pattern Recognition (CVPR), pp. 2360–2367 (2010)
15. Felzenszwalb, P.F., Girshick, R.B., McAllester, D.: Cascade object detection with deformable part models. In: IEEE Conference on Computer Vision and Pattern Recognition (2010)
16. Figueira, D., Bazzani, L., Minh, H., Cristani, M., Bernardino, A., Murino, V.: Semi-supervised multi-feature learning for person re-identification. In: International Conference on Advanced Video and Signal-based Surveillance (2013)
17. Figueiredo, M., Jain, A.: Unsupervised learning of finite mixture models. IEEE Trans. Pattern Anal. Mach. Intell. **24**(3), 381–396 (2002)
18. Forssén, P.E.: Maximally stable colour regions for recognition and matching. In: IEEE Conference on Computer Vision and Pattern Recognition (2007)
19. Gheissari, N., Sebastian, T.B., Tu, P.H., Rittscher, J., Hartley, R.: Person reidentification using spatiotemporal appearance. In: IEEE Conference on Computer Vision and Pattern Recognition, vol. 2, pp. 1528–1535 (2006)
20. Gray, D., Brennan, S., Tao, H.: Evaluating appearance models for recognition, reacquisition, and tracking. In: IEEE International Workshop on Performance Evaluation for Tracking and Surveillance (2007)
21. Gray, D., Tao, H.: Viewpoint invariant pedestrian recognition with an ensemble of localized features. In: European Conference on Computer Vision (2008)
22. Hamdoun, O., Moutarde, F., Stanciulescu, B., Steux, B.: Person re-identification in multi-camera system by signature based on interest point descriptors collected on short video sequences. In: IEEE International Conference on Distribuited Smart Cameras, pp. 1–6 (2008)
23. Hirzer, M., Roth, P.M., Kostinger, M., Bischof, H.: Relaxed pairwise learned metric for person re-identification. In: European Conference on Computer Vision, Lecture Notes in Computer Science, vol. 7577, pp. 780–793 (2012)
24. Isard, M., MacCormick, J.: Bramble: a bayesian multiple-blob tracker. In: IEEE International Conference on Computer Vision, vol. 2, pp. 34–41 (2001)
25. Javed, O., Shafique, K., Rasheed, Z., Shah, M.: Modeling inter-camera space-time and appearance relationships for tracking accross non-overlapping views. Comput. Vis. Image Underst. **109**, 146–162 (2007)
26. Jojic, N., Frey, B., Kannan, A.: Epitomic analysis of appearance and shape. In: International Conference on Computer Vision **1**, 34–41 (2003)
27. Jojic, N., Perina, A., Cristani, M., Murino, V., Frey, B.: Stel component analysis: modeling spatial correlations in image class structure. In: IEEE Conference on Computer Vision and Pattern Recognition, pp. 2044–2051 (2009)
28. Kailath, T.: The divergence and Bhattacharyya distance measures in signal selection. IEEE Trans. Commun. **15**(1), 52–60 (1967)
29. Kasturi, R., Goldgof, D., Soundararajan, P., Manohar, V., Garofolo, J., Bowers, R., Boonstra, M., Korzhova, V., Zhang, J.: Framework for performance evaluation of face, text, and vehicle detection and tracking in video: data, metrics, and protocol. IEEE Trans. Pattern Anal. Mach. Intell. **31**, 319–336 (2009)
30. Kohler, W.: The task of gestalt psychology. Princeton University Press, Princeton (1969)
31. Levinshtein, A., Dickinson, S., Sminchisescu, C.: Multiscale symmetric part detection and grouping. In: International Conference on Computer Vision (2009)
32. Lin, Z., Davis, L.S.: Learning pairwise dissimilarity profiles for appearance recognition in visual surveillance. In: International Symposium on Advances in Visual Computing, pp. 23–34 (2008)
33. Makris, D., Ellis, T., Black, J.: Bridging the gaps between cameras. In: IEEE Computer Society Conference on Computer Vision and Pattern Recognition, vol. 2, pp. II-205–II-210 (2004)
34. Mignon, A., Jurie, F.: PCCA: a new approach for distance learning from sparse pairwise constraints. In: IEEE Conference on Computer Vision and Pattern Recognition, pp. 2666–2672 (2012)

35. Minh, H.Q., Bazzani, L., Murino, V.: A unifying framework for vector-valued manifold regularization and multi-view learning. In: Proceedings of the 30th International Conference on Machine Learning (2013)
36. Nakajima, C., Pontil, M., Heisele, B., Poggio, T.: Full-body person recognition system. Pattern Recogn. Lett. **36**(9), 1997–2006 (2003)
37. Ojala, T., Pietikainen, M., Harwood, D.: A comparative study of texture measures with classification based on featured distributions. Pattern Recogn. **29**(1), 51–59 (1996)
38. Okuma, K., Taleghani, A., de Freitas, N., Little, J.J., Lowe, D.G.: A boosted particle filter: multitarget detection and tracking. In: European Conference on Computer Vision, vol. 1, pp. 28–39 (2004)
39. Pham, N.T., Huang, W.M., Ong, S.H.: Probability hypothesis density approach for multi-camera multi-object tracking. In: Asian Conference on Computer Vision, vol. 1, pp. 875–884 (2007)
40. Pilet, J., Strecha, C., Fua, P.: Making background subtraction robust to sudden illumination changes. In: European Conference on Computer Vision, pp. 567–580 (2008)
41. Prosser, B., Zheng, W., Gong, S., Xiang, T.: Person re-identification by support vector ranking. In: British Machine Vision Conference (2010)
42. Rahimi, A., Dunagan, B., Darrel, T.: Simultaneous calibration and tracking with a network of non-overlapping sensors. In: IEEE Computer Society Conference on Computer Vision and Pattern Recognition, vol. 1, pp. 187–194 (2004)
43. Reisfeld, D., Wolfson, H.J., Yeshurun, Y.: Context-free attentional operators: the generalized symmetry transform. Int. J. Comput. Vision **14**(2), 119–130 (1995)
44. Riklin-Raviv, T., Sochen, N., Kiryati, N.: On symmetry, perspectivity, and level-set-based segmentation. IEEE Trans. Pattern Recogn. Mach. Intell. **31**(8), 1458–1471 (2009)
45. Salvagnini, P., Bazzani, L., Cristani, M., Murino, V.: Person re-identification with a ptz camera: an introductory study. In: International Conference on Image Processing (2013)
46. Satta, R., Fumera, G., Roli, F., Cristani, M., Murino, V.: A multiple component matching framework for person re-identification. In: Proceedings of the 16th International Conference on Image Analysis and Processing, pp. 140–149 (2011)
47. Schwartz, W.R., Davis, L.S.: Learning discriminative appearance-based models using partial least squares. In: Proceedings of the 22nd Brazilian Symposium on Computer Graphics and Image Processing (2009)
48. Sivic, J., Zitnick, C.L., Szeliski, R.: Finding people in repeated shots of the same scene. In: Proceedings of the British Machine Vision Conference (2006)
49. Stauffer, C., Grimson, W.E.L.: Adaptive background mixture models for real-time tracking. In: IEEE Computer Society Conference on Computer Vision and Pattern Recognition, vol. 2, pp. 252–259 (1999)
50. Taylor, G.W., Sigal, L., Fleet, D.J., Hinton, G.E.: Dynamical binary latent variable models for 3d human pose tracking, pp. 631–638. IEEE Computer Society Conference on Computer Vision and Pattern Recognition (2010)
51. Unal, G., Yezzi, A., Krim, H.: Information-theoretic active polygons for unsupervised texture segmentation. Int. J. Comput. Vision **62**(3), 199–220 (2005)
52. Wang, X., Doretto, G., Sebastian, T.B., Rittscher, J., Tu, P.H.: Shape and appearance context modeling. In: International Conference on Computer Vision, pp. 1–8 (2007)
53. Wu, Y., Minoh, M., Mukunoki, M., Lao, S.: Set based discriminative ranking for recognition. In: European Conference on Computer Vision, pp. 497–510. Springer (2012)
54. Zheng, W.S., Gong, S., Xiang, T.: Associating groups of people. In: British Conference on Machine Vision (2009)
55. Zheng, W.S., Gong, S., Xiang, T.: Reidentification by relative distance comparison. IEEE Trans. Pattern Anal. Mach. Intell. **35**(3), 653–668 (2013)

Chapter 4
Re-identification by Covariance Descriptors

Sławomir Bąk and François Brémond

Abstract This chapter addresses the problem of appearance matching, while employing the covariance descriptor. We tackle the extremely challenging case in which the same nonrigid object has to be matched across disjoint camera views. Covariance statistics averaged over a Riemannian manifold are fundamental for designing appearance models invariant to camera changes. We discuss different ways of extracting an object appearance by incorporating various training strategies. Appearance matching is enhanced either by discriminative analysis using images from a single camera or by selecting distinctive features in a covariance metric space employing data from two cameras. By selecting only essential features for a specific class of objects (e.g., humans) without defining *a priori* feature vector for extracting covariance, we remove redundancy from the covariance descriptor and ensure low computational cost. Using a feature selection technique instead of learning on a manifold, we avoid the over-fitting problem. The proposed models have been successfully applied to the person re-identification task in which a human appearance has to be matched across nonoverlapping cameras. We carry out detailed experiments of the suggested strategies, demonstrating their pros and cons *w.r.t.* recognition rate and suitability to video analytics systems.

4.1 Introduction

The present work addresses the re-identification problem that consists in appearance matching of the same subject registered by nonoverlapping cameras. This task is particularly hard due to camera variations, different lighting conditions, different color

S. Bąk (✉) · F. Brémond
INRIA, Sophia Antipolis, France
e-mail: slawomir.bak@inria.fr

F. Brémond
e-mail: francois.bremond@inria.fr

S. Gong et al. (eds.), *Person Re-Identification*,
Advances in Computer Vision and Pattern Recognition,
DOI: 10.1007/978-1-4471-6296-4_4,

responses, and different camera viewpoints. Moreover, we focus on nonrigid objects (i.e., humans) that change their pose and orientation contributing to the complexity of the problem.

In this work, we design two methods for appearance matching across nonoverlapping cameras. One particular aspect is a choice of an image descriptor. A good descriptor should capture the most distinguishing characteristics of an appearance, while being invariant to camera changes. We offer to describe an object appearance by using the *covariance descriptor* [26] as its performance is found to be superior to other methods (Sect. 4.3).

By averaging descriptors on a Riemannian manifold, we incorporate information from multiple images. This produces *mean Riemannian covariance* (Sect. 4.3.2) that yields a compact and robust representation.

Having an effective descriptor, we design efficient strategies for appearance matching. The first method assumes a predefined appearance model (Sect. 4.4.2), introducing discriminative analysis, which can be performed online. On the other hand, the second technique learns an appearance representation during an offline stage, guided by an entropy-driven criterion (Sect. 4.4.3). This removes redundancy from the descriptor and ensures low computational cost.

We carry out detailed experiments of proposed methods (Sect. 4.5), while investigating their pros and cons *w.r.t.* recognition rate and suitability to video analytics systems.

4.2 Related Work

Recent studies have focused on the appearance matching problem in the context of pedestrian recognition. Person re-identification approaches concentrate either on *metric learning* regardless of the representation choice, or on *feature modeling*, while producing a distinctive and an invariant representation for appearance matching.

Metric learning approaches use training data to search for strategies that combine given features maximizing inter-class variation whilst minimizing intra-class variation. These approaches do not pay too much attention to a feature representation. In the result, metric learning techniques use very simple features such as color histograms or common image filters [10, 21, 30]. Moreover, for producing robust metrics, these approaches usually require hundreds of training samples (image pairs with the same person/object registered by different cameras). It raises numerous questions about practicability of these approaches in a large camera network.

Instead, feature-oriented approaches concentrate on an invariant representation, which should handle view point and camera changes. However, these approaches usually do not take into account discriminative analysis [5, 6, 14]. In fact, learning using a sophisticated feature representation is very hard or even unattainable due to a complex feature space.

It is relevant to mention that both approaches proposed in this work belong more to feature-oriented approaches as they employ the covariance descriptor [26]. The covariance matrix can be seen as a meta descriptor that can fuse efficiently different types of features and their modalities. This descriptor has been extensively used in the literature for different computer vision tasks.

In [27] covariance matrix is used for designing a robust human detection algorithm. Human appearance is modeled by a dense set of covariance features extracted inside a detection window. Covariance descriptor is computed from subwindows with different sizes sampled from different locations. Then, a boosting mechanism selects the best regions characterizing a human silhouette.

Unfortunately, using covariance matrices, we also influence significantly computational complexity. This issue has been addressed in [28]. The covariance matrices of feature subsets rather than the full feature vector, provide similar performance while significantly reducing the computation load.

Covariance matrix has also been successfully applied to tracking. In [23] object deformations and appearance changes were handled by a model update algorithm using the Lie group structure of the positive definite matrices.

The first approach which employs the covariance descriptor for appearance matching across nonoverlapping cameras is [2]. In this work, an HOG-based detector establishes the correspondence between body parts, which are matched using a spatial pyramid of covariance descriptors.

In [22] we can find biologically inspired features combined with the similarity measure of covariance descriptors. The new descriptor is not represented by the covariance matrix but by a distance vector computed using the similarity measure between covariances extracted at different resolution bands. This method shows promising results not only for person re-identification but also for face verification.

Matching groups of people by covariance descriptor is the main topic of [7]. It is shown that contextual cues coming from group of people around a person of interest can significantly improve the re-identification performance. This contextual information is also kept by the covariance matrix.

In [4] the authors use *one-against-all* learning scheme to enhance distinctive characteristic of covariances for a specific individual. As covariances do not live on Euclidean space, binary classification is performed on a Riemannian manifold. Tangent planes extracted from positive training data points are used as a classification space for a boosting algorithm. Similarly, in [19] discriminative models are learned by a boosting scheme. However, covariance matrices are transformed into *Sigma Points* to avoid learning on a manifold, which often produces a over-fitted classifier.

Although discriminative approaches show promising results, they are usually computationally intensive, which is unfavorable in practice. In general, discriminative methods are also accused of nonscalability. It may be noted that an extensive learning phase is necessary to extract discriminative signatures at every instant when a new person is added to the set of existing signatures. This updating step makes these approaches very difficult to apply in a real-world scenario.

In this work, we overcome the mentioned issues twofold: (1) by offering an efficient discriminative analysis, which can be performed online even in a large camera network or (2) by an offline learning stage, which learns a general model for appearance matching. Using a feature selection technique instead of learning on a manifold, we avoid the over-fitting problem.

4.3 Covariance Descriptor

In [26], covariance of d-features has been proposed to characterize an image region. The descriptor encodes information on feature variances inside the region, their correlations with each other and their spatial layout. It can fuse different types of features, while producing a compact representation. The performance of the covariance descriptor is found to be superior to other methods, as rotation and illumination changes are absorbed by the covariance matrix.

Covariance matrix can be computed from any type of images such as a one dimensional intensity image, three channel color image or even other types of images, e.g., infrared.

Let I be an image and F be a d-dimensional feature image extracted from I

$$F(x, y) = \phi(I, x, y), \tag{4.1}$$

where function ϕ can be any mapping, such as color, intensity, gradients, filter responses, *etc*. For a given rectangular region $\text{Reg} \subset F$, let $\{f_k\}_{k=1...n}$ be the d-dimensional feature points inside Reg (n is the number of feature points, e.g., the number of pixels). We represent region Reg by the $d \times d$ covariance matrix of the feature points

$$C_{\text{Reg}} = \frac{1}{n-1} \sum_{k=1}^{n} (f_k - \mu)(f_k - \mu)^T, \tag{4.2}$$

where μ is the mean of the points.

Such a defined positive definite and symmetric matrix can be seen as a tensor. The main problem is that such a defined tensor space is a manifold that is not a vector space with the usual additive structure (does not lie on Euclidean space). Hence, many usual operations, such as *mean* or *distance*, need a special treatment. Therefore, the covariance manifold is often specified as Riemannian to determine a powerful framework using tools from differential geometry [24].

4.3.1 Riemannian Geometry

A manifold is a topological space which is locally similar to a Euclidean space. It means that every point on the m-dimensional manifold has a neighborhood homeomorphic to an open subset of the m-dimensional space $\Re^m$. Performing operations on the manifold involves choosing a metric.

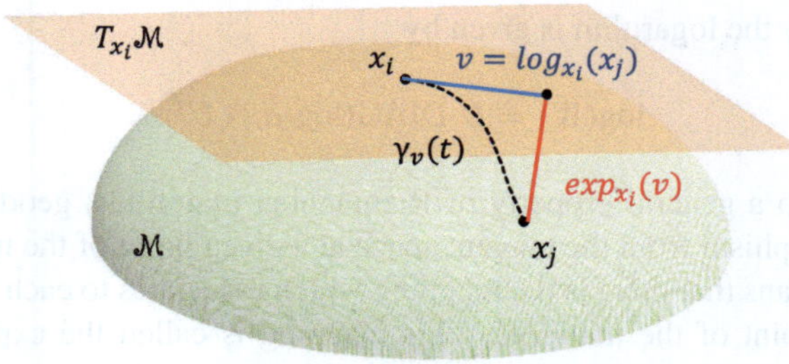

Fig. 4.1 An example of a two-dimensional manifold. We show the tangent plane at x_i, together with the exponential and logarithm mappings related to x_i and x_j [16]

Specifying manifold as Riemannian gives us Riemannian metric. This automatically determines a powerful framework to work on the manifold by using tools from differential geometry [24]. Riemannian manifold $\mathcal{M}$ is a differentiable manifold in which each tangent space has an inner product which varies smoothly from point to point. Since covariance matrices can be represented as a connected Riemannian manifold, we apply operations such as *distance* and *mean* computation using this differential geometry.

Figure 4.1 shows an example of a two-dimensional manifold, a smooth surface living in $\Re^3$. Tangent space $T_x\mathcal{M}$ at x is the vector space that contains the tangent vectors to all 1-D curves on $\mathcal{M}$ passing through x. Riemannian metric on manifold $\mathcal{M}$ associates to each point $x \in \mathcal{M}$, a differentiable varying inner product $\langle \cdot, \cdot \rangle_x$ on tangent space $T_x\mathcal{M}$ at x. This induces a norm of tangent vector $v \in T_x\mathcal{M}$ such that $\|v\|_x^2 = \langle v, v \rangle_x$. The minimum length curve over all possible smooth curves $\gamma_v(t)$ on the manifold between x_i and x_j is called *geodesic*, and the length of this curve stands for geodesic distance $\rho(x_i, x_j)$.

Before defining geodesic distance, let us introduce the exponential and the logarithm functions, which take as an argument a square matrix. The exponential of matrix W can be defined as the series

$$\exp(W) = \sum_{k=0}^{\infty} \frac{W^k}{k!}. \tag{4.3}$$

In the case of symmetric matrices, we can apply some simplifications. Let $W = UDU^T$ be a diagonalization, where U is an orthogonal matrix, and $D = DIAG(d_i)$ is the diagonal matrix of the eigenvalues. We can write any power of W in the same way $W^k = UD^kU^T$. Thus

$$\exp(W) = U\,\mathrm{DIAG}(\exp(d_i))\,U^T, \tag{4.4}$$

and similarly the logarithm is given by

$$\log(W) = U\ \mathrm{DIAG}(\log(d_i))\ U^T. \tag{4.5}$$

According to a general property of Riemannian manifolds, geodesics realize a local diffeomorphism from the tangent space at a given point of the manifold to the manifold. It means that there is the mapping which associates to each tangent vector $v \in T_x\mathcal{M}$ a point of the manifold. This mapping is called the exponential map, because it corresponds to the usual exponential in some matrix groups.

The exponential and logarithmical mappings have the following expressions [24]:

$$\exp_\Sigma(W) = \Sigma^{\frac{1}{2}} \exp\left(\Sigma^{-\frac{1}{2}} W \Sigma^{-\frac{1}{2}}\right)\Sigma^{\frac{1}{2}}, \tag{4.6}$$

$$\log_\Sigma(W) = \Sigma^{\frac{1}{2}} \log\left(\Sigma^{-\frac{1}{2}} W \Sigma^{-\frac{1}{2}}\right)\Sigma^{\frac{1}{2}}, \tag{4.7}$$

where

$$\Sigma^{\frac{1}{2}} = \exp\left(\frac{1}{2}(\log(\Sigma))\right). \tag{4.8}$$

Given tangent vector $v \in T_{x_i}\mathcal{M}$, there exists a unique geodesic $\gamma_v(t)$ starting at x_i (see Fig. 4.1). The exponential map $\exp_{x_i} : T_{x_i}\mathcal{M} \rightarrow \mathcal{M}$ maps tangent vector v to the point on the manifold that is reached by this geodesic. The inverse mapping is given by logarithm map denoted by $\log_{x_i} : \mathcal{M} \rightarrow T_{x_i}\mathcal{M}$. For two points x_i and x_j on manifold $\mathcal{M}$, the tangent vector to the geodesic curve from x_i to x_j is defined as $v = \overrightarrow{x_i x_j} = \log_{x_i}(x_j)$, where the exponential map takes v to the point $x_j = \exp_{x_i}(\log_{x_i}(x_j))$. The Riemannian distance between x_i and x_j is defined as $\rho(x_i, x_j) = \| \log_{x_i}(x_j)\|_{x_i}$. It is relevant to note that an equivalent form of geodesic distance can be given in terms of generalized eigenvalues [13].

The distance between two symmetric positive definite matrices C_i and C_j can be expressed as

$$\rho(C_i, C_j) = \sqrt{\sum_{k=1}^{d} \ln^2 \lambda_k(C_i, C_j)}, \tag{4.9}$$

where $\lambda_k(C_i, C_j)_{k=1\ldots d}$ are the generalized eigenvalues of C_i and C_j, determined by

$$\lambda_k C_i x_k - C_j x_k = 0, \quad k = 1 \ldots d \tag{4.10}$$

and $x_k \neq 0$ are the generalized eigenvectors.

We have already mentioned that we are more interested in extracting covariance statistics from several images rather than from a single image. Then, having a suitable metric, we can define *mean Riemannian covariance*.

4.3.2 Mean Riemannian Covariance

Let $C_1, \ldots, C_N$ be a set of covariance matrices. The Karcher or Fréchet mean is the set of tensors minimizing the sum of squared distances. In the case of tensors, the manifold has a nonpositive curvature, so there is a unique mean value μ:

$$\mu = \arg \min_{C \in \mathcal{M}} \sum_{i=1}^{N} \rho^2(C, C_i). \tag{4.11}$$

As the mean is defined through a minimization procedure, we approximate it by the intrinsic Newton gradient descent algorithm. The following mean value at estimation step $t + 1$ is given by:

$$\mu_{t+1} = \exp_{\mu_t} \left[\frac{1}{N} \sum_{i=1}^{N} \log_{\mu_t}(C_i) \right], \tag{4.12}$$

where $\exp_{\mu_t}$ and $\log_{\mu_t}$ are mapping functions (see Eqs. 4.6 and 4.7). This iterative gradient descent algorithm usually converges very fast (in experiments five iterations were sufficient, which is similar to [24]). This mean value is referred to as *mean Riemannian covariance* (MRC).

MRC versus volume covariance Covariance matrix could be directly computed from a video by merging feature vectors from many frames into a single content (similarly to 3D descriptors, i.e., 3D HOG). Then, this covariance could be seen as *mean covariance*, describing characteristics of the video. Unfortunately, such solution disturbs time dependencies (time order of features is lost). Further, context of the features might be lost and at the same time some features will not appear in the covariance.

Figure 4.2 illustrates the case, in which edge features are lost during computation of the volume covariance. This is a consequence of loosing information that the feature appeared in specific time. Computing volume covariance, order of the feature appearances and their spatial correlations can be lost by merging feature distribution in time. This clearly shows that MRC holds much more information than covariance computed directly from the volume.

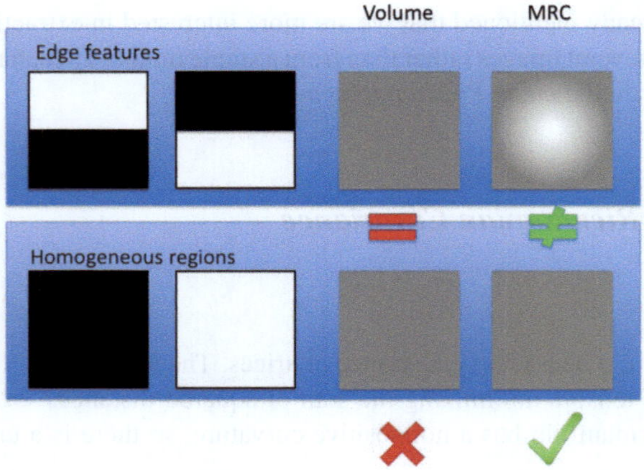

Fig. 4.2 Difference between covariance computed directly from the video content (volume covariance) and MRC. Volume covariance looses information on edge features and can not distinguish two given cases—two edge features (*first row*) from two homogeneous regions (*second row*). MRC holds information on the edges, being able to differentiate both cases

4.4 Efficient Models for Human Re-Identification

In this section we focus on designing efficient models for appearance matching. These models are less computationally expensive than boosting approaches [4, 19], while enhancing distinctive and descriptive characteristics of an object appearance. We propose two strategies for appearance extraction: (1) by using a hand-designed model which is enhanced by a fast discriminative analysis (Sect. 4.4.2) and (2) by employing machine learning technique that selects the most accurate features for appearance matching (Sect. 4.4.3).

4.4.1 General Scheme for Appearance Extraction

The input of the appearance extraction algorithm is a set of cropped images obtained by human detection and tracking results corresponding to a given person of interest (see Fig. 4.3). Color dissimilarities caused by variations in lighting conditions are minimized by applying *histogram equalization* [20]. This technique maximizes the entropy in each color channel (RGB) producing more camera-independent color representation. Then, the normalized image is scaled into a fixed size $W \times H$.

From such scaled and normalized images, we extract covariance descriptors from image subregions and we compute MRC-s (Sect. 4.3.2). Every image subregion (its size and position) as well as features from which covariance is extracted is determined by a model. The final appearance representation is referred to as a *signature*.

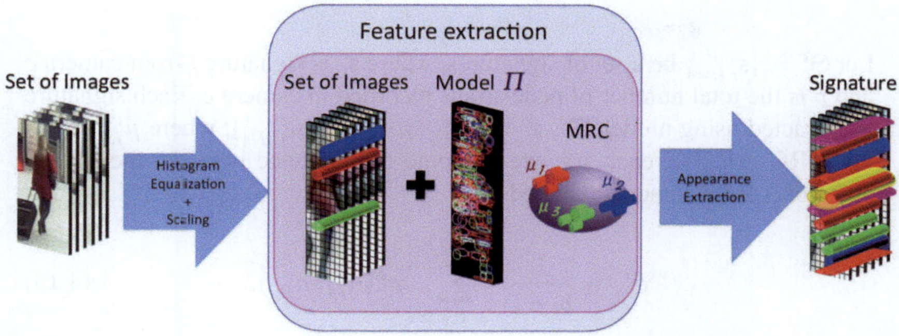

Fig. 4.3 Appearance extraction: features are determined using model Π for computing *mean Riemannian covariances* (MRC), which stand for the final appearance representation—*signature*

4.4.2 MRCG Model

Mean Riemannian covariance grid (MRCG) [3] has been designed to deal with low resolution images and a crowded environment where more specialized techniques (e.g., based on background subtraction) might fail. It combines a dense descriptor philosophy [9] with the effectiveness of MRC descriptor.

MRCG is represented by a dense grid structure with overlapping spatial square subregions (*cells*). First, such dense representation makes the signature robust to partial occlusions. Second, as the grid structure, it contains relevant information on spatial correlations between MRC *cells*, which is essential to carrying out discriminative power of the signature. MRC *cell* describes statistics of an image square subregion corresponding to the specific position in the grid structure. In case of MRCG, we assume a fixed size of *cells* and a fixed feature vector for extracting covariances. Let Π be a model which is actually represented by a set of MRC cells. This model is enhanced by using our discriminative analysis, which weights each cell depending on its distinctive characteristics. These weights are referred to as MRC *discriminants*.

MRC Discriminants

The goal of using discriminants is to identify the relevance of MRC *cells*. We present an efficient way to enhance discriminative features, improving matching accuracy. By employing *one-against-all* learning schema, we highlight distinctive features for a particular individual. The main advantage of this method is its efficiency. Unlike [4], by using simple statistics on Riemannian manifold we are able to enhance features, without applying any time-consuming training process.

Let $\mathfrak{S}^c = \{\mathfrak{s}_i^c\}_{i=1}^p$ be a set of signatures, where $\mathfrak{s}_i^c$ is signature i from camera c and p is the total number of pedestrians recorded in camera c. Each signature is extracted using model Π : $\mathfrak{s}_i^c = \{\mu_{i,1}^c, \mu_{i,2}^c, \ldots, \mu_{i,|\Pi|}^c\}$, where $\mu_{i,j}^c$ stands for MRC *cell*. For each $\mu_{i,j}^c$ we compute the variance between the human signatures from camera c defined as

$$\sigma_{i,j}^c = \frac{1}{p-1} \sum_{k=1;\, k \neq i}^{p} \rho^2(\mu_{i,j}^c, \mu_{k,j}^c). \tag{4.13}$$

In the result, for each human signature $\mathfrak{s}_i^c$, we obtain the vector of discriminants related to our MRC *cells*, $d_i^c = \{\sigma_{i,1}^c, \sigma_{i,2}^c, \ldots, \sigma_{i,|\Pi|}^c\}$. This idea is similar to methods derived from text retrieval where a frequency of *terms* is used to weight relevance of a *word*. As we do not want to quantize covariance space, we use $\sigma_{i,j}^c$ of MRC *cell* to extract its relevance. The MRC is assumed to be more significant when its variance is larger in the class of humans. Here, it is a kind of "*killing two birds with one stone*": (1) it is obvious that the most common patterns belong to the background (the variance is small) and (2) the patterns which are far from the rest are at the same time the most discriminative (the variance is large).

We thought about normalizing the $\sigma_{i,j}^c$ by the variance *within the class* (similarly to Fisher's linear discriminants, we could compute the variance of covariances related to a given cell). However, the results have shown that such normalization does not improve matching accuracy. We believe that it is due to the fact that a given number of images per individual is not sufficient for obtaining the reliable variance of MRC *within the class*.

Scalability Discriminative approaches are often accused of nonscalability. It is true that in the most of these approaches an extensive learning phase is necessary to extract discriminative signatures. This makes these approaches very difficult to apply in a real-world scenario wherein every minute new people appear.

Fortunately, proposing MRC discriminants, we employ a very simple discriminative method which is able to perform in a real world system. It is true that every time when a new signature is created we have to update all signatures in the database. However, for 10,000 signatures, the update takes less than 30 s. Moreover, we do not expect more than such a number of signatures in the database as the re-identification approaches are constrained to *one day period* (the strong assumption about the same clothes). Further, one alternative solution might be to use a fixed *reference dataset*, which can be used as training data for discriminating new signatures.

4.4.3 COSMATI Model

In the previously presented model, we assumed *a priori* the size of MRC cells, the grid layout and the feature vector, from which covariance is extracted. However, depending on image resolution and image characteristics (object class), we could use different feature vectors extracted from different image regions. Moreover, it may happen that different regions of the object appearance ought to be matched using various feature vectors to obtain a distinctive representation. Then, we actually can formulate the appearance matching problem as the task of learning a model that selects the most descriptive features for a specific class of objects. This approach is referred to as *COrrelation-based Selection of covariance MATrIces* (COSMATI) [1].

In contrast to the previous model and to the most of state-of-the-art approaches [4, 19, 26], we do not limit our covariance descriptor to a single feature vector. Instead of defining *a priori* feature vector, we use a machine learning technique to select features that provide the most descriptive apperance representation. The following sections describe our feature space and the learning, by which the appearance model for matching is generated.

Feature Space

Let $L = \{R, G, B, I, \nabla_I, \theta_I, \dots\}$ be a set of feature layers, in which each layer is a mapping such as color, intensity, gradients and filter responses (texture filters, i.e., Gabor, Laplacian, or Gaussian). Instead of using covariance between all of these layers, which would be computationally expensive, we compute covariance matrices of a few relevant feature layers. These relevant layers are selected depending on the region of an object (see Sect. 4.4.3). In addition, let layer $\mathfrak{D}$ be a distance between the center of an object and the current location. Covariance of distance layer $\mathfrak{D}$ and three other layers $l (l \in L)$ form our descriptor, which is represented by a 4×4 covariance matrix. By using distance $\mathfrak{D}$ in every covariance, we keep a spatial layout of feature variances, which is rotation invariant. State-of-the-art techniques very often use pixel location (x, y) instead of distance $\mathfrak{D}$, yielding better description of an image region. Conversely, among our detail experimentation, using $\mathfrak{D}$ rather than (x, y), we did not decrease the recognition accuracy in the general case, while decreasing the number of features in the covariance matrix. This discrepancy may be due to the fact that we hold spatial information twofold, (1) by location of a rectangular subregion from which the covariance is extracted and (2) by $\mathfrak{D}$ in covariance matrix. We constraint our covariances to the combination of four features, ensuring computational efficiency. Also, bigger covariance matrices tend to include superfluous features which can clutter the appearance matching. 4×4 matrices provide sufficiently descriptive correlations while keeping low computational time needed for calculating generalized eigenvalues during distance computation.

Fig. 4.4 A meta covariance feature space. Example of three different covariance features. Every covariance is extracted from a region (P), *distance* layer ($\mathfrak{D}$) and three channel functions (e.g., bottom covariance feature is extracted from region P_3 using layers: $\mathfrak{D}$, I-intensity, ∇_I-gradient magnitude and θ_I-gradient orientation)

Different combinations of three feature layers produce different kinds of covariance descriptors. By using different covariance descriptors, assigned to different locations in an object, we are able to select the most discriminative covariances according to their positions. The idea is to characterize different regions of an object by extracting different kinds of features (e.g., when comparing human appearances, edges coming from shapes of arms and legs are not discriminative enough in most cases as every instance possess similar features). Taking into account this phenomenon, we minimize redundancy in an appearance representation by an entropy-driven selection method.

Let us define index space $\mathbb{Z} = \{(P, l_i, l_j, l_k) : P \in \mathbf{P}; l_i, l_j, l_k \in L\}$, of our meta covariance feature space $\mathfrak{C}$, where $\mathbf{P}$ is a set of rectangular subregions of the object; and l_i, l_j, l_k are color/intensity or filter layers. Meta covariance feature space $\mathfrak{C}$ is obtained by mapping $\mathbb{Z} \rightarrow \mathfrak{C} : cov_P(\mathfrak{D}, l_i, l_j, l_k)$, where $cov_P(\phi)$ is the covariance descriptor [26] of features ϕ: $cov_P(\phi) = \frac{1}{|P|-1} \sum_{k \in P} (\phi_k - \mu)(\phi_k - \mu)^T$. Figure 4.4 shows different feature layers as well as examples of three different types of covariance descriptor. The dimension $n = |\mathbb{Z}| = |\mathfrak{C}|$ of our meta covariance feature space is the product of the number of possible rectangular regions by the number of different combinations of feature layers.

Learning in a Covariance Metric Space

Let $\mathbf{a}_i^c = \{\mathbf{a}_{i,1}^c, \mathbf{a}_{i,2}^c, \ldots \mathbf{a}_{i,m}^c\}$ be a set of relevant observations of an object i in camera c, where $\mathbf{a}_{ij}^c$ is a n-dimensional vector composed of all possible covariance features extracted from image j of object i in the n-dimensional meta covariance feature space $\mathfrak{C}$. We define the distance vector between two samples $\mathbf{a}_{i,j}^c$ and $\mathbf{a}_{k,l}^{c'}$ as follows:

$$\delta(\mathbf{a}_{i,j}^c, \mathbf{a}_{k,l}^{c'}) = \left[\rho(\mathbf{a}_{i,j}^c[z], \mathbf{a}_{k,l}^{c'}[z])\right]_{z \in \mathbb{Z}}^T, \tag{4.14}$$

where ρ is the geodesic distance between covariance matrices [13], and $\mathbf{a}^{c}_{i,j}[z]$, $\mathbf{a}^{c'}_{k,l}[z]$ are the corresponding covariance matrices (the same region P and the same combination of layers). The index z is an iterator of $\mathfrak{C}$.

We cast the appearance matching problem into the following *distance learning* problem. Let δ^{+} be distance vectors computed using pairs of relevant samples (of the same people captured in different cameras, $i = k, c \neq c'$) and let δ^{-} be distance vectors computed between pairs of related irrelevant samples ($i \neq k, c \neq c'$). Pairwise elements δ^{+} and δ^{-} are distance vectors, which stand for positive and negative samples, respectively. These distance vectors define a *covariance metric space*. Given δ^{+} and δ^{-} as training data, our task is to find a general model of appearance to maximize matching accuracy by selecting relevant covariances and thus defining a distance.

Learning on a manifold This is a difficult and unsolved challenge. Methods [4, 27] perform classification by regression over the mappings from the training data to a suitable tangent plane. By defining tangent plane over the Karcher mean of the positive training data points, we can preserve a local structure of the points. Unfortunately, models extracted using means of the positive training data points tend to over-fit. These models concentrate on tangent planes obtained from training data and do not have generalization properties. We overcome this issue by employing a feature selection technique for identifying the most salient features. Based on the hypothesis: "*A good feature subset is one that contains features highly correlated with (predictive of) the class, yet uncorrelated with (not predictive of) each other*" [18], we build our appearance model using covariance features chosen by a *correlation-based feature selection.*

Correlation-based Feature Selection (CFS) [18] is a filter algorithm that ranks feature subsets according to a correlation-based evaluation function. This evaluation function favors feature subsets which contain features highly correlated with the class and uncorrelated with each other. In the *metric learning* problem, we define positive and negative class by δ^{+} and δ^{-}, as relevant and irrelevant pairs of samples.

Further, let feature $f_z = \delta[z]$ be characterized by a distribution of the zth elements in distance vectors δ^{+} and δ^{-}. The feature-class correlation and the feature-feature inter-correlation are measured using a symmetrical uncertainty model [18]. As this model requires nominal valued features, we discretize f_z using the method of Fayyad and Irani [11]. Let X be a nominal valued feature obtained by discretization of f_z (discretization of distances).

We assume that a probabilistic model of X can be formed by estimating the probabilities of the values $x \in X$ from the training data. The information content can be measured by entropy $H(X) = -\sum_{x \in X} p(x) \log_2 p(x)$. A relationship between features X and Y can be given by $H(X|Y) = -\sum_{y \in Y} p(y) \sum_{x \in X} p(x|y) \log_2 p(x|y)$. The amount by which the entropy of X decreases reflects additional information on X provided by Y and is called the *information gain* (*mutual information*) defined as $Gain = H(X) - H(X|Y) = H(Y) - H(Y|X) = H(X) + H(Y) - H(X, Y)$.

Even if the *information gain* is a symmetrical measure, it is biased in favor of features with more discrete values. Thus, the symmetrical uncertainty r_{XY} is used to overcome this problem $r_{XY} = 2 \times \left[Gain/\big(H(X) + H(Y)\big)\right]$.

Having the correlation measure, a subset of features $\mathfrak{S}$ is evaluated using function $\mathfrak{M}(\mathfrak{S})$ defined as

$$\mathfrak{M}(\mathfrak{S}) = \frac{k\,\overline{r_{cf}}}{\sqrt{k + k\,(k+1)\,\overline{r_{ff}}}}, \tag{4.15}$$

where k is the number of features in subset $\mathfrak{S}$, $\overline{r_{cf}}$ is the average feature-class correlation and $\overline{r_{ff}}$ is the average feature-feature inter-correlation

$$\overline{r_{cf}} = \frac{1}{k} \sum_{f_z \in \mathfrak{S}} r_{cf_z}, \quad \overline{r_{ff}} = \frac{2}{k\,(k-1)} \sum_{\substack{f_i, f_j \in \mathfrak{S} \\ i<j}} r_{f_i f_j}, \tag{4.16}$$

where c is the class, or relevance feature, which is $+1$ on δ^+ and -1 on δ^-. The numerator in Eq. 4.15 indicates the predictive ability of subset $\mathfrak{S}$ and the denominator stands for redundancy among the features (for details of $\mathfrak{M}(\mathfrak{S})$, and interested readers please refer to, the interested reader is pointed to [18]).

Equation 4.15 is the core of CFS, which ranks feature subsets in the search space of all possible feature subsets. Since exhaustive enumeration of all possible feature subsets is prohibitive in most cases, a heuristic search strategy has to be applied. We have investigated different search strategies, among which *best first search* [25] performs the best.

Best first search is an Artificial Intelligence search strategy that allows backtracking along the search path. Our *best first* starts with no feature and progresses forward through the search space adding single features. The search terminates if T consecutive subsets show no improvement over the current best subset (we set $T = 5$ in experiments). By using this stopping criterion we prevent the best first search from exploring the entire feature subset search space. Figure 4.5 illustrates CFS method. Let Π be the output of CFS that is the feature subset of $\mathfrak{C}$. This feature subset Π forms a model that is used for appearance extraction and matching.

4.4.4 Appearance Matching

Let $\mathfrak{s}_a^c$ and $\mathfrak{s}_b^{c'}$ be the object signatures. The signature consists of *mean Riemannian covariance* matrices extracted using set Π. The similarity between two signatures $\mathfrak{s}_a^c$ and $\mathfrak{s}_b^{c'}$ is defined as

$$S(\mathfrak{s}_a^c, \mathfrak{s}_b^{c'}) = \frac{1}{|\Pi|} \sum_{i \in \Pi} \frac{\sigma_{a,i}^c + \sigma_{b,i}^{c'}}{\max(\rho(\mu_{a,i}^c, \mu_{b,i}^{c'}), \varepsilon)}, \tag{4.17}$$

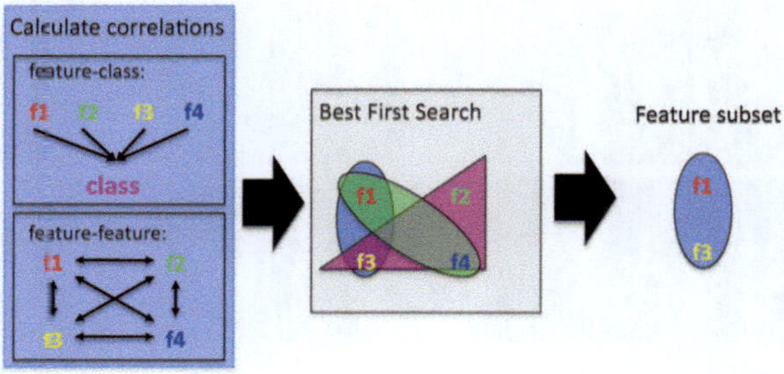

Fig. 4.5 Correlation-based feature selection. *Best first search* evaluates different feature subsets using feature-class and feature-feature correlations (Eq. 4.15). The best feature subset stands for model Π that is used for appearance extraction and matching

where ρ is a geodesic distance, $\mu^{c}_{a,i}$ and $\mu^{c'}_{b,i}$ are mean covariance matrices extracted using covariance feature $i \in \Pi$ and $\varepsilon = 0.1$ is introduced to avoid the denominator approaching to zero. $\sigma^{c}_{a,i}$ and $\sigma^{c'}_{b,i}$ are discriminants of the corresponding MRC-s (see Sect. MRC Discriminants). If discriminants have not been computed, then the nominator is set to 1 ($\sigma^{c}_{a,i} + \sigma^{c'}_{b,i} = 1$). Using the average of similarities computed on feature set Π the appearance matching becomes robust to noise.

4.5 Experiments

In this section, we mostly focus on comparing MRCG with COSMATI model. We carry out experiments on three i-LIDS datasets[1]: i-LIDS-MA [4], i-LIDS-AA [4] and i-LIDS-119 [29]. These datasets have been extracted from the 2008 i-LIDS Multiple-Camera Tracking Scenario (MCTS) dataset for evaluating the re-identification task. The results are analyzed in terms of recognition rate, using the *cumulative matching characteristic* (CMC) [17] curve. The CMC curve represents the expectation of finding the correct match in the top n matches (see Fig. 4.6). We also employ a quantitative scalar appraisal of CMC curve by computing the *normalized area under curve* (nAUC) value.

4.5.1 Experimental Setup

Comparing the proposed models we keep the experimental settings presented in [1, 3]. *Model_Name*$^{+}$ means that signatures were enhanced by using discriminative analysis (Sect. 4.4.2). It should be noted that this discriminative analysis can be applied to MRCG as well as to COSMATI model.

---

[1] The Image Library for Intelligent Detection Systems (i-LIDS) is the UK government's benchmark for Video Analytics (VA) systems.

Fig. 4.6 Example of the person re-identification on i-LIDS-MA. The left-most image is the probe image. The remaining images are the top 20 matched gallery images. The red boxes highlight the correct matches

MRCG Model

Every human image is scaled into a fixed size of 64 × 192 pixels (size of the grid). We extract the MRC *cells* of 16 × 16 pixels, on a fixed grid of eight pixels step (it gives in total 161 *cells*). The feature vector consists of 11 features:

$$\left[x, y, R_{xy}, G_{xy}, B_{xy}, \nabla^R_{xy}, \theta^R_{xy}, \nabla^G_{xy}, \theta^G_{xy}, \nabla^B_{xy}, \theta^B_{xy}\right], \tag{4.18}$$

where x and y are pixel location, R_{xy}, G_{xy}, B_{xy} are RGB channel values and ∇ and θ correspond to gradient magnitude and orientation in each channel, respectively.

COSMATI Model

Feature space We scale every human image into a fixed size window of 64 × 192 pixels. The set of rectangular subregions **P** is produced by shifting 32×8 and 16×16 pixel regions with eight pixels step (up and down). It gives $|\mathbf{P}| = 281$ overlapping rectangular subregions. We set $L = \{(l, \nabla_l, \theta_l)_{l=I,R,G,B}, G_{i=1\ldots4}, \mathcal{N}, \mathfrak{L}\}$, where I, R, G, B refer to intensity, red, green, and blue channel, respectively; ∇ is the gradient magnitude; θ corresponds to the gradient orientation; G_i are Gabor's filters with parameters $\gamma, \theta, \lambda, \sigma^2$ set to $(0.4, 0, 8, 2)$, $(0.4, \frac{\pi}{2}, 8, 2)$, $(0.8, \frac{\pi}{4}, 8, 2)$ and $(0.8, \frac{3\pi}{2}, 8, 2)$, respectively; $\mathcal{N}$ is a gaussian and $\mathfrak{L}$ is a laplacian filter. A learning process involving all possible combinations of three layers would not be computationally tractable (229,296 covariances to consider in Sect. 4.4.3). Thus instead, we experimented with different subsets of combinations and selected a reasonably efficient one. Among all possible combinations of the three layers, we choose 10 combinations ($C_{i=1\ldots10}$) to separate color and texture features, while ensuring an inexpensive computation. We set C_i to (R, G, B), $(\nabla_R, \nabla_G, \nabla_B)$, $(\theta_R, \theta_G, \theta_B)$, (I, ∇_I, θ_I), (I, G_3, G_4), $(I, G_2, \mathfrak{L})$, $(I, G_2, \mathcal{N})$, $(I, G_1, \mathcal{N})$, $(I, G_1, \mathfrak{L})$, (I, G_1, G_2), respectively. Note that we add to every combination C_i layer $\mathfrak{D}$, thus generating our final 4 × 4 covariance descriptors. The dimension of our meta covariance feature space is $n = |\mathfrak{C}| = 10 \times |\mathbf{P}| = 2810$.

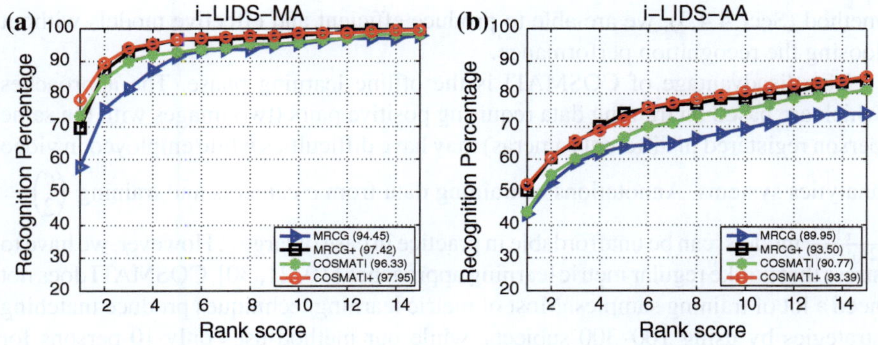

Fig. 4.7 Performance comparison. Evaluation of COSMATI is performed using the models learned on i-LIDS-MA. nAUC values are presented within parentheses. (a) p = 30, (b) p = 100

Learning and testing Let us assume that we have $(p+q)$ individuals seen from two different cameras. For every individual, m images from each camera are given. We take q individuals to learn our model, while p individuals are used to set up the gallery set. We generate positive training examples by comparing m images of the same individual from one camera with m images from the second camera. Thus, we produce $|\delta^+| = q \times m^2$ positive samples. Pairs of images coming from different individuals stand for negative training data, thus producing $|\delta^-| = q \times (q-1) \times m^2$ negative samples.

4.5.2 Results

i-LIDS-MA [4] This dataset consists of 40 individuals extracted from two nonoverlapping camera views. For each individual a set of 46 images is given. The dataset contains in total $40 \times 2 \times 46 = 3680$ images. For each individual we randomly select $m = 10$ images. Then, we randomly select $q = 10$ individuals to learn COSMATI models. The evaluation is performed on the remaining $p = 30$ individuals. We evaluate MRCG and COSMATI on the same sets of people. Every signature is used as a query to the gallery set of signatures from the other camera. This procedure has been repeated 10 times to obtain averaged CMC curves.

We compare COSMATI with MRCG in Fig. 4.7a. The best performance is achieved by COSMATI$^+$. It appears that discriminative analysis has more significant impact on MRCG than on COSMATI model. This result may be due to the fact that COSMATI already selects distinctive representation for appearance matching. We can also note that MRCG$^+$ achieves similar recognition rate as COSMATI$^+$. However, it is relevant to mention that COSMATI is significantly faster than MRCG, as it uses small covariance matrices (4×4). The experiment bears out that by designing the efficient feature space (Sect. 4.4.3) and employing the effective selection method (Sect. 4.4.3), we are able to produce efficient and effective models without loosing the recognition performance.

The disadvantage of COSMATI is the offline learning phase. The approaches which are based on training data requiring positive pairs (two images with the same person registered in different cameras) may have difficulties while employed in video analytics systems. Annotations of training data from c cameras and training $\binom{c}{2} = \frac{c!}{2!(c-2)!}$ models, can be unaffordable in practice in case of large c. However, we have to stress that unlike regular metric learning approaches [10, 21, 30], COSMATI does not need a lot of training samples. Most of metric learning techniques produce matching strategies by using 100–300 subjects, while our method uses only 10 persons for obtaining an effective and efficient model.

i-LIDS-AA [4] This dataset contains 100 individuals automatically detected and tracked in two cameras. Cropped images are noisy, which makes the dataset more challenging (e.g., detected bounding boxes are not accurately centered around the people, only part of the people is detected due to occlusion). For minimizing misalignment issues, we employ *discriminatively trained deformable part models* [12, 15], which slightly improve detection accuracy. COSMATI is evaluated using the models learned on i-LIDS-MA. Figure 4.7b illustrates the results. Although, data are noisy, we can observe the same trends as in the case of evaluating on i-LIDS-MA data.
i-LIDS-119 [29] For comparing MRCG and COSMATI models with state-of-the-art techniques, we select i-LIDS-119 data. This dataset is extensively used in the literature for testing the person re-identification approaches. It consists of 119 individuals with 476 images. The dataset is very challenging since there are many occlusions and often only the top part of the person is visible. As only few images are given per individual, we extract signatures using maximally $N = 2$ images.

In Fig. 4.8 we compare MRCG$^+$ and COSMATI with LCP [4], CPS [8], SDALF [5] and HPE [6]. In case of COSMATI, we have used models learned on i-LIDS-MA to evaluate our approach on the full dataset of 119 individuals.

COSMATI performs the best among all considered methods. We believe that it is due to the informative appearance representation obtained by CFS technique (Sect. 4.4.3). It clearly shows that a combination of the strong covariance descriptor with the efficient selection method produces distinctive models for the appearance matching problem. For more extensive evaluation and competitive results of COSMATI and of MRCG, the interested reader is pointed to [1] and [3], respectively.

Computational speed The level of performance achieved by COSMATI comes with a significant computational gain *w.r.t.* MRCG. In our experiments, for $q = 10$ and $m = 10$ we generate $|\delta^+| = 1000$ and $|\delta^-| = 9000$ training samples. Learning on 10,000 samples takes around 20 min on Intel quad-core 2.4 GHz. COSMATI model is composed of 150 covariance features in average, which is similar to MRCG (161 cells). Comparing the time calculation of generalized eigenvalues (distance computation) of 4×4 covariance with 11×11 covariance, we always reach 10–15 speedup depending on the hardware architecture. In the result, we can expect the same speedup while retrieving signatures in video analytics systems.

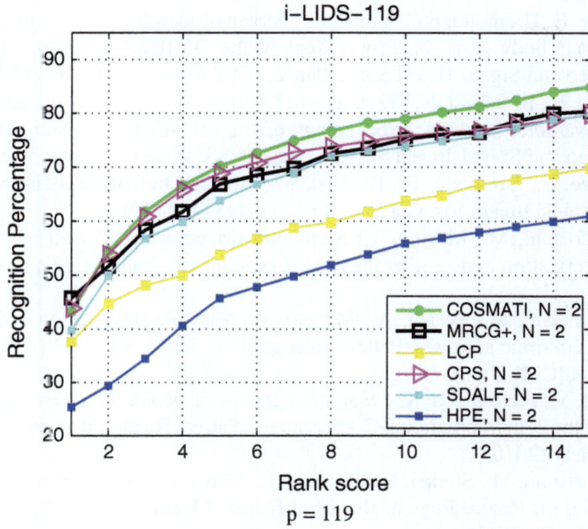

Fig. 4.8 Comparison with state-of-the-art approaches on i-LIDS-119 dataset: LCP [4], CPS [8], SDALF [5] and HPE [6]

4.6 Conclusions

This chapter presented two strategies for appearance matching, while employing covariance statistics averaged over a Riemannian manifold. We discussed different ways of extracting an object appearance by incorporating various training strategies. We showed that by applying efficient discriminative analysis, we are able to improve re-identification accuracy. Further, we demonstrated that by introducing an offline learning stage, we can characterize an object appearance in a more efficient and distinctive way. In the future, we plan to integrate the notion of motion in our recognition framework. This will allow to distinguish individuals using their shape characteristics and to extract only the features which surely belong to foreground region.

Acknowledgments This work has been supported by VANAHEIM, ViCoMo and PANORAMA European projects.

References

1. Bak, S., Charpiat, G., Corvee, E., Bremond, F., Thonnat, M.: Learning to match appearances by correlations in a covariance metric space. In: Proceedings of the 12th European Conference on Computer Vision, IEEE Computer Society (2012)

2. Bak, S., Corvee, E., Bremond, F., Thonnat, M.: Person re-identification using spatial covariance regions of human body parts. In: Proceedings of the 7th IEEE International Conference on Advanced Video and Signal-Based Surveillance, IEEE Computer Society (2010)
3. Bak, S., Corvee, E., Bremond, F., Thonnat, M.: Multiple-shot human re-identification by mean riemannian covariance grid. In: Proceedings of the 8th IEEE International Conference on Advanced Video and Signal-Based Surveillance, AVSS. IEEE Computer Society (2011)
4. Bak, S., Corvee, E., Bremond, F., Thonnat, M.: Boosted human re-identification using riemannian manifolds. Image Vis. Comput. **30**(6–7), 443–452 (2012)
5. Bazzani, L., Cristani, M., Murino, V.: Symmetry-driven accumulation of local features for human characterization and re-identification. Comput. Vis. Image Underst. **117**(2), 130–144 (2013)
6. Bazzani, L., Cristani, M., Perina, A., Murino, V.: Multiple-shot person re-identification by chromatic and epitomic analyses. Pattern Recogn. Lett. **33**(7), 898–903 (2012) (Special Issue on Awards from ICPR 2010).
7. Cai, Y., Takala, V., Pietikainen, M.: Matching groups of people by covariance descriptor. In: Proceedings of the 20th International Conference on Pattern Recognition, pp. 2744–2747. IEEE Computer Society (2010)
8. Cheng, D.S., Cristani, M., Stoppa, M., Bazzani, L., Murino, V.: Custom pictorial structures for re-identification. In: Proceedings of the 22nd British Machine Vision Conference, pp. 68.1-68.11. BMVA Press (2011)
9. Dalal, N., Triggs, B.: Histograms of oriented gradients for human detection. In: Proceedings of the 18th Conference on Computer Vision and Pattern Recognition, pp. 886–893. IEEE Computer Society (2005)
10. Dikmen, M., Akbas, E., Huang, T.S., Ahuja, N.: Pedestrian recognition with a learned metric. In: Proceedings of the 10th Asian Conference on Computer Vision, pp. 501–512. IEEE Computer Society (2010)
11. Fayyad, U.M., Irani, K.B.: Multi-interval discretization of continuous-valued attributes for classification learning. In: Proceedings of the International Joint Conference on Uncertainty in AI, IJCAI, pp. 1022–1027 (1993)
12. Felzenszwalb, P.F., Girshick, R.B., McAllester, D., Ramanan, D.: Object detection with discriminatively trained part based models. IEEE Trans. Pattern Anal. Mach. Intell. **32**(9), 1627–1645 (2010)
13. Förstner, W., Moonen, B.: A metric for covariance matrices. In: Quo vadis geodesia ...?, Festschrift for Erik W. Grafarend on the occasion of his 60th birthday, TR Deptadtment of Geodesy and Geoinformatics, Stuttgart University (1999)
14. Gheissari, N., Sebastian, T.B., Hartley, R.: Person reidentification using spatiotemporal appearance. In: Proceedings of the 19th Conference on Computer Vision and Pattern Recognition, CVPR, pp. 1528–1535. IEEE Computer Society (2006)
15. Girshick, R.B., Felzenszwalb, P.F., McAllester, D.: Discriminatively trained deformable part models, release 5. http://people.cs.uchicago.edu/rbg/latent-release5/
16. Goh, A., Vidal, R.: Unsupervised riemannian clustering of probability density functions. In: Proceedings of the 2008 European Conference on Machine Learning and Knowledge Discovery in Databases—Part I, ECML PKDD, pp. 377–392. Springer, Berlin (2008)
17. Gray, D., Brennan, S., Tao, H.: Evaluating Appearance Models for Recognition, Reacquisition, and Tracking. In: Proceedings of the IEEE International Workshop on Performance Evaluation for Tracking and Surveillance, PETS. IEEE Computer Society (2007)
18. Hall, M.A.: Correlation-based Feature Subset Selection for Machine Learning. Ph.D. thesis, Department of Computer Science, University of Waikato (1999)
19. Hirzer, M., Beleznai, C., Roth, P.M., Bischof, H.: Person re-identification by descriptive and discriminative classification. In: Proceedings of the 17th Scandinavian Conference on Image Analysis, pp. 91–102. Springer, Berlin (2011)
20. Hordley, S.D., Finlayson, G.D., Schaefer, G., Tian, G.Y.: Illuminant and device invariant colour using histogram equalisation. Pattern Recognit. **38**(2), 179–190 (2005)

21. Köstinger, M., Hirzer, M., Wohlhart, P., Roth, P.M., Bischof, H.: Large scale metric learning from equivalence constraints. In: Proceedings of the 25th Conference on Computer Vision and Pattern Recognition, pp. 2288–2295 (2012)
22. Ma, B., Su, Y., Jurie, F.: Bicov: a novel image representation for person re-identification and face verification. In: Proceedings of the 23rd British Machine Vision Conference (2012)
23. Oncel, F.P., Porikli, F., Tuzel, O., Meer, P.: Covariance tracking using model update based on lie algebra. In: Proceedings of the 19th Conference on Computer Vision and Pattern Recognition (2006)
24. Pennec, X., Fillard, P., Ayache, N.: A riemannian framework for tensor computing. Int. J. Comput. Vis. **66**(1), 41–66 (2006)
25. Rich, E., Knight, K.: Artificial Intelligence. McGraw-Hill Higher Education (1991)
26. Tuzel, O., Porikli, F., Meer, P.: Region covariance: a fast descriptor for detection and classification. In: Proceedings of the 9th European Conference on Computer Vision, pp. 589–600. Springer (2006)
27. Tuzel, O., Porikli, F., Meer, P.: Pedestrian detection via classification on riemannian manifolds. IEEE Trans. Pattern Anal. Mach. Intell. **30**(10), 1713–1727 (2008)
28. Yao, J., Odobez, J.M.: Fast human detection from joint appearance and foreground feature subset covariances. Comput. Vis. Image Underst. **115**(3), 1414–1426 (2011)
29. Zheng, W.S., Gong, S., Xiang, T.: Associating groups of people. In: Proceedings of the 20th British Machine Vision Conference, BMVC Press (2009)
30. Zheng, W.S., Gong, S., Xiang, T.: Person re-identification by probabilistic relative distance comparison. In: Proceedings of the 24th Conference on Computer Vision and Pattern Recognition, pp. 649–656. IEEE Computer Society (2011)

Chapter 5
Attributes-Based Re-identification

Ryan Layne, Timothy M. Hospedales and Shaogang Gong

Abstract Automated person re-identification using only visual information from public-space CCTV video is challenging for many reasons, such as poor resolution or challenges involved in dealing with camera calibration. More critically still, the majority of clothing worn in public spaces tends to be non-discriminative and therefore of limited disambiguation value. Most re-identification techniques developed so far have relied on low-level visual-feature matching approaches that aim to return matching gallery detections earlier in the ranked list of results. However, for many applications an initial probe image may not be available, or a low-level feature representation may not be sufficiently invariant to viewing condition changes as well as being discriminative for re-identification. In this chapter, we show how mid-level "semantic attributes" can be computed for person description. We further show how this attribute-based description can be used in synergy with low-level feature descriptions to improve re-identification accuracy when an attribute-centric distance measure is employed. Moreover, we discuss a "zero-shot" scenario in which a visual probe is unavailable but re-identification can still be performed with user-provided semantic attribute description.

5.1 Introduction

Person re-identification, or *inter-camera entity association*, is the task of recognising an individual in diverse scenes obtained from non-overlapping cameras. In particular, for surveillance applications performed over space and time, an individual disappear-

R. Layne (✉) · T. M. Hospedales · S. Gong
Queen Mary University of London, London, UK
e-mail: rlayne@eecs.qmul.ac.uk

T. M. Hospedales
e-mail: tmh@eecs.qmul.ac.uk

S. Gong
e-mail: sgg@eecs.qmul.ac.uk

S. Gong et al. (eds.), *Person Re-Identification*,
Advances in Computer Vision and Pattern Recognition,
DOI: 10.1007/978-1-4471-6296-4_5, © Springer-Verlag London 2014

ing from one view would need to be differentiated from numerous possible targets and matched in one or more other views at different locations and time. Potentially each view may be taken from a different angle, featuring different static and dynamic lighting conditions, degrees of occlusion and other view-specific variables.

Relying on manual re-identification in large camera networks is prohibitively costly and inaccurate. Operators are often assigned more cameras to monitor than what is optimal and manual matching can be prone to attentive gaps [19]. Moreover, baseline human performance is determined by individual operator's experience amongst other factors. It is difficult to transfer this expertise directly between operators without knowledge being affected by operator-bias [45].

As public space camera networks have grown quickly in recent years, there has also been an increasing interest in the computer vision community for developing automated re-identification solutions. These efforts have primarily focused on two strategies: (i) developing feature representations which are discriminative for identity, yet invariant to view angle and lighting [4, 12, 37] and (ii) learning methods to discriminatively optimise parameters of a re-identification model [50]. Until now, automated re-identification remains largely an unsolved problem due to the underlying challenge that most visual features are either insufficiently discriminative for cross-view entity association, especially with low resolution images, or insufficiently robust to viewing condition changes.

In this chapter, we take inspiration from the operating procedures of human experts [8, 33, 43] and recent research in attribute learning for classification [21] in order to introduce a new mid-level *semantic attribute* representation.

When performing person re-identification, human experts rely upon matching appearance or functional attributes that are discrete and unambiguous in interpretation, such as hair-style, shoe-type or clothing-style [33]. This is in contrast to the continuous and more ambiguous quantities measured by contemporary computer vision based re-identification approaches using visual features such as colour and texture [4, 12, 37]. This attribute-centric representation is similar to a description provided verbally to a human operator, e.g. by an eye-witness. We call this task attribute-profile identification, or *zero-shot re-identification*. Furthermore, we will show in our study that humans and computers have important differences in attribute-centric re-identification. In particular descriptive attributes that are favoured by humans may not be the most *useful* or *computable* for fully automated re-identification because of variance in the ability of computer vision techniques to detect each attribute and variability in how discriminative each attribute is across the entire population.

This approach of measuring similarity between attributes rather than within the feature-space has two advantages: (i) it allows re-identification (from a probe image) and identification (from a verbal description) to be performed in the same representational space and (ii) as attributes provide a very different type of information to low-level features, which can be considered as a separate modality, they can be fused together with low-level features to provide more accurate and robust re-identification.

5.2 Problem Definitions

5.2.1 The Re-identification Problem

Contemporary approaches to re-identification typically exploit low-level features (LLFs) such as colour [29], texture, spatial structure [4], or combinations thereof [3, 13, 37], because they can be relatively easily and reliably measured, and provide a reasonable level of inter-person discrimination together with inter-camera invariance.

Once a suitable representation has been obtained, nearest-neighbour [4] or model-based matching algorithms such as support-vector ranking [37] may be used for re-identification. In each case, a distance metric (e.g. Euclidean or Bhattacharyya) must be chosen to measure the similarity between two samples. There is now a body of work on discriminatively optimising re-identification models or distance metrics [2, 15, 47, 50] as well as discriminatively learning the low-level features themselves [24]. Other complementary aspects of the re-identification problem have also been pursued to improve performance, such as improving robustness by combining multiple frames worth of features along a trajectory tracklet [3], between sets [48], in a group [46], and learning the topology of camera networks by learning inter-camera activity correlations [27] in order to reduce matching search space and hence reduce false-positives.

5.2.2 Attributes as Representation

Attribute-based modelling has recently been exploited to good effect in object [21] and action [11, 25] recognition. To put this in context: in contrast to low-level features or high-level classes or identities, attributes provide the mid-level *description* of both classes and instances. There are various unsupervised (e.g. PCA or topic-models) or supervised (e.g. neural networks) modelling approaches which produce data-driven mid-level representations. These techniques aim to project the data onto a basis set defined by the assumptions of the particular model (e.g. maximisation of variance, likelihood or sparsity). In contrast, attribute learning focuses on representing data instances by projecting them onto a basis set defined by domain-specific axes which are semantically meaningful to humans. Recent work in this area has also examined the exploitation of the constantly growing semantic web in order to automatically retrieve visual data correlating to relevant metatext [10] and vice-versa for visual retrieval using metatext queries [38].

Semantic attribute representations have various benefits: (i) In re-identification, a single pair of images may be available for each target—which can be seen as a challenging case of "one-shot" learning. In this case attributes can be more powerful than low-level features [21, 25, 41] because they provide a form of transfer learning as attributes are learned from a larger dataset *a priori*; (ii) they can be used synergistically in conjunction with raw data for greater effectiveness [25] and (iii) they are a suitable representation for direct human interaction, therefore allowing searches to be specified, initialised or constrained using human-labelled attribute-profiles [20, 21, 41].

5.2.3 Attributes for Identification

One view of attributes is as a type of transferable context [49] in that they provide auxiliary information about an instance to aid in (re-)identification. Here they are related to the study of soft-biometrics, which aims to enhance biometric identification performance with ancillary information [9, 18]. High-level features such as ethnicity, gender, age or indeed identity itself would be the most useful to us for re-identification. However, soft biometrics are exceptionally difficult to reliably compute in typical surveillance video as visual information is often impoverished and individuals are often at "stand-off distances" as well as in unconstrained or unknown viewing angles.

Alternatively attributes can be used for semantic attribute-profile identification (c.f. zero-shot learning [21]), in which early research has aimed to retrieve people matching a verbal attribute description from a camera network [43]. However, this has only been illustrated on relatively simple data with a small set of similarly-reliable facial attributes. We will illustrate in this study that one of the central issues for exploiting attributes for general automated (re)-identification is dealing with their unequal and variable informativeness and reliability of measurement from raw imagery data.

In this chapter, we move towards leveraging semantic mid-level attributes for automated person identification and re-identification. Specifically, we make four main contributions as follows. In Sect. 5.3.1, we introduce an ontology of attributes based on a subset from a human expert defined larger set [33]. These were selected for being relatively more reliable to compute whilst also discriminative for identification in typical populations. We evaluate our ontology from the perspective of both human-centric and automation-centric purposes and discuss considerations for successful ontology selection. In Sect. 5.3.6 we show how to learn an attribute-space distance metric to optimally weight attributes for re-identification, and do so in a synergistic way with low-level features. We evaluate our model in Sect. 5.4 and show significantly improved re-identification performance compared to conventional feature-based techniques on the two largest benchmark datasets. In the subsequent sections, we provide additional analysis and insight into the results, including contrast against zero-shot re-identification from attribute-profile descriptions.

5.3 Computing Attributes for Re-identification

5.3.1 Ontology Selection

The majority of recent work on attributes looks to human expertise in answer to the question as to which attributes to learn. Typically, ontology selection is performed manually prior to research or via learning from existing metadata [5]. Hand-picked ontologies can be broadly categorised as top-down and bottom-up. In the top-down case, ontology selection may be predicated on the knowledge of experienced human domain-experts. In the latter, it may be based on the intuition of vision researchers, based on factors such as how detectable an attribute might be with available methods or data availability.

For the purposes of automated re-identification, we are concerned with descriptions that permit us to reliably discriminate; that is to say, we wish to eliminate identity ambiguity between individuals. Ontology selection therefore is guided by two factors: *computability* and *usefulness*. That is, *detectable* attributes, which can be detected reliably using current machine learning methods and available data [11], and *discriminative* (informative) attributes which, if known, would allow people to be effectively disambiguated [28].

The notion of discriminative attributes encompasses a nuance. Humans share a vast prior pool of potential attributes and experience. If required to describe a person in a way which uniquely identifies them against a gallery of alternatives, they typically choose a short description in terms of the rare attributes which uniquely discriminate the target individual (e.g. imperial moustache). In contrast, in the ideal discriminative ontology of attributes for automated processing, each attribute should be uncorrelated with all others, and should occur in exactly half of the population (e.g. male vs. female). In this way, no one attribute can distinguish a person uniquely, but together they effectively disambiguate the population: a "binary search" strategy. There are two reasons for this: constraining the ontology size and training data requirement.

Ontology size: Given a "binary search" ontology, any individual can be uniquely identified among a population of n candidates with only an $O(\log(n))$ sized attribute ontology or description. In contrast, the single rare-attribute strategy favoured by people means that while a person may be identified with a short length 1 attribute description, an ontology size and computation size $O(n)$ may be required to describe, interpret and identify this person.

Training data: Given a "binary search" ontology, each training image may be re-used and be (equally) informative for all n attributes (attributes are typically positive for half the images). In contrast, the single rare-attribute strategy would require an infeasible n times as much training data, because different data would be needed for each attribute (e.g. finding a significant number of wearers of imperial moustaches) to train the detectors). In practice, rare attributes do not have enough training data to learn good classifiers, and are thus not reliably *detectable*. A final consideration is the visual subtlety of the attributes, which humans may be able to easily pick out

Table 5.1 Our attribute ontology for re-identification

Redshirt	Blueshirt	Lightshirt
Darkshirt	Greenshirt	Nocoats
Not light dark jeans colour	Dark bottoms	Light bottoms
Hassatchel	Barelegs	Shorts
Jeans	Male	Skirt
Patterned	Midhair	Darkhair
Bald	Has handbag carrier bag	Has backpack

based on their lifetime of experience but which would require prohibitive amounts of training data as well as feature/classifier engineering for machines to detect.

Whether or not a particular ontology is detectable and discriminative cannot therefore be evaluated prior to examination of representative data. However, given a putative ontology and a representative and annotated training set, the detectability of the ontology can be measured by the test performance of the trained detectors whilst the discriminativeness of the ontology can be measured by the mutual information (MI) between the attributes and person identity. The question of how to trade off discriminativeness and detectability when selecting an ontology on the basis of maximum predicted performance is not completely clear [22, 23]. However, we will take some steps to address this issue in Sect. 5.3.6.

5.3.2 Ontology Creation and Data Annotation

Given the considerations discussed in the previous section, we select our ontology jointly based on four criteria. (i) We are informed by the operational procedures of human experts [33] as well as (ii) prioritising suitable findings from [22, 23, 38, 44], (iii) whether the ontology is favourably distributed in the data (binary search) and (iv) those which are likely to be detectable (sufficient training data and avoiding subtlety).

Specifically, we define the following space of $N_a = 21$ binary attributes (Table 5.1). Ten of these attributes are related to colour, one to texture and the remaining ten are related to soft biometrics. Figure 5.1 shows a visual example of each attribute.[1]

Human annotation of attribute labels is costly in terms of both time and human effort. Due to the semantic nature of the attributes, accurate labelling can be especially challenging for cases where data are visually impoverished. Typically problems can arise where (i) ontology definition allows for ambiguity between members of the ontology and (ii) boundary cases are difficult for an annotator to binarily classify with confidence. These circumstances can be natural places for subjective labelling errors [42].

---

[1] We provide our annotations here: http://www.eecs.qmul.ac.uk/~rlayne/

Fig. 5.1 Positive instances of our ontology from (*top*) the VIPeR and (*bottom*) the PRID datasets

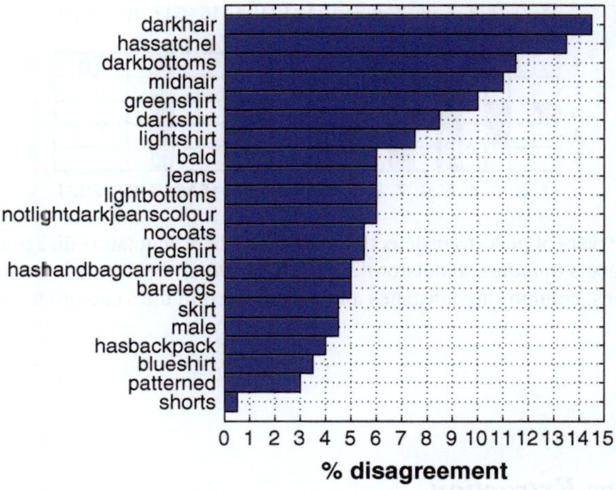

Fig. 5.2 Annotation disagreement error frequencies for two annotators on PRID

To investigate the significance of this issue, we independently double-annotated the PRID dataset [15] for our attribute ontology. Figure 5.2 illustrates frequency of label disagreements for each attribute in the PRID dataset measured as the Hamming distance between all annotations for that attribute across the dataset.

For attributes such as *shorts* or *gender*, uncertainty and therefore error is low. However, attributes whose boundary cases may be less well globally agreed upon can be considered to have the highest relative error between annotators. For example, in Fig. 5.2 attributes *hassatchel* and *darkhair* are most disagreed upon since lighting variations make determining darkness of hair difficult in some instances and satchel refers to a wide variety of rigid or non-rigid containers held in multiple ways. This means that attributes such as *darkhair* and *hassatchel* may effectively be subject to a significant rate of label noise [51] in the training data and hence perform poorly. This adds another source of variability in reliability of attribute detection which will have to be accounted for later. Figure 5.3 illustrates pairs of individuals in the PRID dataset whose shared attribute-profiles were the most disagreed upon. The figure highlights the extent of noise that can be introduced through semantic labelling errors, a topic we will revisit later in Sect. 5.3.6.

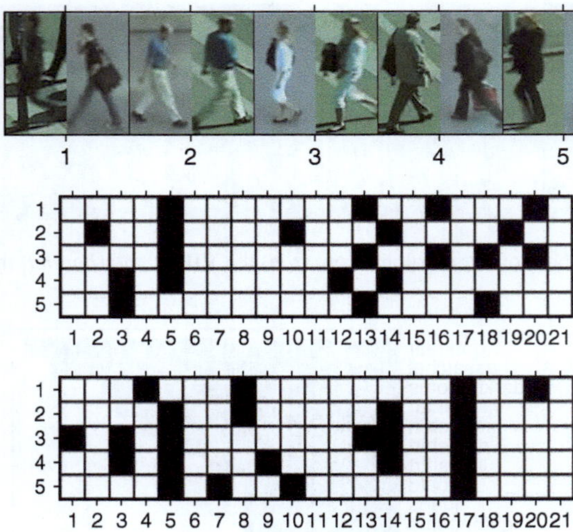

Fig. 5.3 Top five pairs of pedestrian detections in PRID where annotators disagreed most (*top row*). Annotator #1's labels (*middle*), annotator #2's labels (*bottom*). Each row is an attribute-profile for a pair of detections, columns are attributes and are arranged in the same order as Fig. 5.2

5.3.3 Feature Extraction

To detect attributes, we first select well-defined and informative low-level features with which to train robust classifiers. We wish to choose a feature which is also typically used for re-identification in order to enable later direct comparison between conventional and attribute-space re-identification in a way which controls for the input feature used. Typical descriptors used for re-identification include the Symmetry Driven Accumulation of Local Features (SDALF) [4] and Ensemble of Localised Features (ELF) [13].

The content of our ontology includes semantic attributes such as jeans, shirt colours, gender. We can infer that the information necessary for humans to distinguish these items is present visually, and wish to select a feature that incorporates information pertaining to colour, texture and spatial information. For our purposes, SDALF fulfils the requirements for our ontology but does not produce positive semi-definite distances, therefore ruling it out for classification using kernel methods. As a result, we therefore exploit ELF.

To that end, we first extract a 2784-dimensional low-level colour and texture feature vector denoted x from each person image I following the method in [37]. This consists of 464-dimensional feature vectors extracted from six equal sized horizontal strips from the image. Each strip uses eight colour channels (RGB, HSV and YCbCr) and 21 texture filters (Gabor, Schmid) derived from the luminance channel. We use the same parameter choices for γ, λ, θ and σ^2 as proposed in [37] for Gabor filter extraction, and for τ and σ for Schmid extraction. Finally, we use a bin size of 16 to quantise each channel.

5.3.4 Attribute Detection

Classifier Training and Attribute Feature Construction

We train Support Vector Machines (SVM) [40] to detect attributes. We use Chang et al.'s LIBSVM [6] and investigate Linear, RBF, χ^2 and Intersection kernels. We select the intersection kernel as it compares closely with χ^2 but is faster to compute.[2]

For each attribute, we perform cross validation to select values for the SVM's slack parameter C from the set $C \in \{-10, \ldots, 10\}$ with increments of $\epsilon = 1$. The SVM scores are probability mapped, so each attribute detector i outputs a posterior $p(a_i|x)$. We follow the standard approach for mapping SVM scores to posterior probabilities [36] as implemented by LIBSVM [6].

Spatial Feature Selection

Since some attributes (e.g. shorts) are highly unlikely to appear outside of their expected spatial location, one might ask whether it is possible to improve performance by discriminatively selecting or weighting the individual strips within the feature vector (Sect. 5.3.3). We experimented with defining a kernel for each strip as well as for the entire image, and training multi-kernel learning SVM using the DOGMA library with *Obscure* as classifiers [34, 35]. This approach discriminatively optimises the weights for each kernel in order to improve classifier performance and has been shown to improve performance when combining multiple features. However in this case, it did not reliably improve on the conventional SVM approach, presumably due to the relatively sparse and imbalanced training data being insufficient to correctly tune the inter-kernel weights.

---

[2] Our experiments on LIBSVM performance versus attribute training time show the intersection kernel as being a good combination of calculation time and accuracy. For example, training the attribute ontology results in 65.4 % mean accuracy with 0.8 h training for the intersection kernel, as compared to the χ^2 kernel (63.8 % with 4.1 h), the RBF kernel (65.9 % with 0.76 h and the linear kernel (61.8 % with 1.2 h) respectively with LIBSVM. Although RBF is computed slightly faster and has similar accuracy, we select the intersection kernel overall, since the RBF kernel would require cross-validating over a second parameter. Providing LIBSVM with pre-built kernels reduces training time considerably in all cases.

Imbalanced Attribute Training

The prevalence of each attribute in a given dataset tends to vary dramatically and some attributes have a limited number of positive examples in an absolute sense as a result. This imbalance can cause discriminative classifiers such as SVMs to produce biased or degenerate results. There are various popular approaches to dealing with imbalanced data [14], such as synthesising further examples from the minority class to improve the definition of the decision boundary, for example using SMOTE [7] or weighting SVM instances or mis-classification penalties [1, 14]. However, neither of these methods outperformed simple subsampling in our case.

To avoid bias due to imbalanced data, we therefore simply train each attribute detector with all the positive training examples of that attribute, and obtain the same number of negative examples by sub-sampling the rest of the data at regular intervals.

Mid-Level Attribute Representation

Given the learned bank of attribute detectors, at test time we generate mid-level features as $1 \times N_a$ sized vectors of classification posteriors which we use to represent the probability that each attribute is present in the detection. Effectively we have projected the high dimensional, low-level features onto a mid-level, low-dimensional semantic attribute space. In particular, each person image is now represented in semantic attribute space by stacking the posteriors from each attribute detector into the N_a dimensional vector: $A(x) = [p(a_1|x), \ldots, p(a_{N_a}|x)]^T$.

5.3.5 Attribute Fusion with Low-Level Features

To use our attributes for re-identification, we can define a distance solely on the attribute space, or use the attribute distance in conjunction with conventional distance between low-level features such as SDALF [4] and ELF [12]. SDALF provides state-of-the-art performance for a non-learning nearest-neighbour (NN) approach while ELF has been widely used by model-based learning approaches [37, 46]. We also use it as the feature for our attribute detectors in Sect. 5.3.3.

We therefore introduce a rather general formulation of a distance metric between two images I_p and I_g which combines both multiple attributes and multiple low-level features as follows:

$$d_{\mathbf{w}^L,\mathbf{w}^A}\left(I_p, I_g\right) = \sum_{l \in LL} w_l^L \, d_l^L \, \left(L_l\left(I_p\right), L_l\left(I_g\right)\right) + d_{\mathbf{w}^A}^A\left(A\left(I_p\right), A\left(I_g\right)\right). \tag{5.1}$$

Here Eq. (5.1) (first term) corresponds to the contribution from a set LL of low-level distance measures, where $L_l(I_p)$ denotes extraction of type l low-level features from image I_p, d_l^L denotes the distance metric defined for low-level feature type l, and w_l^L is a weighting factor for each feature type l. Eq. (5.1) (second term) corresponds to the contribution from our attribute-based distance metrics. Where $A(I_p)$ denotes the attribute encoding of image I_p. For the attribute-space distance we experiment with two metrics: weighted $L1$ (Eq. (5.2)) and weighted Euclidean (Eq. (5.3)).

$$d^A_{\mathbf{w}^A}(I_p, I_g) = (\mathbf{w}^A)^T \left|\left(A(x_p) - A(x_g)\right)\right|, \tag{5.2}$$

$$d^A_{\mathbf{w}^A}(I_p, I_g) = \sqrt{\sum_i w^A_i \left(p(a_i|x_p) - p(a_i|x_g)\right)^2}. \tag{5.3}$$

5.3.6 Attribute Selection and Weighting

As discussed earlier, all attributes are not equal due to variability in how reliably they are measured due to imbalance, subtlety (detectability) and how informative they are about identity (discriminability). How to account for variable detectability and discriminability of each attribute ($\mathbf{w}^A$), and how to weight attributes relative to low-level features ($\mathbf{w}^{LL}$) are important challenges, which we discuss now.

Exhaustively searching the N_a dimensional space of weights directly to determine attribute selection and weighting is computationally intractable. However, we can re-formulate the re-identification task as an optimisation problem and apply standard optimisation methods [32] to search for a good configuration of weights.

Importantly, we only search $|\mathbf{w}^A| = N_a = 21$ parameters for the within-attribute-space metric $d^A_{\mathbf{w}^A}(\cdot, \cdot)$. and one or two parameters for weighting attributes relative to low-level features. In contrast to previous learners for low-level features [37, 47, 50], which must optimise 100s or 1,000s of parameters, this gives us considerable flexibility in terms of computation requirement of the objective.

An interesting question is therefore what is the ideal criterion for optimisation. Previous studies have considered optimising, e.g. relative rank [37] and relative distance [15, 50]. While effective, these metrics are indirect proxies for what the re-identification application ultimately cares about, which is the average rank of the true match to a probe within the gallery set, which we call Expected Rank (ER). That is, how far does the operator have to look down the list before finding the target. See Sect. 5.4 for more discussion.

We introduce the following objective for ER:

$$ER = \frac{1}{|P|} \sum_{p \in P} \sum_{g \in G} L_{\mathbf{w}}\left(D_{pp}, D_{pg}\right) + \lambda \parallel \mathbf{w} - \mathbf{w}_0 \parallel, \tag{5.4}$$

where D_{pg} is the matrix of distances (Eqs. (5.1)) from probe image p to gallery image g; L is a loss function, which can penalise the objective according to the relative distance of the true match D_{pp} versus false matches D_{pg}; and $\mathbf{w}_0$ is a regulariser bias with strength λ. To complete the definition of the objective, we define the loss function L as in Eq. (5.5). That is, imposing a penalty every time a false match is ranked ahead of the true match. ($\mathbf{I}$ is an indicator function which returns 1 when the parameter is

Algorithm 1 Attributes-based re-identification

Training
for all Attribute **do**
  Subsample majority class to length of minority class
  Cross-validate to obtain parameter C that gives best average accuracy.
  Retrain SVM on all training data with selected C
end for
Determine inter and intra-attribute weighting $\mathbf{w}$ by minimising Eq. (5.4).

Testing (Re-identification)
for all Person $\mathbf{x}_g \in$ gallery set **do**
  Classify each attribute a
  Stack attribute posteriors into person signature $A(\mathbf{x}_g)$.
end for
for all Person $\mathbf{x}_p \in$ probe set **do**
  Classify each attribute a
  Stack attribute posteriors into person signature $A(\mathbf{x}_p)$.
  Compute distance to gallery set fusing attribute and LLF cues with weight $\mathbf{w}$. (Eq. (5.1))
  Nearest-neighbour re-identification in gallery according to their similarity to person $\mathbf{x}_p$.
end for

true.) The overall objective (Eq. (5.4)) thus returns the ER of the true match. This is now a good objective, because it directly reflects the relevant end-user metric for effectiveness of the system. However it is hard to efficiently optimise because it is non-smooth: a small change to the weights $\mathbf{w}$ may have exactly zero change to the ER (the optimisation surface is piece-wise linear). We therefore soften this loss-function using a sigmoid, as in Eq. (5.6), which is now smooth and differentiable. This finally allows efficient gradient-based optimisation with Newton [26] or conjugate-gradient methods [32].

$$L_{\mathbf{w}}^{HardRank,ER} = \mathbf{I}\left(d_{pp} - d_{pg} > 0\right). \tag{5.5}$$

$$L_{\mathbf{w}}^{Sigmoid,ER} = \sigma\left(d_{pp} - d_{pg}\right). \tag{5.6}$$

We initialise $\mathbf{w}_{initial} = 1$. To prevent over fitting, we use regularisation parameters $\mathbf{w}_0$=1, and $\lambda = 0.2$ (i.e. everything is assumed to be equal a priori) and set the sigmoid scale to $k = 32$. Finally for fusion with low-level features (Eq. (5.1)), we use both SDALF and ELF.

In summary, this process uses gradient-descent to search for a setting of weights $\mathbf{w}$ for each LLF and for each attribute (Eq. (5.1)) that will (locally) minimise the ER within the gallery of the true match to each probe image (Eq. (5.4)). See Algorithm 1 for an overview of our complete system.

5.4 Experiments

5.4.1 Datasets

We select two challenging datasets with which to validate our model, VIPeR [12] and PRID [15]. VIPeR contains 632 pedestrian image pairs from two cameras with different viewpoint, pose and lighting. Images are scaled to 128×48 pixels. We follow [4, 12] in considering Cam B as the gallery set and Cam A as the probe set. Performance is evaluated by matching each test image in Cam A against the Cam B gallery.

PRID is provided as both multi-shot and single-shot data. It consists of two camera views overlooking an urban environment from a distance and from fixed viewpoints. As a result PRID features low pose variability with the majority of people captured in profile. The first 200 shots in each view correspond to the same person, however the remaining shots only appear once in the dataset. To maximise comparability with VIPeR, we use the single-shot version and use the first 200 shots from each view. Images are scaled to 128×64 pixels.

For each dataset, we divide the available data into training, validation and test partitions. We initially train classifiers and produce attribute representations from the training portion, and then optimise the attribute weighting as described in Sect. 5.3.6 using the validation set. We then retrain the classifiers on both the training and validation portions, while re-identification performance is reported on the held out test portion.

We quantify re-identification performance using three standard metrics and one less common one metric. The standard re-identification metrics are performance at rank n, cumulative matching characteristic (CMC) curves and normalised area under the CMC curve [4, 12]. Performance at rank n reports the probability that the correct match occurs within the first n ranked results from the gallery. The CMC curve plots this value for all n, and the nAUC summarises the area under the CMC curve (so perfect nAUC is 1.0 and chance nAUC is 0.5).

We additionally report ER, as advocated by Avraham et al. [2] as CMC Expectation. The ER reflects the mean rank of the true matches and is a useful statistic for our purposes; in contrast to the standard metrics, lower ER scores are more desirable and indicate that on average the correct matches are distributed more toward the lower ranks. (So perfect ER is 1 and random ER would be half the gallery size). In particular, ER has the advantage of a highly relevant practical interpretation: it is the average number of returned images the operator will have to scan before reaching the true match.

We compare the following re-identification methods: (1) SDALF [4] using code provided by the authors (note that SDALF is already shown to decisively outperform [13]); (2) ELF: Prosser et al.'s [37] spatial variant of ELF [12] using Strips of ELF; (3) Attributes: Raw attribute based re-identification (Euclidean distance); (4) Optimised Attribute Re-identification (OAR): our Optimised Attribute based Re-identification method with weighting between low-level features and within attributes learned by directly minimising the ER (Sect. 5.3.6).

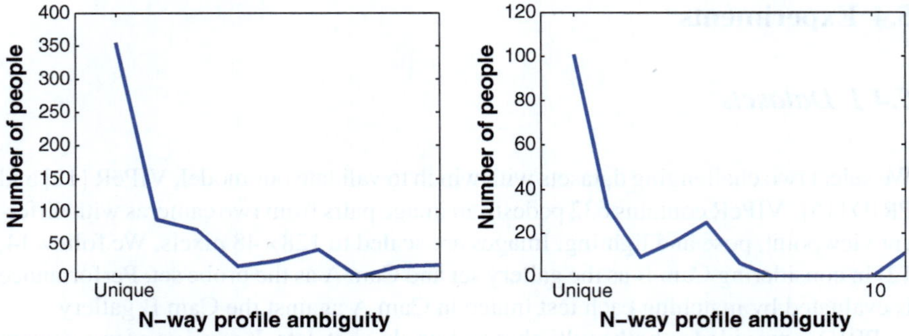

Fig. 5.4 Uniqueness of attribute descriptions in a population, **i** VIPeR and **ii** PRID. The peak around unique shows that most people are uniquely identifiable by attributes

5.4.2 Attribute Analysis

We first analyse the intrinsic discriminative potential of our attribute ontology independently of how reliably detectable the attributes are (assuming perfect detectability). This analysis plays provides an upper bound of performance that would be obtainable with sufficiently advanced attribute detectors. Fig. 5.6 reports the prevalence of each attribute in the datasets. Many attributes have prevalence near to 50 %, which is reflected in their higher MI with person identity. As we discussed earlier this is a desirable property because it means each additional attribute known can potentially halve the number of possible matches. Whether this is realised or not depends on if attributes are correlated/redundant, in which case each additional redundant attribute provides less marginal benefit. To check this we compute the correlation coefficient between all attributes, and found that the average inter-attribute correlation was only 0.07. We therefore expect the attribute ontology to be effective.

Figure 5.4 shows a histogram summarising how many people are uniquely identifiable solely by attributes and how many would be confused to a greater or lesser extent. The peak around unique/unambiguous shows that a clear majority of people can be uniquely or otherwise near-uniquely identified by their attribute-profile alone, while the tail shows that there are a small number of people with very generic profiles. This observation is important; near-uniqueness means that approaches which rank distances between attribute-profiles are still likely to feature the correct match high enough in the ranked list to be of use to human operators.

The CMC curve (for gallery size $p = 632$) that would be obtained assuming perfect attribute classifiers is shown in Fig. 5.5. This impressive result (nAUC near a perfect score of 1.0) highlights the potential for attribute-based re-identification. Also shown are the results with only the top five or 10 attributes (sorted by MI with identity), and a random 10 attributes. This shows that: (i) as few as 10 attributes are sufficient if they are good (i.e. high MI) and perfectly detectable, while five is too few and (ii) attributes with high MI are significantly more useful than low MI (always present or absent) attributes (Fig. 5.6).

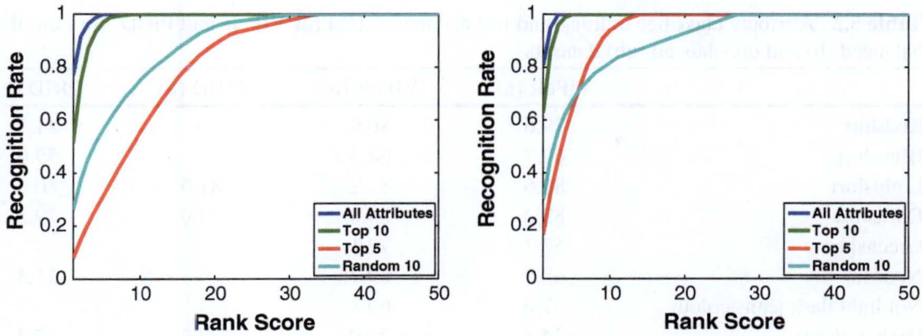

Fig. 5.5 Best-case (assuming perfect attribute detection) re-identification using attributes with highest *n* ground-truth MI scores, **i** VIPeR and **ii** PRID

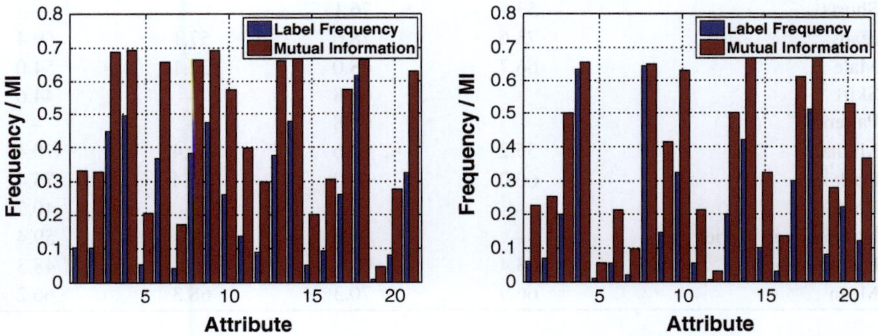

Fig. 5.6 Attribute occurrence frequencies and Attribute MI scores in VIPeR (*left*) and PRID (*right*)

5.4.3 Attribute Detection

Given the analysis of the intrinsic effectiveness of the ontology in the previous section, the next question is whether the selected attributes can indeed be detected or not. Attribute detection on both VIPeR and PRID achieves reasonable levels on both balanced and unbalanced datasets as seen in Table 5.2. (dash indicates failure to train due to insufficient data). For all datasets, a minimum of nine classifiers can be trained on unbalanced PRID, and 16 on unbalanced VIPeR, in both cases some attribute classifiers are unable to train due to extreme class imbalances or data sparsity. Average accuracies for these datasets are also reasonable; 66.9 % and 68.3 % respectively. The benefit of sub-sampling negative data for attribute learning is highlighted in the improvement for the balanced datasets. Balancing in this case increases the number of successfully trained classifiers to 20 for balanced VIPeR and

Table 5.2 Attribute classifier training and test accuracies (%) for VIPeR and PRID, for both the balanced (b) and unbalanced (ub) datasets

	VIPeR (u)	VIPeR (b)	PRID (u)	PRID (b)
Redshirt	79.6	80.9	–	41.3
Blueshirt	62.7	68.3	–	59.6
Lightshirt	80.6	82.2	81.6	80.6
Darkshirt	82.2	84.0	79.0	79.5
Greenshirt	57.3	72.1	–	–
Nocoats	68.5	69.7	–	31.3
Not light dark jeans colour	57.6	69.1	–	–
Dark bottoms	74.4	75.0	72.2	67.3
Light bottoms	75.3	74.7	76.0	74.0
Hassatchel	–	56.0	51.9	55.0
Barelegs	60.4	74.4	–	50.2
Shorts	53.1	76.1	–	–
Jeans	73.6	78.0	57.1	69.4
Male	66.7	68.0	52.1	54.0
Skirt	–	68.8	–	44.6
Patterned	–	60.8	–	–
Midhair	55.2	64.6	69.4	70.4
Dark hair	60.0	60.0	75.4	75.4
Bald	–	–	–	40.2
Has handbag carrier bag	–	54.5	–	59.4
Has backpack	63.4	68.6	–	48.3
Mean	66.9	70.3	68.3	66.2

16 on balanced PRID with mean accuracies rising to 70.3 % for VIPeR. Balancing slightly reduces classification performance on PRID to an average of 66.2 %.

5.4.4 Using Attributes to Re-identify

Given the previous analysis of discriminability and detectability of the attributes, we now address the central question of attributes for re-identification. We first consider vanilla attribute re-identification (no weighting or fusion; $\mathbf{w}^L = 0$, $\mathbf{w}_a = 1$ in Eq. (5.1)). The re-identification performance of attributes alone is summarised in Table 5.3 in terms of ER. There are a few interesting points to note: (i) In most cases using $L2$ NN matching provides lower ER scores than $L1$ NN matching. (ii) On VIPeR and PRID, SDALF outperforms the other low-level features, and outperforms our basic attributes in VIPeR. (iii) Although the attribute-centric re-identification uses the *same low-level input features* (ELF), and the same $L1/L2$ NN matching strategy, attributes decisively outperform raw ELF. We can verify that this large difference is due to the semantic attribute space rather than the implicit dimensionality reduction effect of attributes by performing Principle Components Analysis (PCA) on ELF

Table 5.3 Re-identification performance, we report ER scores for VIPeR (left, gallery size $p = 316$) and PRID (right, gallery size $p = 100$) and compare different features and distance measures against our balanced attribute-features prior to fusion and weight selection.

VIPeR	$L1$		$L2$
ELF [37]	84.3		72.1
ELF PCA	85.3		74.5
Raw attributes	34.4		37.8
SDALF [4]		44.0	
Random chance		158	
PRID	$L1$		$L2$
ELF	28.2		37.0
ELF PCA	32.7		38.1
Raw attributes	24.1		24.4
SDALF [4]		31.8	
Random chance		50	

Smaller values indicate better re-identification performance

to reduce its dimensionality to the same as our attribute space ($N_a = 21$). In this case the re-identification performance is still significantly worse than the attribute-centric approach (See Table 5.3). The improvement over raw ELF is thus due to the attribute-centric approach.

5.4.5 Re-identification with Optimised Attributes

Given the promising results for vanilla attribute re-identification in the previous section, we finally investigate whether our complete model (including discriminative optimisation of weights to improve ER) can further improve performance. Figure 5.7 and Table 5.4 summarise final re-identification performance. In each case, optimising the attributes with the distance metric and fusing with low-level SDALF and ELF improves re-identification uniformly compared to using attributes or low-level features alone. Our approach improves ER by 38.3 and 35 % on VIPeR, and 38.8 and 46.5 % on PRID for the balanced and unbalanced cases vs. SDALF and 66.9, 65.1, 77.1 and 80 % versus ELF features.

Critically for re-identification scenarios, the most important rank 1 accuracies are improved convincingly. For VIPeR, OAR improves 40 % over SDALF in the balanced case, and 33.3 % for unbalanced data. For PRID, OAR improves by 30 and 36.6 %. As in the case of ER, rank is uniformly improved, indicating the increased likelihood that correct matches appear more frequently at earlier ranks using our approach.

The learned weights for fusion between our attributes and low-level features indicate that SDALF is informative and useful for re-identification on both datasets. In contrast, ELF is substantially down-weighted to 18 % compared to SDALF on PRID

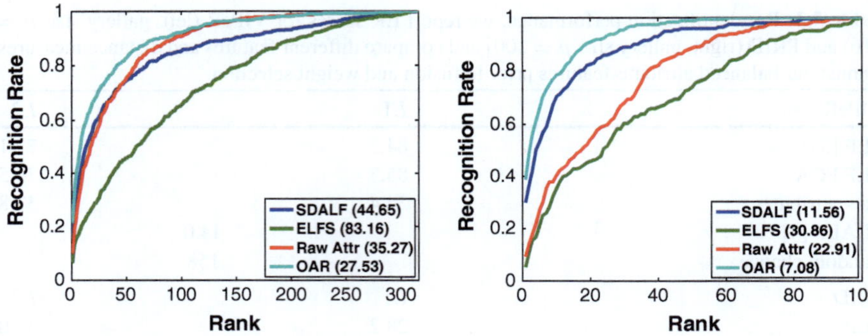

Fig. 5.7 Final attribute re-identification CMC plots for **i** VIPeR and **ii** PRID, gallery sizes $p = 316$, $p = 100$. ER is given in parentheses

Table 5.4 Final attribute re-identification performance

VIPeR	ER	Rank 1	Rank 5	Rank10	Rank25	nAUC
Farenzena et al. [4]	44.7	15.3	34.5	44.3	61.6	0.86
Prosser et al. [37]	83.2	6.5	16.5	21.0	30.9	0.74
Raw attributes (b)	35.3	10.0	26.3	39.6	58.4	0.89
OAR (b)	27.5	21.4	41.5	55.2	71.5	0.94
Raw attributes (u)	40.4	6.5	23.9	34.8	55.9	0.88
OAR (u)	29.0	19.6	39.7	54.1	71.2	0.91
PRID	ER	Rank 1	Rank 5	Rank10	Rank25	nAUC
Farenzena et al.	11.6	30.0	53.5	70.5	86.0	0.89
Prosser et al.	30.9	5.5	21.0	35.5	52.0	0.70
Raw attributes (b)	22.9	9.5	27.0	40.5	60.0	0.78
OAR (b)	7.1	39.0	66.0	78.5	93.5	0.93
Raw attributes (u)	20.8	8.5	28.5	44.0	69.0	0.80
OAR (u)	6.2	41.5	69.0	82.5	95.0	0.95

We report ER scores [2] (lower scores indicate that overall, an operator will find the correct match appears lower down the ranks), Cumulative Match Characteristic (CMC) and normalised Area-Under-Curve (nAUC) scores (higher is better, the maximum nAUC score is one). We further report accuracies for our approach using unbalanced data for comparison

and on VIPeR, disabled entirely. This makes sense because SDALF is at least twice as effective as ELF for VIPeR (Table 5.3).

The intra-attribute weights (Fig. 5.8) are relatively even on PRID but more varied on VIPeR where the highest weighted attributes (*jeans, hasbackpack, nocoats, midhair, shorts*) are weighted at 1.43, 1.20, 1.17, 1.10 and 1.1; while the least informative attributes are *barelegs, lightshirt, greenshirt, patterned* and *hassatchel* which are weighted to 0.7, 0.7, 0.66, 0.65 and 0.75. Jeans is one of the attributes that is detected most accurately and is most common in the datasets, so its weight is expected to be high. However the others are more surprising, with some of the most accurate attributes such as *darkshirt* and *lightshirt* weighted relatively low (0.85 and

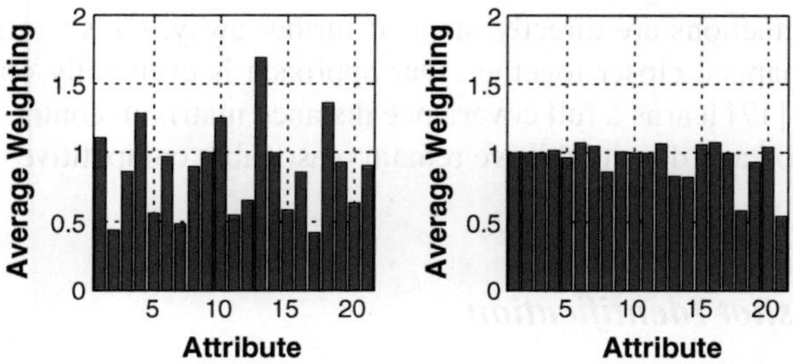

Fig. 5.8 Final attribute feature weights for VIPeR (*left*) and PRID (*right*)

Table 5.5 Comparison of results between our OAR method and other state-of-art results for the VIPeR dataset

VIPeR	Rank 1	Rank 10	Rank 20	Rank 50	nAUC
OAR	21.4	55.2	71.5	82.9	0.92
Hirzer et al.[16]	22.0	63.0	78.0	93.0	
Farenzena et al.[4]	9.7	31.7	46.5	66.6	0.82
Hirzer et al.[17]	27.0	69.0	83.0	95.0	-
Avraham et al.[2]	15.9	59.7	78.3	-	-
Zheng et al.[47, 50]	15.7	53.9	70.1	-	-
Prosser et al.[37]	14.6	50.9	66.8	-	-

0.7). For PRID, *darkshirt, skirt, lightbottoms, lightshirt* and *darkbottoms* are most informative (1.19, 1.04, 1.02 and 1.03); *darkhair, midhair, bald, jeans* are the least (0.78, 0.8, 0.92, 0.86).

Interestingly, the most familiar indicators which might be expected to differentiate good versus bad attributes are not reflected in the final weighting. Classification accuracy, annotation error (label noise) and MI are not significantly correlated with the final weighting, meaning that some unreliably detectable and rare/low MI attributes actually turn out to be *useful* for re-identification with low ER; and vice versa. Moreover, some of the weightings vary dramatically between dataset, for example, the attribute *jeans* is the strongest weighted attribute on VIPeR, however it is one of the lowest on PRID despite being reasonably accurate and prevalent on both datasets. These two observations both show (i) the necessity of jointly learning a combined weighting for all the attributes, (ii) doing so with a relevant objective function (such as ER) and (iii) learning a model which is adapted for the statistics of each given dataset/scenario.

In Table 5.5, we compare our approach with the performance other methods as reported in their evaluations. In this case, the cross-validation folds are not the same, so the results are not exactly comparable, however they should be indicative. Our approach performs comparably to [16] and convincingly compared to [4, 47, 50] and [37]. Both [17] and [2] exploit pairwise learning; in [2] a binary classifier is trained on correct and incorrect pairs of detections in order to learn the projection from one camera to another, in [17] incorrect (i.e. matches that are nearer to the probe than the true match) detections are directly mapped further away whilst similar but correct matches are mapped closer together. Our approach is eventually outperformed by [17], however [17] learns a full covariance distance matrix in contrast to our simple diagonal matrix, and despite this we remain reasonably competitive.

5.4.6 Zero-shot Identification

In Sect. 5.4.2 we showed that with perfect attribute detections, highly accurate re-identification is possible. Even with merely 10 attributes, near-perfect re-identification can be performed. Zero-shot identification is the task of generating an attribute-profile either manually or from a different modality of data, then matching individuals in the gallery set via their attributes. This is highly topical for surveillance: consider the case where a suspect is escaping through a public area surveilled by CCTV. The authorities in this situation may have enough information build a semantic-attribute-profile of the suspect using attributes taken from eyewitness descriptions.

In zero-shot identification (a special case of re-identification), we replace the probe image with a manually specified attribute description. To test this problem setting, we match the ground truth attribute-profiles of probe persons against their inferred attribute-profiles in the gallery as in [43].

An interesting question one might ask is whether this is expected to be better or worse than conventional attribute-space re-identification based on attributes detected from a probe *image*. One might expect zero-shot performance to be better because we know that in the absence of noise, attribute re-identification performs admirably (Sect. 5.4.2 and Fig. 5.5)—and there are two sources of noise (attribute detection inaccuracies in the probe and target images) of which the former noise source has been removed in the zero-shot case. In this case, a man-in-the-loop approach to querying might be desirable, even if a probe image is available. That is, the operator could quickly indicate the ground-truth attributes for the probe image and search based on this (noise-free) ground-truth.

Table 5.6 shows re-identification performance for both datasets. Surprisingly, while the performance is encouraging, it is not as compelling as when the profile is constructed by our classifiers, *despite the elimination of the noise on the probe images*.

This significant difference between the zero-shot case we outline here and the conventional case we discuss in the previous section turns out to be because of *noise correlation*. Intuitively, consider that if someone with a hard-to-classify hairstyle is classified in one camera with some error ($p(a_{hair}|\mathbf{x}) - a_{hair}^{true}$), then this person might also be classified in another camera with an error *in the same direction*. In this case, using the ground-truth attribute in one camera will actually be detrimental to re-identification performance (Fig. 5.9).

To verify this explanation, we perform Pearson's product-moment correlation analysis on the error (difference between ground-truth labels and the predicted attributes) between the probe and gallery sets. The average cross-camera error correlation coefficient is 0.93 in VIPeR and 0.97 in PRID, and all of the correlation coefficients were statistically significant ($p < 0.05$).

Table 5.6 Zero-shot re-identification results for VIPeR and PRID

	Exp Rank	Rank 1	Rank 5	Rank 10	Rank 25
VIPER (u)	50.1	6.0	17.1	26.0	48.1
VIPER (b)	54.8	5.4	14.9	25.3	44.9
PRID (u)	19.2	8.0	29.0	47.0	73.0
PRID (b)	26.1	3.0	16.0	32.0	62.0

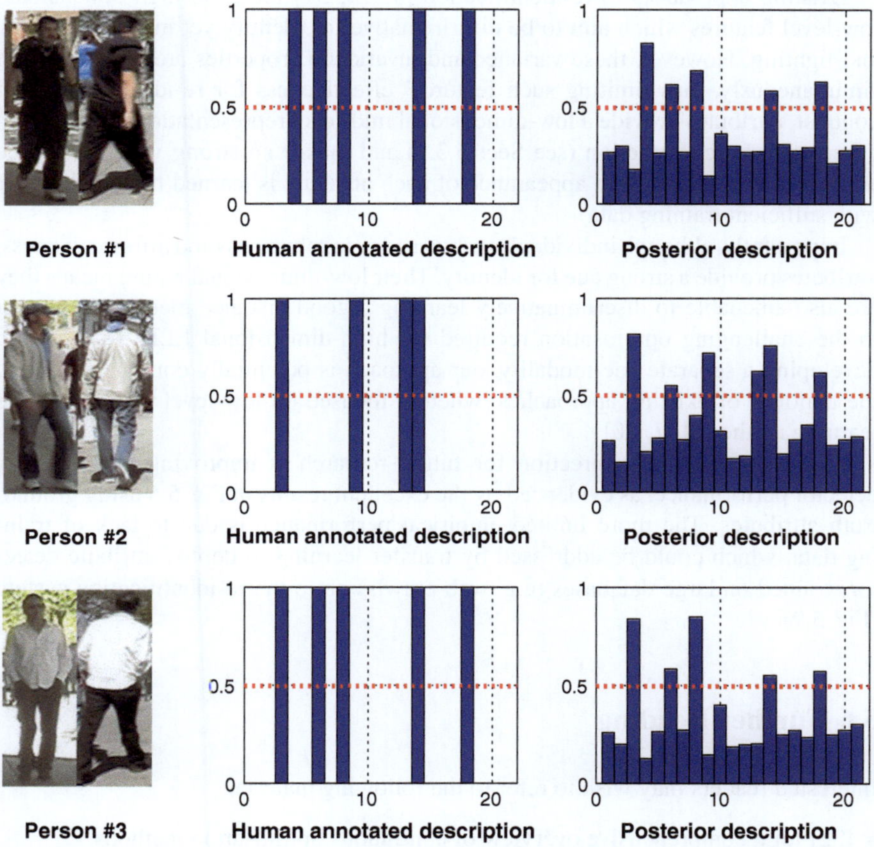

Fig. 5.9 Success cases for zero-shot re-identification on VIPeR. The *left* column shows two probe images; **i** is the image annotated by a human operator and **ii** is the correct rank #1 match as selected by our zero-shot re-identification system. The human-annotated probe descriptions (*middle*) and the matched attribute-feature gallery descriptions (*right*) are notably similar for each person; the attribute detections from the gallery closely resemble the human-annotated attributes (particularly those above *red* line)

Although these results show that man-in-the-loop zero-shot identification—if intended to replace a probe image—may not always be beneficial, it is still evident that zero-shot performs reasonably in general and is a valuable capability for the case where descriptions are verbal rather than extracted from a visual example.

5.5 Conclusions

We have shown how mid-level attributes trained using semantic cues from human experts [33] can be an effective representation for re-identification and (zero-shot) identification. Moreover, this provides a different modality to standard low-level features and thus synergistic opportunities for fusion.

Existing approaches to re-identification [4, 12, 37] focus on high-dimensional low-level features which aim to be discriminative for identity yet invariant to view and lighting. However, these variance and invariance properties are hard to obtain simultaneously, thus limiting such features' effectiveness for re-identification. In contrast, attributes provide a low-dimensional mid-level representation which is discriminative by construction (see Sect. 5.3.1) and makes no strong view invariance assumptions (variability in appearance of each attribute is learned by the classifier with sufficient training data)

Importantly, although individual attributes vary in robustness and informativeness, attributes provide a strong cue for identity. Their low-dimensional nature means they are also amenable to discriminatively learning a good distance metric, in contrast to the challenging optimisation required for high-dimensional LLFs [47, 50]. In developing a separate cue-modality, our approach is potentially complementary to the majority of existing approaches, whether focused on low-level features [4], or learning methods [47, 50]

The most promising direction for future research is improving the attribute-detector performance, as evidenced by the excellent results in Fig. 5.5 using ground-truth attributes. The more limited empirical performance is due to lack of training data, which could be addressed by transfer learning to deploy attribute detectors trained on large databases (e.g. web-crawls) on to the re-identification system (Fig. 5.9).

5.6 Further Reading

Interested readers may wish to refer to the following material:

- [32] for a comprehensive overview of continuous optimisation methods.
- [31] for detailed exposition and review of contemporary features and descriptors.
- [30] discusses classifier training and machine learning methods.
- [39] for trends on surveillance hardware development.

Acknowledgments The authors shall express their deep gratitude to Colin Lewis of the UK MOD SA(SD) who made this work possible and to Toby Nortcliffe of the UK Home Office CAST for providing human operational insight. We also would like to thank Richard Howarth for his assistance in labelling datasets.

References

1. Akbani, R., Kwek, S., Japkowicz, N.: Applying support vector machines to imbalanced datasets. In: European Conference on Machine Learning (2004)
2. Avraham, T., Gurvich, I., Lindenbaum, M., Markovitch, S.: Learning implicit transfer for person re-identification. In: European Conference on Computer Vision, First International Workshop on Re-identification, Florence (2012)
3. Bazzani, L., Cristani, M., Perina, A., Murino, V.: Multiple-shot person re-identification by chromatic and epitomic analyses. Pattern Recogn. Lett. **33**(7), 898–903 (2012)
4. Bazzani, L., Cristani, M., Murino, V.: Symmetry-driven accumulation of local features for human characterization and re-identification. Comput. Vis. Image Underst. **117**(2), 130–144 (2013)
5. Berg, T.L., Berg, A.C., Shih, J.: Automatic attribute discovery and characterization from noisy web data. In: European Conference on Computer Vision (2010)
6. Chang, C.C., Lin, C.J.: LIBSVM: a library for support vector machines. In: ACM Trans. Intell. Syst. Technol. **2**(3), 27:1–27:27 (2011)
7. Chawla, N.V., Bowyer, K.W., Hall, L.O.: SMOTE : synthetic minority over-sampling technique. J. Artif. Intell. Res. **16**, 321–357 (2002)
8. Cheng, D., Cristani, M., Stoppa, M., Bazzani, L.: Custom pictorial structures for re-identification. In: British Machine Vision Conference (2011)
9. Dantcheva, A., Velardo, C., Dángelo, A., Dugelay, J.L.: Bag of soft biometrics for person identification. Multimedia Tools Appl. **51**(2), 739–777 (2011)
10. Ferrari, V., Zisserman, A.: Learning visual attributes. In: Neural Information Processing Systems (2007)
11. Fu, Y., Hospedales, T., Xiang, T., Gong, S.: Attribute learning for understanding unstructured social activity. In: European Conference on Computer Vision, Florence (2012)
12. Gray, D., Brennan, S., Tao, H.: Evaluating appearance models for recognition, reacquisition, and tracking. In: IEEE International Workshop on Performance Evaluation for Tracking and Surveillance, vol. 3 (2007)
13. Gray, D., Tao, H.: Viewpoint invariant pedestrian recognition with an ensemble of localized features. In: European Conference on Computer Vision, Marseille (2008)
14. He, H., Garcia, E.A.: Learning from imbalanced data. In: IEEE Transactions on Data and Knowledge Engineering, vol. 21 (2009)
15. Hirzer, M., Beleznai, C., Roth, P., Bischof, H.: Person re-identification by descriptive and discriminative classification. In: Scandinavian Conference on Image analysis (2011)
16. Hirzer, M., Roth, P.M., Bischof, H.: Person re-identification by efficient impostor-based metric learning. In: IEEE International Conference on Advanced Video and Signal-Based Surveillance (2012)
17. Hirzer, M., Roth, P.M., Martin, K., Bischof, H., Köstinger, M.: Relaxed pairwise learned metric for person re-identification. In: European Conference on Computer Vision, Florence (2012)
18. Jain, A.K., Dass, S.C., Nandakumar, K.: Soft biometric traits for personal recognition systems. In: International Conference on Biometric Authentication, Hong Kong (2004)
19. Keval, H.: CCTV Control room collaboration and communication: does it Work? In: Human Centred Technology Workshop (2006)
20. Kumar, N., Berg, A., Belhumeur, P.: Describable visual attributes for face verification and image search. IEEE Trans. Pattern Anal. Mach. Intell. **33**(10), 1962–1977 (2011)
21. Lampert, C.H., Nickisch, H., Harmeling, S.: Learning to detect unseen object classes by between-class attribute transfer. In: IEEE Conference on Computer Vision and Pattern Recognition (2009)
22. Layne, R., Hospedales, T.M., Gong, S.: Person re-identification by attributes. In: British Machine Vision Conference (2012)
23. Layne, R., Hospedales, T.M., Gong, S.: Towards person identification and re-identification with attributes. In: European Conference on Computer Vision, First International Workshop on Re-identification, Florence (2012)

24. Liu, C., Gong, S., Loy, C.C., Lin, X.: Person re-identification: what features are important? In: European Conference on Computer Vision, First International Workshop on Re-identification, Florence (2012)
25. Liu, J., Kuipers, B.: Recognizing human actions by attributes. In: IEEE Conference on Computer Vision and Pattern Recognition pp. 3337–3344 (2011)
26. Liu, D., Nocedal, J.: On the limited memory method for large scale optimization. Math. Program. B **45**(3), 503–528 (1989)
27. Loy, C.C., Xiang, T., Gong, S.: Time-Delayed Correlation Analysis for Multi-Camera Activity Understanding. Int. J. Comput. Vision **90**(1), 106–129 (2010)
28. Mackay, D.J.C.: Information Theory, Inference, and Learning Algorithms, 4th edn. Cambridge University Press, Cambridge (2003)
29. Madden, C., Cheng, E.D., Piccardi, M.: Tracking people across disjoint camera views by an illumination-tolerant appearance representation. Mach. Vis. Appl. **18**(3–4), 233–247 (2007)
30. Murphy, K.P.: Machine Learning: A Probabilistic Perspective. MIT Press, Cambridge, MA, (2012)
31. Nixon, M.S., Aguado, A.S.: Feature Extraction and Image Processing for Computer Vision, 3rd edn. Academic Press, Waltham (2012)
32. Nocedal, J., Wright, S.: Numerical Optimization, 2nd edn. Springer-Verlag, Newyork (2006)
33. Nortcliffe, T.: People Analysis CCTV Investigator Handbook. Home Office Centre of Applied Science and Technology, UK Home Office (2011)
34. Orabona, F., Jie, L.: Ultra-fast optimization algorithm for sparse multi kernel learning. In: International Conference on Machine Learning (2011)
35. Orabona, F.: DOGMA: a MATLAB toolbox for online learning (2009)
36. Platt, J.C.: Probabilities for SV machines. In: Advances in Large Margin Classifiers. MIT Press, Cambridge (1999)
37. Prosser, B., Zheng, W.S., Gong, S., Xiang, T.: Person re-identification by support vector ranking. In: British Machine Vision Conference (2010)
38. Satta, R., Fumera, G., Roli, F.: A general method for appearance-based people search based on textual queries. In: European Conference on Computer Vision, First International Workshop on Re-Identification (2012)
39. Schneiderman, R.: Trends in video surveillance give dsp an apps boost. IEEE Signal Process. Mag. **6**(27), 6–12 (2010)
40. Schölkopf, B., Smola, A.J.: Learning with kernels: Support Vector Machines, Regularization, Optimization, and Beyond. MIT Press, Cambridge, MA (2002)
41. Siddiquie, B., Feris, R.S., Davis, L.S.: Image ranking and retrieval based on multi-attribute queries. In: IEEE Conference on Computer Vision and Pattern Recognition (2011)
42. Smyth, P.: Bounds on the mean classification error rate of multiple experts. Pattern Recogn. Lett. **17**, 1253–1257 (1996)
43. Vaquero, D.A., Feris, R.S., Tran, D., Brown, L., Hampapur, A., Turk, M.: Attribute-based people search in surveillance environments. In: IEEE International Workshop on the Applications of Computer Vision, Snowbird, Utah (2009)
44. Walt, C.V.D., Barnard, E.: Data characteristics that determine classifier performance. In: Annual Symposium of the Pattern Recognition Association of South Africa (2006)
45. Williams, D.: Effective CCTV and the challenge of constructing legitimate suspicion using remote visual images. J. Invest. Psychol. Offender Profil. **4**(2), 97–107 (2007)
46. Zheng, W.S., Gong, S., Xiang, T.: Associating groups of people. In: British Machine Vision Conference (2009)
47. Zheng, W.S., Gong, S., Xiang, T.: Person re-identification by probabilistic relative distance comparison. In: IEEE Conference on Computer Vision and Pattern Recognition (2011)
48. Zheng, W.S., Gong, S., Xiang, T.: Transfer re-identification : from person to set-based verification. In: IEEE Conference on Computer Vision and Pattern Recognition (2012)
49. Zheng, W.S., Gong, S., Xiang, T.: Quantifying and Transferring Contextual Information in Object Detection. IEEE Trans. Pattern Anal. Mach. Intell. **1**(8), 762–777 (2011)

50. Zheng, W.S., Gong, S., Xiang, T.: Re-identification by Relative Distance Comparison. IEEE Trans. Pattern Anal. Mach. Intell. **35**(3), 653–668 (2013)
51. Zhu, X., Wu, X.: Class Noise vs. Attribute Noise: A Quantitative Study of Their Impacts. Artif. Intell. Rev. **22**(1), 177–210 (2004)

Chapter 6
Person Re-identification by Attribute-Assisted Clothes Appearance

Annan Li, Luoqi Liu and ShuichengYan

Abstract Person re-identification across nonoverlapping camera views is a challenging computer vision task. Due to the often low video quality and high camera position, it is difficult to get clear human faces. Therefore, clothes appearance is the main cue to re-identify a person. It is difficult to represent clothes appearance using low-level features due to its nonrigidity, but daily clothes have many characteristics in common. Based on this observation, we study person re-identification by embedding middle-level clothes attributes into the classifier via a latent support vector machine framework. We also collect a large-scale person re-identification dataset, and the effectiveness of the proposed method is demonstrated on this dataset under open-set experimental settings.

6.1 Introduction

Person re-identification is a computer vision task of matching people across nonoverlapping camera views in a multicamera surveillance system. It has a wide range of applications and great commercial value. However, it still remains an unsolved

---

These authors Annan Li and Luoqi Liu contributed equally to this work.

A. Li (✉) · L. Liu · S. Yan
Department of Electrical and Computer Engineering at National University of Singapore, Singapore, Singapore
e-mail: lia@i2r.a-star.edu.sg
e-mail: li.annan.cn@gmail.com

L. Liu
e-mail: liuluoqi@nus.edu.sg

S. Yan
e-mail: eleyans@nus.edu.sg

S. Gong et al. (eds.), *Person Re-Identification*,
Advances in Computer Vision and Pattern Recognition,
DOI: 10.1007/978-1-4471-6296-4_6, © Springer-Verlag London 2014

problem because of the low video quality and variations of viewpoint, pose, and illumination [19].

In a video surveillance system, to get a wider perspective range, cameras are usually installed at positions much higher than the height of people. High camera position leads to longer sight distance and also limitation in viewpoint, which makes it difficult to get clear human faces. Therefore, the appearance of a person is most influenced by clothes. However, representing clothes appearance is still a challenging problem. Most person re-identification approaches represent clothes appearance by low-level texture descriptors [1, 9, 11, 15, 20, 24–26], which are similar to the feature representation for rigid object recognition, e.g., face recognition. Since human body is nonrigid, in person re-identification such feature representation methods cannot achieve as good performance as they can get in face recognition. Low-level features are not sufficient for person re-identification.

As artificialities, clothes can vary in any way. However in daily life, people usually wear ordinary clothes, which vary in color and texture patterns but sill have many similar characteristics in style. Therefore, it is possible to describe properties of daily clothes by some attributes with middle-level semantic meaning. Consequently, using middle-level clothes attributes may improve the performance of person re-identification. Based on the above observations, we define several clothes attributes and propose a new person re-identification method by embedding clothes attributes into person re-identification via a latent Support Vector Machines (SVM) framework [6, 23].

Person re-identification methods can be evaluated in two ways, i.e., as a closed-set identification problem and as an open-set verification problem, respectively. The former usually treats people appearing in one camera as the gallery set, and the people appearing in other cameras as the probe set. And it requires that no new persons appear in the probe set. In other words, it requires closed environments. However in real-world scenarios, most video surveillance systems are installed in open environments. Therefore, it is reasonable to evaluate person re-identification methods under an open-set experimental setting. Due to the limitation of publicly available datasets, many works on person re-identification are evaluated in closed-set experimental settings [1, 9, 11, 15, 20, 25]. To address this problem, we collect a large-scale dataset which contains more than 1,000 videos. The proposed method is evaluated under open-set experimental settings using this dataset.

In this chapter we study the person re-identification problem with robustness. Specifically, the content includes three main points:

- A part-based human appearance representation approach.
- A person re-identification method by embedding into the discriminant classifier by a latent SVM framework.
- A person re-identification benchmark, including a large-scale database, an evaluation method based on open-set experimental settings and results of the proposed method.

The rest of this chapter is organized as follows. Section 6.2 gives a brief description of related work. The proposed person re-identification approach is described in Sect. 6.3. Section 6.4 describes the large-scale database we have collected. Section 6.5 shows the experimental results. Conclusions and discussions are given in Sect. 6.6.

6.2 Related Work

In recent years, person re-identification has attracted growing attention in the field of computer vision and pattern recognition. As one of the subproblems in multicamera surveillance, a brief literature survey can be found in a recent review on intelligent multicamera video surveillance [19].

Following the categorization in [19], research works on person re-identification can be roughly divided into two categories, i.e., *feature* and *learning*. Gheissari et al. [7] proposed to use local motion features to re-identify people across camera views. In this approach, correspondence between body parts of different persons is obtained through space-time segmentation. Color and edge histograms are extracted on these body parts. Person re-identification is performed by matching the body parts based on the features and correspondence. Wang et al. [20] proposed shape and appearance context for person re-identification. The shape and appearance context computes the co-occurrence of shape words and visual words. In this approach, human body is segmented into L parts using the shape context and a learned shape dictionary. Each part is further partitioned into M subregions by a spatial kernel. The histogram of visual words is extracted on each subregion. Consequently, the $L \times M$ histograms are used as visual features for person re-identification. Bazzani et al. [1] represented the appearance of a pedestrian by combining three kinds of features, i.e., weighted color histograms, maximally stable color regions, and recurrent highly structured patches, respectively. The above-mentioned features are sampled according to the symmetry and asymmetry axes obtained from silhouette segmentation.

Besides exploring better handcrafted features, learning discriminant models on low-level visual features is another popular way to tackle the problem of person re-identification. Gray and Tao [9] used AdaBoost to select an optimal ensemble of localized features for pedestrian recognition. They claimed that the learned feature is robust to viewpoint change. Schwartz and Davis [16] used partial least squares to perform person re-identification. In this work, high-dimensional original features are projected into a low-dimensional discriminant latent subspace learned by Partial Least Squares. Person re-identification is performed in the latent subspace. Prosser et al. [15] treated person re-identification as a pairwise ranking problem. And they used ranking SVM to learn the ranking model. In recent years, metric learning has become popular in person re-identification. Zheng et al. [25] proposed a probabilistic relative distance comparison model. The proposed model maximizes the probability that the distance between a pair of true match is smaller than that between an incorrect match pair. Therefore, the learned metric can achieve good results in person re-identification. Besides the above-mentioned methods, Zheng et al. [26] extended the person re-identification approach in [15, 25] to set-based verification by transfer learning. Hirzer et al. [11] proposed a more efficient metric learning approach for person re-identification. In this approach, a discriminative Mahalanobis metric learning model was used. With some relaxations, this model can be efficient, and faster than previous approaches.

Appearance-based human analysis and recognition is a hot topic in the computer vision literature. Clothes information is utilized implicitly or explicitly in these works. For example, Tapaswi et al. [17] used clothes appearance to assist recognizing actors in TV Series. However, the clothes analysis in such works is quite simple. Compared with other human-related research topics, the number of research works on clothes analysis is quite small. Recently, Yamaguchi et al. [21] conducted a study on parsing clothes in fashion photographs. In this work, a fashion clothes dataset including images and corresponding is crawled from a fashion website. The authors proposed to use super-pixel segmentation and a conditional random fields model to segment clothes and predict the attributes labels. However, most of the clothes in this dataset are high-level concepts such as jacket and coat, which are difficult to distinguish. Besides [21], Liu et al. [14] conducted a study on cross-scenario clothing retrieval. In this work, clothes images are retrieved across daily clothing images and online store images. The appearance of clothes is represented by local features extracted on the aligned body parts. Cross-scenario clothing retrieval is performed by indirectly matching the appearance via an auxiliary set. In this work, 15 are defined to assist the retrieval.

The work of Vaquero et al. [18] was the first to introduce middle-level attributes in human recognition. However, the attributes used in this work are mainly facial attributes. Recently, Layne et al. [12, 13] utilized in person re-identification. Fifteen attributes are defined and predicted in their work. The similarity between two sets of predicted attributes is measured by Mahalanobis distance. Distance between two people is measured by a weighted sum of the attribute distance and the distance given by [1].

6.3 Person Re-identification with by Latent SVM

The process of re-identifying a person in the video surveillance system usually includes three necessary steps: human detection, visual feature representation, and classification. As shown in Fig. 6.1, our method contains two more steps. The first one is body part detection, which provides better alignment. The second one is the embedding of into the classifier.

The details of body part detection are described in Sect. 6.3.1. The definitions and estimation are given in Sect. 6.3.2. In Sect. 6.3.3, we describe how to embed into the classifier using a latent SVM framework.

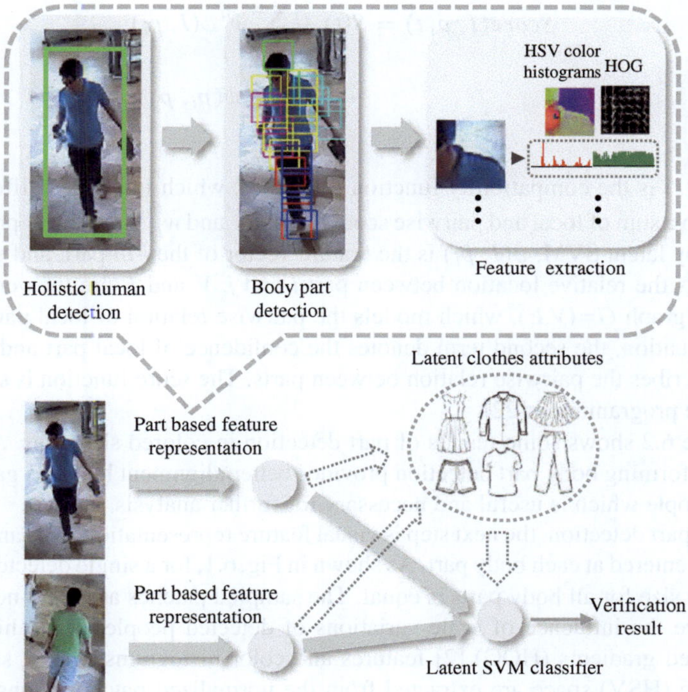

Fig. 6.1 The flowchart of person re-identification

6.3.1 Body Part Detection and Feature Representation

In the proposed method, the input of body part detection is an initial bounding box of holistic human body. We obtain the bounding box by a deformable part model-based cascade detector [5]. Then, we perform body part detection using the method of Yang and Ramanan [22]. In Yang and Ramanan's approach, human body is represented by K local parts. Candidates of local parts are obtained by the deformable part model-based detector [6]. These local part candidates produce many configuration candidates. Part locations are estimated by selecting the configuration with the best score.

The configurations are categorized into several types according to pose difference. Denote $i \in \{1, \ldots, K\}$ as the index of local parts, $p_i \in \{1, \ldots, L\}$ and $t_i \in \{1, \ldots, T\}$ as the position and type of the *i-th* part. The score of a configuration for type t at position p in image I is given by:

$$Score(I, p, t) = S(t) + \sum_{i \in V} w_i^{t_i} \phi(I, p_i) + \sum_{ij \in E} w_{ij}^{t_i, t_j} \psi(p_i, p_j). \tag{6.1}$$

Here, $S(t)$ is the compatibility function for type t, which integrates different part types into a sum of local and pairwise scores [22]. $w_i^{t_i}$ and $w_{ij}^{t_i,t_j}$ are linear projections learned by latent SVM. $\phi(I, p_i)$ is the feature vector of the *i-th* part, and $\psi(p_i, p_j)$ represents the relative location between part i and j. V and E are the vertices and edges of graph $G=(V,E)$, which models the pairwise relation of local parts. In the above equation, the second term denotes the confidence of local part and the third term describes the pairwise relation between parts. The score function is solved via quadratic programming [22].

Figure 6.2 shows some results of part detection in colored skeletons. As can be seen, performing body part detection provides better alignment between gallery and probe people which is useful and necessary for further analysis.

After part detection, the next step is visual feature representation. We sample local patches centered at each body part. As shown in Fig. 6.1, for a single detected person, the patch size for all body parts is equal. The sampled patches are scale-normalized to remove the influence of scale variations of detected people. Then, histograms of oriented gradients (HOG) [2] features and color histograms in hue, saturation, and value (HSV) space are extracted from the normalized patches. Consequently, the appearance of a person is described by a feature vector, which is obtained by concatenating features of all aligned parts. To lower the computational cost, the dimension of the feature vectors is reduced by principal component analysis (PCA).

6.3.2 Definition and Estimation

As in most popular person re-identification methods, the feature representation approach described in the previous subsection reflects low-level visual features. However, identifying people is a high-level task. There is a semantic gap between low-level feature representation and high-level task are recognized as middle-level description of people. Therefore, embedding into the identification provides a possible way to bridge the semantic gap.

In the literature of computer vision, are obtained two approaches. In the work of Yamaguchi et al. [21], the are crawled from the fashion website. Although such a data acquisition method can provide plenty of attributes, these attributes are not suitable for person re-identification. For example, in [21], *jacket* and *coat* are both annotated, but they are high-level clothes concepts and difficult to distinguish in low-quality surveillance videos. For the person re-identification task, the attributes should be visually separable in the surveillance video scenario.

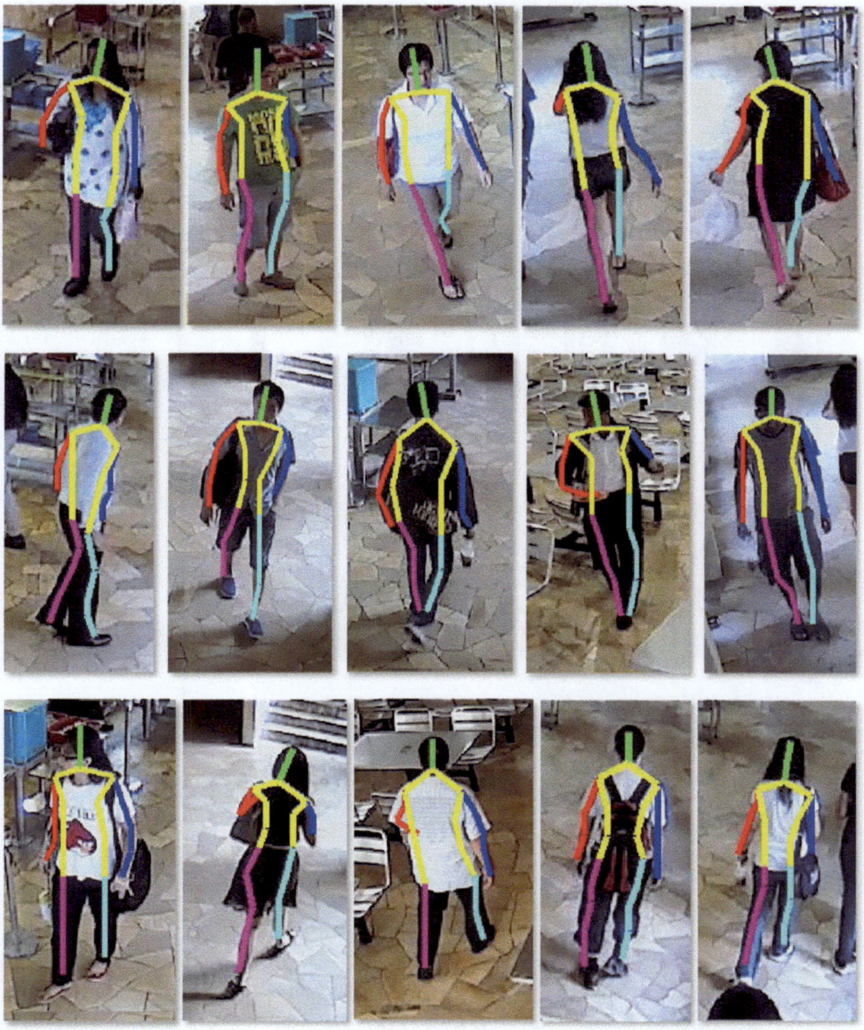

Fig. 6.2 Some results of body part detection illustrated in skeletons

Besides mining Web data, another approach to obtain is manual annotation. Layne et al. [13] annotated 15 binary-valued and utilized them in person re-identification. In this work, we define 11 kinds, to describe the appearance of people. Each kind of attributes has 2–5 values. The combinations of our attributes are more than those in [13]. Details of the attribute definitions are shown in Fig. 6.3.

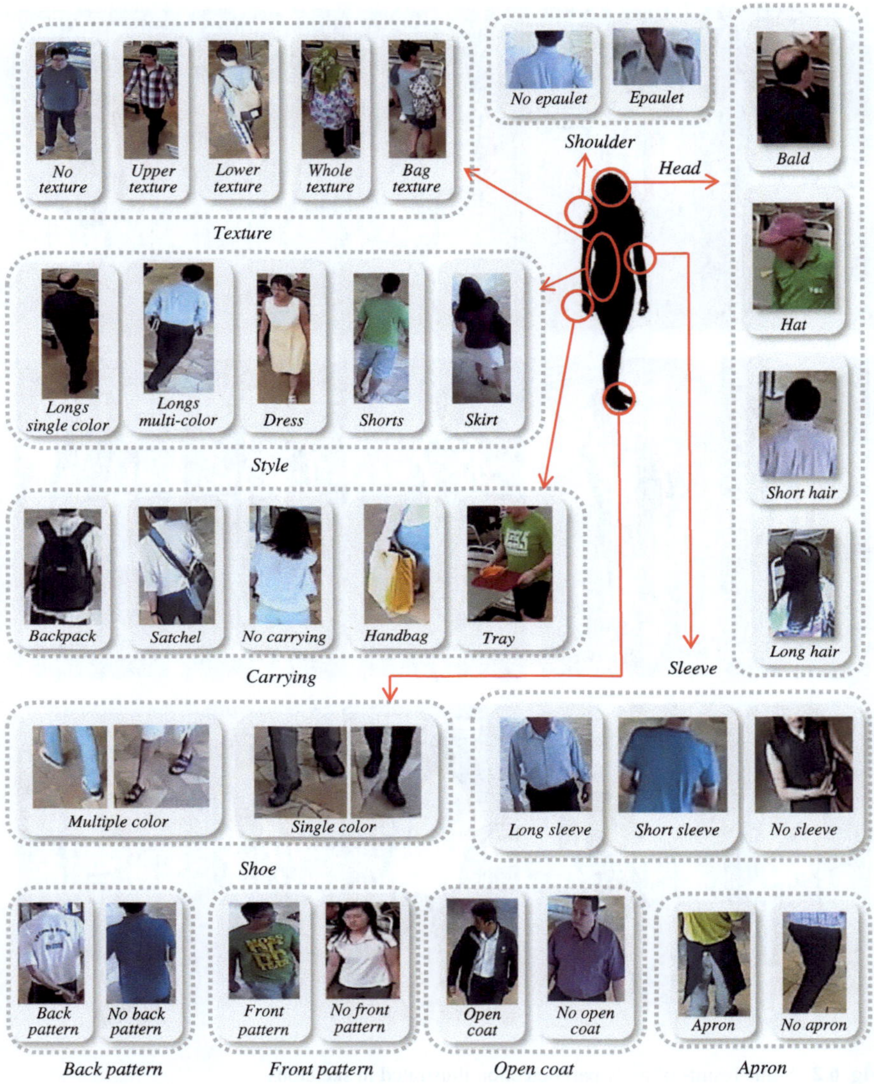

Fig. 6.3 The definition of clothing attributes

6.3.3 Person Re-identification with by Latent SVM

In this subsection, we first describe the problem of image set-based person re-identification in an open-set scenario in Open-Set Person Re-identification as a Binary Classification Problem. Then, we show how to address the above-mentioned problem by a latent SVM-based, assisted person re-identification method. The details of

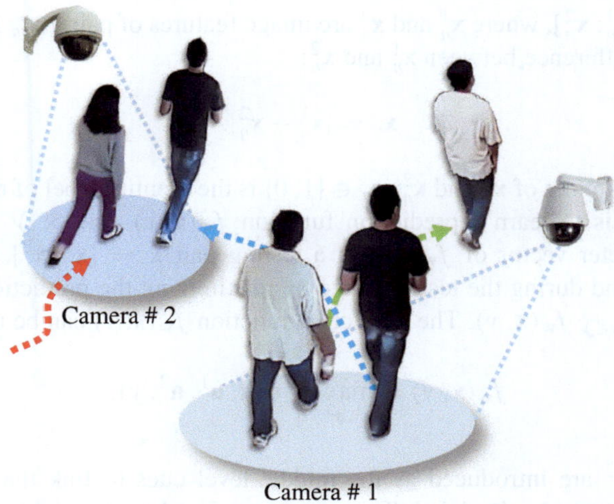

Fig. 6.4 The illustration of the open-set person re-identification problem

the method include: the objective, the potential function, the optimization, and the inference. They are described in The Latent SVM-Based Model, Potential Functions, Optimization, and Inference, respectively.

Open-Set Person Re-identification as a Binary Classification Problem

Person re-identification can be considered as a closed-set identification problem or an open-set verification problem. As shown in Fig. 6.4, in many scenarios, people appearing in one camera do not necessarily appear in another camera. And a camera view may include people never appearing in other cameras either. Therefore, it is better to treat person re-identification as a verification problem.

In the verification problem, the testing set is open. It can be formulated as a binary classification problem whether a pair of samples belong to the same person. In this work, we tackle this problem by a binary latent SVM classifier. The input is the absolute value of the differences between a pair of testing samples. The output is the probability of their belonging to the same person.

Re-identification is an application in the video surveillance system. Therefore, multiple images of a person are usually available. The problem can be further formulated as a set-to-set classification problem. To simplify the problem, the binary latent SVM classifier is trained and tested on single image pairs. Based on the similarities between image pairs, the similarity between a pair of image sets is measured by a set-to-set metric, for example, the Hausdorff distance [10].

The Latent SVM-Based Model

As described above, we formulate the person re-identification problem as a binary classification problem on image pairs. The training set is represented as a set of N sample tuples $\{(\mathbf{x}_n, \mathbf{a}_n^1, \mathbf{a}_n^2, y_n)\}_{n=1}^N$. Here $\mathbf{x}_n \in \mathbb{R}^s$ is the low-level feature vector of image pairs $[\mathbf{x}_n^1; \mathbf{x}_n^2]$, where $\mathbf{x}_n^1$ and $\mathbf{x}_n^2$ are image features of pair n. $\mathbf{x}_n$ is the absolute value of the difference between $\mathbf{x}_n^1$ and $\mathbf{x}_n^2$:

$$\mathbf{x}_n = |\mathbf{x}_n^1 - \mathbf{x}_n^2|. \tag{6.2}$$

$\mathbf{a}_n^1, \mathbf{a}_n^2 \in \mathbb{R}^{N_a}$ are the of $\mathbf{x}_n^1$ and $\mathbf{x}_n^2$. $y_n \in \{1, 0\}$ is the identity label of the image pair.

The target is to learn a prediction function $f_{\mathbf{w}}(\mathbf{x}, y) : \mathcal{X} \times \mathcal{Y} \rightarrow \mathbb{R}$, where $\mathbf{w}$ is a parameter vector of $f_{\mathbf{w}}$. Given a testing pair $\mathbf{x} = [\mathbf{x}^1; \mathbf{x}^2]$, identity label y^* can be found during the testing stage by maximizing the prediction function as $y^* = \arg\max_{y \in \mathcal{Y}} f_{\mathbf{w}}(\mathbf{x}, y)$. The prediction function $f_{\mathbf{w}}(\mathbf{x}, y)$ can be modeled as:

$$f_{\mathbf{w}}(\mathbf{x}, y) = \max_{\mathbf{a}^1, \mathbf{a}^2} \mathbf{w}^T \psi(\mathbf{x}, \mathbf{a}^1, \mathbf{a}^2, y). \tag{6.3}$$

The $\mathbf{a}^1$ and $\mathbf{a}^2$ are introduced as the middle-level cues to link the original low-level features and identity label. They are treated as latent variables in the whole formulation and are automatically predicted in both training and testing. The ground truth labels of clothing attributes are implicitly used to obtain the predictor in training, while no ground truth label is available in testing. $\mathbf{w}^T \psi(\mathbf{x}, \mathbf{a}^1, \mathbf{a}^2, y)$ is defined as follows:

$$\begin{aligned} \mathbf{w}^T \psi(\mathbf{x}, \mathbf{a}^1, \mathbf{a}^2, y) = {} & \mathbf{w}^{y^T} \phi_y(\mathbf{x}, y) \\ & + \sum_{i \in \mathcal{A}} \mathbf{w}_i^{1^T} \phi_a(\mathbf{x}, a_i^1) \\ & + \sum_{i \in \mathcal{A}} \mathbf{w}_i^{2^T} \phi_a(\mathbf{x}, a_i^2) \\ & + \sum_{i \in \mathcal{A}} \mathbf{w}_i^{ay^T} \phi_{ay}(a_i^1, a_i^2, y), \end{aligned} \tag{6.4}$$

where $\mathcal{A}$ is the attribute set. The first term on the right corresponds to the binary classifier which directly makes decision from the original feature. The second and third terms correspond to attributes prediction. The role of the fourth term is to transform the influence of to identity classification.

The parameter vector $\mathbf{w}$ of $\mathbf{w}^T \psi(\mathbf{x}, \mathbf{a}^1, \mathbf{a}^2, y)$ can be solved within the latent SVM framework:

$$\begin{aligned} & \arg\min_{\mathbf{w}, \xi} \beta \|\mathbf{w}\|^2 + \sum_{n=1}^{N} \xi_n \\ & s.t. \max_{\mathbf{a}^1, \mathbf{a}^2} \mathbf{w}^T \psi(\mathbf{x_n}, \mathbf{a}^1, \mathbf{a}^2, y_n) - \max_{\mathbf{a}^1, \mathbf{a}^2} \mathbf{w}^T \psi(\mathbf{x_n}, \mathbf{a}^1, \mathbf{a}^2, y) \\ & \geq \Delta(y_n, y) - \xi_n, \ \forall n, \ y_n \neq y, \end{aligned} \tag{6.5}$$

where β is a coefficient of ℓ_2 regularizer $\|\mathbf{w}\|^2$, and ξ_n is the slack variable for the n-th training sample. $\Delta(y_n, y)$ is a loss function defined as

$$\Delta_{0/1}(y_n, y) = \begin{cases} 1 & y_n \neq y \\ 0 & otherwise. \end{cases} \tag{6.6}$$

Potential Functions

The potential functions of Eq. (6.4) are defined as follows:

The term $\mathbf{w}^{y^T} \phi_y(\mathbf{x}, y)$ is a linear model to predict the identity of an image pair from the low-level features. We can use linear SVM [4] to estimate the parameters of this potential function. The mapping response $\phi_y(\mathbf{x}, y)$ is represented as the confidence vector of SVM.

The term $\mathbf{w}_i{}^T \phi_a(\mathbf{x}, a_i)$ is a linear model that represents the prediction of the i-th attribute from the low-level features. Similarly, we can also train an SVM classifier [4] to model this potential function and output the SVM confidence score as the function value of $\phi_a(\mathbf{x}, a_i)$.

The term $\mathbf{w}_i^{ay^T} \phi_{ay}(a_i^1, a_i^2, y)$ is a linear model for the i-th attribute and identity y. This potential integrates the relationship between attributes and identity. When a_i^1 and a_i^2 have the same label, x^1 and x^2 are more likely to be the same person (y is equal to 1). When a_i^1 and a_i^2 have different labels, x^1 and x^2 are more likely to be different persons (y is equal to 0). $\phi_{ay}(a_i^1, a_i^2, y)$ is a sparse vector of dimension $|\mathcal{A}_i| \times 2$, where $|\mathcal{A}_i|$ is the possible number of values for attributes a_i.

Optimization

The latent SVM formulation can be solved by the nonconvex cutting plane method [3] which minimizes the Lagrange form of Eq. (6.5), $\min_{\mathbf{w}} L(\mathbf{w}) = \beta\|\mathbf{w}\|^2 + \sum_{n=1}^{N} R^n(\mathbf{w})$. $R^n(\mathbf{w})$ is a hinge loss function defined as:

$$\begin{aligned} R^n(\mathbf{w}) = &\max_y(\Delta(y_n, y) + \max_{\mathbf{a}^1,\mathbf{a}^2} \mathbf{w}^T \psi(\mathbf{x}_n, \mathbf{a}^1, \mathbf{a}^2, y) \\ &- \max_{\mathbf{a}^1,\mathbf{a}^2} \mathbf{w}^T \psi(\mathbf{x}_n, \mathbf{a}^1, \mathbf{a}^2, y_n). \end{aligned} \tag{6.7}$$

The cutting plane method iteratively builds an increasingly accurate piecewise quadratic approximation of $L(\mathbf{w})$ based on its subgradient $\partial_{\mathbf{w}} L(\mathbf{w})$.

Define:

$$\begin{aligned} \{\mathbf{a}_n^{1*}, \mathbf{a}_n^{2*}\} &= \arg\max_{a^1,a^2} \mathbf{w}^T \psi(\mathbf{x}_n, \mathbf{a}^1, \mathbf{a}^2, y), \forall n, \\ \{\mathbf{a}_n^1, \mathbf{a}_n^2\} &= \arg\max_{\mathbf{a}^1,\mathbf{a}^2} \mathbf{w}^T \psi(\mathbf{x}_n, \mathbf{a}^1, \mathbf{a}^2, y_n), \forall n, \\ y_n^* &= \arg\max_y \Delta(y_n, y) + \mathbf{w}^T \psi(\mathbf{x}_n, \mathbf{a}_n^1, \mathbf{a}_n^2, y). \end{aligned} \tag{6.8}$$

The subgradient $\partial_{\mathbf{w}} L(\mathbf{w})$ can be calculated as

$$\partial_{\mathbf{w}} L(\mathbf{w}) = 2\beta\mathbf{w} + \sum_{n=1}^{N} \psi(\mathbf{x}_n, \mathbf{a}_n^{1*}, \mathbf{a}_n^{2*}, y_n^*) - \sum_{n=1}^{N} \psi(\mathbf{x}_n, \mathbf{a}_n^1, \mathbf{a}_n^2, y_n). \tag{6.9}$$

Given the subgradient $\partial_{\mathbf{w}} L(\mathbf{w})$, $L(\mathbf{w})$ can be minimized by the cutting plane method.

Inference

Given a testing pair $\mathbf{x} = [\mathbf{x}^1; \mathbf{x}^2]$ and specific y, the identity score can be inferred over latent variable $\mathbf{a}^1$ and $\mathbf{a}^2$ as $f_{\mathbf{w}}(\mathbf{x}, y) = \max_{\mathbf{a}^1,\mathbf{a}^2} \mathbf{w}^T \psi(\mathbf{x}, \mathbf{a}^1, \mathbf{a}^2, y)$. Then the identity is obtained as the predicted label with the highest score

$$y^* = \arg\max_{y}\{\max_{\mathbf{a}^1,\mathbf{a}^2} \mathbf{w}^T \psi(\mathbf{x}, \mathbf{a}^1, \mathbf{a}^2, y)\}. \tag{6.10}$$

6.4 Database

6.4.1 The NUS-Canteen Database

In the literature of person re-identification, VIPeR [8], i-LIDS [24], and ETHZ [16] are the most frequently used datasets. In the VIPeR dataset, there are 632 pedestrian image pairs captured from two cameras. The publicly available subset of i-LIDS [24] has 476 images corresponding to 119 people, in which each person has 4 images. In the ETHZ dataset, which is originally proposed for pedestrian detection, videos are captured from moving cameras. The dataset contains three video sequences, each containing 83, 35, and 28 persons, respectively. Corresponding numbers of images are 4,857, 1,961, and 1,762.

The publicly available datasets are limited in sample numbers and camera views. To address this problem, we collected and annotated a large-scale person re-identification dataset. As shown in Fig. 6.5, the raw videos are captured from 10 cameras installed at a university canteen. The canteen has roofs but no inclosure wall. Therefore, it can be considered as a semi-outdoor scenario. The illumination is influenced by both controlled lights and sunlight. There are multiple entrances in the canteen, which the cameras cannot completely cover. It is a typical open environment for person re-identification.

We have annotated 1,129 short videos. Each video corresponds to one person and contains 12–61 frames. 74.31% of the videos have 61 frames. There are 215 people annotated in our dataset. Each person has 2–19 videos and appears in 1–6 camera(s). The detailed statistics of the data are shown in Fig. 6.6. On average, one person corresponds to more than five videos and appears in more than three cameras. The number of samples and camera views are much bigger than the above-mentioned datasets. It provides a better benchmark for person re-identification.

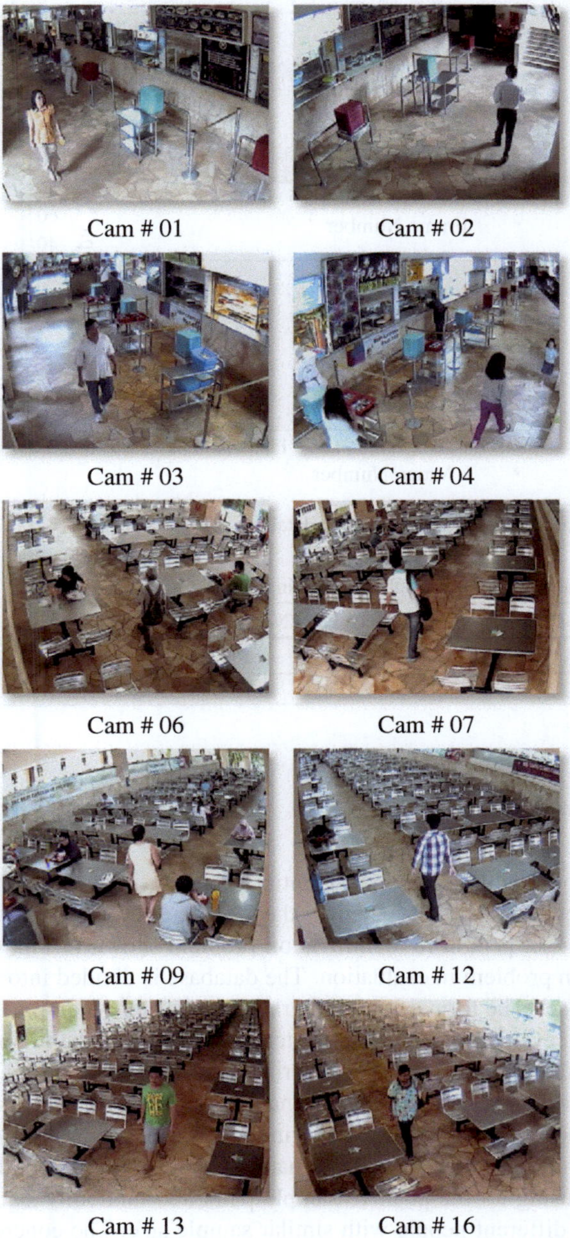

Fig. 6.5 Example frames in NUS-canteen database

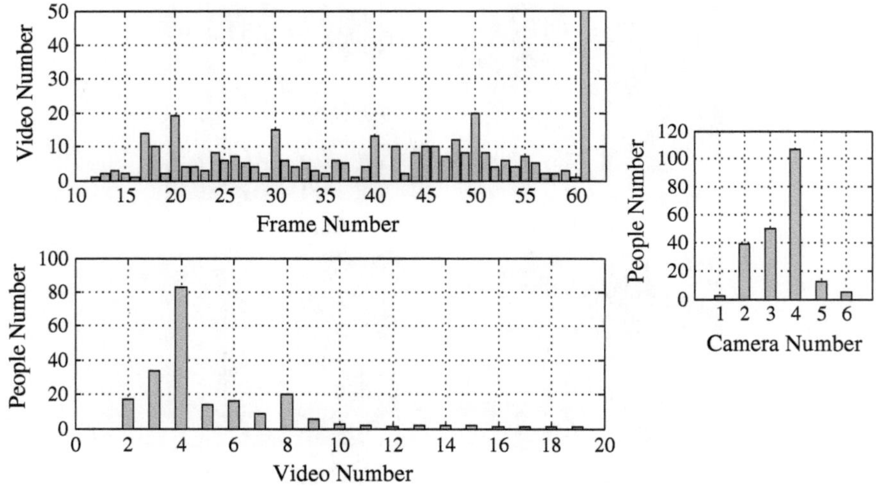

Fig. 6.6 The statistics of University-canteen database

Table 6.1 The sample and pair number of training and testing set

	Subject No.	Video No.	Pairs No. (Same people)	Pairs No. (Different people)
Training Set	100	514	1512	4884
Testing Set	115	615	1889	4918

6.4.2 Evaluation

In many previous studies, person re-identification is usually treated as a closed-set identification problem and evaluated by the CMC curves [1, 9, 11, 15, 20, 25]. To tackle the open-set person re-identification problem shown in Fig. 6.4, we treat it as a verification problem in evaluation. The database is divided into the training set and the testing set. The former is used to train the person re-identification model, the latter is used for evaluation. There is no intersection between them. Since we treat the person re-identification as an open-set verification problem, the training and testing are performed on sample pairs. The pair number of different people is much bigger than that of the same person. It brings imbalance between positive and negative samples, which may lead to bias in evaluation. Thus, we construct the training and testing set by using all the positive sample pairs and randomly sampling a subset of the pairs of different people with similar sample size. The concrete numbers of training and testing set are shown in Table 6.1.

The performance of a person re-identification approach is measured by Receiver Operating Characteristic (ROC) curves.

6.5 Experiments

The experiments are organized into three parts. In Sect. 6.5.1, we make comparisons between the holistic feature representation and the proposed part-based approach. In Sect. 6.5.2, we show the prediction accuracy. The proposed latent SVM-based clothes assisted person re-identification method is validated in 6.5.3.

6.5.1 Holistic Versus Part-Based Feature Representation

As shown in Fig. 6.1, the first step of person re-identification is the holistic pedestrian detection. For this step, we use the detector in [5]. Some detection results are shown in Fig. 6.7. In the next step, we perform body part detection using the approach in [22].

In the holistic feature representation, we first normalize the detection bounding boxes to 48×128 pixels and divide them into 3×8 grid of nonoverlapping patches of 16×16 pixels. HOG and color histogram features are extracted from each patch. The size of HOG cell is set to 4 and the color histograms are quantified into 16 bins in each channel. Consequently, each patch is represented by a 48-dimensional color feature and 124-dimensional HOG feature vector. The total length of HOG and color feature vector is 2,976 and 1,152.

As defined in [22], the human body is divided into 26 local parts. In the experiments, the body parts are normalized to 32×32 pixels. The size of the HOG cell is set to 8. The color histograms are quantified into 16 bins in each channel. Consequently, the length of color and HOG feature vectors extracted from a body part are 48 and 496, respectively. Then we have a 1,248-dimensional color vector and a 12,896-dimensional HOG vector for each sample. Since the dimensions are too high, we reduce the dimension of HOG and color vectors to 1,000 by principal component analysis (PCA). The combination of color and HOG feature is simply done by concatenating them into one vector. In the experiments, we apply two kinds of set-to-set metric, i.e., the Hausdorff distance and the average Euclidean distance, respectively. Let X and Y be two sets of feature vectors. Their Hausdorff distance $d_{Hausdorff}(X, Y)$ is given by

$$\begin{aligned} d_{Hausdorff}(X, Y) &= max(h(X, Y), h(Y, X)) \\ h(X, Y) &= \max_{x \in X}(\min_{y \in Y}(d(x, y))), \end{aligned} \tag{6.11}$$

where $d(x, y)$ denotes the Euclidean distance between feature vector x and y.

The performance comparisons between holistic and part-based feature representation are shown in Fig. 6.8. As can be seen, no matter what type of visual features are used, the part-based feature representation is obviously better than the holistic feature representation. Performing PCA also enhances the representation power. Based on

Fig. 6.7 Examples of detected people

the PCA feature, integrating color and HOG feature also improves the performance. The experimental results show that the proposed part-based feature representation is very effective. We also find that simply using average Euclidean distance gets better performance than Hausdorff distance.

The part-based, PCA-enhanced, color + HOG feature representation achieves the best results. We use this feature in the next experiments.

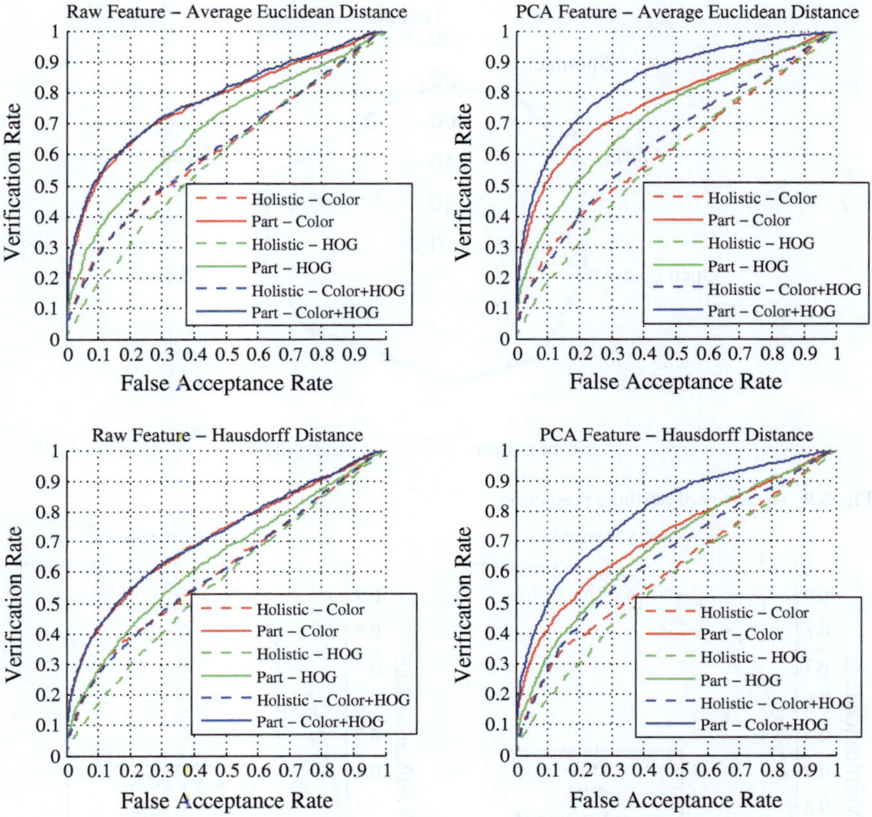

Fig. 6.8 Performance comparisons between holistic and part-based feature representations

6.5.2 Prediction

In this chapter we embed middle-level into person re-identification. The clothing attributes are treated as latent variables and automatically predicted in both training and testing phases. This part corresponds to the second and third terms in Eq. (6.4). The prediction accuracy is an important factor that influences the results. Thus, it is better to show the prediction accuracy before analyzing the clothing attribute-assisted person re-identification approach.

The prediction accuracy on the testing set is illustrated in Fig. 6.9. The prediction accuracy of multivalued attributes is lower than that of binary-valued attributes. There is still much room for improvement.

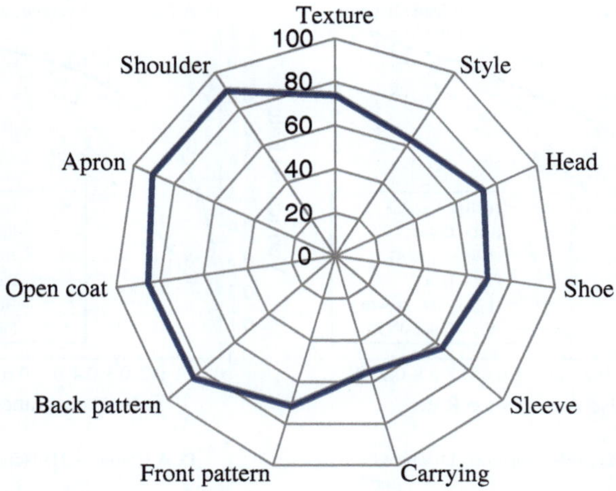

Fig. 6.9 Accuracy of attribute prediction

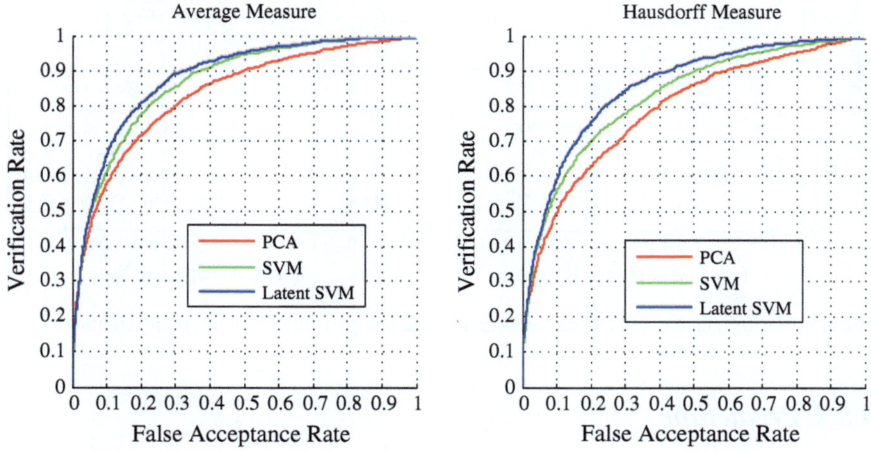

Fig. 6.10 Performance comparisons between SVM and latent SVM

6.5.3 SVM Versus Latent SVM

To validate the effectiveness of embedding into person re-identification, we make comparisons between the SVM classifier and the latent SVM classifier. The former does not use any information and corresponds to the first term of Eq. (6.4). The latter integrates attributes information and corresponds to all the four terms of Eq. (6.4). In the experiments we use the linear SVM classifier [4].

The performance comparisons are shown in Fig. 6.10. Besides the ROC curves of SVM and latent SVM models, we also plot the results of original PCA features as a baseline. As can be seen, the latent SVM outperforms SVM using both average metric and Hausdorff metric. The experimental results prove that embedding can improve the performance of person re-identification. It also proves that the proposed latent SVM model is effective. Considering that the accuracy of attributes prediction is not high, the performance of latent SVM can be further enhanced by improving the prediction.

6.6 Conclusions

In this chapter, we describe how to use middle-level information to assist person re-identification. The assistance is performed by embedding as latent variables into the classifier via a latent SVM framework. As a necessary preprocessing step, a body part-based feature representation approach is also proposed. The experimental results demonstrate the effectiveness of both the feature representation approach and the latent SVM-based, assisted person re-identification method.

References

1. Bazzani, L., Cristani, M., Murino, V.: Symmetry-driven accumulation of local features for human characterization and re-identification. Comput. Vision. Image Underst. **117**(2), 130–144 (2013)
2. Dalal, N., Triggs, B.: Histograms of Oriented Gradients for Human Detection. In: IEEE Conference on Computer Vision and, Pattern Recognition, pp. 886–893 (2005)
3. Do, T., Artières, T.: Large margin training for hidden Markov models with partially observed states. In: International Conference on, Machine Learning, pp. 265–272 (2009)
4. Fan, R., Chang, K., Hsieh, C., Wang, X., Lin, C.: LIBLINEAR: a library for large linear classification. J. Mach. Learn. Res. **9**, 1871–1874 (2008)
5. Felzenszwalb, P., Girshick, R., McAllester, D.: Cascade object detection with deformable part models. In: IEEE Conference on Computer Vision and, Pattern Recognition, pp. 2241–2248 (2010)
6. Felzenszwalb, P.F., Girshick, R.B., McAllester, D., Ramanan, D.: Object detection with discriminatively trained part based models. IEEE Trans. Pattern Anal. Mach. Intell. **32**(9), 1627–1645 (2010)
7. Gheissari, N., Sebastian, T., Hartley, R.: Person reidentification using spatiotemporal appearance. In: IEEE Conference on Computer Vision and, Pattern Recognition, pp. 1528–1535 (2006)
8. Gray, D., Brennan, S., Tao, H.: Evaluating appearance models for recognition, reacquisition, and tracking. In: IEEE International Workshop on Performance Evaluation of Tracking and Surveillance (PETS) (2007)
9. Gray, D., Tao, H.: Viewpoint invariant pedestrian recognition with an ensemble of localized features. In: European Conference on Computer Vision, pp. 262–275 (2008)
10. Hausdorff, F.: Dimension und äußeres Maß. Mathematische Annalen **79**(1–2), 157–179 (1918)
11. Hirzer, M., Roth, P., Köstinger, M., Bischof, H.: Relaxed pairwise learned metric for person re-identification. In: European Conference on Computer Vision, pp. 780–793 (2012)
12. Layne, R., Hospedales, T., Gong, S.: Person re-identification by attributes. In: British Machine Vision Conference, vol. 2, p. 3 (2012)

13. Layne, R., Hospedales, T.M., Gong, S.: Towards person identification and re-identification with attributes. In: European Conference on Computer Vision, First International Workshop on Re-Identification, pp. 402–412 (2012)
14. Liu, S., Song, Z., Liu, G., Xu, C., Lu, H., Yan, S.: Street-to-Shop: cross-scenario clothing retrieval via parts alignment and auxiliary set. In: IEEE Conference on Computer Vision and, Pattern Recognition, pp. 3330–3337 (2012)
15. Prosser, B., Zheng, W., Gong, S., Xiang, T.: Person re-identification by support vector ranking. In: British Machine Vision Conference, pp. 21.1-21.11 (2010)
16. Schwartz, W.R., Davis, L.S.: Learning discriminative appearance-based models using partial least squares. In: Brazilian Symposium on, Computer Graphics and Image Processing, pp. 322–329 (2009)
17. Tapaswi, M., Bauml, M., Stiefelhagen, R.: "Knock! Knock! Who is it?" Probabilistic person identification in TV-series. In: IEEE Conference on Computer Vision and, Pattern Recognition, pp. 2658–2665 (2012)
18. Vaquero, D., Feris, R., Tran, D., Brown, L., Hampapur, A., Turk, M.: Attribute-based people search in surveillance environments. In: Workshop on the Applications of Computer Vision (2009)
19. Wang, X.: Intelligent multi-camera video surveillance: a review. Pattern Recogn. Lett. **34**(1), 3–19 (2013)
20. Wang, X., Doretto, G., Sebastian, T., Rittscher, J., Tu, P.: Shape and appearance context modeling. In: International Conference on Computer Vision, pp. 1–8 (2007)
21. Yamaguchi, K., Kiapour, M., Ortiz, L., Berg, T.: Parsing clothing in fashion photographs. In: IEEE Conference on Computer Vision and, Pattern Recognition, pp. 3570–3577 (2012)
22. Yang, Y., Ramanan, D.: Articulated pose estimation with flexible mixtures of parts. In: IEEE Conference on Computer Vision and, Pattern Recognition, pp. 1385–1392 (2011)
23. Yu, C.N.J., Joachims, T.: Learning structural SVMs with latent variables. In: International Conference on, Machine Learning, pp. 1169–1176 (2009)
24. Zheng, W., Gong, S., Xiang, T.: Associating groups of people. In: British Machine Vision Conference (2009)
25. Zheng, W., Gong, S., Xiang, T.: Person re-identification by probabilistic relative distance comparison. In: IEEE Conference on Computer Vision and, Pattern Recognition, pp. 649–656 (2011)
26. Zheng, W., Gong, S., Xiang, T.: Transfer re-identification: from person to set-based verification. In: IEEE Conference on Computer Vision and, Pattern Recognition, pp. 2650–2657 (2012)

Chapter 7
Person Re-identification by Articulated Appearance Matching

Dong Seon Cheng and Marco Cristani

Abstract Re-identification of pedestrians in video-surveillance settings can be effectively approached by treating each human figure as an articulated body, whose pose is estimated through the framework of Pictorial Structures (PS). In this way, we can focus selectively on similarities between the appearance of body parts to recognize a previously seen individual. In fact, this strategy resembles what humans employ to solve the same task in the absence of facial details or other reliable biometric information. Based on these insights, we show how to perform single image re-identification by matching signatures coming from articulated appearances, and how to strengthen this process in multi-shot re-identification by using Custom Pictorial Structures (CPS) to produce improved body localizations and appearance signatures. Moreover, we provide a complete and detailed breakdown-analysis of the system that surrounds these core procedures, with several novel arrangements devised for efficiency and flexibility. Finally, we test our approach on several public benchmarks, obtaining convincing results.

7.1 Introduction

Human re-identification (re-id) consists in recognizing a person in different locations over various non-overlapping camera views. We adopt the common assumption that individuals do not change their clothing within the observation period, and that finer

D. S. Cheng (✉)
Department of Computer Science and Engineering, Hankuk University of Foreign Studies,
81, Oedae-ro, Mohyeon-myeon, Cheoin-gu, Yongin-si Gyeonggi-do 449-791, South Korea
e-mail: cheng_ds@hufs.ac.kr

M. Cristani
Department of Computer Science, University of Verona,
Ca' Vignal 2, Verona 37134, Italy
e-mail: marco.cristani@univr.it

S. Gong et al. (eds.), *Person Re-Identification*,
Advances in Computer Vision and Pattern Recognition,
DOI: 10.1007/978-1-4471-6296-4_7, © Springer-Verlag London 2014

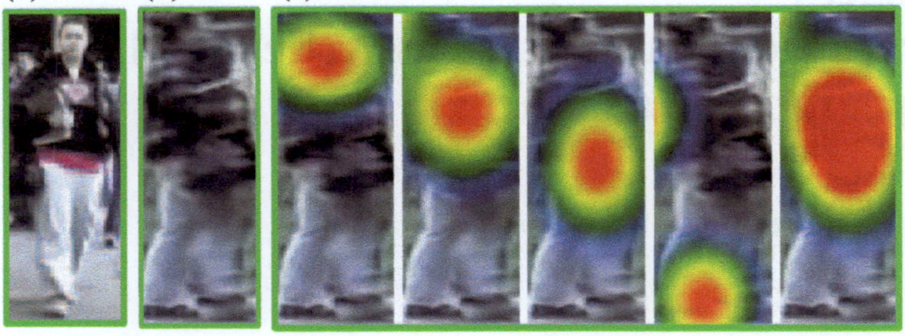

Fig. 7.1 Re-id performed by a human subject: (**a**) the test probe, (**b**) the correct match in the gallery, and (**c**) the fixation heat maps from eye-tracking over consecutive 1s intervals—the hotter the color, the longer the time spent looking at that area

biometric cues (face, fingerprint, gait, etc..) are unavailable: we consider, that is, only *appearance-based* re-id.

In this chapter, we present an extensive methodology for person re-id through articulated appearance matching, based on Pictorial Structures (PS) [16], and its variant Custom Pictorial Structures (CPS) [9], to decompose the human appearance into body parts for pose estimation and signature matching. In the PS framework of [1], the parts are initially located by general part detectors, and then a full body pose is inferred by solving their kinematic constraints. In this work, we propose a novel type of part detector, fast to train and to use, based on the *histogram of oriented gradients* (HOG) [10] features and a *linear discriminant analysis* (LDA) [24] classifier. Moreover, we use the *belief propagation* algorithm to infer MAP body configurations from the kinematic constraints, represented as a tree-shaped factor graph.

More in general, our proposal takes inspiration from how humans approach appearance-based re-id. As we showed in [9], monitoring subjects performing re-id confirmed a tendency to scan for salient (structurally known) parts of the body, looking for part-to-part correspondences (we reproduce a sample of the study in Fig. 7.1). We think that encoding and exploiting the human appearance per parts is a convenient strategy for re-id, and PS is particularly well suited to this task. In particular, we exploit the conventional PS fitting on separate individual images for *single-shot* re-id, which consists in matching pairs probe/gallery of images for each subject. Our approach aims to obtain robust signatures from features extracted from the segmented parts.

Secondly, for *multi-shot* re-id, where each subject has multiple images distributed between probe set and gallery set, we can use the extra information to improve the re-id process in two ways: by improving the PS fitting using the CPS algorithm [9] that iteratively performs appearance modeling and pose estimation, and by using *set-matching* to compute distances between probe set and gallery set.The rationale of CPS is that the local appearance of each part should be relatively consistent among images of the same subject, and hence it is possible to build an appearance model. Thus, localizing parts can be enhanced by evaluating the similarity to the model.

Our goal in this work is to crystallize the use of PS for re-id with a complete and detailed breakdown of the stages in our process. We intend to introduce several novel arrangements devised for efficiency and flexibility, with an eye towards future extensions. In particular, we introduce a new class of part detectors based on HOG features and linear discriminant analysis to feed the PS pose estimation algorithm, and a new color histogram technique to extract feature vectors. Experiments have been carried out on many publicly available datasets (iLIDS, ETHZ1,2,3, VIPeR, CAVIAR4REID) with convincing results in all modalities.

The chapter is organized as follows: we analyze related work in Sect. 7.2; we provide an overview of our approach in Sect. 7.3 and all the details in Sect. 7.4; we provide details about the training of our part detectors in Sect. 7.5; and we discuss the experiments in Sect. 7.6. Finally, Sect. 7.7 wraps up with remarks and future perspectives.

7.2 State of the Art

Pictorial structures: The literature on PS is large and multifaceted. Here, we briefly review the studies that focus on the appearance modeling of body parts. We can distinguish two types of approaches: the single-image and multiple-image methods. In the former case, a PS processes each image individually. In [29], a two-step image parsing procedure is proposed, which enriches an edge-based model by adding chromatic information. In [12], a learning strategy estimates relations between body parts and a shared color-based appearance model is used to deal with occlusions. A recent, very successful, strategy is to build *deformable part models* (DPM) [15, 38] with many small pieces, less subject to warps and distortions.

In the other case, several images of the same person are available, but very few methods deal with this situation. In [30], two approaches for building PS have been proposed for tracking applications. A top-down approach automatically builds people models starting by convenient key poses detections; a bottom-up method groups together candidate body parts found along the considered sequence exploiting spatio-temporal reasoning. This technique shares some similarities with our approach, but it requires a high number of temporally consecutive frames (50–100). In our setting, few ($\leq$5), unordered images are instead expected. In a photo-tagging context, PS are grown over face detections to recognize few people [35], modeling the parts with Gaussian distributions in the color space.

Re-identification: Appearance-based techniques for re-id can be organized in two groups of methods: *learning-based* and *direct* approaches. In the former, a dataset is split into training and test sets, with the training individuals used to learn features and/or strategies for combining features to achieve high re-id accuracy, and the test ones used as validation. Direct methods are instead pure feature extractors.

An orthogonal classification separates the *single-shot* and the *multi-shot* techniques. As learning-based methods, an ensemble of discriminant localized features and classifiers is selected by boosting in [22]. In [25], pairwise dissimilarity profiles between individuals are learned and adapted for nearest-neighbor classification. Similarly, in [33], a high-dimensional signature formed by multiple features is projected onto a low-dimensional discriminant space by Partial Least Squares reduction. Contextual visual information is exploited in [39], enriching a bag-of-word-based descriptor by features derived from neighboring people, assuming that people stay together across different cameras. Bak et al. [3] casts re-id as a binary classification problem (one vs. all), while [28, 40] as a relative ranking problem in a higher dimensional feature space, where true and wrong matches become more separable. In [17], re-id is cast as a semi-supervised single-shot recognition problem where multiple features are fused at the classification output level, using the multi-view learning approach of [26]. Finally, re-id is cast as a Multiple Instance Learning in [32], where in addition a method for synthetically augmenting the training dataset is presented.

As direct methods, a spatio-temporal local feature grouping and matching is proposed in [20]: a decomposable triangulated graph is built that captures the spatial distribution of the local descriptions over time. In [36], images are segmented into regions and their color spatial relationship acquired with co-occurrence matrices. In [23], interests points (SURF) are collected in subsequent frames and matched. Symmetry and asymmetry perceptual attributes are exploited in [7, 14], based on the idea that features closer to the bodies' axes of symmetry are more robust against scene clutter. Covariance features, originally employed for pedestrian detection, are tailored in [4] for re-id, extracted from coarsely located body parts; later on, such descriptors are embedded into a learning framework in [2]. In [8], epitomic analysis is used to collapse a set of images into a small collage of overlapped patches containing the essence of textural, shape and appearance properties. To be brief, in addition to color, a large number of features types is employed for re-id: textures [14, 22, 28, 33], edges [33], Haar-like features [3], interest points [20] and image regions [14, 22, 36]. The features, when not collected densely, can be extracted from horizontal stripes, triangulated graphs, concentric rings [39], symmetry-driven structures [7, 14], and localized patches [4]. Very recently, other modalities and sensors (such as RGB-D cameras) are used to extract 3D soft-biometric cues from depth images: this avoids the constraint that people must be dressed in the same way during a re-id session [6]. Another unconventional application considers Pan-Tilt-Zoom cameras, where distances between signatures are also computed across different scales [31]. For an extensive review on re-id methods, please see [11].

7.3 Overview of our Approach

This section gives an overview of our re-id process, which is summarized in Fig. 7.2. Implementation details of each stage can be found later, in Sect. 7.4. The method is based on obtaining good pedestrian segmentations from which effective re-id

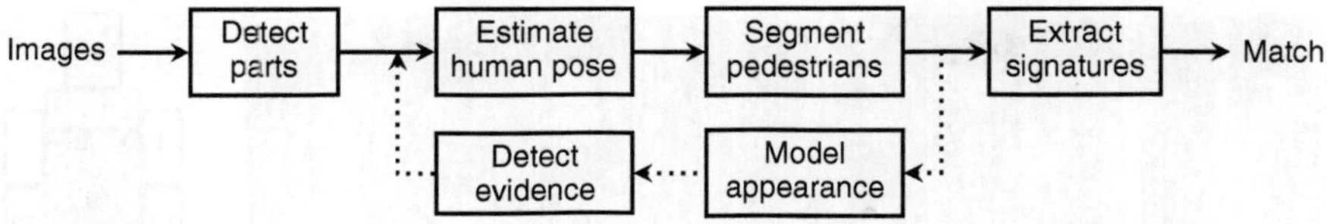

Fig. 7.2 Diagram of the stages in our approach. In single-shot mode, the estimated pose of the articulated human figure is used to segment the image and extract the features, joined into a signature. In multi-shot mode, with multiple images for each pedestrian, we model the common appearance of its parts, and thus refine the part detections with additional evidence, to be able to improve the pose estimation

signatures can be extracted. The basic idea is that we can segment accurately after we estimate the pose of the human figure within each image, and this pose estimation can be performed with PS.

The single-shot modality

Every image is processed individually to retrieve a feature vector that acts as its signature. By calculating distances between signatures, we can match a given probe image against a set of gallery images, ranking them from lowest to highest distance, and declaring the rank-1 gallery to be our guess for the identity of the probe.

Our proposed approach tries to increase the effectiveness of the signatures by filtering out as much of the background scene as possible, and by decomposing a full pedestrian figure into semantically reasonable body parts (like head, torso, arms and legs) in such a way that we can compose a full signature by joining part signatures. This increases the robustness of the method to partial (self)occlusions and changes in local appearance, like the presence of bags, different looks between frontal, back and side views, and imperfections in the pose estimation. Figure 7.3 (left) shows two cases from the VIPeR experiment, illustrating several aspects of the problems just mentioned. It is clear that the segmentations provide a good filtering of the background scene, even when they do not perfectly isolate the pedestrian figure.

However, the decomposition into parts is not sufficient to overcome persistent dataset-wise occlusions or poor image resolution. For example, the iLIDS dataset is made up of images taken from airport cameras, and an overwhelming number of pedestrians are captured with several bags, backpacks, trolleys and other occluding objects (including other different pedestrians). In this challenging situation, legs and arms are often hidden and their discriminating power is greatly reduced. Therefore, our approach is to balance the contributions of each part through a weight that indicates, percentage wise, its importance with respect to the torso, which remains the main predictor (see Fig. 7.3 (right) for an example and Sect. 7.4.4 for details).

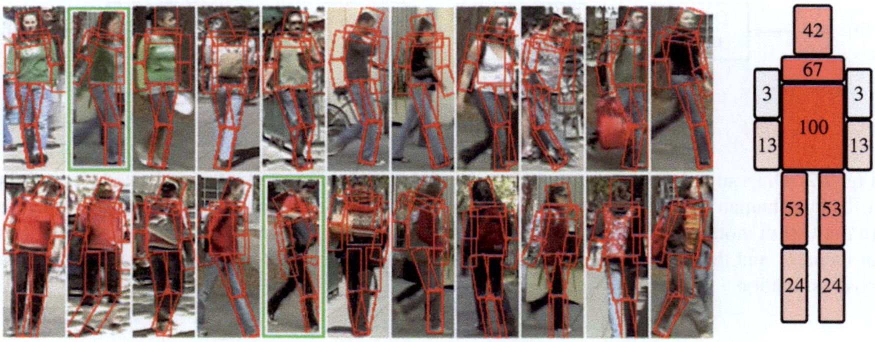

Fig. 7.3 (*Left*) Two illustrative lineups in single-shot re-id from the VIPeR experiments: the leftmost image is the probe and the rest are gallery images sorted by increasing distance from the probe. The correct match is shown with a green outline. (*Right*) Model of the articulated human figure, with percentages and color intensities proportional to the importance of a part in the VIPeR experiment

The multi-shot modality

Multi-shot re-id is performed when probe and gallery sets are made of multiple images for each subject. We can exploit this situation in two ways: firstly, by using *set matching* (the minimal distance across all pairs) when comparing signatures, so that the most unlike matches are discarded; secondly, by improving the pose estimations based on the appearance evidence. We create this evidence by building an appearance model of each pedestrian and using it to localize his parts with greater accuracy than just by using the generalized part detectors. Then, we feed this information back into the PS algorithm to compute new pose estimations, and hence segmentations. This process can be repeated until we reach a satisfactory situation.

In the end, our goal is to reinforce a coherent image of pedestrians, such that we can compute more robust signatures. Then, with multiple signatures available, the most natural way to match a probe set to the gallery sets is to find the closest pairs: this potentially matches frontal views with frontal views, side views with side views, occluding bags with occluding bags, and so on.

7.4 Details of our Approach

We now give a detailed description of the stages in our re-id approach, with a critical review of our previous method [9], where we adapted Andriluka's publicly available PS code to perform articulated pose estimation. Here instead, we developed a new and completely independent system with a novel part detector and our own implementation of the PS algorithm.

7.4.1 Part Detection

In [1], the authors use discriminatively trained part detectors to feed their articulated pose estimation process. In particular, their part detectors densely sample a shape context descriptor that captures the distribution of locally normalized gradient orientations in a log-polar histogram. With 12 bins for the location and eight bins for the gradient orientation, they obtain 96 dimensional descriptors. Then, they concatenate the histograms of all shape context descriptors falling inside the bounding box of a part. During detection, many positions, scales, and orientations of parts are scanned in a sliding window fashion. All color images are converted to gray-scale before feature extraction.

To classify the feature vectors, they train an AdaBoost classifier [19] using as weak learners simple decision stumps that test histogram bins against a threshold. More formally, given a feature vector $\mathbf{x}$, there are $t = 1, \ldots, T$ stump functions $h_t(\mathbf{x}) = \text{sign}(\xi_t(x_{n(t)} - \varphi_t))$, where φ_t is a threshold, ξ_t is a label equal to ± 1, and $n(t)$ is the index of the bin chosen by the stump. Training the AdaBoost classifier results in a strong classifier $H_i(\mathbf{x}) = \text{sign}(\sum_t \alpha_{i,t} h_t(\mathbf{x}))$ for each part i, where $\alpha_{i,t}$ are the learned weights of the weak classifiers.

During training, each annotated part is scaled and rotated to a canonical pose prior to learning, and the same process is applied during testing of candidate parts. The negative feature vectors come from sampling the image regions outside the objects, and the classifiers are then re-trained with a new training set augmented with false positives from the initial round. The classifier outputs are then converted into pseudo-probabilities by interpreting the normalized classifier margin as follows:

$$f_i(\mathbf{x}) = \frac{\sum_t \alpha_{i,t} h_t(\mathbf{x})}{\sum_t \alpha_{i,t}} \tag{7.1}$$

$$\tilde{p}(\mathbf{d}_i | \mathbf{l}_i) = \max(f_i(\mathbf{x}(\mathbf{l}_i)), \varepsilon_0), \tag{7.2}$$

where $\mathbf{x}(\mathbf{l}_i)$ is the feature vector for part configuration $\mathbf{l}_i$, and $\varepsilon_0 = 10^{-4}$ is a cutoff threshold. Even if the authors claim it works well, this simple conversion formula in fact produces poorly calibrated probabilities, as it is known that AdaBoost with decision stumps sacrifices the margin of the easier cases to obtain larger margins on cases close to the decision surface [34]. Our experience suggests that it produces weak and sparse candidate part configurations, because the decision boundary is assigned probability zero (not 0.5 as you would expect) and the weak margins (none of which approach 1) are linearly mapped to probabilities. A better choice would be to calibrate the predictions using Platt scaling [27].

The HOG-LDA Detector

Histograms of oriented gradients (HOG) features for pedestrian detection were first introduced by Dalal and Triggs in [10]. They proved to be efficient and effective

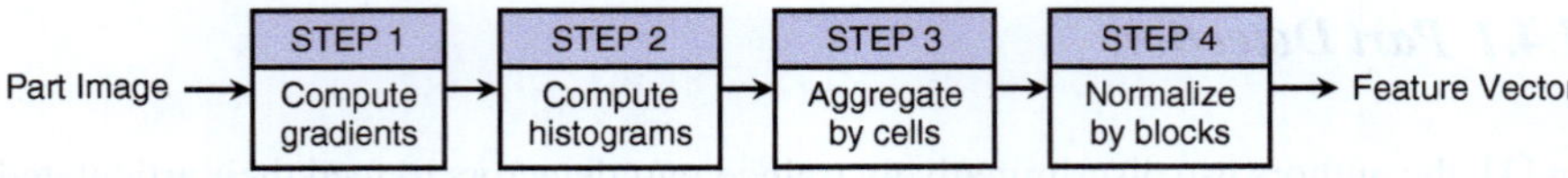

Fig. 7.4 Overview of the HOG feature extraction

for object detection, not only pedestrians, both as wholes and as collection of parts [38]. The HOG features are usually combined with a linear SVM classifier, but [24] shows that an opportunely trained *linear discriminant analysis* (LDA) classifier can be competitive while being faster, and easier, to train and test.

Calculating the HOG features requires a series of steps, shown summarized in Fig. 7.4. At each step, Dalal and Triggs experimentally show that certain choices produce better results than others, and they call the resultant procedure *the default detector* (HOG-dd). Like other recent implementations [15], we largely operate the same choices, but also introduce some tweaks.

Step 1. Here, we assume the input is an image window of canonical size for the body part we are considering. Like in HOG-dd, we directly compute the gradients with the masks $[-1, 0, 1]$. For color images, each RGB color channel is processed separately, and pixels assume the gradient vector with the largest norm. While it does not take full advantage of the color information, it is better than discarding it like in the Andriluka's detector.

Step 2. Next, we turn each pixel gradient vector into an histogram by quantizing its orientation into 18 bins. The orientation bins are evenly spaced over the range 0–180° so each bin spans 10°. For pedestrians there is no a-priori light/dark scheme between foreground and background (due to clothes and scenes) that justifies the use of the "signed" gradients with range 0–360°: in other words, we use the contrast insensitive version [15]. To reduce aliasing, when an angle does not fall squarely in the middle of a bin, its gradient magnitude is split linearly between the neighboring bin centers. The outcome can be seen as a sparse image with 18 channels, which is further processed by applying a spatial convolution, to spread the votes to four neighboring pixels [37].

Step 3. We then spatially aggregate the histograms into cells made by 7×7 pixel regions, by defining the feature vector at a cell to be the sum of its pixel-level histograms.

Step 4. As in the HOG-dd, we group cells into larger blocks and contrast normalize each block separately. In particular, we concatenate features from 2×2 contiguous cells into a vector $\mathbf{v}$, then normalize it as $\tilde{\mathbf{v}} = \min(\mathbf{v}/||\mathbf{v}||, 0.2)$, L2 norm followed by clipping. This produces 36-dimensional feature vectors for each block. The final feature vector for the whole part image is obtained by concatenating the vectors of all the blocks.

When the initial part image is rotated such that its orientation is not aligned with the image grid, the default approach is to normalize this situation by counter-rotating

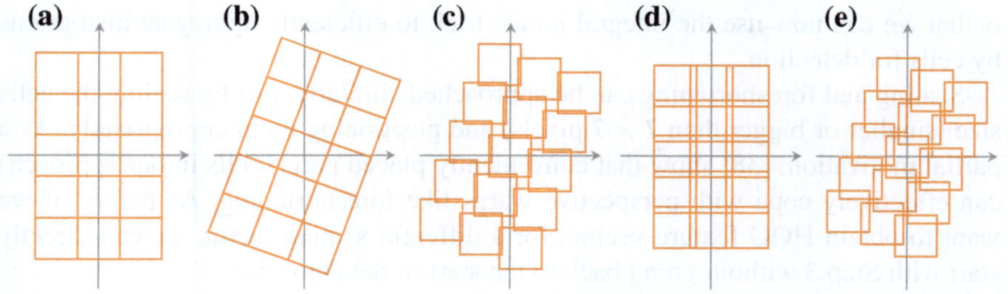

Fig. 7.5 Rotation approximation for a part defined by a matrix of 5 × 3 cells. From *left* to *right*: **a** default configuration with disjointed cells, **b** clockwise rotation by 20°, **c** approximation by non-rotated cells, **d** tighter configuration with cells overlapping by one pixel on each side, **e** rotation approximation of the tighter configuration. This approximation allows us to use the integral image technique

the entire image (or the bounding box of the part) before processing it as a canonical window. This can be computationally expensive during training, where image parts have all sorts of orientations, and during testing, even if we limit the number of detectable angles. Furthermore, dealing with changes in the scaling factor of the human figures and the foreshortening of limbs introduces additional computational burdens. In the following, we introduce a novel approximation method that manages to speed up the detection process.

Rotation and Scaling Approximation

Let p be a body part defined by a matrix of $M_p \times N_p$ cells (see Fig. 7.5). Rotating this part by θ degrees away from the vertical orientation creates two problems: how to compute the histograms in Step 2, and how to aggregate them by cells in Step 3. Step 1 can compute gradients regardless of the rotation and Step 4 does not care after we have the cell aggregates.

The first problem arises because we need to collect a histogram of the gradient angles with respect to the axis of the rotated part, and they are instead expressed with respect to the image grid. We propose our first approximation: with a fine enough binning of the histograms (our resolution of 10° is double the HOG-dd), we can approximate the "rotated" histograms by circularly shifting the bin counts of the neutral histograms of $-r_\theta$ places, where $r_\theta = \text{round}(\theta/10°)$. This operation is much more efficient than re-computing the features after counter-rotating the source image, and can be performed fast for all the rotation angles we are interested in.

We solve the second problem by approximating the rotated cells with no rotation at all. As can be seen in Fig. 7.5, this leaves quite large holes in the covering of the part image, which is only partially mitigated by the spatial convolution in Step 2 that spreads bin votes around. Our solution is to use a tighter packing of the cells, overlapped by one pixel on each side, so that they leave much smaller holes even at the worst angle for this approximation. The main purpose of avoiding rotated cells is that we can now use the integral image trick to efficiently aggregate histograms by cells for detection.

Scaling and foreshortening can be approached similarly, just by scaling the cells size (smaller or bigger than 7×7 pixels) and positioning them appropriately. As a partial motivation, [38] show that conveniently placed parts (cells in our approach) can effectively cope with perspective warps like foreshortening. As before, if we want to obtain HOG feature vectors for a different scaling factor, we can directly start with Step 3 without going back to the start of the algorithm.

Efficient Detection

Detection of a given part from a new image is usually performed with a sliding window approach: a coarse or fine grid of detection points is selected, and the image is tested at each point by the detector, once for every orientation angle and scale allowed for the part (we usually are not interested in all angles or scales for pedestrians). This means extracting HOG feature vectors for many configurations of position, orientation, scale, and all the approximations introduced so far make this task very efficient, especially when we use the integral image technique.

In fact, at the end of Step 2, instead of providing the gradient histograms, we compute their integral image, so that all sums in Step 3 can be performed in constant time for each cell, in every configuration we wish for. If the resolution of the orientation angles matches the one in the histograms binning, we expect the least amount of information loss to happen in the approximations.

The last component of our fast detection algorithm is the LDA classifier. As shown in [24], LDA models can be trained almost trivially, and with little or no loss in performance compared to SVM classifiers. An LDA model classifies a given feature vector $\mathbf{x}_i$ as a part p instead of background if

$$\mathbf{w}_p^t \mathbf{x}_i - c_p > 0 \tag{7.3}$$

where

$$\mathbf{w}_p = \mathbf{S}^{-1}(\mathbf{m}_p - \mathbf{m}_{bg}) \tag{7.4}$$

$$c_p = \mathbf{w}_p^t(\mathbf{m}_p + \mathbf{m}_{bg})/2. \tag{7.5}$$

The background mean $\mathbf{m}_{bg}$ and the common covariance $\mathbf{S}$ are trained from many images including different objects and scenes, and $\mathbf{m}_p$ is trained from feature vectors extracted from annotated images (see left Fig. 7.6).

Furthermore, given the scores $f_i = \mathbf{w}_p^t \mathbf{x}_i - c$, we retrieve well calibrated probability values $p(\mathbf{x}_i)$ using the Platt scaling method [27], where

$$p(\mathbf{x}_i) = \frac{1}{1 + \exp(Af_i + B)} \tag{7.6}$$

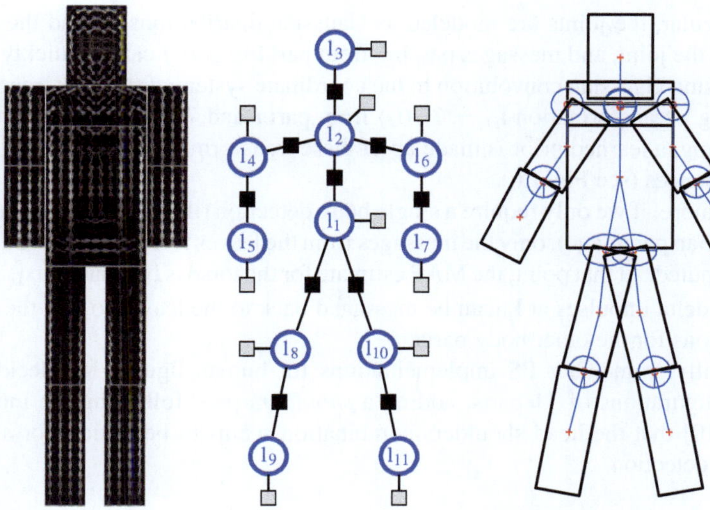

Fig. 7.6 (*left*) Composite image showing the positive weights in all the model weights $\mathbf{w}_p$ after training: each block shows the gradients that vote positively towards a part identification, with brighter colors in proportion to the vote strength. (*center*) Factor graph of the kinematic prior model for pose estimation. (*right*) Learned model of the relative position and rotation of the parts, including spatial localization covariances of the joints

and the parameters A and B are found using maximum likelihood estimation as

$$\arg\min_{A,B}\{-\sum_i y_i \log p(\mathbf{x}_i) + (1 - y_i)\log(1 - p(\mathbf{x}_i))\} \tag{7.7}$$

using the calibration set (f_i, y_i) with labels $y_i \in \{0, 1\}$.

7.4.2 Pose Estimation

After the part detectors independently scan an input image, giving us image evidence $D = \{\mathbf{d}_p\}$, it is time to detect full body configurations, denoted as $L = \{\mathbf{l}_p\}$, where $\mathbf{l}_p = (x_p, y_p, \vartheta_p, s_p)$ encodes position, orientation and scale of part p, respectively.In PS, the posterior of L is modeled as $p(L|D) \propto p(D|L)p(L)$, where $p(D|L)$ is the image likelihood and $p(L)$ is a prior modeling the links between parts. The latter is also called the *kinematic prior* because it can be seen as a system of masses (parts) and springs (joints) that rule the body's motions.

In fact, we can represent the prior as a factor graph (see Fig. 7.6), where we have two types of factors: the detection maps $p(\mathbf{d}_p|\mathbf{l}_p)$ (gray boxes) and the joints $p(\mathbf{l}_i|\mathbf{l}_j)$ (black boxes). This graph is actually a tree with the torso $p = 1$ as root, which means that we can use standard (non loopy) belief propagation to get the MAP estimates.

In particular, the joints are modeled as Gaussian distributions around the mean location of the joint, and messages passing from part i to part j can be quickly computed by using Gaussian convolution in the coordinate system of the joint, reachable by applying a transformation $\mathbf{l}_{ij} = T_{ij}(\mathbf{l}_i)$ from part i and $T_{ji}^{-1}(\mathbf{l}_{ij})$ towards part j. After training, a learned prior is made up of these transformations together with the joint covariances (see Fig. 7.6).

Furthermore, if we only require a single body detection (the default situation with one pedestrian per image), only the messages from the leaves to the root must be accurately computed. At that point, the MAP estimate for the torso is $\hat{\mathbf{l}}_1 = \arg\max_{\mathbf{l}_1} p(\mathbf{l}_1)$, and single delta impulses at $\hat{\mathbf{l}}_p$ can be messaged back to the leaves to find the MAP configurations for the other body parts.

Differently from other PS implementations for human figures, we decided to create configurations of 11 parts, adding a *shoulders* part, following the intuition of Dalal [10] that the head-shoulders combination seems to be critical for a good pedestrian detection.

7.4.3 Pedestrian Segmentation

To obtain well discriminating signatures, it is crucial to filter out as much of the background scene as possible, which is a potential source of spurious matches. After computing the pose estimation, we retrieve a segmentation of the image into separate body part regions, depending on the position and orientiation within the full body configuration. We encode such information in the form of image masks: thus, we get 11 body part masks and a combined set-union full body mask. Early experiments on removing the residual background within the masks, like cutting the pixels proximal, in location and appearance, to the known background, resulted in worse performances. In fact, the limited size of the images, usually cropped close to the pedestrian, and the cluttered scenes made figure/background inference difficult.

7.4.4 Feature Extraction

Having the masks, the task is to identify feature extraction methods that provide discriminating and robust signatures. As in our previous work [9], we rely on two proven techniques: color histograms and *maximally stable color regions* (MSCR) [18]. We experimented on several different variants of color histograms, both in our previous work and in this one: it is our experience that each dataset is suited to certain methods rather than others, with no method clearly outperforming the rest.

However, we reached a good compromise with a variant that separates shades of gray from colored pixels. We first convert all pixel values (r, g, b) to the HSV color space (h, s, v), and then we perform the following selections: all pixels with

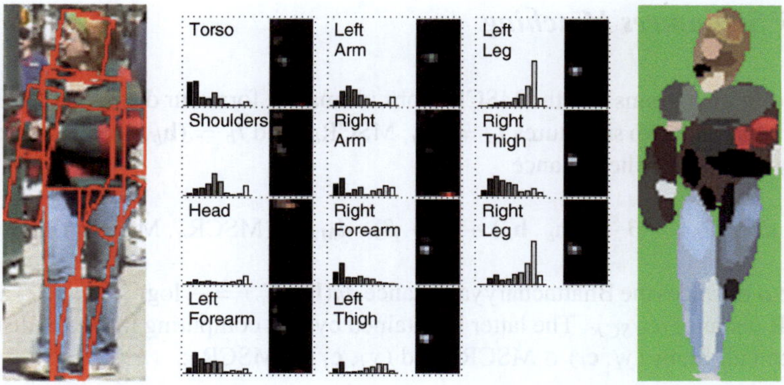

Fig. 7.7 (*Left*) A sample image from VIPeR with parts segmentation. (*Center*) Color histogram features, shown here separately for the 11 parts, each comprising of a histogram of grays and a histogram of colors. (*Right*) Blobs from the MSCR operator

value $v < \tau_{black}$ are counted in the bin of blacks, all remaining pixels with saturation $s < \tau_{gray}$ are counted in the gray bins according to their value v, all remaining pixels are counted in the color bins according to their hue-saturation coordinates (h, s).

We basically count the dark and unsaturated pixels separately from the others, and we ignore the brightness of the colored pixels, counting only their chromaticity in a 2D histogram (see Fig. 7.7). This procedure is also tweaked in several ways to improve speed and accuracy: the HSV channels are quantized into [20, 10, 10] levels, the votes are (bi)linearly interpolated into the bins to avoid aliasing, the residual chromaticity of the gray pixels is counted into the color histograms with a weight proportional to their saturation s. The image regions of each part are processed separately and provide a combined grays-colors histogram (GC histogram in short) which is vectorized and normalized. We then multiply each of these histograms by the part relevance weights λ_p (shown for example in Fig. 7.3 (right)), and then concatenate and normalize to form a single feature vector. Moreover, we allow the algorithm to adapt to particular camera settings by varying the importance of grays versus colors with a weight w_G, which can be tuned for each dataset.

Independently, the full body masks are used to constrain the extraction of the MSCR blobs. The MSCR operator detects a set of blob regions by looking at successive steps of an agglomerative clustering of image pixels. Each step groups neighboring pixels with similar color within a threshold that represents the maximal chromatic distance between colors. Those maximal regions that are stable over a range of steps become MSCR blobs. As in [14], we create a signature $\text{MSCR} = \{(y_i, \mathbf{c}_i) | i = 1, \dots, N\}$ containing the height and color of the N blobs. The algorithm is setup in a way that provides many small blobs and avoids creating ones too big (see Fig. 7.7). The rationale is that we want to localize details of the pedestrians appearance, which is more accurate for small blobs.

7.4.5 Signatures Matching

The color histograms and the MSCR blobs ultimately form our desired image signatures. Matching two signatures $I_a = (\mathbf{h}_a, \text{MSCR}_a)$ and $I_b = (\mathbf{h}_b, \text{MSCR}_b)$ is carried out by calculating the distance

$$d(I_a, I_b) = \beta \cdot d_h(\mathbf{h}_a, \mathbf{h}_b) + (1 - \beta) \cdot d_{MSCR}(\text{MSCR}_a, \text{MSCR}_b), \tag{7.8}$$

where β balances the Bhattacharyya distance $d_h(\mathbf{h}_a, \mathbf{h}_b) = -\log(\sqrt{\mathbf{h}_a}^t \sqrt{\mathbf{h}_b})$ and the MSCR distance d_{MSCR}. The latter is obtained by first computing the set of distances between all blobs $(y_i, \mathbf{c}_i) \in \text{MSCR}_a$ and $(y_j, \mathbf{c}_j) \in \text{MSCR}_b$:

$$v_{ij} = \gamma \cdot d_y(y_i, y_i) + (1 - \gamma) \cdot d_{lab}(\mathbf{c}_i, \mathbf{c}_j), \tag{7.9}$$

where γ balances the height distance $d_y = |y_i - y_j|/H$ and the color distance $d_{lab} = \|labcie(\mathbf{c}_i) - labcie(\mathbf{c}_j)\|/200$, which is the Euclidean distance in the LABCIE color space. Then, we compute the sets $M_a = \{(i, j)|v_{ij} \leq v_{ik}\}$ and $M_b = \{(i, j)|v_{ij} \leq v_{kj}\}$ of minimum distances from the two point of views, and finally obtain their average:

$$d_{MSCR}(\text{MSCR}_a, \text{MSCR}_b) = \frac{1}{|M_a \cup M_b|} \sum_{(i,j) \in M_a \cup M_b} v_{ij}. \tag{7.10}$$

The normalization factor H for the height distance is set to the height of the images in the dataset, while the parameters β and γ are tuned through cross-validation.

Additionally, we have experimented with different distances than the Bhattacharyya, like Hellinger, L1, L2, Mahalanobis, χ^2, but performances were inferior.

7.4.6 Multi-Shot Iteration

In multi-shot mode, we use CPS to improve the segmentations before extracting the features. This is a two-step iterative process that alternates between setting/updating the appearance model for the parts and updating the pose estimations. At the first iteration, we start with the conventional PS fittings, fed by the general part detectors. We thus collect all the part regions in the given images, normalize the different orientations, and stack them to estimate their common appearance. In particular, CPS employs a Gaussian model $\mathcal{N}(\mu_k, \sigma_k)$ in RGB space for all pixels k. In order to reinforce the statistics, the samples are extended by including spatial neighbors of similar color by performing k-means segmentation on each sub-image t and including the neighbors of k that belong to the same segment. The resulting Gaussian distribution is thus more robust to noise.

In the lead up to the second step of the iteration, these Gaussian models are used to evaluate the original images, scoring each location for similarity, providing thus *evidence maps* $p(\mathbf{e}_p|\mathbf{l}_p)$. This process can be efficiently performed using FFT-based Gaussian convolutions. Then, these maps must be combined with the part detections to feed the PS algorithm. Differently from [9], we experimented with different ways to combine them. It is our experience that maps that are too sparse and poorly populated generate pose estimations that rely on the default configuration in the kinematic prior. A fusion rule based on multiplication of probabilities (the default approach in a Bayesian update setting) tends to reduce the maps to isolated peaks. We thus propose a fusion rule based on the probability rule for union, which provides richer, but still selective, maps:

$$p(\mathbf{f}_p|\mathbf{l}_p) = p(\mathbf{d}_p|\mathbf{l}_p) + p(\mathbf{e}_p|\mathbf{l}_p) - p(\mathbf{d}_p|\mathbf{l}_p)p(\mathbf{e}_p|\mathbf{l}_p), \tag{7.11}$$

where the resulting $p(\mathbf{f}_p|\mathbf{l}_p)$ is then used in place of $p(\mathbf{d}_p|\mathbf{l}_p)$ in the pose estimation algorithm of Sect. 7.4.2. Experimentally, CPS converges after 4–5 iterations, and we can finally extract signatures like in the single-shot case. As for the matching, when we compare M probe signatures of a given subject against N gallery signatures of another one, we simply calculate all the possible $M \times N$ single-shot distances, and keep the smallest one.

7.5 Training

Training was performed on the PARSE[1], the PASCAL VOC2010[13], and the INRIA Person[2] databases. PARSE consists of 305 images of people in various poses that can be mirrored to generate 610 training images. The database also provides labels for each image, in the form of locating 14 body points of interest. From these points it is possible to retrieve configurations of body parts to train the PS models, and our setup is described in Table 7.1. PASCAL and INRIA are used to generate negative cases: PASCAL has 17,125 images containing all sorts of objects, including human figures of different sizes; INRIA Person has a negative training set of 1,218 non-person images.In particular, as in [24], all the images in PASCAL were used to extract the background model for the HOG-LDA detectors, while the first 200 annotated images in PARSE (mirrored to 400) were used to compute the foreground models for the parts. The remaining 105 images (mirrored to 210) and parts randomly drawn from INRIA Person's negative set were used to train the Platt calibration parameters. The PS kinematic model was trained on PARSE.

[1] http://phoenix.ics.uci.edu/software/pose/

[2] http://pascal.inrialpes.fr/data/human/

Table 7.1 Setup of the HOG-LDA detectors: configuration of the body parts used in our approach, with the canonical size in pixels and in number of cells. Detected orientations angles are −30, −20, −10, 0, 10, 20, 30°

Parts	Size (pixels)	Size (cells)	Codenames
Torso	43 × 31	7 × 5	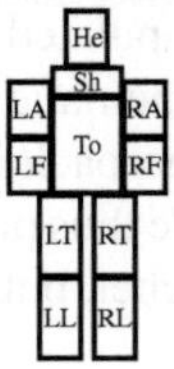
Shoulders	13 × 31	2 × 5	
Head	25 × 19	4 × 3	
2 × Arms	25 × 13	4 × 2	
2 × Forearms	25 × 13	4 × 2	
2 × Thighs	37 × 13	6 × 2	
2 × Legs	27 × 13	6 × 2	

7.6 Experiments

In this section we present the experimental evaluation of our approach and we compare our results to those at the state of the art. The main performance report tool for re-id is the Cumulative Matching Characteristic (CMC) curve, which plots the cumulative expectation of finding the correct match in the first n matches. Higher curves represent better performances, and hence it is also possible to compare results at-a-glance by computing the normalized area under curve (nAUC) value, indicated on the graphs within parentheses after the name when available. What follows is a detailed explanation of the experiments we performed on these datasets: VIPeR, iLIDS, ETHZ, CAVIAR for re-id.

Experimental Setup: The HOG-LDA detectors scan images once every four pixels and interpolate the results in between. The PS algorithm discards torso, head, shoulders detections below 50, 40, 30 respectively. Only one scale is evaluated in each dataset since the images are normalized. The calibration parameters γ, β, w_G, and the part weights $\{\lambda_p\}$ are tuned by cross-validation on half the dataset for single-shot, and one or more tuning partitions for multi-shot.

VIPeR Dataset [21]: This dataset contains 632 pedestrian image pairs taken from arbitrary viewpoints under varying illumination conditions. Each image is 128×48 pixels and presents a centered unoccluded human figure, although cropped short at the feet in some side views. In the literature, results on VIPeR are typically produced by mediating over ten runs, each consisting in a partition of randomly selected 316 image pairs. Our approach handily outperforms our previous result (BMVC in the figures), as well as SDALF [14], PRSVM [28], and ELF [21], setting the rank-1 matching rate at 26 %, and exceeding 61 % at rank-10 (see Fig. 7.8 (left)). We note that weights for arms are very low, due to the fact that pose estimation is unable to correctly account for self-occlusions in side views, which abound in this dataset.

iLIDS Dataset: The iLIDS MCTS videos have been captured at a busy airport arrival hall [39]: the dataset consists of 119 pedestrians with 479 images that we normalize to 64×192 pixels. The images come from non-overlapping cameras, subject to quite large

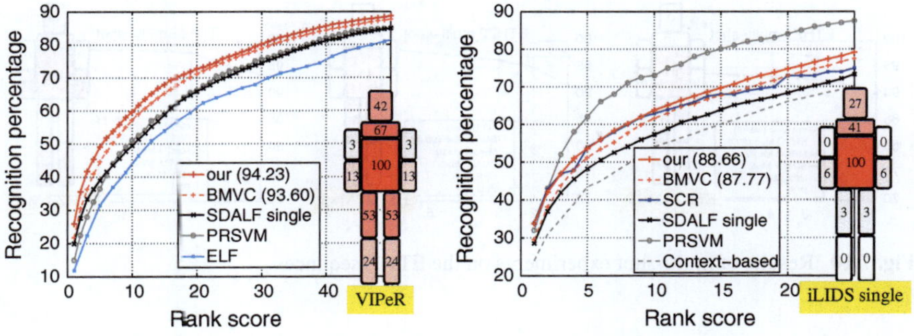

Fig. 7.8 Results of single-shot experiments on VIPeR (*left*) and iLIDS (*right*). Also shown on the puppets are the corresponding part weights: note how the legs in iLIDS are utterly useless because of too many occlusions by bags and trolleys

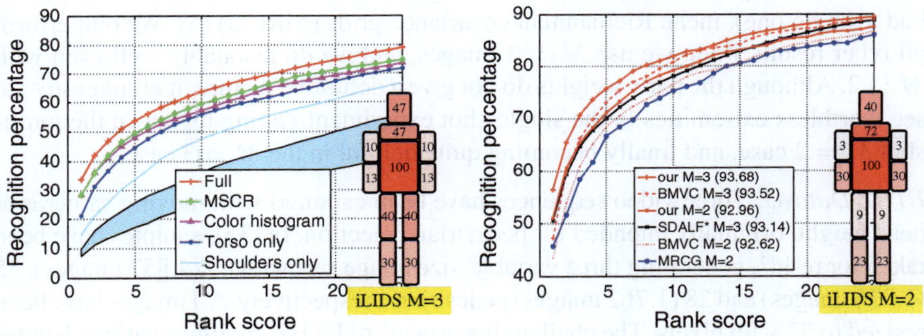

Fig. 7.9 (*Left*) Breakdown of our iLIDS multi-shot experiment showing the performance of the full distance, only the MSCR, only the color histograms, separately for the torso and shoulders parts (the shaded region contains the other parts curves). (*Right*) Comparison with the state of the art for multi-shot on iLIDS

illumination changes and occlusions. On average, each individual has four images, with some ones having only two. In the single-shot case, we reproduce the same experimental settings of [14, 39]: we randomly select one image for each pedestrian to build the gallery set, while all the remaining images (360) are used as probes. This is repeated 10 times, and the average CMC is displayed in Fig. 7.8 (right): we outperform all methods except for PRSVM [28], where the comparison is slightly unfair due to a completely different validation setup (learning-based). We do well compared to a covariance-based technique (SCR) [4] and the Context-based strategy of [39], which is also learning-based.

As for the multi-shot case, we follow a multi-vs-multi matching policy introduced in [14], where both probe and gallery sets have groups of M images per individual. We obtain our best result with $M = 3$, shown in Fig. 7.9 (left): the full distance combines individually good performances of the MSCR and color histogram distances detailed in Sect. 7.4.5; also of note is that torso and shoulders are far more reliable than the

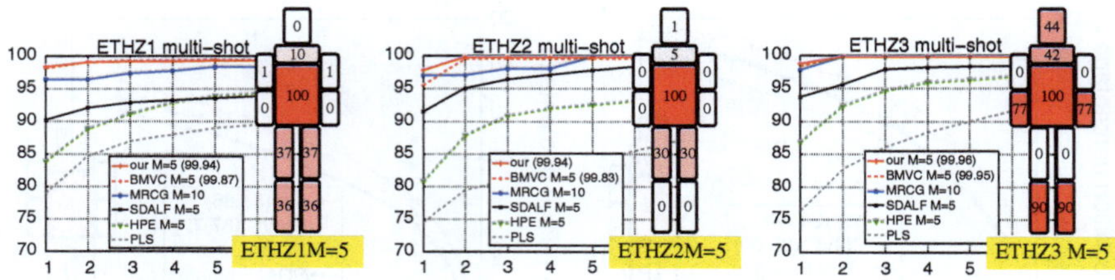

Fig. 7.10 Results of multi-shot experiments on the ETHZ sequnces

other parts, even though, the high importance given to thighs and legs (see puppet) indicates a good support role in difficult matches.

In Fig. 7.9 (right), we compare our multi-shot results with SDALF* (obtained in the multi-vs-single modality $M = 3$, where galleries had three signatures and probes had a single one), mean Riemannian covariance grids (MRCG) [5]. We outperform all other results when we use $M = 3$ images, and we do resonably well even with $M = 2$. Although the parts weights do not give a definitive picture, it is suggestive to see worthless extremities in the single-shot experiment getting higher in the multi-shot $M = 2$ case, and finally becoming quite helpful in the $M = 3$ case.

ETHZ Dataset: Three video sequences have been captured with moving cameras at head height, originally intended for pedestrian detection. In [33], samples have been taken for re-id[3], generating three variable-size image sets with 83 (4.857 images), 35 (1.936 images) and 28 (1.762 images) pedestrians, respectively. All images have been resized to 32×96 pixels. The challenging aspects of ETHZ are illumination changes and occlusions, and while the moving camera provides a good range of variations in people's appearances, the poses are rather few. Nevertheless, our approach is very close to obtaining perfect scores with $M = 5$. See Fig. 7.10 for a comparison with MCRG, SDALF and HPE. Note how the part weights behave rather strangely in ETHZ3: since the part weights are tuned on a particular tuning subset of the dataset, if this happens to give perfect re-id on a wide range of values for the parameters, then it is highly likely that they turn up some unreasonable values. In fact, checking the breakdown of the performances, it is apparent that the torso alone is able to re-id at 99.85 %.

CAVIAR for re-id Dataset: CAVIAR4REID[4] has been introduced in [9] to provide a challenging real-world setup. The images have been cropped from CAVIAR video frames recorded by two different cameras in an indoor shopping center in Lisbon. Of the 72 different individuals identified (with images varying from 17×39 to 72×144), 50 are captured by both views and 22 from only one camera. In our experiments, we reproduce the original setup: focusing only on the 50 double-camera subjects, we select M images from the first camera for the probe set and M images from the

[3] http://www.umiacs.umd.edu/~schwartz/datasets.html

[4] Available at http://www.re-identification.net/.

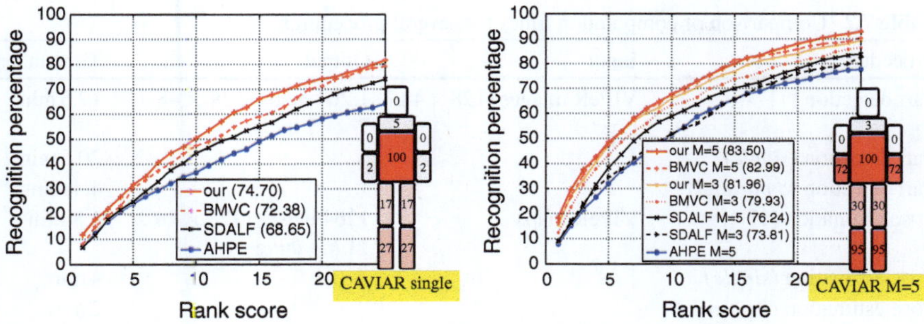

Fig. 7.11 Results of single-shot and multi-shot experiments on CAVIAR4REID

second camera as the gallery set, and then perform multi-shot re-id (called Camera-based multi-vs-multi, or CMvsM in short). All images are resized to 32×96 pixels. Both in single-shot and multi-shot, we outperform our previous results, SDALF (see Fig. 7.11) and AHPE [8]. The part weights suggest relatively poor conditions, due to low resolution and low appearance specificity, even though the pose estimation is fine thanks to decent contrast and low clutter.

Computation Time: All experiments were run on a machine with one CPU (2.93 Ghz, 8 cores) and 8 GB of RAM. The implementation was done in MATLAB (except for the MSCR algorithm), using the facilities of the Parallel Computing toolbox to take advantage of the multi-core architecture. To establish a baseline, experiments on the VIPeR dataset with our approach initially require for each of the 1,264 images: part detection to extract probability maps of size $128 \times 48 \times N_r \times N_p$ ($N_r = 7$ number of orientation angles, $N_p = 11$ number of parts), pose estimation, and feature extraction. Then, we calculate distances between all probes and galleries to produce a 632×632 matrix, and compute the matching and associated CMC curves for 10 trial runs of randomly chosen 316 subjects. The time taken by the last step is negligible since it is simply a matter of selecting and sorting distances, and can be safely ignored in this report.

We took the publicly available C++ source code of [1] and compiled it under Windows (after suitable adjustments), to compare against our approach: its part detection with SHAPE descriptors and AdaBoost is faster than our pure MATLAB code, while the pose estimation is slower because it provides full marginal posteriors (useful in other contexts than re-id) against our MAP estimates. We also report the speed of our approach when activating eight parallel workers in MATLAB, noting that the C++ implementation can also run parallel processes. The time taken by distance calculations heavily depends on the distance being used: Bhattacharyya, Hellinger and L2 can be fully vectorized and take less than 1 s, χ^2 and L1 are slower, and distances like Earth Mover Distance are basically unpractical.

Running the full experiment on VIPeR takes approximately 30 min in single-thread mode, and 12 min using eight parallel workers (see Table 7.2). Training the background model for the HOG-LDA detectors takes approximately 3 h but it is done

Table 7.2 Comparison of computation times for several procedures

Procedure	Input	Output	Time taken
Part detection [1]	VIPeR images (128 × 48 *pixels*)	1,264 maps (128 × 48 × 7 × 11 *mats*)	12.5 min
Part detection *(single)*			20.5 min
Part detection *(8 parallel)*			4.4 min
Pose estimation [1]	VIPeR maps	1264 masks (128 × 48 × 11 *bin images*)	6.8 min
Pose estimation *(single)*			4 min
Pose estimation *(8 parallel)*			2 min
GC extraction	VIPeR images+masks	1264 hists (210 × 11 *mats*)	11–13 sec
MSCR extraction	VIPeR images+masks	1264 blobs lists	30 sec
MSCR dist. calculation	VIPeR blobs	632 × 632 mat	3.5–5 min

once for all detectors (even future ones for different parts or objects, as detailed in [24]), and negligible time for the foreground models. The kinematic prior estimation is also practically instantaneous.

7.7 Conclusions

When we approach the problem of person re-id from a human point of view, it is reasonable to exploit our prior knowledge about person appearances: that they are decomposable into articulated parts, and that matching can be carried out per part as well as on the whole. Thus, we proposed a framework for estimating the local configuration of body parts using PS, introducing novel part detectors that are easy and fast to train and to apply.

In our methodology and experimentation, we strove to devise discriminating and robust signatures for re-id. We currently settled on color histograms and MSCR features because of their speed and accuracy, but the overall framework is not dependent on them, and could be further enhanced. In fact, we plan to publicly release the source code of our system[5] as an incentive for more comparative discussions.

Acknowledgments This work was supported by Hankuk University of Foreign Studies Research Fund of 2013.

[5] Available at http://san.hufs.ac.kr/~chengds/software.html.

References

1. Andriluka, M., Roth, S., Schiele, B.: Pictorial structures revisited: people detection and articulated pose estimation. In: IEEE Conference on Computer Vision and Pattern Recognition, pp. 1014–1021 (2009)
2. Bak, S., Charpiat, G., Corvee, E., Bremond, F., Thonnat, M.: Learning to match appearances by correlations in a covariance metric space. In: European Conference on Computer Vision, pp. 806–820 (2012)
3. Bak, S., Corvee, E., Bremond, F., Thonnat, M.: Person re-identification using haar-based and DCD-based Signature. In: 2nd Workshop on Activity Monitoring by Multi-Camera Surveillance Systems (2010)
4. Bak, S., Corvee, E., Bremond, F., Thonnat, M.: Person re-identification using spatial covariance regions of human body parts. In: 7th IEEE International Conference on Advanced Video and Signal-Based Surveillance (2010)
5. Bak, S., Corvee, E., Bremond, F., Thonnat, M.: Multiple-shot human re-identification by mean riemannian covariance grid. In: 8th IEEE International Conference on Advanced Video and Signal-Based Surveillance, pp. 179–184 (2011)
6. Barbosa, I.B., Cristani, M., Del Bue, A., Bazzani, L., Murino, V.: Re-identification with rgb-d sensors. In: Fusiello, A., Murino, V., Cucchiara, R. (eds.) European Conference on Computer Vision: Workshops and Demonstrations. Lecture Notes in Computer Science, vol. 7583, pp. 433–442. Springer, Berlin, Heidelberg (2012)
7. Bazzani, L., Cristani, M., Murino, V.: Symmetry-driven accumulation of local features for human characterization and re-identification. Comput. Vis. Image Underst. **117**(2), 130–144 (2013)
8. Bazzani, L., Cristani, M., Perina, A., Murino, V.: Multiple-shot person re-identification by chromatic and epitomic analyses. Pattern Recogn. Lett. **33**(7), 898–903 (2012)
9. Cheng, D.S., Cristani, M., Stoppa, M., Bazzani, L., Murino, V.: Custom pictorial structures for re-identification. In: British Machine Vision Conference, pp. 1–11 (2011)
10. Dalal, N., Triggs, B.: Histograms of oriented gradients for human detection. In: IEEE Conference on Computer Vision and Pattern Recognition, vol. 1, pp. 886–893 (2005)
11. Doretto, G., Sebastian, T., Tu, P., Rittscher, J.: Appearance-based person reidentification in camera networks: problem overview and current approaches. J. Ambient Intell. Hum. Comput. **2**(2), 127–151 (2011)
12. Eichner, M., Ferrari, V.: Better appearance models for pictorial structures. In: British Machine Vision Conference (2009)
13. Everingham, M., Van Gool, L., Williams, C.K.I., Winn, J., Zisserman, A.: The PASCAL Visual Object Classes Challenge 2010 (VOC2010) Results (2010)
14. Farenzena, M., Bazzani, L., Perina, A., Murino, V., Cristani, M.: Person re-Identification by symmetry-driven accumulation of local features. In: IEEE Conference on Computer Vision and Pattern Recognition (2010)
15. Felzenszwalb, P.F., Girshick, R.B., McAllester, D., Ramanan, D.: Object detection with discriminatively trained part-based models. IEEE Trans. Pattern Anal. Mach. Intell. **32**(9), 1627–1645 (2010)
16. Felzenszwalb, P.F., Huttenlocher, D.P.: Pictorial structures for object recognition. Int. J. Comput. Vision **61**(1), 55–79 (2005)
17. Figueira, D., Bazzani, L., Minh, H., Cristani, M., Bernardino, A., Murino, V.: Semi-supervised multi-feature learning for person re-identification. In: International Conference on Advanced Video and Signal-Based Surveillance (2013)
18. Forssén, P.E.: Maximally stable colour regions for recognition and matching. In: IEEE Conference on Computer Vision and Pattern Recognition (2007)
19. Freund, Y., Schapire, R.: A decision-theoretic generalization of on-line learning and an application to boosting. J. Comput. Syst. Sci. **55**(1), 119–139 (1997)

20. Gheissari, N., Sebastian, T.B., Tu, P.H., Rittscher, J., Hartley, R.: Person reidentification using spatiotemporal appearance. In: IEEE Conference on Computer Vision and Pattern Recognition, vol. 2, pp. 1528–1535 (2006)
21. Gray, D., Brennan, S., Tao, H.: Evaluating appearance models for recognition, reacquisition and tracking. In: IEEE International Workshop on Performance Evaluation for Tracking and Surveillance (2007)
22. Gray, D., Tao, H.: Viewpoint invariant pedestrian recognition with an ensamble of localized features. In: European Conference on Computer Vision, pp. 262–275 (2008)
23. Hamdoun, O., Moutarde, F., Stanciulescu, B., Steux, B.: Person re-identification in multi-camera system by signature based on interest point descriptors collected on short video sequences. In: ACM/IEEE International Conference on Distributed Smart Cameras, pp. 1–6 (2008)
24. Hariharan, B., Malik, J., Ramanan, D.: Discriminative decorrelation for clustering and classification. In: European Conference on Computer Vision, pp. 459–472 (2012)
25. Lin, Z., Davis, L.: Learning pairwise dissimilarity profiles for appearance recognition in visual surveillance. In: 4th International Symposium on Advances in Visual Computing (2008)
26. Minh, H.Q., Bazzani, L., Murino, V.: A unifying framework for vector-valued manifold regularization and multi-view learning. In: 30th International Conference on Machine Learning (2013)
27. Platt, J.: Probabilistic outputs for support vector machines and comparisons to regularized likelihood methods. Adv. Large Margin Classif. **10**(3), 61–74 (1999)
28. Prosser, B., Zheng, W., Gong, S., Xiang, T.: Person re-identification by support vector ranking. In: British Machine Vision Conference (2010)
29. Ramanan, D.: Learning to parse images of articulated bodies. In: Advances in Neural Information Processing Systems, pp. 1129–1136 (2007)
30. Ramanan, D., Forsyth, D.A., Zisserman, A.:Tracking people by learning their appearance. IEEE Trans. Pattern Anal. Mach. Intell. **29**(1), 65–81 (2007)
31. Salvagnini, P., Bazzani, L., Cristani, M., Murino, V.: Person re-identification with a ptz camera: an introductory study. In: IEEE International Conference on Image Processing (2013)
32. Satta, R., Fumera, G., Roli, F., Cristani, M., Murino, V.: A multiple component matching framework for person re-identification. In: 16th International Conference on Image Analysis and Processing, ICIAP'11, pp. 140–149. Springer-Verlag, Berlin, Heidelberg. http://dl.acm.org/citation.cfm?id=2042703.2042719 (2011)
33. Schwartz, W., Davis, L.: Learning discriminative appearance-based models using partial least squares. In: XXII SIBGRAPI 2009 (2009)
34. Shapire, R., Freund, Y., Bartlett, P., Lee, W.: Boosting the margin: a new explanation for the effectiveness of voting methods. Ann. Stat. **26**(5), 1651–1686 (1998)
35. Sivic, J., Zitnick, C.L., Szeliski, R.: Finding people in repeated shots of the same scene. In: British Machine Vision Conference (2006)
36. Wang, X., Doretto, G., Sebastian, T.B., Rittscher, J., Tu, P.H.: Shape and appearance context modeling. In: IEEE International Conference on Computer Vision, pp. 1–8 (2007)
37. Wang, X., Han, T.X., Yan, S.: An hog-lbp human detector with partial occlusion handling. In: IEEE International Conference on Computer Vision, pp. 32–39 (2009)
38. Yang, Y., Ramanan, D.: Articulated human detection with flexible mixtures-of-parts. IEEE Trans. Pattern Anal. Mach. Intell. **PP**(99), 1–1 (2012)
39. Zheng, W., Gong, S., Xiang, T.: Associating groups of people. In: British Machine Vision Conference (2009)
40. Zheng, W.S., Gong, S., Xiang, T.: Reidentification by relative distance comparison. IEEE Trans. Pattern Anal. Mach. Intell. **35**(3), 0653–668 (2013)

Chapter 8
One-Shot Person Re-identification with a Consumer Depth Camera

Matteo Munaro, Andrea Fossati, Alberto Basso, Emanuele Menegatti and Luc Van Gool

Abstract In this chapter, we propose a comparison between two techniques for one-shot person re-identification from soft biometric cues. One is based upon a descriptor composed of features provided by a skeleton estimation algorithm; the other compares body shapes in terms of whole point clouds. This second approach relies on a novel technique we propose to warp the subject's point cloud to a standard pose, which allows to disregard the problem of the different poses a person can assume. This technique is also used for composing 3D models which are then used at testing time for matching unseen point clouds. We test the proposed approaches on an existing RGB-D re-identification dataset and on the newly built *BIWI RGBD-ID* dataset. This dataset provides sequences of RGB, depth, and skeleton data for 50 people in two different scenarios and it has been made publicly available to foster advancement in this new research branch.

M. Munaro (✉) · A. Basso · E. Menegatti
Intelligent Autonomous Systems Laboratory, University of Padua, Via Gradenigo 6a, 35131 Padua, Italy
e-mail: munaro@dei.unipd.it

A. Basso
e-mail: bassoal1@dei.unipd.it

E. Menegatti
e-mail: emg@dei.unipd.it

A. Fossati · L. Van Gool
Computer Vision Laboratory, ETH Zurich, Sternwartstrasse 7, 8092 Zurich, Switzerland
e-mail: fossati@vision.ee.ethz.ch

L. Van Gool
e-mail: vangool@vision.ee.ethz.ch

S. Gong et al. (eds.), *Person Re-Identification*,
Advances in Computer Vision and Pattern Recognition,
DOI: 10.1007/978-1-4471-6296-4_8,

8.1 Introduction

The task of identifying a person that is in front of a camera has plenty of important practical applications: Access control, video surveillance, and people tracking are a few examples of such applications.

The computer vision problem that we tackle in this chapter is inside the branch of noninvasive and noncooperative biometrics. This implies no access to more reliable and discriminative data such as the DNA sequence and fingerprints, but simply rely on the input provided by a cheap consumer depth camera.

We decided to take advantage of a depth-sensing device to overcome a few shortcomings intrinsically present in standard video-based re-identification. These include for example noninvariance to different viewpoints and lighting conditions, in addition to being very sensitive to clothing appearance. On the other hand, the known disadvantages of consumer depth cameras, i.e., sensitivity to solar infrared light and limited functioning range, do not usually constitute a problem in standard re-identification scenarios.

The set of features that we adopt to identify a specific person are commonly known as soft biometrics. This means that each feature alone is not a univocal identifier for a certain subject. Still, the combination of several soft biometrics features can show a very good discriminative performance even within large sets of persons.

We take into account both skeleton lengths and the global body shape to be able to describe a subject's identity. Moreover, we extract also facial features for comparison purposes. All the necessary information is collected using a single device, namely a Microsoft Kinect. Given that a body shape can vary also because of the different poses the subject can assume, we warp every point cloud back to a standard pose before comparing them.

Both the approaches we propose in this chapter aim at a one-shot re-identification. After a training phase, during which the classifier parameters or the training models are learned for each of the subjects in the dataset, the system is able to estimate the ID label of detected people separately for each input frame, in real time. To improve the robustness in the estimation, the output of multiple consecutive frames can be easily integrated, for example using a voting scheme.

The contributions of this chapter are three-fold: First we propose a novel technique for exploiting skeleton information to transform persons' point clouds to a standard pose in real time. Moreover, we explain how to use this transformed point clouds for composing 3D models of moving people which can be used for re-identification by means of an ICP matching with new test clouds and we compare this approach with feature-based approaches which classify skeleton and face descriptors. Finally, we present a novel biometrics RGB-D dataset including 50 subjects: For each subject, we provide a sequence including standard video input, depth input, a segmentation mask, and the skeleton as provided by the Kinect SDK. Additionally, the dataset includes several labeled testing sequences collected in a different scenario.

8.2 State of the Art

As cheap depth-sensing devices have started appearing in the market only very recently, the literature in this specific field is quite limited. We will first introduce several vision-based soft biometrics approaches, and then analyze in more detail a few depth-based identification techniques.

The integration of multimodal cues for person identification is an active research topic since the 1990s [14]. For example, in [10] the authors integrate a voice-based system with face recognition using hyper basis function networks. The concept of information fusion in biometrics has been methodically studied in [22], in which the authors propose several different architectures to combine multiple vision-based modalities: Fusion can happen at the feature extraction level, which usually consists in concatenating the input feature vectors. Otherwise there can be fusion at the matching score level, by combining the scores of the different subsystems, or at the decision level, i.e., each subsystem takes a decision and then decisions are combined, for example through a majority voting scheme.

Most vision-based systems fall in the category of soft biometrics, which are defined to be a set of characteristics that provide some biometric information, but are not able to individually authenticate the person, mainly due to lack of distinctiveness and permanence [15].

Vision-based biometrics systems can be either collaborative, as for example iris recognition or fingerprint analysis, or noncollaborative. We will mainly focus on noncollaborative traits, as they are more generally applicable and easier to process: Face-based identification is a deeply studied topic in the computer vision literature [34]. Efforts have been spent in making it more robust to different alignment and illumination conditions [28], and to small training set sizes [35]. The problem has also been tackled in a real-time setup [1, 16] and from a 3D perspective [7]. Another type of vision-based analysis that has been used for people identification is gait recognition [21, 31], which can be either model-based, i.e., a skeleton is first fitted to the data, or model-free, for example by analyzing directly silhouettes. This is by definition a soft biometrics, as it is in general not discriminative enough to identify a subject, but can be very powerful if combined with other traits. Finally, visual techniques have also been proposed, which try to re-identify a subject based on a global appearance model [4, 5, 29]. The intrinsic drawback of such approaches is that they can only be applied to tracking scenarios and are not suitable for long time-span recognition.

As mentioned above, due to the very recent availability of cheap depth-sensing devices, only a few works exist that focused on identification using such multimodal input. In [19], it is shown that anthropometric measures are discriminative enough to obtain a 97 % accuracy on a population of 2,000 subjects. The authors apply Linear Discriminant Analysis to very accurate laser scans to obtain such performance. Also the authors of [26] studied a similar problem. They in fact used a few anthropometric features, manually measured on the subjects, as a preprocessing pruning step to make face-based identification more efficient and reliable. In [23], the authors have recently proposed an approach which uses the input provided by a network of Kinect cameras: the depth data in their case are only used for segmentation, while their re-identification techniques purely rely on appearance-based features. The authors of [2] propose a method that relies only on depth data, by extracting a signature for each subject. Such signature includes features extracted from the skeleton as the lengths of a few limbs and the ratios of some of these lengths. In addition, geodesic distances on the body shape between some pairs of body joints are considered. The choice of the most discriminative features is based upon experiments carried on a validation dataset. The signatures are extracted from a single training frame for each subject, which renders the framework quite prone to noise, and weighted Euclidean distances are used to compute distances between signatures. The weights of the different feature channels are simply estimated through an exhaustive grid search. The dataset used in the chapter has also been made publicly available, but this does not contain facial information of the subjects, in contrast with the dataset proposed within this chapter. Also Kinect Identity [17], the software running on the Kinect for XBox360, uses multimodal data, namely the subject's height, a face descriptor, and a color model of the user's clothing to re-identify a player during a gaming session. In this case, though, the problem is simplified as such re-identification only covers a very short time span and the number of different identities is usually very limited.

8.3 Datasets

With the recent availability of cheap depth sensors, a lot of effort in the computer vision community has been put into collecting novel datasets. In particular, several groups have proposed databases of human motions, usually making available skeleton and depth data in conjunction with regular RGB input [18, 20, 25, 30, 32, 33]. Nonetheless, the vast majority of these are focusing on human activity analysis and action recognition, and for this reason they are generally composed by many gestures performed by few subject.

On the other hand, the problem we tackle in this chapter is different and requires data relative to many different subjects, while the number of gestures is not crucial. From this perspective, only a dataset has been proposed so far [2]. It consists of 79 different subjects collected in four different scenarios. The collected information, for each subject and for each scenario, includes five RGB frames (in which the face has been blurred), the foreground segmentation mask, the extracted skeleton, the corresponding 3D mesh, and an estimation of the ground plane. This dataset contains very few frames for each subject, thus machine learning approaches can be hardly tested because of the little data available for training a person classifier. Moreover, the faces of the recorded subjects have been blurred for privacy reasons, making the comparison with a baseline built upon face recognition impossible.

8.3.1 *BIWI RGBD-ID Dataset*

To perform more extensive experiments on a larger amount of data we also collected our own RGB-D Identification dataset called *BIWI RGBD-ID*.[1] It consists of video sequences of 50 different subjects, performing a certain routine of motions in front of a Kinect, such as a rotation around the vertical axis, several head movements, and two walks toward the camera. The dataset includes synchronized RGB images (captured at the highest resolution possible with the Kinect, i.e. 1,280 $\times$ 960 pixels), depth images, persons' segmentation maps, and skeletal data (as provided by the Kinect SDK), in addition to the ground plane coordinates. These videos have been acquired at about 10fps and last about one minute for every subject.

Moreover, we have collected 56 testing sequences with 28 subjects already present in the dataset. These have been collected on a different day and therefore most subjects are dressed differently. These sequences are also shot in different locations than the studio room where the training dataset had been collected. For every person in the testing set, a *Still* sequence and a *Walking* sequence have been collected. In the *Walking* video, every person performs two walks frontally and two other walks diagonally with respect to the Kinect.

8.4 Approach

The framework we have designed allows to identify a subject standing in front of a consumer depth camera, taking into account a single input frame. To achieve this goal, we consider two different approaches. In the former, a descriptor is computed from the body skeleton information provided by the Microsoft Kinect SDK [24] and fed to a pretrained classifier. In the latter, we compare people point clouds by means of the fitness score obtained after an Iterative Closest Point (ICP) [6] registration. For tackling the problem of different poses people can have, we exploit the skeleton information for transforming a person point cloud to a standard pose before applying ICP.

8.4.1 *Feature-Based Re-identification*

In this section, our feature-based approach to person re-identification is described. In a first phase, as a subject is detected in front of the depth-sensing device, the descriptor is extracted from the input channels. Our feature extraction step relies on the body skeleton obtained through the Kinect SDK, since the data are already available and computation is optimized.

[1] The *BIWI RGBD-ID* dataset can be downloaded at: http://robotics.dei.unipd.it/reid.

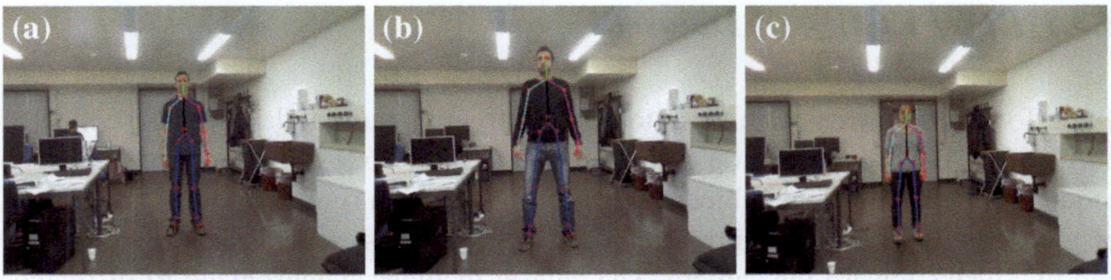

Fig. 8.1 Pictorial illustration of the skeleton features composing the skeleton descriptor

Skeleton Descriptor

The extraction of skeleton-based information is substantially the computation of a few limb lengths and ratios, using the 3D location of the body joints provided by the skeletal tracker. We extended the set of skeleton features used in [2], in order to collect measurements from all the human body. In particular, we extract the following 13 distances:

(a) head height,
(b) neck height,
(c) neck to left shoulder distance,
(d) neck to right shoulder distance,
(e) torso to right shoulder distance,
(f) right arm length,
(g) left arm length,
(h) right upper leg length,
(i) left upper leg length,
(j) torso length,
(k) right hip to left hip distance,
(l) ratio between torso length and right upper leg length (j/h),
(m) ratio between torso length and left upper leg length (j/i).

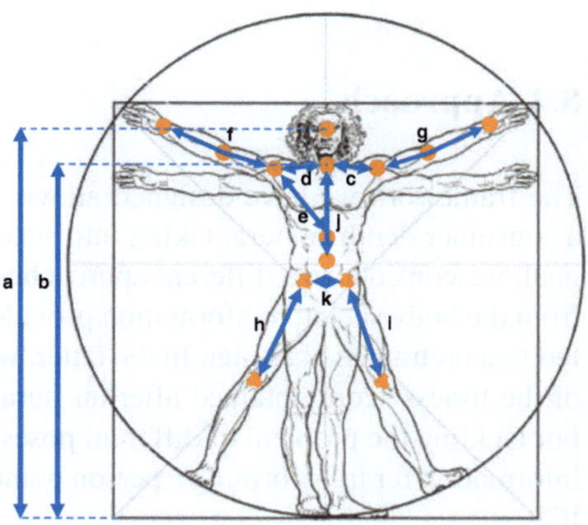

All these distances are concatenated into a single skeleton descriptor $\mathbf{x}_S$. In Fig. 8.1, the skeleton computed with Microsoft Kinect SDK is reported for three very different people of our dataset, while in Figs. 8.2 and 8.3, we show how the value of some skeleton features varies along time when these people are still and walking, respectively. We also report the average standard deviation of these features for the people of the two testing sets. As expected, the heights of the head and the neck from the ground are the most discriminative features. What is more interesting is that the standard deviation of these features doubles for the walking test set with respect to the test set where people are still, thus suggesting that the skeleton joint positions are better estimated when people are static and frontal.

When a person is seen from the side or from the back, Microsoft's skeletal tracking algorithm [24] does not provide correct estimates because it is based on a random

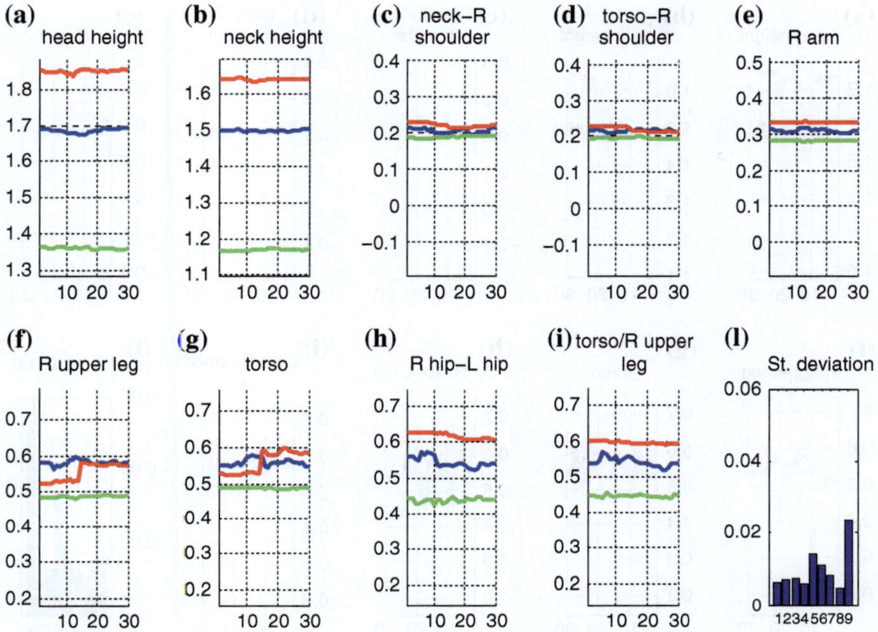

Fig. 8.2 **a–i** Estimated skeleton features for some frames of the *Still* test sequence for the three subjects of Fig. 8.1. Those subjects are represented by *blue*, *red*, and *green curves*, respectively. In **l**, the standard deviation of these features is reported

forest classifier which has been trained with examples of frontal people only. For this reason, in this work, we discard frames with at least one not tracked joint.[2] Then, we keep only those where a face is detected [27] in the proximity of the head joint position. This kind of selection is needed for discarding also those frames where the person is seen from the back, which come with a wrong skeleton estimation.

Classification

For classifying the descriptor presented in the previous section, we tested four different classification approaches. The first method compares descriptors extracted from the testing dataset with those of the training dataset by means of a Nearest-Neighbor classifier based on the Euclidean distance. The second one consists in learning the parameters of a Support Vector Machine (SVM) [11] for every subject of the training dataset. As SVMs are originarily designed for binary classification, these classifiers are trained in a One-vs-All fashion: For a certain subject i, the descriptors computed on that subject are considered as positive samples while the descriptors computed on all the subjects except i are considered as negative samples.

The One-vs-All approach requires all the training procedure to be performed again if a new person is inserted in the database. This need makes the approach

[2] Microsoft's SDK provides a flag for every joint stating if it is tracked, inferred, or not tracked.

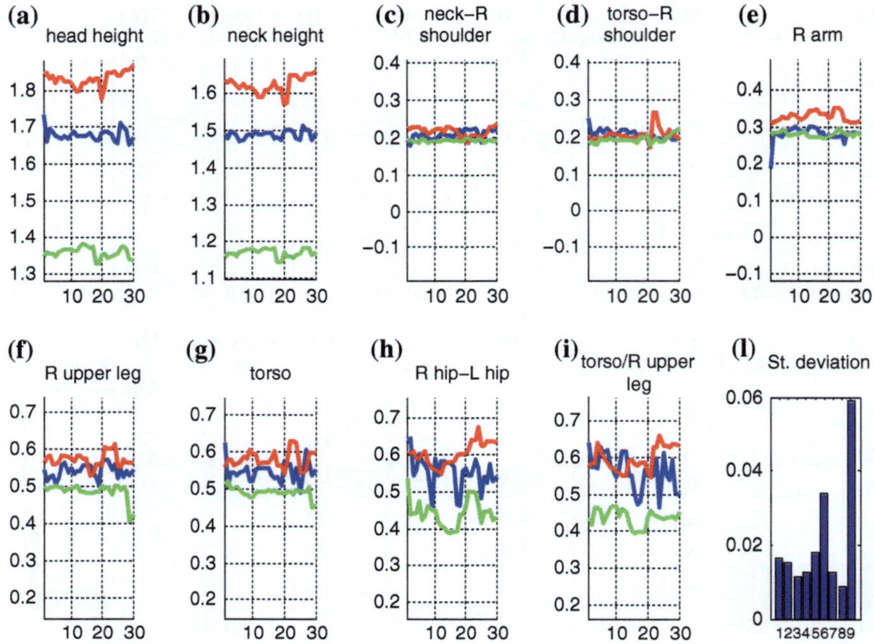

Fig. 8.3 **a–i** Estimated skeleton features for some frames of the *Walking* test sequence for the three subjects of Fig. 8.1. Those subjects are represented by *blue*, *red*, and *green curves*, respectively. In **l**, the standard deviation of these features is reported

not suitable for a scenario where new people are inserted online for a subsequent re-identification. For this purpose, we also trained a *Generic SVM* which does not learn how to distinguish a specific person from all the others, but it learns how to understand if two descriptors have been extracted from the same person or not. The positive training examples which are fed to this SVM are of the form

$$\text{pos} = \left| d_1^i - d_2^i \right|, \tag{8.1}$$

where d_1^i and d_2^i are descriptors extracted from two frames containing the same subject i, while the negative examples are of the form

$$\text{neg} = \left| d_1^i - d_2^j \right|, \tag{8.2}$$

where d_1^i and d_2^j are descriptors extracted from frames containing different subjects. At testing time, the current descriptor d_{test} is compared to the training descriptors d_k^i of every subject i by using this Generic SVM for classifying the vector $|d_{\text{test}} - d_k^i|$ and the test descriptor is associated to the class for which the maximum SVM confidence is obtained.

Finally, we tested also a Naive Bayes approach: as a training stage, we computed mean and standard deviation of a normal distribution for every descriptor feature and for every person of the training dataset; at testing time, we used these data to calculate the likelihood with which a new descriptor could belong to each person in the training set.

8.4.2 Point Cloud Matching

The skeleton descriptor explained in Sect. 8.4.1 provides information about the characteristic lengths of the human body. However, it does not take into account many shape traits which are important for discriminating people with similar body lengths. In this section, we propose a process which takes the whole point cloud shape into account for the re-identification task. In particular, given two persons' point clouds, we try to align them and then compute a similarity score between the two. As a fitness score, we compute the average distance of the points of a cloud to the nearest points of the other cloud. If $\mathcal{P}_1$ and $\mathcal{P}_2$ are two point clouds, the fitness score of $\mathcal{P}_2$ with respect to $\mathcal{P}_1$ is then

$$f_{2\rightarrow 1} = \sum_{p_i \in \mathcal{P}_2} \left\| p_i - q_i^* \right\|, \tag{8.3}$$

where q_i^* is defined as

$$q_i^* = \arg\min_{q_j \in \mathcal{P}_1} \left\| p_i - q_j \right\|. \tag{8.4}$$

It is worth noticing that this fitness score is not symmetric, that is $f_{2\rightarrow 1} \neq f_{1\rightarrow 2}$.

For what concerns the alignment, the position and orientation of a reference skeleton joint, e.g., the hip center, is used to perform a rough alignment between the clouds to compare. Then, that alignment is refined by means of an ICP-based registration, which should converge in few iterations if the initial alignment is good enough. When the input point clouds have been aligned with this process, the fitness score between them should be minimum, ideally zero if they coincide or if $\mathcal{P}_2$ is contained in $\mathcal{P}_1$.

For the purpose of re-identification, this procedure can be used to compare a testing point cloud with the point clouds of the persons in the training set and to select the subject whose point cloud has the minimum fitness score when matched with the testing cloud. However, for this approach to work well, a number of problems should be taken into account, such as the quality of the depth estimates and the different poses people can assume.

(a)

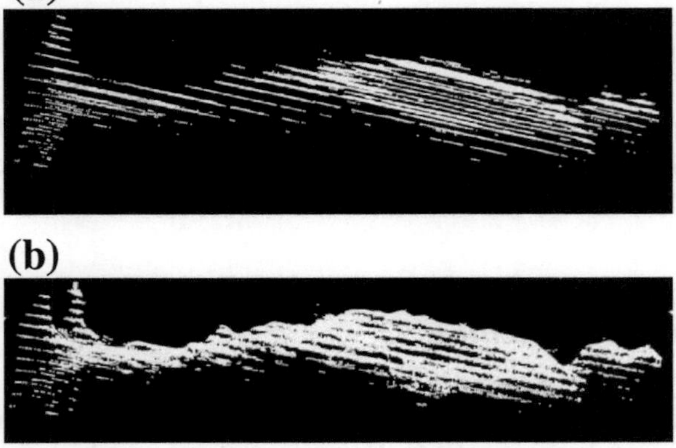

(b)

Fig. 8.4 **a** Raw person point cloud at 3 m of distance from the Kinect and **b** point cloud after the preprocessing step

Point Cloud Smoothing

3D point clouds acquired with consumer depth sensors have good resolution but the depth quantization step increasing quadratically with the distance does not allow to obtain smooth people point clouds beyond two meters from the sensor. In Fig. 8.4a, the point cloud of a person three meters from the sensor is reported. It can be noticed that the point cloud results are divided into slices produced by the quantization steps. As a preprocessing step, we improve the person point cloud by applying a voxel grid filter and a Moving Least Squares surface reconstruction method to obtain a smoothing, as reported in Fig. 8.4b.

Point Cloud Transformation to Standard Pose

The point cloud matching technique we described is derived from the 3D object recognition research, where objects are supposed to undergo rigid transformations only. However, when dealing with moving people, the rigidity assumption does not hold any more, because people are articulated and they can appear in a very large number of different poses, thus these approaches would be doomed to fail.

Bronstein et al. [8, 9] tackle this problem by applying an isometric embedding which allows to get rid of pose variability (extrinsic geometry) by warping shapes to a canonical form where geodesic distances are replaced by Euclidean ones. In this space, an ICP matching is applied to estimate similarity between shapes. However, a geodesic masking which retains the same portion of every shape is needed for this method to work well. In particular, for matching people's shape, a complete and accurate 3D scan has to be used, thus partial views cannot be matched with a full model because they could lead to very different embeddings. Moreover, this approach needs to solve a complicated optimization problem, thus requiring several seconds to complete.

For these reasons, we studied a new technique which exploits the information provided by the skeleton for efficiently transforming people point clouds to a standard pose before applying the matching procedure. This result is obtained by rototranslating each body part according to the positions and orientations of the skeleton joints and links given by Microsoft's skeletal tracking algorithm.

A preliminary operation consists in segmenting the person's point cloud into body parts. Even if Microsoft's skeletal tracker estimates this segmentation as a first step and then derives the joints position, it does not provide to the user the result of the depth map labeling into body parts. For this reason, we implemented the reverse procedure for obtaining the segmentation of a person point cloud into parts by starting from the 3D positions of the body joints. In particular, we assign every point cloud point to the nearest body link. For a better segmentation of the torso and the arms, we added two further fictitious links between the hips and the shoulders.

Once we performed the body segmentation, we warp the pose assumed by the person to a new pose, which is called *standard pose*. The standard pose makes the point clouds of all the subjects directly comparable, by imposing the same orientation between the links. On the other hand, the joints/links position is person-dependent and is estimated from a valid frame of the person and then kept fixed. This approach allows the standard pose skeleton to adapt to the different body lengths of the subjects. This transformation consists in rototranslating the points belonging to every body part according to the corresponding skeleton link position and orientation.[3] In particular, every body part is rotated according to the corresponding link orientation and translated according to its joints coordinates. If Q_c is the quaternion representing the orientation of a link in the current frame given by the skeleton tracker and Q_s is the one expressing its orientation in standard pose, the whole rotation to apply can be computed as

$$R = Q_s\,(Q_c)^{-1}, \tag{8.5}$$

while the full transformation applied to a point p can be synthesized as

$$p' = T_{V_s}\left(R\left(T_{V_c}^{-1}\,(p)\right)\right), \tag{8.6}$$

where T_{V_c} and T_{V_s} are the translation vectors of the corresponding skeleton joint at the current frame and in the standard pose, respectively.

As the standard pose, we chose a typical frontal pose of a person at rest. In Fig. 8.5, we report two examples of person's point clouds before and after the transformation to standard pose. For the point cloud before the transformation, the body segmentation is shown with colors, while the points with RGB texture are reported for the transformed point cloud.

It is worth noting that the process of rotating each body part according to the skeleton estimation can have two negative effects on the point cloud: some body parts can intersect with each other and some gaps can appear around the joint centers.

[3] It is worth noting that all the links belonging to the torso have the same orientation, as the hip center.

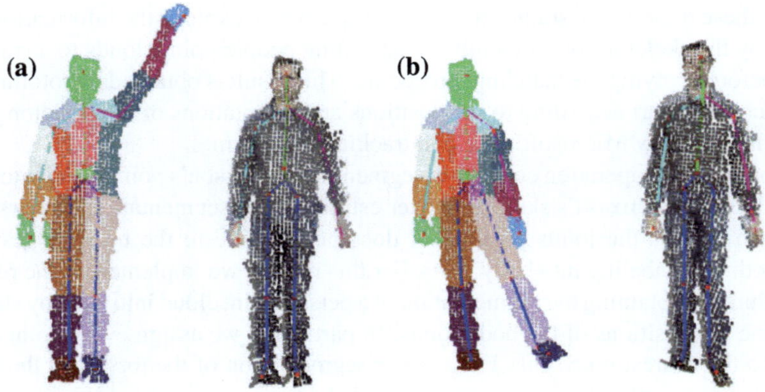

Fig. 8.5 Two examples (a, b) of standard pose transformation. On the *left* of every example, the body segmentation is shown with colors, and on the *right*, the RGB texture is applied to the transformed point cloud

However, the parts intersection is tackled by voxel grid filtering the transformed point cloud, while the missing points do not represent a problem for the matching phase, since a test point cloud is considered to perfectly match a training point cloud if it is fully contained in it, as explained in Sect. 8.4.2.

Creation of Point Cloud Models

The transformation to standard pose is not only useful because it allows to compare people clouds disregarding their initial pose, but also because more point clouds belonging to the same moving person can be easily merged to compose a wider person model. In Fig. 8.6, a single person's point cloud (a) is compared with the model we obtained by merging together some point clouds acquired from different points of view and transformed to standard pose. It can be noticed how the union cloud is denser and more complete with respect to the single one. We also show, in Fig. 8.6c and d, a side view of the person model when no smoothing is performed and when the smoothing of Sect. 8.4.2 is applied. Our approach is not focused on obtaining realistic 3D models for computer graphics, but on creating 3D models which can be useful for the re-identification task. In fact, these models can be used as a reference for matching new test point clouds with the people database. In particular, a point cloud model is created for every person from a sequence of frames where the person is turning around. Then, a new testing cloud can be transformed to standard pose and compared with all the persons' models by means of the approach described in Sect. 8.4.2. Given that, with Microsoft's skeletal tracker, we do not obtain valid frames if the person is seen from the back, we can only obtain 180° people models.

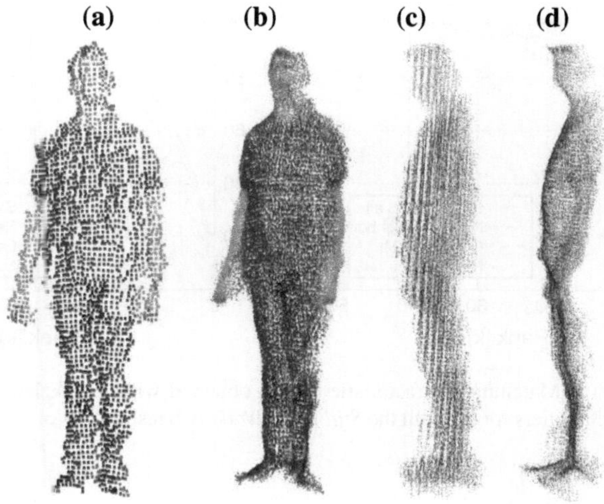

Fig. 8.6 **a** A single person's point cloud and **b** the point cloud model obtained by merging together several point clouds transformed to standard pose. Person's point cloud model **c** before and **d** after the smoothing described in Sect. 8.4.2

8.5 Experiments

In this section, we report the experiments we carried out with the techniques described in Sect. 8.4. For evaluation purposes, we compute *Cumulative Matching Characteristic* (CMC) *Curves* [13], which are commonly used for evaluating re-identification algorithms. For every k from 1 to the number of training subjects, these curves express the mean person recognition rate computed when considering a classification to be correct if the ground truth person appears among the subjects who obtained the k best classification scores. The typical evaluation parameters for these curves are the *rank-1* recognition rate and the *normalized Area Under Curve* (nAUC), which is the integral of the CMC. In this work, the recognition rates are separately computed for every subject and then averaged to obtain the final recognition rate.

8.5.1 Tests on the BIWI RGBD-ID Dataset

We present here some tests we performed on the *BIWI RGBD-ID* dataset. For the feature-based re-identification approach of Sect. 8.4.1, we extracted frame descriptors and trained the classifiers on the 50 sequences of the training set and we used them to classify the *Still* and *Walking* sequences of the 28 people of the testing set. In Fig. 8.7, we report the CMCs obtained on the *Still* and *Walking* testing sets when classifying the skeleton descriptor with the four classifiers described in Sect. 8.4.1. The best classifier for this kind of descriptor proved to be the Nearest Neighbor,

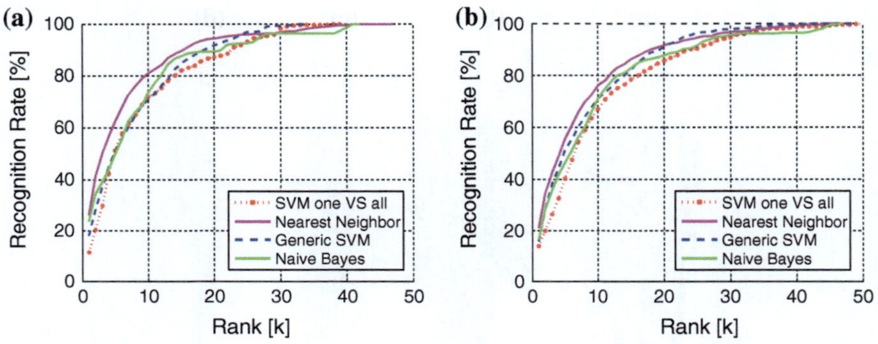

Fig. 8.7 Cumulative Matching Characteristic curves obtained with the skeleton descriptor and different types of classifiers for the both the *Still* **a** and *Walking* **b** testing sets of the *BIWI RGBD-ID* dataset

which obtained a rank-1 recognition rate of 26.6 % and a nAUC of 89.7 % for the testing set where people are still and 21.1 and 86.6 % respectively, for the testing set with walking people.

For testing the point cloud matching approach of Sect. 8.4.2, we built one point cloud model for every person of the training set by merging together point clouds extracted from their training sequences and transformed to standard pose. At every frame, a new cloud is added and a voxel grid filter is applied to the union result for re-sampling the cloud and limiting the number of points. At the end, we exploit a moving least squares surface reconstruction method for obtaining a smoothing. At testing time, every person's cloud is transformed to standard pose, aligned, and compared to the 50 persons' training models and classified according to the minimum fitness score $f_{\text{test}\rightarrow\text{model}}$ obtained. It is worth noticing that the fitness score reported in Eq. 8.3 correctly returns the minimum score (zero) if the test point cloud is contained in the model point cloud, while it would return a different score if the test cloud would only partially overlap the model. Also for this reason, we chose to build the persons' models described above, i.e., by having training models covering 180° while the test point clouds are smaller and for this reason only cover portions of the training point clouds. In Fig. 8.8, we compare the described method with a similar matching method which does not exploit the point cloud transformation to standard pose. For the testing set with still people, the differences are small because people are often in the same pose, while, for the walking test set, the transformation to standard pose outperforms the method which does not exploit it, reaching a rank-1 performance of 22.4 % against 7.4 % and a nAUC of 81.6 % against 64.3 %.

We compare the main approaches we described in Fig. 8.9. As a reference, we report also the results obtained with a face recognition technique. This technique extracts the subject's face from the RGB input using a standard face detection algorithm [27]. To increase the computational speed and decrease the number of false positives, the search region is limited to a small neighborhood of the 2D location of the head, as provided by the skeletal tracker. Once the face has been detected,

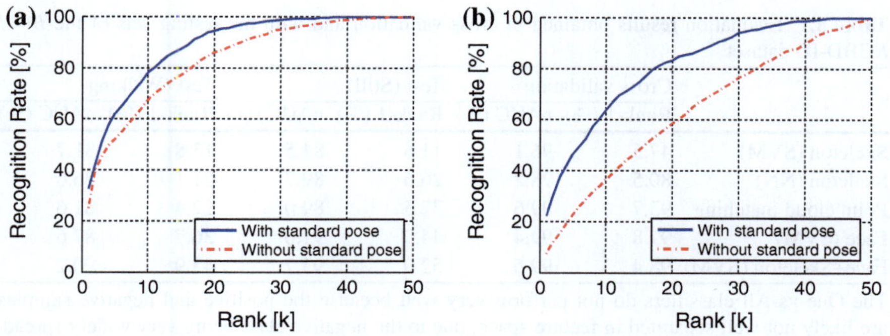

Fig. 8.8 Cumulative Matching Characteristic curves obtained with the point cloud matching approach with and without transformation to standard pose on the testing sets of the *BIWI RGBD-ID* dataset. **a** Still. **b** Walking

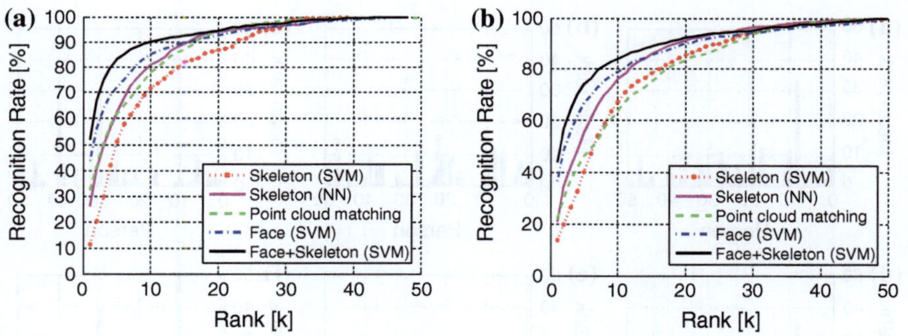

Fig. 8.9 Cumulative Matching Characteristic curves obtained with the main approaches described in this chapter for the *BIWI RGBD-ID* dataset. **a** Still. **b** Walking

a real-time method to extract the 2D location of 10 fiducials points is applied [12]. Finally, SURF descriptors [3] are computed at the location of the fiducials and concatenated forming a single vector. Unlike the skeleton descriptor, the face descriptor provided the best results with the One-vs-All SVM classifier, reaching 44 % of rank-1 for the *Still* testing set and 36.7 % for the *Walking* set. An advantage of the SVM classification is that descriptors referring to different features can be easily fused by concatenating them and leaving to the classifier the task to learn the suitable weights. We report, as an example, the results obtained with the concatenation of the face and skeleton descriptors which are then classified with the One-vs-All SVM approach. This method allows to further gain 8 % of rank-1 for the *Still* test set and 7.2 % for the *Walking* test set. In Table 8.1, all the numerical results are reported, together with those obtained by executing a three-fold cross validation on the training videos where two folds were used for training and one for testing. In the remaining experiments, all the training videos were used for training and all the testing data were used for testing. The point cloud matching technique performs slightly better than

Table 8.1 Evaluation results obtained in cross validation and with the testing sets of the *BIWI RGBD-ID* dataset

	Cross validation		Test (Still)		Test (Walking)	
	Rank-1 (%)	nAUC (%)	Rank-1 (%)	nAUC (%)	Rank-1 (%)	nAUC (%)
Skeleton (SVM)	47.5	96.1	11.6	84.5	13.8	81.7
Skeleton (NN)	80.5	98.2	26.6	89.7	21.1	86.6
Point cloud matching	93.7	99.6	32.5	89.0	22.4	81.6
Face (SVM)	97.8	99.4	44.0	91.0	36.7	87.6
Face+Skeleton (SVM)	98.4	99.5	52.0	93.7	43.9	90.2

The One-vs-All classifiers do not perform very well because the positive and negative samples are likely not well separated in feature space, due to the negative class being very widely spread. Although it is possible that pairwise classifiers may perform better, this would lead to a very large number of classifiers, which may be impractical given the number of classes. This nonseparability at the category level is supported by the good performance of the nearest neighbor classifier, which further suggests that there are overlaps among categories, but locally some classification is possible

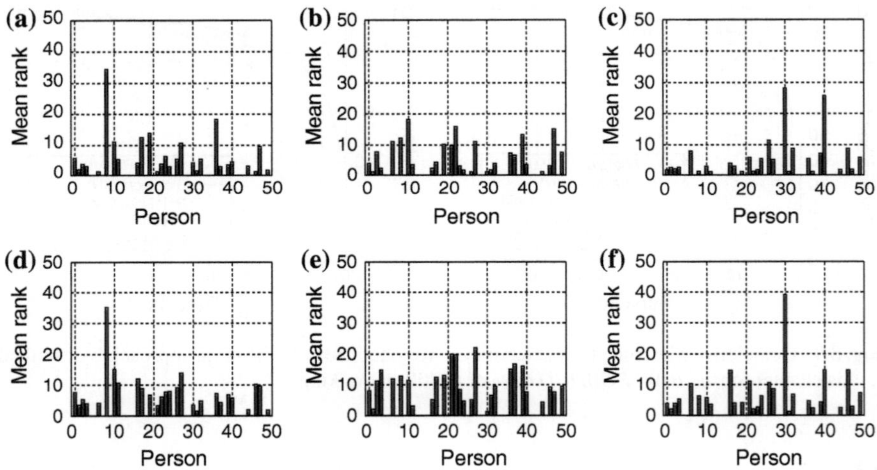

Fig. 8.10 Mean ranking histograms obtained with different techniques for every person of the *Still* (*top row*) and *Walking* (*bottom row*) test sets of the *BIWI RGBD-ID* dataset. **a** Skeleton (NN). **b** Point cloud matching. **c** Face (SVM). **d** Skeleton (NN). **e** Point cloud matching. **f** Face (SVM)

the skeleton descriptor classification for the *Still* test set and slightly worse for the *Walking* test set, thus proving to be useful too for the re-identification task.

For analyzing how the re-identification performance changes for the different people of our dataset, we report in Fig. 8.10 the histograms of the mean ranking for every person of the testing dataset, which is the average ranking at which the correct person is classified. The missing values in the x axis are due to the fact that not all the training subjects are present in the testing set. It can be noticed that there is a correspondence between the mean ranking obtained in the *Still* testing set and that obtained in the *Walking* test set. On the contrary, it is also clear that

Table 8.2 Evaluation results on the *RGB-D Person Re-Identification* dataset

Training	Testing	[2]		Ours (NN)		Ours (Generic SVM)	
		Rank-1	nAUC (%)	Rank-1 (%)	nAUC (%)	Rank-1 (%)	nAUC (%)
Collaborative	Walking1	N/A	90.1	7.8	81.1	5.3	79.0
Collaborative	Walking2	13(%)	88.9	4.8	81.3	4.1	78.6
Collaborative	Backwards	N/A	85.6	4.6	78.8	3.6	76.0
Walking1	Walking2	N/A	91.8	28.6	89.9	35.7	92.8
Walking1	Backwards	N/A	88.7	17.8	82.7	18.5	90.6
Walking2	Backwards	N/A	87.7	13.2	84.1	22.3	91.6

different approaches lead to mistakes on different people, thus showing to be partially complementary.

8.5.2 Tests on the RGB-D Person Re-identification Dataset

As explained in Sect. 8.3, the *RGB-D Person Re-identification* dataset is the only other public dataset for person re-identification using RGB-D data. Unfortunately, there are only a few examples available for each of the subjects, which makes the use of many machine learning techniques, including SVMs trained with a One-vs-All approach, quite complicated. However, given that the Generic SVM described in Sect. 8.4.1 is one for all the subjects, we had enough examples to train it correctly. In Table 8.2, we compare the results reported in [2] with our results obtained when classifying the skeleton descriptor with the Nearest Neighbor and the Generic SVM. Unfortunately, the authors of [2] report performances only in terms of normalized Area Under Curve (nAUC) of the Cumulative Matching Characteristic (CMC) curve, thus their rank-1 scores are not available except for one result that can be inferred from a figure. The classification of our skeleton descriptor with the Generic SVM performed better than [2] and that of our Nearest Neighbor classifier for the tests which do not involve the *Collaborative* set, where people walk with open arms. We also tested the geodesic features the authors propose, but they did not provide substantial improvement to the skeleton alone. We did not test the point cloud matching and the face recognition techniques on this dataset because the links orientation information was not provided and the face in the RGB image was blurred.

8.5.3 Multiframe Results

The re-identification methods we described in this work are all based on a one-shot re-identification from a single test frame. However, when more frames of the same person are available, the results obtained for each frame can be merged to obtain a sequence-wise result. In Table 8.3, we compare on our dataset the single-frame rank-1 performances with what can be obtained with a simple multiframe reasoning,

Table 8.3 Rank-1 results with the single-frame and the multiframe evaluation for the testing sets of the *BIWI RGBD-ID* dataset

	Cross validation		Test (Still)		Test (Walking)	
	Single (%)	Multi (%)	Single (%)	Multi (%)	Single (%)	Multi (%)
Skeleton (SVM)	47.5	66.0	11.6	10.7	13.8	17.9
Skeleton (NN)	80.5	100	26.6	32.1	21.1	39.3
Point cloud matching	93.7	100	32.5	42.9	22.4	39.3
Face (SVM)	97.8	100	44.0	57.1	36.7	57.1
Face+Skeleton (SVM)	98.4	100	52.0	67.9	43.9	67.9

Table 8.4 Runtime performance of the algorithms used for the point cloud matching method

	Time (ms)
Face detection	42.19
Body segmentation	3.03
Transformation to standard pose	0.41
Filtering and smoothing	56.35
ICP and fitness scores computation	254.34

that is by associating each test sequence to the subject voted by the highest number of frames. On average, this voting scheme allows to obtain a performance improvement of about 8–10 %. The Nearest Neighbor classification of the skeleton descriptor for the *Walking* test set seems to benefit most from this approach, thus its rank-1 almost doubles. The best performance is again obtained with the SVM classification of the combined face and skeleton descriptors, which reaches 67.9 % of rank-1 for both the testing sets.

8.5.4 Runtime Performance

The feature-based re-identification method of Sect. 8.4.1 exploits information which is already precomputed by Microsoft Kinect SDK and classification methods which takes less than a millisecond to classify one frame; thus, the runtime performance is only limited by the sensor frame rate and by the face detection algorithm used to select frames with a valid skeleton, which runs at more than 20fps with a C++ implementation on a standard workstation with an Intel Core i5-3570k@3.40GHz processor.

In Table 8.4, the runtime of the single algorithms needed for the point cloud matching method of Sect. 8.4.2 is reported. The most demanding operation is the matching between the test point cloud transformed to standard pose and the models of every subject in the training set, which takes 250 ms for performing 50 comparisons. The overall frame rate is then about 2.8fps, which suggests that this approach could also be used in a real-time scenario with further optimization and with a limited number of people in the database.

8.6 Conclusions and Directions for Future Work

In this chapter, we have compared two different techniques for one-shot person re-identification with soft biometric cues obtained through a consumer depth sensor. The skeleton information is used to build a descriptor which can then be classified with standard machine learning techniques. Moreover, we also proposed to identify subjects by comparing their global body shape. For this purpose, we described how to warp point clouds to a standard pose in order to allow a rigid comparison based on a typical ICP fitness score. We also proposed to use this transformation for obtaining a 3D body model which can be used for re-identification from a series of point clouds of the subject while moving freely.

We tested the proposed algorithms on a publicly available dataset and on the newly created *BIWI RGBD-ID* dataset, which contains 50 training videos and 56 testing sequences with synchronized RGB, depth, and skeleton data. Experimental results show that both the skeleton and the shape information can be used for effectively re-identifying subjects in a noncollaborative scenario, because similar results have been obtained with these two approaches.

As future work, we envision to study techniques for combining skeleton classification and point cloud matching results into a common single re-identification framework.

Acknowledgments The authors would like to thank all the people at the BIWI laboratory of ETH Zurich who took part in the *BIWI RGBD-ID* dataset.

References

1. Apostoloff, N., Zisserman, A.: Who Are You? - Real-time Person Identification. In: British Machine Vision Conference (2007)
2. Barbosa, B.I., Cristani, M., Del Bue, A., Bazzani, L., Murino, V.: Re-identification with rgb-d sensors. In: First International Workshop on Re-identification (2012)
3. Bay, H., Ess, A., Tuytelaars, T., Van Gool, L.: Speeded-up robust features (surf). Comput. Vis. Image Underst. **110**(3), 346–359 (2008)
4. Bazzani, L., Cristani, M., Murino, V.: Symmetry-driven accumulation of local features for human characterization and re-identification. Comput. Vis. Image Underst. **117**(2), 130–144 (2013)
5. Bedagkar-Gala, A., Shah, S.: Multiple person re-identification using part based spatio-temporal color appearance model. In: Computational Methods for the Innovative Design of Electrical Devices'11, pp. 1721–1728 (2011)
6. Besl, P.J., McKay, N.: A method for registration of 3-d shapes. IEEE Trans. Pattern Anal. Mach. Intell. **14**, 239–256 (1992)
7. Bowyer, K.W., Chang, K., Flynn, P.: A survey of approaches and challenges in 3d and multi-modal 3d + 2d face recognition. Comput. Vis. Image Underst. **101**(1), 1–15 (2006)
8. Bronstein, A.M., Bronstein, M.M., Kimmel, R.: Three-dimensional face recognition. Int. J. Comput. Vision **64**, 5–30 (2005)
9. Bronstein, A.M., Bronstein, M.M., Kimmel, R.: Topology-invariant similarity of nonrigid shapes. Int. J. Comput. Vision **81**, 281–301 (2009)

10. Brunelli, R., Falavigna, D.: Person identification using multiple cues. IEEE Trans. Pattern Anal. Mach. Intell. **17**(10), 955–966 (1995)
11. Cortes, C., Vapnik, V.N.: Support-vector networks. Machine Learning **20**(3), 273–297 (1995)
12. Dantone, M., Gall, J., Fanelli, G., Gool, L.V.: Real-time facial feature detection using conditional regression forests. In: IEEE Conference on Computer Vision and Pattern Recognition (2012)
13. Gray, D., Tao, H.: Viewpoint invariant pedestrian recognition with an ensemble of localized features. In: European Conference on Computer Vision, vol. 5302, pp. 262–275 (2008)
14. Hong, L., Jain, A., Pankanti, S.: Can multibiometrics improve performance? In: Proceedings IEEE Workshop on Automatic Identification Advanced Technologies, pp. 59–64 (1999)
15. Jain, A.K., Dass, S.C., Nandakumar, K.: Can soft biometric traits assist user recognition? In: Proceedings of SPIE, Biometric Technology for Human Identification **5404**, 561–572 (2004)
16. Lee, S.U., Cho, Y.S., Kee, S.C., Kim, S.R.: Real-time facial feature detection for person identification system. In: Machine Vision and Applications, pp. 148–151 (2000)
17. Leyvand, T., Meekhof, C., Wei, Y.C., Sun, J., Guo, B.: Kinect identity: Technology and experience. Computer **44**(4), 94–96 (2011)
18. Li, W., Zhang, Z., Liu, Z.: Action recognition based on a bag of 3d points. In: IEEE International Workshop on CVPR for Human Communicative Behavior Analysis (in conjunction with CVPR 2010), San Francisco (2010)
19. Ober, D., Neugebauer, S., Sallee, P.: Training and feature-reduction techniques for human identification using anthropometry. In: Fourth IEEE International Conference on Biometrics: Theory Applications and Systems (BTAS), pp. 1–8 (2010)
20. Ofli, F., Chaudhry, R., Kurillo, G., Vidal, R., Bajcsy, R.: A comprehensive multimodal human action database. In: Proceeding of the IEEE Workshop on Applications on Computer Vision (2013)
21. Preis, J., Kessel, M., Werner, M., Linnhoff-Popien, C.: Gait recognition with kinect. In: Proceedings of the First Workshop on Kinect in Pervasive Computing (2012)
22. Ross, A., Jain, A.: Information fusion in biometrics. Pattern Recogn. Lett. **24**, 2115–2125 (2003)
23. Satta, R., Pala, F., Fumera, G., Roli, F.: Real-time appearance-based person re-identification over multiple Kinect cameras. In: VisApp (2013)
24. Shotton, J., Fitzgibbon, A.W., Cook, M., Sharp, T., Finocchio, M., Moore, R., Kipman, A., Blake, A.: Real-time human pose recognition in parts from single depth images. In: IEEE Conference on Computer Vision and Pattern Recognition, pp. 1297–1304 (2011)
25. Sung, J., Ponce, C., Selman, B., Saxena, A.: Unstructured human activity detection from rgbd images. In: International Conference on Robotics and Automation (2012)
26. Velardo, C., Dugelay, J.L.: Improving identification by pruning: A case study on face recognition and body soft biometric. In: International Workshop on Image and Audio Analysis for Multimedia Interactive Services, pp. 1–4 (2012)
27. Viola, P.A., Jones, M.J.: Robust real-time face detection. In: International Conference on Computer Vision, p. 747 (2001)
28. Wagner, A., Wright, J., Ganesh, A., Zhou, Z., Ma, Y.: Towards a practical face recognition system: Robust registration and illumination by sparse representation. In: IEEE Conference on Computer Vision and Pattern Recognition, pp. 597–604 (2009)
29. Wang, S., Lewandowski, M., Annesley, J., Orwell, J.: Re-identification of pedestrians with variable occlusion and scale. In: International Conference on Computer Vision Workshops, pp. 1876–1882 (2011)
30. Wang, J., Liu, Z., Wu, Y., Yuan, J.: Mining actionlet ensemble for action recognition with depth cameras. In: IEEE Conference on Computer Vision and Pattern Recognition (2012)
31. Wang, C., Zhang, J., Pu, J., Yuan, X., Wang, L.: Chrono-gait image: A novel temporal template for gait recognition. In: Proceedings of the 11th European Conference on Computer Vision, pp. 257–270 (2010)
32. Wolf, C., Mille, J., Lombardi, E., Celiktutan, O., Jiu, M., Baccouche, M., Dellandrea, E., Bichot, C.E., Garcia, C., Sankur, B.: The liris human activities dataset and the icpr 2012 human activities recognition and localization competition. Tech. Rep. RR-LIRIS-2012-004 (2012)

33. Zhang, H., Parker, L.E.: 4-dimensional local spatio-temporal features for human activity recognition. In: IEEE/RSJ International Conference on Intelligent Robots and Systems, pp. 2044–2049 (2011)
34. Zhao, W., Chellappa, R., Phillips, P.J., Rosenfeld, A.: Face recognition: A literature survey. ACM Comput. Surv. **35**(4), 399–458 (2003)
35. Zhu, P., Zhang, L., Hu, Q., Shiu, S.: Multi-scale patch based collaborative representation for face recognition with margin distribution optimization. In: European Conference on Computer Vision, pp. 822–835 (2012)

Chapter 9
Group Association: Assisting Re-identification by Visual Context

Wei-Shi Zheng, Shaogang Gong and Tao Xiang

Abstract In a crowded public space, people often walk in groups, either with people they know or with strangers. Associating a group of people over space and time can assist understanding an individual's behaviours as it provides vital visual context for matching individuals within the group. This seems to be an 'easier' task compared with person re-identification due to the availability of more and richer visual content in associating a group; however, solving this problem turns out to be rather challenging because a group of people can be highly non-rigid with changing relative position of people within the group and severe self-occlusions. In this work, the problem of matching/associating groups of people over large space and time gaps captured in multiple non-overlapping camera views is addressed. Specifically, a novel people group representation and a group matching algorithm are proposed. The former addresses changes in the relative positions of people in a group and the latter uses the proposed group descriptors for measuring the similarity between two candidate images. Based on group matching, we further formulate a method for matching individual person using the group description as visual context. These methods are validated using the 2008 i-LIDS Multiple-Camera Tracking Scenario (MCTS) dataset on multiple camera views from a busy airport arrival hall.

W.-S. Zheng (✉)
Sun Yat-sen University, Guangzhou, China
e-mail: wszheng@ieee.org

S. Gong
Queen Mary University of London, London, UK
e-mail: sgg@eecs.qmul.ac.uk

T. Xiang
Queen Mary University of London, London, UK
e-mail: txiang@eecs.qmul.ac.uk

S. Gong et al. (eds.), *Person Re-Identification*,
Advances in Computer Vision and Pattern Recognition,
DOI: 10.1007/978-1-4471-6296-4_9,

9.1 Introduction

Object recognition has been a focus of computer vision research for the past five decades. In recent years, the focus of object recognition has shifted from recognising objects captured in isolation against clean background under well-controlled lighting conditions to a more challenging but also potentially more useful problem of recognising objects under occlusion against cluttered background with drastic view angle and illumination changes, known as 'recognition in the wild'. In particular, the problem of person re-identification or tracking between disjoint views has received increasing interest [1–6], which aims to match a person observed at different non-overlapping locations observed in different camera views. Typically, person re-identification is addressed by detecting and matching the visual appearance of isolated (segmented) individuals. In this work, we go beyond the conventional individual person re-identification by framing the re-identification problem in the context of associating groups of people in proximity across different camera views. We call this the *group association* problem. Moreover, we also consider how to explore a group of people as non-stationary visual context for assisting individual centred person re-identification within a group. This is often the condition under which re-identification needs be performed in a public space such as transport hubs.

In a crowded public space, people often walk in groups, either with people they know or with strangers. To be able to associate the same group of people over different camera views at different locations can bring about two benefits: (1) Matching a group of people over large space and time can be extremely useful in understanding and inferring longer term association and more holistic behaviour of a group of people in public space. (2) It can provide vital visual context for assisting the match of individuals as the appearance of a person often undergoes drastic change across camera views caused by lighting and view angle variations. Most significantly, people appearing in public space are prone to occlusions by others nearby. These viewing conditions make person re-identification in crowded spaces an extremely difficult problem. On the other hand, groups of people are less affected by occlusion which can provide a richer context and reduce ambiguity in discriminating an individual against others. This is illustrated by examples shown in Fig. 9.1a where each of the six groups of people consists of one or two people in dark clothing. Based on visual appearance alone, it is difficult if not impossible to distinguish them in isolation. However, when they are considered in context by associating groups of people they appear together, it becomes much clearer that all candidates highlighted by red boxes are different people. Figure 9.1b shows examples of cases where matching groups of people together seems to be easier than matching individuals in isolation due to the changes in the appearance of people in different views caused by occlusion or change of body posture. We consider that the group context is more robust against these changes and more consistent over different views, and thus should be exploited for improving the matching of individual person.

However, associating groups of people introduces new challenges: (1) Compared to an individual, the appearance of a group of people is highly non-rigid and the

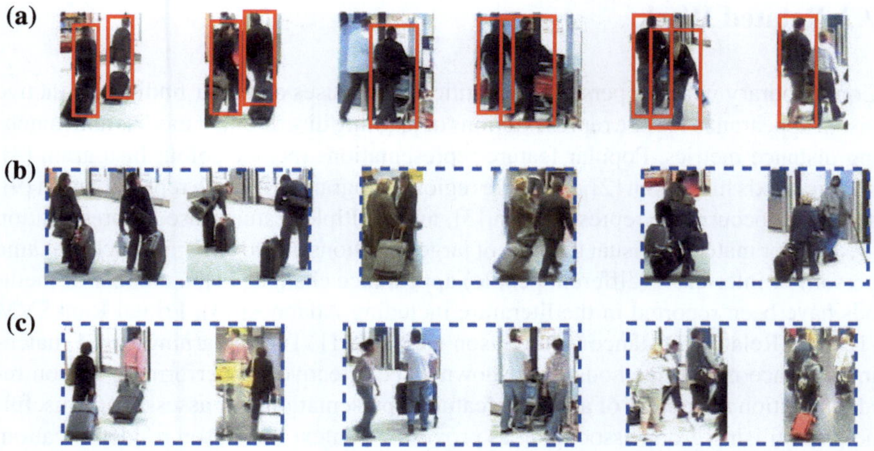

Fig. 9.1 Advantages from and challenges in associating groups of people versus person re-identification in isolation

relative positions of the members can change significantly and frequently. (2) Although occlusion by other objects is less of an issue, self-occlusion caused by people within the group remains a problem which can cause changes in group appearance. (3) Different from a relatively stable shape of every upright person which has similar aspect ratio, the aspect ratio of the shapes of different groups of people can be very different. Some difficult examples are shown in Fig. 9.1c.

Due to these challenges, conventional representations and matching methodsfor person re-identification are not suitable for solving the group association problem, because they are designed for person in isolation rather than in a group. In this work, a novel people group representation is presented based on two ratio-occurrence descriptors. This is in order to make the representation robust against within-group position changes. Given this group representation, a group matching algorithm is formulated to achieve group association robustness against both changes in relative positions of people within a group and variations in illumination and viewpoint across different camera views. In addition, a new person re-identification method is introduced by utilising associated group of people as visual context to improve the matching of individuals across camera views. This group association model is validated using 2008 i-LIDS Multiple-Camera Tracking Scenario (MCTS) dataset captured by multiple camera views from a busy airport arrival hall [7].

The remaining sections are as follows. Section 9.2 overviews related work and assesses this work in context. Section 9.3 describes how the visual appearance of a group of people can be represented for robust group matching. Section 9.4 introduces the metric we used to measure the similarity between two group images and Sect. 9.5 formulates a method for utilising a group of people as contextual cues for individual person re-identification within the group. Section 9.6 presents experimental validation on these methods and Sect. 9.7 concludes the chapter.

9.2 Related Work

Contemporary work on person re-identification focuses on either finding distinctive visual appearance feature representations or learning discriminant models and matching distance metrics. Popular feature representations include colour histogram [4], principal axis histogram [2], rectangle region histogram [8], graph representation [9], spatial co-occurrence representation [3], and multiple feature based representation [4, 10]. For matching visual features of large variations due to either intra-class (same person) or inter-class (different people) appearance change [11], a number of methods have been reported in the literature including Adaboost [4], Primal RankSVM [12] and Relative Distance Comparison (RDC) [11]. These learning-based matching distance metric methods have shown to be effective for performing person re-identification regardless of a chosen feature representation. For assessing the usefulness of utilising group association as proximity context for person re-identification, the RankSVM-based matching method [12] is adopted in this work for matching individuals in a group.

The concept of exploiting contextual information for object recognition has been extensively studied in the literature. Most existing context modelling works require manual annotation/labelling of contextual information. Given both the annotated target objects and contextual information, one of the most widely used methods is to model the co-occurrence of context and object. Torralba et al. [13], Rabinovich et al. [14], Felzenszwalb et al. [15] and Zheng et al. [16] model how a target object category co-occurs frequently with other object categories (e.g. a person carrying a bag, a tennis ball with a tennis bracket) or where the target objects tend to appear (e.g. a TV in a living room). Besides co-occurrence information, spatial relationship between objects and context has also been explored for context modelling. Spatial relationships are typically modelled using Markov Random Field (MRF) or Conditionally Random Field (CRF) [17–19], or other graphical models [20]. These models incorporate the spatial support of target object against other objects either from the same category or from different categories and background, such as a boat on a river/sea or a car on a road. Based on a similar principle, Hoim et al. [21] and Bao et al. [22] proposed to infer the interdependence of object, 3D spatial geometry and the orientation and position of camera as context; and Galleguillos et al. [23] inferred the contextual interactions at pixel, region and object levels and combined them together using a multi-kernel learning algorithm [23, 24]. Comparing to those works, this work has two notable differences: (1) we focus on the problem of intra-category identification (individual re-identification whilst all people look alike) rather than inter-category classification (differentiating different object classes between for instance cars and bicycles); (2) we are specifically interested in exploring a group of people as non-stationary proximity context to assist in the matching of one of the individuals in the group.

There are other related works on crowd detection and analysis [25–28] and group activity recognition [29, 30]. However, those works are not concerned with group association over space and time, either within the same camera view or across different views. A preliminary version of this work was reported in [31].

9.3 Group Image Representation

Given a gallery set and a probe set of images of different groups of people, we aim to design and construct suitable group image descriptors for matching gallery images with any probe image of a group of people.

9.3.1 From Pixel to Local Region-Based Feature Representation

Similar to [3, 32], we first assign a label to each pixel of a given group image **I**. The label can be a simple colour or a visual word index of colour together with gradient information. Due to the change in camera view and varying positions and motions of a group of people, we consider that integration of local rotational invariant features and colour density information is better for constructing visual words for indexing. In particular, we extract SIFT features [33] (a 128-dimensional vector) for each RGB channel at each pixel with a surrounding support region (12×12 in our experiment). We also obtain an average RGB colour vector of pixel over a support region (3×3), where the colour vector is normalised to $[0, 1]^3$. The SIFT vector and colour vector are then concatenated for each pixel for representation, which we call the SIFT+RGB feature. The SIFT+RGB features are quantised into n clusters by K-means and a code book $\mathcal{A}$ of n visual words $\mathbf{w}_1, ..., \mathbf{w}_n$ is built. Finally, an appearance label image is built by assigning a visual word index to the corresponding SIFT+RGB feature at each pixel of the group image. In order to remove background information, background subtraction is first performed. Then, only features extracted for foreground pixels are used to construct visual words for group image representation.[1]

To represent the distribution of visual words of any image, a single histogram of visual words, which we call the holistic histogram, can be considered [34]. However autoedited1, this representation loses all spatial distribution information about the visual words. One way to alleviate this problem is to divide the image into grid blocks and concatenate the histograms of blocks one by one, for instance similar to [35]. However, this representation will still be sensitive to the appearance changes in situations when people swap their positions in a group (Fig. 9.1c). Moreover, corresponding image grid positions between two group images are not always guaranteed to represent foreground regions, thus such a hard-wired grid block-based representation is not suitable.

Considering such characteristics of group images, we consider to represent a group image by constructing two local region-based descriptors: a *center rectangular ring*

---

[1] This step is omitted when continuous image sequences are not available.

ratio-occurrence descriptor which aims to describe the ratio information of visual words within and between different rectangular ring regions, and a *block based ratio-occurrence descriptor* for exploring more specific local spatial information between visual words that could be stable. These two descriptors are combined to form a group image representation. These two descriptors are motivated by the observation that whilst global spatial relationships between people within a group can be highly unstable, local spatial relationships between small patches within a local region is more stable, e.g. within the bounding box of a person.

9.3.2 Center Rectangular Ring Ratio-Occurrence (CRRRO)

Rectangular ring regions are considered approximately rotationally invariant. An efficient integral computation of visual words histogram is also available [32]. Given both, we define a holistic rectangular ring structure expanding from the centre of a group image. The ℓ rectangular rings divide a group image into ℓ non-overlapped regions $P_1, \cdots, P_\ell$ from inside to outside. Every rectangular ring is $0.5 \cdot N/\ell$ and $0.5 \cdot M/\ell$ thick along the vertical and horizontal directions respectively (see Fig. 9.2a with $\ell = 3$), where the group image is of size $M \times N$. Such a partitioning of a group image is especially useful for describing a pair of people because the distribution of constituent patches of each person in each ring is likely to be more stable against changes in relative positions between the two people over different viewpoints or scales (Fig. 9.3).

After a partition of any image for representation, a common approach to constructing a codebook is to concatenate the histogram of visual words from each ring. However, this ignores any spatial relationships existing between visual words from different ring-zones of a partition. We consider retaining such spatial relationships critical, thus we introduce a notion of *intra-* and *inter-ratio-occurrence maps* as follows. For each ring-region P_i, a histogram $\mathbf{h}_i$ is built, where $\mathbf{h}_i(a)$ indicates the frequency (occurrence) of visual word $\mathbf{w}_a$. Then for P_i, an *intra ratio-occurrence map* $\mathbf{H}_i$ is defined as

$$\mathbf{H}_i(a, b) = \frac{\mathbf{h}_i(a)}{\mathbf{h}_i(a) + \mathbf{h}_i(b) + \varepsilon}, \tag{9.1}$$

where ε is a very small positive value in order to avoid $0/0$. $\mathbf{H}_i(a, b)$ then represents the ratio-occurrence between words $\mathbf{w}_a$ and $\mathbf{w}_b$ within the region.

In order to capture any spatial relationships between visual words within and outside region P_i, we further define another two ratio occurrence maps for ring-region P_i as follows:

$$\mathbf{g}_i = \sum_{j=1}^{i-1} \mathbf{h}_j, \quad \mathbf{s}_i = \sum_{j=i+1}^{\ell} \mathbf{h}_j,$$

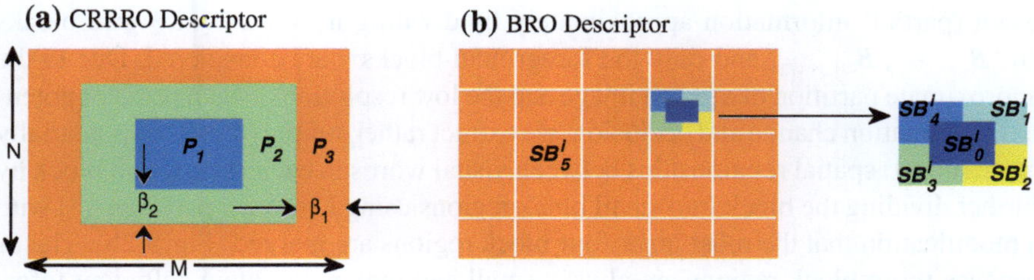

Fig. 9.2 Partition of a group image by two descriptors. *Left* the center rectangular ring ratio-occurrence descriptor ($\beta_1 = M/2\ell, \beta_2 = N/2\ell, \ell = 3$); *Right* the block based ratio-occurrence descriptor ($\gamma = 1$), where *white lines* show the grids of the image

Fig. 9.3 An illustration of a group of people against *dark* background

where $\mathbf{g}_i$ represents the distribution of visual words enclosed by the rectangular ring P_i and $\mathbf{s}_i$ represents the distribution of visual words outside P_i, where we define $\mathbf{g}_1 = \mathbf{0}$ and $\mathbf{s}_\ell = \mathbf{0}$. Then two *inter ratio-occurrence maps* $\mathbf{S}_i$ and $\mathbf{G}_i$ are formulated as follows:

$$\mathbf{G}_i(a, b) = \frac{\mathbf{g}_i(a)}{\mathbf{g}_i(a) + \mathbf{h}_i(b) + \varepsilon}, \quad \mathbf{S}_i(a, b) = \frac{\mathbf{s}_i(a)}{\mathbf{s}_i(a) + \mathbf{h}_i(b) + \varepsilon}. \tag{9.2}$$

Therefore, for each ring-region P_i, we construct a triplet representation $\mathbf{T}_r^i = \{\mathbf{H}_i, \mathbf{S}_i, \mathbf{G}_i\}$, and a group image is represented by a set $\{\mathbf{T}_r^i\}_{i=1}^{\ell}$.

We show in the experiments that this group image representation using a set of triplet intra- and inter-ratio occurrence maps gives better performance for associating groups of people than that of using a conventional concatenation based representation.

9.3.3 Block-Based Ratio-Occurrence (BRO)

The CRRRO descriptor introduced above still cannot cope well with large non-center-rotational changes in people's positions within a group. It also does not utilise any local structure information that may be more stable or consistent across different views of the same group, e.g. certain parts of a person can be visually more consistent than others. As we do not make any assumptions on people in a group being well segmented due to self-occlusion, we revisit a group image to explore patch (partial) information approximately by dividing it into $\omega_1 \times \omega_2$ grid blocks $B_1, B_2, \cdots, B_{\omega_1 \times \omega_2}$, and only the foreground blocks[2] are considered. Due to the approximate partition of a group image and the low resolution of each patch or potential illumination change and occlusion, we extract rather simple (therefore potentially more robust) spatial relationships between visual words in each foreground block by further dividing the block into small block regions using L-shaped partition [3] with a modification that the most inner four block regions are merged (Fig. 9.2b). This is because those block regions are always small and may not contain sufficient information. As a result, we obtain $4\gamma + 1$ block regions within each block B_i denoted by $SB_0^i, \cdots, SB_{4\gamma}^i$ for some positive integer γ.

For associating groups of people over different views, we first note that not all blocks B_i appear in the same position in the group images. For example, a pair of people may swap their positions resulting in the blocks corresponding to those foreground pixels changing their positions in different images. Also, there may be other visually similar blocks in the same group image. Hence, describing local matches only based on features within block B_i could not be distinct enough. To reduce this ambiguity, region $SB_{4\gamma+1}^i$, which is the image portion outside block B_i (see Fig. 9.2b with $\gamma = 1$). Therefore, for each block B_i, we partition the group image into $SB_0^i, SB_1^i, \cdots, SB_{4\gamma}^i$ and $SB_{4\gamma+1}^i$. We show in the experiments that including such complementary region $SB_{4\gamma+1}^i$ would significantly enhance matching performance.

Similar to the CRRRO descriptor, for each block B_i, we learn an intra ratio-occurrence map $\mathbf{H}_j^i$ between visual words in each block region SB_j^i. Similarly, we explore an inter ratio-occurrence map $\mathbf{O}_j^i$ between different block regions SB_j^i. Since the size of each block region in block B_i would always be relatively much smaller than the complementary region $SB_{4\gamma+1}^i$, the ratio information between them will be sensitive to noise. Consequently, we consider two simplified inter ratio-occurrence maps $\mathbf{O}_j^i$ between block B_i and its complementary region $SB_{4\gamma+1}^i$ formulated as follows:

$$\mathbf{O}_1^i(a, b) = \frac{\mathbf{t}_i(a)}{\mathbf{t}_i(a) + \mathbf{z}_i(b) + \varepsilon}, \quad \mathbf{O}_2^i(a, b) = \frac{\mathbf{z}_i(a)}{\mathbf{z}_i(a) + \mathbf{t}_i(b) + \varepsilon}, \tag{9.3}$$

where $\mathbf{z}_i$ and $\mathbf{t}_i$ are the histograms of visual words of block B_i and image region $SB_{4\gamma+1}^i$, respectively. Then, each block B_i is represented by $\mathbf{T}_b^i = \{\mathbf{H}_j^i\}_{j=0}^{4\gamma+1} \bigcup \{\mathbf{O}_j^i\}_{j=1}^{2}$, and a group image is represented by a set $\{\mathbf{T}_b^i\}_{i=1}^m$ where m is the amount of foreground blocks B_i.

To summarise, two local region-based group image descriptors, CRRRO and BRO, are specially designed and constructed for associating images of groups of people. Due to highly unstable positions of people within a group and likely partial occlusions among them, these two descriptors explore the inter-person spatial relational information in a group and the likely local patch (partial) information for each person respectively.

[2] A foreground block is defined as an image block with more than 70 % pixels being foreground.

9.4 Group Image Matching

We match two group images $\mathbf{I}_1$ and $\mathbf{I}_2$ by combining the distance metrics of the two proposed descriptors as follows:

$$d(\mathbf{I}_1, \mathbf{I}_2) = d_r\left(\{\mathbf{T}_r^i(\mathbf{I}_1)\}_{i=1}^{\ell}, \{\mathbf{T}_r^{i'}(\mathbf{I}_2)\}_{i'=1}^{\ell}\right) + \alpha \cdot d_b\left(\{\mathbf{T}_b^i(\mathbf{I}_1)\}_{i=1}^{m_1}, \{\mathbf{T}_b^{i'}(\mathbf{I}_2)\}_{i'=1}^{m_2}\right), \quad \alpha \geq 0, \tag{9.4}$$

where $\{\mathbf{T}_r^i(\mathbf{I}_1)\}_{i=1}^{\ell}$ indicates the center rectangular ring ratio-occurrence descriptor for group image $\mathbf{I}_1$ whilst $\{\mathbf{T}_b^i(\mathbf{I}_1)\}_{i=1}^{m_1}$ is for the block based descriptor.

For d_r, the L_1 norm metric is used to measure the distance between each corresponding ratio-occurrence map and d_r is obtained by averaging these distances. Note that L_1 norm metric is more robust and tolerant to noise as compared to Euclidean metric [36]. For d_b, since the spatial relationship between patches is not constant in different images of the same group and also not all the patches in one group image can be matched with those in another, it is inappropriate to directly measure the distance between the corresponding patches (blocks) of two group images. To address this problem, we assume that for each pair of group images, there exist at most k pairs of matched local patches between two images. We then define d_b as a *top k-match metric* where k is a positive integer as follows:

$$d_b\left(\{\mathbf{T}_b^i(\mathbf{I}_1)\}_{i=1}^{m_1}, \{\mathbf{T}_b^{i'}(\mathbf{I}_2)\}_{i'=1}^{m_2}\right) = \min_{\mathbf{C},\mathbf{D}} \left\{k^{-1} \cdot ||\mathbf{AC} - \mathbf{BD}||_1\right\}, \quad \mathbf{A} \in \mathbb{R}^{q \times m_1}, \mathbf{B} \in \mathbb{R}^{q \times m_2}, \mathbf{C} \in \mathbb{R}^{m_1 \times k}, \mathbf{D} \in \mathbb{R}^{m_2 \times k}, \tag{9.5}$$

where the ith (i'th) column of matrix $\mathbf{A}$ ($\mathbf{B}$) is the vector representation of $\mathbf{T}_b^i(\mathbf{I}_1)$ ($\mathbf{T}_b^{i'}(\mathbf{I}_2)$). Each column $\mathbf{c}_j$ ($\mathbf{d}_j$) of $\mathbf{C}$ ($\mathbf{D}$) is an indicator vector in which only one entry is 1 and the others are zeros, and the columns of $\mathbf{C}$ ($\mathbf{D}$) are orthogonal. Note that m_1 and m_2, the number of foreground blocks in two group images, may be unequal. Generally, directly solving Eq. (9.5) is hard. Note that $\min_{\mathbf{C},\mathbf{D}}\left\{||\mathbf{AC} - \mathbf{BD}||_1\right\} \leq \sum_{j=1}^{k} \min_{\mathbf{c}_j,\mathbf{d}_j}\left\{||\mathbf{Ac}_j - \mathbf{Bd}_j||_1\right\}$ where $\{\mathbf{c}_j\}$ and $\{\mathbf{d}_j\}$ are sets of orthogonal indicator vectors. We therefore approximate the k-match metric value as follows: the most matched patches $\mathbf{a}_{i_1}$ and $\mathbf{b}_{i'_1}$ are first found by finding the smallest L_1 distance between columns of $\mathbf{A}$ and $\mathbf{B}$. We then remove $\mathbf{a}_{i_1}$ and $\mathbf{b}_{i'_1}$ from $\mathbf{A}$ and $\mathbf{B}$ respectively and find the next most matched pair. This procedure repeats until the top k matched patches are found.

9.5 Exploring Group Context in Person Re-identification

9.5.1 Re-identification by Ranking

Person re-identification can be casted as a ranking problem [11, 12, 37], by which the problem is further addressed either in terms of feature selection or matching distance metric learning. This approach aims to learn a set of most discriminant and robust features, based on which a weighted L1 norm distance is used to measure the similarity between a pair of person images.

More specifically, person re-identification by ranking the relevance of their image features can be formulated as follows: There exist a set of relevance scores $\lambda = \{r_1, r_2, \cdots, r_\rho\}$ such that $r_\rho \succ r_{\rho-1} \succ \cdots \succ r_1$ where ρ is the number of scores and $\succ$ indicates the order. Most commonly, this problem only has two relevance considerations: relevant and related irrelevant observation feature vectors, that is, the correct and incorrect (but may still be visually similar) matches. Given a dataset $X = \{(\mathbf{x}_i, y_i)\}_{i=1}^m$ where $\mathbf{x}_i$ is a multi-dimensional feature vector representing the appearance of a person captured in one view, y_i is its label and m is the number of training samples. Each vector $\mathbf{x}_i (\in R^d)$ has an associated set of relevant observation feature vectors $\mathbf{d}_i^+ = \{\mathbf{x}_{i,1}^+, \mathbf{x}_{i,2}^+, \cdots, \mathbf{x}_{i,m^+(\mathbf{x}_i)}^+\}$ and irrelevant observation feature vectors $\mathbf{d}_i^- = \{\mathbf{x}_{i,1}^-, \mathbf{x}_{i,2}^-, \cdots, \mathbf{x}_{i,m^-(\mathbf{x}_i)}^-\}$ corresponding to correct and incorrect matches from another camera view, respectively. Here $m^+(\mathbf{x}_i)$ and $m^-(\mathbf{x}_i)$ are the respective numbers of relevant and irrelevant observations for query $\mathbf{x}_i$. In general, $m^+(\mathbf{x}_i) << m^-(\mathbf{x}_i)$ because there are likely only a few instances of correct match and many incorrect matches. The goal of ranking any paired image relevance is to learn function δ for all pairs of $(\mathbf{x}_i, \mathbf{x}_{i,j}^+)$ and $(\mathbf{x}_i, \mathbf{x}_{i,j'}^-)$ such that the relevant score $\delta(\mathbf{x}_i, \mathbf{x}_{i,j}^+)$ is larger than $\delta(\mathbf{x}_i, \mathbf{x}_{i,j'}^-)$.

Here, we seek to compute the score δ in terms of the pairwise sample $(\mathbf{x}_i, \mathbf{x}_{i,j})$ by a linear function $\mathbf{w}$ as follows:

$$\delta(\mathbf{x}_i, \mathbf{x}_{i,j'}) = \mathbf{w}^\top |\mathbf{x}_i - \mathbf{x}_{i,j}|, \tag{9.6}$$

where $|\mathbf{x}_i - \mathbf{x}_{i,j}| = \left[|\mathbf{x}_i(1) - \mathbf{x}_{i,j}(1)|, \cdots, |\mathbf{x}_i(d) - \mathbf{x}_{i,j}(d)|\right]^\top$. We call $|\mathbf{x}_i - \mathbf{x}_{i,j}|$ the absolute difference vector.

Note that for a query feature vector $\mathbf{x}_i$, we wish to have the following rank relationship for a relevant feature vector $\mathbf{x}_{i,j}^+$ and a related irrelevant feature vector $\mathbf{x}_{i,j'}^-$

$$\mathbf{w}^\top (|\mathbf{x}_i - \mathbf{x}_{i,j}^+| - |\mathbf{x}_i - \mathbf{x}_{i,j'}^-|) > 0. \tag{9.7}$$

Let $\hat{\mathbf{x}}_s^+ = |\mathbf{x}_i - \mathbf{x}_{i,j}^+|$ and $\hat{\mathbf{x}}_s^- = |\mathbf{x}_i - \mathbf{x}_{i,j'}^-|$. Then, by going through all samples $\mathbf{x}_i$ in the dataset X, we obtain a corresponding set of these pairwise relevant difference vectors denoted by $P = \{(\hat{\mathbf{x}}_s^+, \hat{\mathbf{x}}_s^-)\}$ where $\mathbf{w}^\top (\hat{\mathbf{x}}_s^+ - \hat{\mathbf{x}}_s^-) > 0$ is expected. A RankSVM model then is defined as the minimisation of the following objective function:

$$\frac{1}{2}\|\mathbf{w}\|^2 + C\sum_{s=1}^{|P|}\xi_s$$
$$\text{s.t. } \mathbf{w}^\top(\hat{\mathbf{x}}_s^+ - \hat{\mathbf{x}}_s^-) \geq 1 - \xi_s,\ s = 1, \cdots, |P|,\ \xi_s \geq 0,\ \ s = 1, \cdots, |P|, \tag{9.8}$$

where C is a parameter that trades margin size against training error.

A computational difficulty in using a SVM to solve the ranking problem is the potentially large size of P. In problems with lots of queries and/or queries represented as feature vectors of high dimensionality, the size of P means that forming the $\hat{\mathbf{x}}_s^+ - \hat{\mathbf{x}}_s^-$ vectors becomes computationally challenging. In the case of person re-identification, the ratio of positive to negative observation samples is $m : m \cdot (m-1)$ and as m increases the size of P can become very large rapidly. Hence, the RankSVM in Eq. (9.8) can be computationally intractable for large-scale constraint problems due to memory usage.

Chapelle and Keerthi [38] proposed primal RankSVM that relaxes the constrained RankSVM and formulated a non-constraint model as follows:

$$\mathbf{w} = \arg\min_{\mathbf{w}} \frac{1}{2}\|\mathbf{w}\|^2 + C\sum_{s=1}^{|P|}\ell\left(0, 1 - \mathbf{w}^\top\left(\hat{\mathbf{x}}_s^+ - \hat{\mathbf{x}}_s^-\right)\right)^2, \tag{9.9}$$

where C is a positive importance weight on the ranking performance and ℓ is the hinge loss function. Moreover, a Newton optimisation method is introduced to reduce the training time of the SVM. Additionally, it removes the need for an explicit computation of the $\hat{\mathbf{x}}_s^+ - \hat{\mathbf{x}}_s^-$ pairs through the use of a sparse matrix. However, in the case of person re-identification the size of the training set can also be a limiting factor. The effort required to construct all the $\hat{\mathbf{x}}_s^+$ and $\hat{\mathbf{x}}_s^-$ for model learning is determined by the ratio of positive to negative samples as well as the feature dimension d. As the number of related observation feature vectors is increased, i.e. more people are observed, the space complexity (memory cost) of creating all the training samples is

$$O\left(\sum_{i=1}^{m} d \cdot m^+(\mathbf{x}_i) \cdot m^-(\mathbf{x}_i)\right), \tag{9.10}$$

where $m^-(\mathbf{x}_i) = m - m^+(\mathbf{x}_i) - 1$ for the problems addressed here.

9.5.2 Re-identification with Group Context

We wish to explore group information for reducing the ambiguity in person re-identification if a person stays in the same group. Suppose a set of L paired samples $\{(\mathbf{I}_p^i, \mathbf{I}_g^i)\}_{i=1}^L$ is given, where $\mathbf{I}_g^i$ is the corresponding group image of the ith person image $\mathbf{I}_p^i$. We introduce a group-contextual-descriptor similar in spirit to the center rectangular ring descriptor introduced in Sect. 9.3.1, with a modification that we expand the rectangular ring structure surrounding each person. This makes the group context person specific, i.e. two people in the same group would have different context. Note that, only context features at foreground pixels are extracted. As a result, the most inner rectangular region P_1 is the bounding box of a person, and for other outer rings, they are $\max\{M - a_1 - 0.5 \cdot M_1, a_1 - 0.5 \cdot M_1\}/(\ell - 1)$ and $\max\{N - b_1 - 0.5 \cdot N_1, b_1 - 0.5 \cdot N_1\}/(\ell - 1)$ thick along the horizontal and vertical directions, where (a_1,b_1) is the centre of region P_1, M and N are width and height of the group image, and M_1 and N_1 are width and height of P_1. In particular, when $\ell = 2$, the rectangular ring structure would divide a group image into two parts: a person-centred bounding box and a surrounding complementary image region.

To integrate group information for person re-identification, in this work, we adopt to combine the distance metric d_p of a pair of person descriptors and the distance metric d_r of the corresponding group context descriptors computed from a probe and gallery image pair to be matched. More specifically, denote the person descriptors of person image $\mathbf{I}_p^1$ and $\mathbf{I}_p^2$ as $\mathbf{P}_1$ and $\mathbf{P}_2$ respectively and denote their corresponding group context descriptors as $\mathbf{T}_1$ and $\mathbf{T}_2$ respectively. Then the distance between two people is computed as:

$$d(\mathbf{I}_p^1, \mathbf{I}_p^2) = d_p(\mathbf{P}_1, \mathbf{P}_2) + \beta \cdot d_r(\mathbf{T}_1, \mathbf{T}_2), \quad \beta \geq 0, \tag{9.11}$$

where d_r is defined in Sect. 9.4 and d_p is formulated as

$$d_p(\mathbf{P}_1, \mathbf{P}_2) = -\delta(\mathbf{P}_1, \mathbf{P}_2) = -\mathbf{w}^\top |\mathbf{P}_1 - \mathbf{P}_2|, \tag{9.12}$$

where $\mathbf{w}$ is learned by RankSVM as described in Sect. 9.5.1.

For making use of group context in assisting person re-identification, we consider the following processing steps:

1. Detect a target person;
2. Extract features for each person and measure its distance from the gallery person images using the ranking distance in Eq. (9.12);
3. Segment the group of people around a detected person;
4. Represent each group of people using the group descriptor described in Sect. 9.3;
5. Measure the distance of each group descriptor from the group images to their corresponding gallery person images using the matching distance given in Sect. 9.4;
6. Combine the two distances using Eq. (9.11).

In computing the group descriptors, we focus on demonstrating the effectiveness of the proposed group descriptors and the group assisted matching model for improving person re-identification. We consider that person detection and the segmentation of groups of people in steps (1) and (3) above are performed using standard techniques readily available.

9.6 Experiments

We conducted extensive experiments using the 2008 i-LIDS MCTS dataset to evaluate the feasibility and performance of the proposed methods for associating groups of people and for person re-identification assisted by group context in a crowded public space.

9.6.1 Dataset and Settings

The i-LIDS MCTS dataset was captured at an airport arrival hall by a multi-camera CCTV network. We extracted image frames captured from two non-overlapping camera views. In total, 64 groups were extracted and 274 group images were cropped. Most of the groups have four images, either from different camera views or from the same camera but captured at different locations at different time. These group images are of different sizes. From the group images, we extracted 476 person images for 119 pedestrians, most of which are with four images. All person images were normalised to 64×128 pixels. Different from other person re-identification datasets [1, 3, 4], these images were captured by non-overlapping camera views, and many of them underwent large illumination change and were subject to occlusion.

For code book learning, additional 80 images (of size 640×480) were randomly selected with no overlap with the dataset described above. As described in Sect. 9.3, the SIFT+RGB features were extracted at each pixel of an image. In our experiments, a code book with 60 visual words (clusters) was built using K-means.

Unless otherwise stated, our descriptors are set as follows. For the CRRRO descriptor, we set $\ell = 3$. For the BRO descriptor, each image was divided into 5×5 blocks, γ was set to 1, and the top 10-match score was computed. The default combination weight α in (Eq. (9.4)) was set to 0.8. For the colour histogram, the number of colour bins was set to 16.

9.6.2 Evaluation of Group Association

We randomly selected one image from each group to build the gallery set and the other group images formed the probe set. For each group image in the probe set, we measured its similarity with each template image in the gallery. The s-nearest correct match for each group image was obtained. This procedure was repeated 10 times and the average cumulative match characteristic (CMC) curve [3] and the synthetic disambiguation rate (SDR) curve [4] were used to measure the performance, where the top 25 matching rates are shown for CMC curve and the SDR curve is able to give an overview of the whole CMC curve from the reacquisition point of view [4].

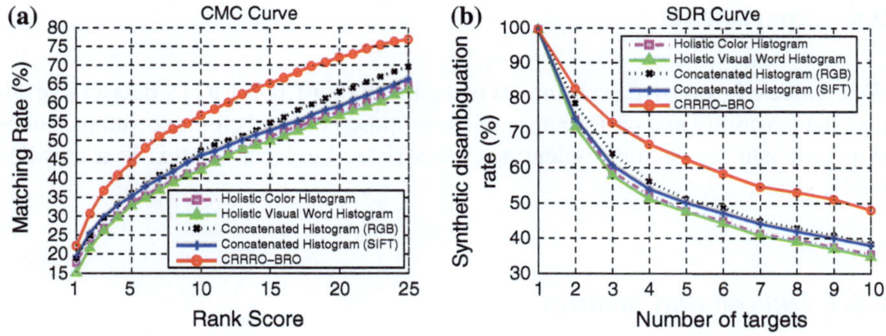

Fig. 9.4 Compare the CMC and SDR *curves* for associating groups of people using the proposed CRRRO–BRO descriptor with those from other commonly used descriptors

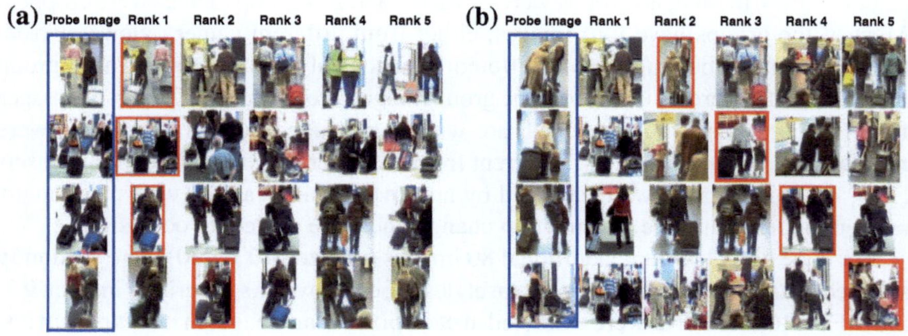

Fig. 9.5 Examples of associating groups of people: the correct matches are highlighted by *red boxes*

The performance of the combined Center Rectangular Ring Ratio-Occurrence and Block based Ratio-Occurrence (CRRRO–BRO) descriptor approach (Eq. (9.4)) is shown in Fig. 9.4. We compare our model with two commonly used descriptors, colour histogram and visual word histogram of SIFT features (extracted at each colour channel) [34], which represent the distributions of colour or visual words of each group image holistically. We also applied these two descriptors to the designed center rectangular ring structure by concatenating the colour or visual word histogram of each rectangular ring. In order to make the compared descriptors scale invariant, the histograms used in the compared methods were normalised [39]. For measurement, the *Chi-square* distance χ^2 [39] was used.

Results in Fig. 9.4 show the proposed CRRRO–BRO descriptor gives the best performance. It always keeps a notable margin to the CMC curve of the second best method, with 44.62 % against 36.14 % and 77.29 % against 69.57 % for rank 5 and 25 matching respectively. Compared to the existing holistic representations and the concatenation of local histograms representations, the proposed descriptor benefits from exploring the ratio information between visual words within and outside each

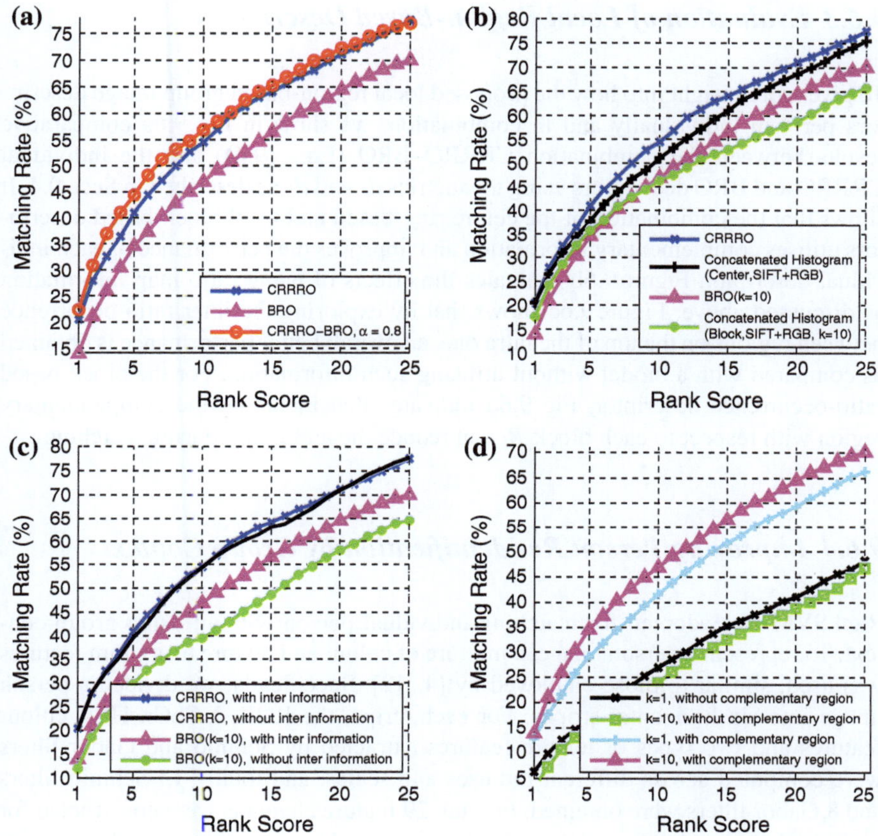

Fig. 9.6 Evaluation of the proposed group image descriptors

local region. Moreover, Fig. 9.6b shows that either using the proposed centre-based or block-based descriptor can still achieve an overall improvement as compared to the concatenated histogram of visual words using SIFT+RGB features (Sect. 9.3) denoted by "Concatenated Histogram (Center, SIFT+RGB)" and "Concatenated Histogram (Block, SIFT+RGB, $k = 10$)" in the figure, respectively. This suggests the ratio maps can provide more information for matching. Finally, Fig. 9.5 shows some examples of associating groups of people using the proposed model (Eq. (9.4)) with $\alpha = 0.8$). It demonstrates that this model is capable of establishing correct matching when there are large variations in people's appearances and their relative positions in a group caused by some challenging viewing conditions, including significantly different view angles and severe occlusions.

9.6.3 Evaluation of Local Region-Based Descriptors

To give more insight into how the proposed local region-based group image descriptors perform individually and in combination, we show in Fig. 9.6a comparative results between the combination CRRRO–BRO (Eq. (9.4)) and the individual CRRRO and BRO descriptors using the metrics d_r and d_b as described in Sect. 9.4. It shows that the combination of the centre ring-based and local block-based descriptors utilises complementary information and improves the performance of each individual descriptor. Figure 9.6b evaluates the effects of using ratio map information as discussed above. Figure 9.6c shows that by exploring the inter ratio-occurrence between regions on the top of the intra one, an overall better performance is obtained as compared with a model without utilising such information. For the block-based ratio-occurrence descriptor, Fig. 9.6d indicates that including the complementary region with respect to each block B_i can reduce the ambiguity during matching.

9.6.4 Improving Person Re-identification by Group Context

RankSVM was adopted for matching individual person without using group context. To represent a person image, a mixture of colour and texture histogram features was used, similar to those employed by [4, 12]. Specifically, we divided a person image into six horizontal stripes. For each stripe, the RGB, YCbCr, HSV colour features and two types of texture features extracted by Schmid and Gabor filters were computed across different radiuses and scales, and totally 13 Schmid filters and 8 Gabor filters were obtained. In total, 29 feature channels were constructed for each stripe and each feature channel was represented by a 16-dimensional histogram vector. The details are given in [4, 12]. Each person image was thus represented by a feature vector in a 2,784-dimensional feature space $\mathcal{Z}$. Since the features computed for this representation include low-level features widely used by existing person re-identification techniques, this representation is considered as generic and representative. With group context, as described in Sect. 9.5.2, a two-rectangular-ring structure is expanded from the centre of the bounding box of each person, and the group matching score is fused with the RankSVM score, where we set $C = 0.005$ in Eq. (9.9).

For evaluating whether there is any benefit to re-identification when using group context information, we randomly selected all images of p people (classes) to set up the test set, and the images of the rest of the people (classes) were used for training. Different values of p were used to evaluate the matching performance of models learned with different amounts of training data. Each test set was composed of a gallery set and a probe set. The gallery set consisted of one image for each person, and the remaining images were used as the probe set. This procedure was repeated 10 times and the average performances of these techniques without and with group context are shown in Fig. 9.7. It is evident that including group context

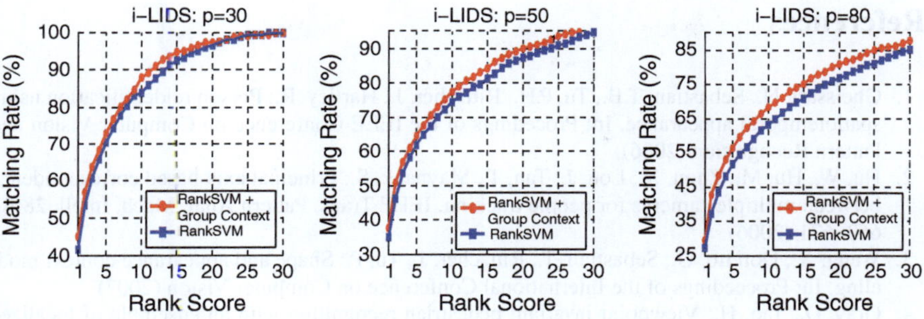

Fig. 9.7 Improving person re-identification using group context

notably improves the matching rate regardless of the choice of different person re-identification techniques. Although RankSVM has been shown in the literature as a very effective method for person re-identification, a clear margin of improvement is consistently achieved over the baseline RankSVM model when group context information is utilised. This suggests that group context helps alleviate the appearance variations due to occlusion, large variations in both view angle and illumination caused by non-overlapping multiple camera views.

9.7 Conclusions

In this work, we considered and addressed the problem of associating groups of people over multiple non-overlapping camera views and formulated local region-based group image descriptors in the form of both a centre rectangular ring and block based ratio-occurrence descriptors. They are designed specifically for the representation of images of groups of people in crowded public spaces. We evaluated their effectiveness using a top k-match distance model. Moreover, we demonstrated the advantages gained from utilising group context information in improving person re-identification under challenging viewing conditions using the 2008 i-LIDS MCTS dataset.

A number of future research directions are identified. First, both the grouping matching method and how the scores of group matching and person matching can benefit from further investigation, e.g. by exploiting more effective distance metric learning methods. Second, the problem of automatically identifying groups, especially groups of people who move together over a sustained period of time, needs to be solved more systematically in order to fully apply the presented method, e.g. by exploiting crowd analysis and modelling crowd flow patterns. Finally, dynamical contextual information inferred from groups can be further used to complement the method presented in this work, which is not utilised in the current model.

References

1. Gheissari, N., Sebastian, T.B., Tu, P.H., Rittscher, J., Hartley, R.: Person reidentification using spatiotemporal appearance. In: Procedings of the IEEE Conference on Computer Vision and Pattern Recognition (2006)
2. Hu, W., Hu, M., Zhou, X., Lou, J., Tan, T., Maybank, S.: Principal axis-based correspondence between multiple cameras for people tracking. IEEE Trans. Pattern Anal. Mach. Intell. **28**(4), 663–671 (2006)
3. Wang, X., Doretto, G., Sebastian, T., Rittscher, J., Tu, P.: Shape and appearance context modeling. In: Proceedings of the International Conference on Computer Vision (2007)
4. Gray, D., Tao, H.: Viewpoint invariant pedestrian recognition with an ensemble of localized features. In: Proceedings of the European Conference on Computer Vision (2008)
5. Javed, O., Rasheed, Z., Shafique, K., Shah, M.: Tracking across multiple cameras with disjoint views. In: Proceedings of the International Conference on Computer Vision (2003)
6. Madden, C., Cheng, E., Piccardi, M.: Tracking people across disjoint camera views by an illumination-tolerant appearance representation. Mach. Vision Appl. **18**(3), 233–247 (2007)
7. HOSDB: Imagery library for intelligent detection systems (i-lids). In: Proceedings of the IEEE Conference on Crime and Security (2006)
8. Dollar, P., Tu, Z., Tao, H., Belongie, S.: Feature mining for image classification. In: Proceedings of the IEEE Conference on Computer Vision and Pattern Recognition (2007)
9. Gheissari, N., Sebastian, T., Hartley, R.: Person reidentification using spatiotemporal appearance. In: Proceedings of the IEEE Conference on Computer Vision and Pattern Recognition (2006)
10. Bazzani, L., Cristani, M., Murino, V.: Symmetry-driven accumulation of local features for human characterization and re-identification. Comput. Vis. Image Underst. **117**(2), 130–144 (2013)
11. Zheng, W., Gong, S., Xiang, T.: Re-identification by relative distance comparison. IEEE Trans. Pattern Anal. Mach. Intell. **35**(3), 653–668 (2013)
12. Prosser, B., Zheng, W., Gong, S., Xiang, T.: Person re-identification by support vector ranking. In: Proceedings of the British Machine Vision Conference (2010)
13. Torralba, A., Murphy, K., Freeman, W., Rubin, M.: Context-based vision system for place and object recognition. In: Proceedings of the International Conference on Computer Vision (2003)
14. Rabinovich, A., Vedaldi, A., Galleguillos, C., Wiewiora, E., Belongie, S.: Objects in context. In: Proceedings of the International Conference on Computer Vision (2007)
15. Felzenszwalb, P., Girshick, R., McAllester, D., Ramanan, D.: Object detection with discriminatively trained part-based models. IEEE Trans. Pattern Anal. Mach. Intell. **32**(9), 1627–1645 (2010)
16. Zheng, W., Gong, S., Xiang, T.: Quantifying and transferring contextual information in object detection. IEEE Trans. Pattern Anal. Mach. Intell. **34**(4), 762–777 (2012)
17. Kumar, S., Hebert, M.: A hierarchical field framework for unified context-based classification. In: Proceedings of the International Conference on Computer Vision (2005)
18. Carbonetto, P., de Freitas, N., Barnard, K.: A statistical model for general contextual object recognition. In: Proceedings of the European Conference on Computer Vision (2004)
19. Galleguillos, C., Rabinovich, A., Belongie, S.: Object categorization using co-occurrence, location and appearance. In : Proceedings of the IEEE Conference on Computer Vision and Pattern Recognition (2008)
20. Gupta, A., Davis, L.S.: Beyond nouns: exploiting prepositions and comparative adjectives for learning visual classifier. In: Proceedings of the European Conference on Computer Vision (2008)
21. Hoiem, D., Efros, A., Hebert, M.: Putting objects in perspective. Int. J. Comput. Vision **80**(1), 3–15 (2008)
22. Bao, S.Y.Z., Sun, M., Savarese, S.: Toward coherent object detection and scene layout understanding. In: Proceedings of the IEEE Conference on Computer Vision and Pattern Recognition, pp. 65–72 (2010)

23. Galleguillos, C., McFee, B., Belongie, S., Lanckriet, G.: Multi-class object localization by combining local contextual interactions. In: Proceedings of the IEEE Conference on Computer Vision and Pattern Recognition (2010)
24. Vedaldi, A., Gulshan, V., Varma, M., Zisserman, A.: Multiple kernels for object detection. In: Proceedings of the IEEE International Conference on Computer Vision (2009)
25. Brostow, G.J., Cipolla, R.: Unsupervised bayesian detection of independent motion in crowds. In: Proceedings of the IEEE Conference on Computer Vision and Pattern Recognition (2006)
26. Arandjelović, O.: Crowd detection from still images. In: Proceedings of the British Machine Vision Conference (2008)
27. Kong, D., Gray, D., Tao, H.: Counting pedestrians in crowds using viewpoint invariant training. In: Proceedings of the British Machine Vision Conference (2005)
28. Rabaud, V., Belongie, S.: Counting crowded moving objects. In: Proceedings of the IEEE Conference on Computer Vision and Pattern Recognition (2006)
29. Gong, S., Xiang, T.: Recognition of group activities using dynamic probabilistic networks. In: Proceedings of the International Conference on Computer Vision (2003)
30. Saxena, S., Brémond, F., Thonnat, M., Ma, R.: Crowd behavior recognition for video surveillance. In: Proceedings of the International Conference on Advanced Concepts for Intelligent Vision Systems (2008)
31. Zheng, W., Gong, S., Xiang, T.: Associating groups of people. In: Proceedings of the British Machine Vision Conference (2009)
32. Savarese, S., Winn, J., Criminisi, A.: Discriminative object class models of appearance and shape by correlatons. In: Proceedings of the IEEE Conference on Computer Vision and Pattern Recognition (2006)
33. Lowe, D.: Distinctive image features from scale-invariant keypoints. Int. J. Comput. Vision **2**(60), 91–110 (2004)
34. Csurka, G., Dance, C., Fan, L., Willamowski, J., Bray, C.: Visual categorization with bags of keypoints. In: Proceedings of the European Conference on Computer Vision, International Workshop on Statistical Learning in Computer Vision (2004)
35. Dalal, N., Triggs, B.: Histograms of oriented gradients for human detection. In: Proceedings of the IEEE Conference on Computer Vision and Pattern Recognition (2005)
36. He, R., Zheng, W.S., Hu, B.G.: Maximum correntropy criterion for robust face recognition. IEEE Trans. Pattern Anal. Mach. Intell. **33**(8), 1561–1576 (2011)
37. Zheng, W., Gong, S., Xiang, T.: Person re-identification by probabilistic relative distance comparison. In: Proceedings of the IEEE Conference on Computer Vision and Pattern Recognition, pp. 649–656. Colorado Springs (2011)
38. Chapelle, O., Keerthi, S.S.: Efficient algorithms for ranking with svms. Inf. Retrieval **13**(3), 201–215 (2010)
39. Fowlkes, C., Belongie, S., Chung, F., Malik, J.: Spectral grouping using the nystrom method. IEEE Trans. Pattern Anal. Mach. Intell. **26**(2), 214–225 (2004)

Chapter 10
Evaluating Feature Importance for Re-identification

Chunxiao Liu, Shaogang Gong, Chen Change Loy and Xinggang Lin

Abstract Person re-identification methods seek robust person matching through combining feature types. Often, these features are assigned implicitly with a single vector of global weights, which are assumed to be universally and equally good for matching all individuals, independent of their different appearances. In this study, we present a comprehensive comparison and evaluation of up-to-date imagery features for person re-identification. We show that certain features play more important roles than others for different people. To that end, we introduce an unsupervised approach to learning a bottom-up measurement of feature importance. This is achieved through first automatically grouping individuals with similar appearance characteristics into different prototypes/clusters. Different features extracted from different individuals are then automatically weighted adaptively driven by their inherent appearance characteristics defined by the associated prototype. We show comparative evaluation on the re-identification effectiveness of the proposed *prototype-sensitive feature importance*-based method as compared to two generic weight-based *global feature importance* methods. We conclude by showing that their combination is able to yield more accurate person re-identification.

C. Liu (✉) · X. Lin
Tsinghua University, Beijing, China
e-mail: lcx08@mails.tsinghua.edu.cn

X. Lin
e-mail: xglin@mail.tsinghua.edu.cn

S. Gong
Queen Mary University of London, London, UK
e-mail: sgg@eecs.qmul.ac.uk

C. C. Loy
The Chinese University of Hong Kong, Hong Kong, China
e-mail: ccloy@ie.cuhk.edu.hk

S. Gong et al. (eds.), *Person Re-Identification*,
Advances in Computer Vision and Pattern Recognition,
DOI: 10.1007/978-1-4471-6296-4_10,

10.1 Introduction

Visual appearance-based person re-identification aims to establish a visual match between two imagery instances of the same individual appearing at different locations and times under unknown viewing conditions which are often significantly different. Solving this problem is non-trivial owing to both very sparse samples of the person of interest, often a single example imagery to compare against, and the unknown viewing condition changes, including visual ambiguities and uncertainties caused by illumination changes, viewpoint and pose variations and inter-object occlusion [16, 27, 28]. In order to cope with sparsity of data and the challenging view conditions, most existing methods [8, 9, 17] combine different appearance features, such as colour and texture, to improve reliability and robustness in person matching. Typically, feature histograms are concatenated and weighted in accordance to their *importance*, i.e. their discriminative power in distinguishing a target of interest from other individuals.

Current re-identification techniques [19, 30, 33, 41] assume implicitly a feature weighting or selection mechanism that is *global*, i.e. a set of generic weights on feature types invariant to a population. That is, to assume a single weight vector (or a linear weight function) that is globally optimal for all people. For instance, one often assumes colour is the most important (intuitively so) and universally a good feature for matching all individuals. In this study, we refer such a generic weight vector as a Global Feature Importance (GFI) measure. They can be learned either through boosting [19], rank learning [33] or distance metric learning [41]. Scalability is the main bottleneck of such approaches as the learning process requires exhaustive supervision on pairwise individual correspondence from a known dataset.

Alternatively, we consider that certain appearance features are more important than others in describing an individual and distinguishing him/her from other people. For instance, colour is more informative to describe and distinguish an individual wearing a textureless bright red shirt, but texture information can be equally or more critical for a person wearing a plaid shirt (Fig. 10.1). It is therefore undesirable to bias all the weights to the features that are universally good for all individuals. Instead, feature weighting should be able to *selectively distribute different weights adaptively according to the informativeness of features given different visual appearance attributes under changing viewing conditions and for different people*. By visual appearance attributes, we refer to conceptually meaningful appearance characteristics of an individual, e.g. dark shirt, blue jeans.

In this study, we first provide a comprehensive review of various feature representations and weighting strategies for person re-identification. In particular, we investigate the roles of different feature types given different appearance attributes and give insights into what features are more important under what circumstances. We show that selecting features specifically for different individuals can yield more robust re-identification performance than feature histogram concatenation with GFI as adopted by [27, 37].

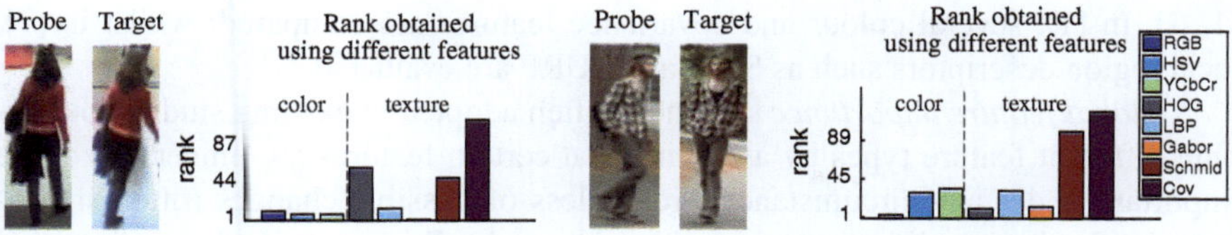

Fig. 10.1 Two examples of a pair of probe image against a target (gallery) image, together with the rank of correct matching by different feature types independently

It is non-trivial to quantify feature importance adaptively driven by specific appearance attributes detected on an individual. A plausible way is to apply supervised attribute learning method, i.e. training a number of attribute detectors to cover an exhaustive set of possible attributes, and then defining feature importance associated to each specific attribute. This method requires expensive annotation and yet the annotation obtained may have low quality due to inevitable visual ambiguity. Previous studies [10, 18, 29] have shown great potential in using unsupervised attributes in various computer vision problems such as object recognition. Despite that the unsupervised attributes are not semantically labelled or explicitly named, they are discriminative and correlated with human attribute perception.

Motivated by the unsupervised attribute studies, we investigate here a random forests-based method to discover *prototypes* in an unsupervised manner. Each prototype reveals a mixture of attributes to describe specific population of persons with similar appearance characteristics, such as wearing colourful shirt and black pants. With the discovered prototypes, we further introduce an approach to quantify the feature importance specific for an individual driven by his/her inherent appearance attributes. We call the discovered feature importance *Prototype-Sensitive Feature Importance* (PSFI). We conduct extensive evaluation using four different person re-identification benchmark datasets, and show that combining prototype-sensitive feature importance with global feature importance can yield more accurate re-identification without any extra supervision cost as compared to existing learning-based approaches.

10.2 Recent Advances

Most person re-identification methods benefit from integrating several types of features [1, 8, 9, 14, 17, 19, 26, 33, 36, 37, 41]. In [17], weighted colour histograms derived from maximally stable colour regions (MSCR) and structured patches are combined for visual description. In [8], histogram plus epitome features are proposed as a human signature. Essentially, they explore the combination of colour and texture properties on the human appearance but with more specific feature types. There are a number of reviews on features and feature evaluation for person re-identification

[1, 7]. In [1], several colour and covariance features are compared; whilst in [7], local region descriptors such as SIFT and SURF are evaluated.

A *global feature importance* scheme is often adopted in existing studies to combine different feature types by assuming that certain features are universally more important under any circumstances, regardless of possible changes (often significant) in viewing conditions between the probe and gallery views and specific visual appearance characteristics of different individuals. Recent advances based on metric learning or ranking [2, 19, 21, 30, 33, 41] can be considered as data-driven *global feature importance mining* techniques. For example, the ranking support vector machines (RankSVM) method [33] converts the person re-identification task from a matching problem into a pairwise binary classification problem (correct match vs. incorrect match), and aims to find a linear function to weight the absolute difference of samples via optimisation given pairwise relevance constraints. The Probabilistic Relative Distance Comparison (PRDC) [41] maximises the probability of a pair of true match having a smaller distance than that of a wrong matched pair. The output is an orthogonal matrix that encodes the global importance of each feature. In essence, the learned global feature importance reflects the stability of each feature component across two cameras. For example, if two camera locations are under significantly different lighting conditions, the colour features will be less important as they are unstable/unreliable. A major weakness of this type of pairwise learning-based methods is their potential limitation on scalability since the supervised learning process requires exhaustive supervision on pairwise correspondence, i.e. the building of a training set is cumbersome as it requires to have for each subject a pair of visual instances. The size of such a pairwise labelled dataset required for model learning is difficult to be scaled up.

Schwartz and Davis [36] propose a feature selection process depending on the feature type and the location. This method, however, requires labelled gallery images to discover the gallery-specific feature importance. To relax such conditions, in this work we investigate a fully unsupervised learning method for adaptive feature importance mining which aims to be more flexible (attribute-driven) without any limitations to a specific gallery set. A more recent study in [34] explores prototype relevance for improving processing time in re-identification. In a similar spirit but from a different perspective, this study investigates salient feature importance mining based on prototype discovery for improving matching accuracy. In [23], a supervised attribute learning method is proposed to describe the appearance for each individual. However, it needs massive human annotation of attributes which is labour-intensive. In contrast, we explore in an unsupervised way to discover the inherent appearance attributes.

10.3 Feature Representation

Different types of visual appearance features have been proposed for person re-identification, including colour histogram [17, 22], texture filter banks [33], shape context [37], covariance [3, 4, 6] and histogram plus epitome [8]. In general, colour information is dominant when the lighting changes are not severe, as colour is more robust to viewpoint changes as compared to other features. Although texture or structure information can be more stable under significant lighting changes, they are sensitive to changes in viewpoint and occlusion. As shown in [8, 17], re-identification matching accuracy can be improved by combining several features so as to gain benefit from different and complementary information captured by different features.

In this study, we investigate a mixture of commonly used colour, structure and texture features for re-identification, similar to those employed in [19, 33], plus a few more additional local structure features. In particular, the following range of imagery features are considered:

- **Colour Histogram**: HSV colour histogram is employed in [8, 17, 36]. Specifically, in [17] they generate a weighted colour histogram according to pixel's location to the vertical symmetry axes of the human body. The intuition is that central pixels should be more robust to pose variations. HSV is effective in describing the bright colours, such as red, but not robust to neutral colour as the hue channel is undefined. An alternative representation is to combine the colour histograms from several complementary colour spaces, such as HSV, RGB, and YCbCr [19, 21, 33, 41].
- **Texture and Structure**: Texture and structure patterns are commonly found on clothes, such as the plaid (see Fig. 10.1) or the stripes (see Fig. 10.5b) on a sweater. Possible texture descriptors include Gabor and Schmid filters [19, 33] or local binary patterns (LBP) [39]. As to structure descriptor, histogram of gradient (HOG) [15] that prevails in human detection is considered in [1, 36, 37]. As these texture and structure features are computed on the intensity image, they play an important role in establishing correspondence when colour information degrades under drastic illumination changes and/or change of camera settings.
- **Covariance**: Covariance feature has been reported to be effective in [4, 5, 20, 24]. It has three advantages: (1) it reflects second-order regional statistical property discarded by histogram; (2) different feature types such as colour and gradient can be readily integrated; (3) it is versatile with no limitation to the region's shape, suggesting its potential to be integrated with most salient region detectors.

In this study, we divide a person image into six horizontal stripes (see Fig. 10.2). This is a generic human body partitioning method that is widely used in existing methods [33, 41] to capture distinct areas of interest. Alternative partitioning schemes, such as symmetry segmentation [8] or pictorial model [14], are also applicable. A total of 33 feature channels including RGB, HSV, YCbCr, Gabor (8 filters), Schmid (13 filters), HOG, LBP and Covariance are computed for each stripe. For the first five types of features, each channel is represented by a 16-dimensional vector. A detailed explanation of computing the former 5 features can be found in [33]. For HOG feature, each strip is further divided into 4×4 pixels cell and each cell is represented by a 9-dimensional gradient histogram, yielding a 36-dimensional feature vector for each strip. For LBP feature, we compute a 59-dimensional local binary pattern histogram on the intensity image. As for covariance feature for a given strip

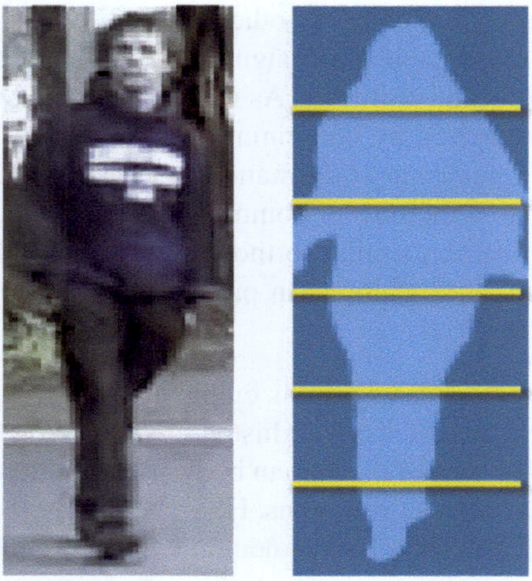

Fig. 10.2 A spatial representation of human body [33, 41] is used to capture visually distinct areas of interest. The representation employs six equal-sized horizontal strips in order to capture approximately the head, upper and lower torso and upper and lower legs

$R \subset I$, let $\{\mathbf{z}_m\}_{m=1\ldots M}$ be the feature vectors extracted from M pixels inside R. The covariance descriptor of region R is derived by

$$\mathbf{C}_R = \frac{1}{M-1}\sum_{m=1}^{M}(\mathbf{z}_m - \mu)(\mathbf{z}_m - \mu)^{\mathsf{T}}$$

where μ denotes the mean vector of $\{\mathbf{z}_m\}$. Here we use the following features to reflect information of each pixel $\mathbf{z} = [H, S, V, I_x, I_y, I_{xx}, I_{yy}]$ where H, S, V are the HSV colour values. The first-order (I_x and I_y) and second-order (I_{xx} and I_{yy}) image derivatives are calculated through the filters $[-1, 0, 1]^{\mathsf{T}}$ and $[-1, 2, -1]^{\mathsf{T}}$, respectively. The subscript x or y denotes the direction for filtering. Thus the covariance descriptor is a 7×7 matrix. While in this form covariance matrix cannot be directly combined with other features to form a single histogram representation. Hence, we follow the approach proposed by [20] to convert the 7×7 covariance matrix $\mathbf{C}$ into *sigma points*, expressed as follows:

$$\mathbf{s}_0 = \mu \tag{10.1}$$

$$\mathbf{s}_i = \mu + \alpha(\sqrt{\mathbf{C}})_i \tag{10.2}$$

$$\mathbf{s}_{i+d} = \mu - \alpha(\sqrt{\mathbf{C}})_i, \tag{10.3}$$

where μ is the mean value of sample data and $(\sqrt{\mathbf{C}})_i$ denotes the i-th column of the covariance matrix square root. Parameter α is a scalar weight for the elements in $\mathbf{C}$ and is set to $\alpha = \sqrt{2}$ for Gaussian data. Thus, the vector form of covariance feature can be obtained by concatenation of all sigma points, in our case resulting in a 105-dimensional vector. Therefore, it allows for integration of other feature channels into one compact feature vector.

10.4 Unsupervised Mining of Feature Importance

Given the range of features included in our feature representation, we consider an unsupervised way to compute and evaluate a bottom-up measurement of feature importance driven by intrinsic appearance of individuals. To that end, we propose a three-step procedure as follows: (1) automatic discovery of feature prototypes by exploiting clustering forests; (2) prototype-sensitive feature importance mining by classification forests; (3) determining the feature importance of a probe image on-the-fly adapting to changes in viewing condition and inherent appearance characteristics of individuals. An overview of the proposed approach is depicted in Fig. 10.3.

Our unsupervised feature importance mining method is formulated based on random forests models, particularly the clustering forests [25] and classification forests [12]. Before introducing and discussing the proposed method, we briefly review the two forests models.

10.4.1 Random Forests

Random forests [12] are a type of decision trees constructed by an ensemble learning process, and can be designed for performing either classification, clustering or regression tasks. Random forests have a number of specific properties that make it suitable for the re-identification problem. In particular

1. It defines the pairwise affinity between image samples by the tree structure itself, therefore, avoiding manual definition of distance function.
2. It selects implicitly optimal features via optimisation of the well-defined information gain function [12]. This feature selection mechanism is beneficial to mitigating noisy or redundant visual features in our representation.
3. It performs empirically well on high-dimensional input data [13], a problem that is typical in person re-identification problem.

In addition to the three aforementioned characteristics, there are other attractive general properties in random forests such as it approximates the Bayes optimal classifier [35], it handles inherently multiple-class problem and it provides probabilistic outputs.

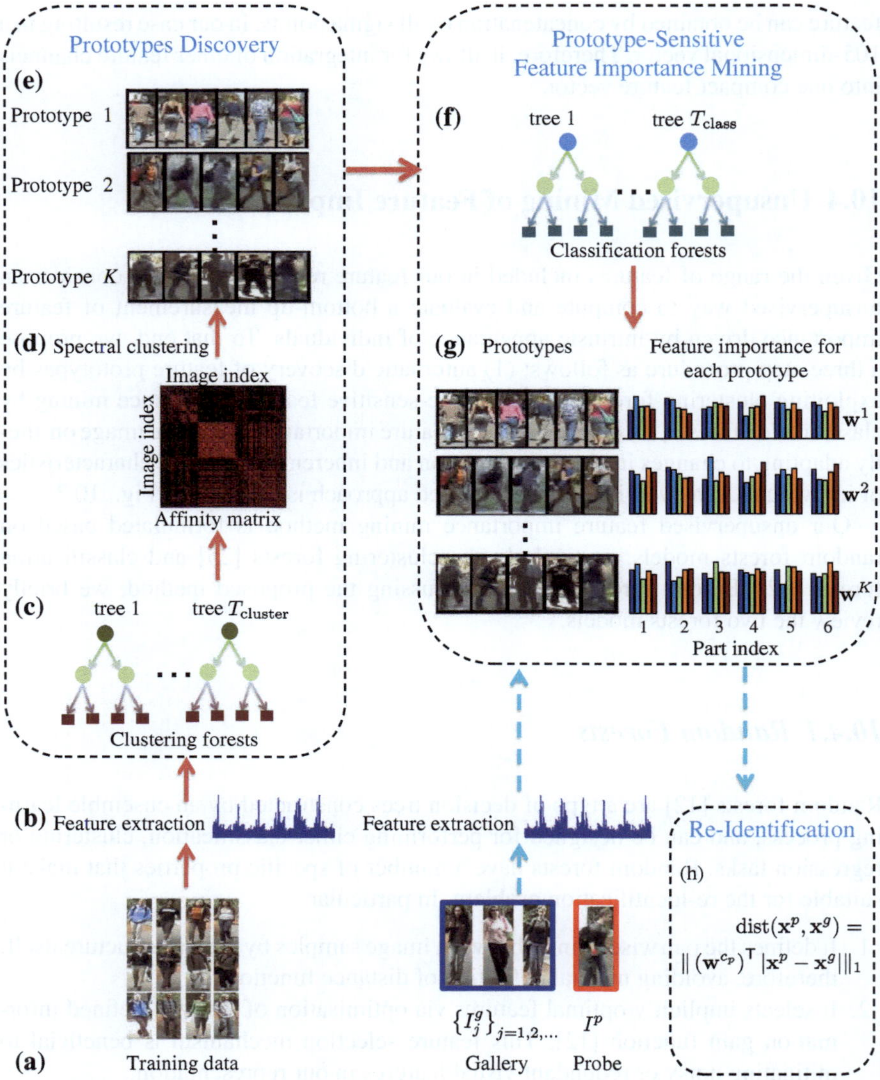

Fig. 10.3 Overview of prototype-sensitive feature importance mining. Training steps are indicated by *red solid arrows* and testing steps are denoted by *blue slash arrows*

Classification Forests

A common type of random forests is the classification forests. Classification forests [12, 35] consists of a set of T_{class} binary decision trees $\mathcal{T}(\mathbf{x}) : \mathcal{X} \to \mathcal{R}^K$, where $\mathcal{X} = \mathbb{R}^D$ is the D-dimensional feature space and $\mathcal{R}^K = [0, 1]^K$ represents the space of class probability distribution over the label space $\mathcal{C} = \{1, \ldots, K\}$. During testing, given an unseen sample $\mathbf{x}^* \in \mathbb{R}^D$, each decision tree produces a posterior $p_t(c|\mathbf{x})$, and a probabilistic output from the forests can be obtained via averaging

$$p(c|\mathbf{x}^*) = \frac{1}{T_{\text{class}}} \sum_{t}^{T_{\text{class}}} p_t(c|\mathbf{x}^*). \tag{10.4}$$

The final class label c^* can be obtained as $c^* = \text{argmax}_c p(c|\mathbf{x}^*)$.

In the learning stage, each decision tree is trained independently from each other using a random subset of training samples, i.e. bagging [12]. Typically, one draws $\frac{2}{3}$ of the original training samples randomly for growing a tree, and reserves the remaining as *out-of-bag* (oob) validation samples. We will exploit these oob samples for computing importance of each feature (Section 10.4.3).

Growing a decision tree involves an iterative node splitting procedure that optimises a binary split function of each internal node. We define the split function as:

$$h(\mathbf{x}, \boldsymbol{\theta}) = \begin{cases} 0 & \text{if } \mathbf{x}_{\boldsymbol{\theta}_1} < \boldsymbol{\theta}_2 \\ 1 & \text{otherwise} \end{cases}. \tag{10.5}$$

The above split function is parameterised by two parameters: (i) a feature dimension $\boldsymbol{\theta}_1 \in \{1, \ldots, D\}$, and (ii) a threshold $\boldsymbol{\theta}_2 \in \mathbb{R}$. Based on the outcome of Eq. (10.5), a sample $\mathbf{x}$ arriving at the split node will be channelled to either the left or right child nodes.

The best parameter $\boldsymbol{\theta}^*$ is chosen by optimising

$$\boldsymbol{\theta}^* = \underset{\boldsymbol{\theta} \in \Theta}{\text{argmax}}\, \Delta\mathcal{I}, \tag{10.6}$$

where Θ is a randomly sampled set of $\{\boldsymbol{\theta}^i\}$. The information gain $\Delta\mathcal{I}$ is defined as follows:

$$\Delta\mathcal{I} = \mathcal{I}_p - \frac{n_l}{n_p}\mathcal{I}_l - \frac{n_r}{n_p}\mathcal{I}_r, \tag{10.7}$$

where p, l and r refer to a splitting node, the left and right child, respectively; n denotes the number of samples at a node, with $n_p = n_l + n_r$. The $\mathcal{I}$ can be computed as either the entropy or Gini impurity [11]. Throughout this paper we use the Gini impurity.

Clustering Forests

In contrast to classification forests, clustering forests does not require any ground truth class labels for learning. Therefore, it is suitable for our problem of unsupervised prototype discovery. Clustering forests consists of T_{cluster} decision trees whose leaves define a spatial partitioning or grouping of the data. Although the clustering forests is an unsupervised model, it can be trained using the classification forests optimisation routine by following the pseudo two-class algorithm proposed in [12, 25]. In particular, in each splitting node we add n_p uniformly distributed pseudo points $\bar{\mathbf{x}} = \{\bar{x}_1, \ldots, \bar{x}_D\}$, with $\bar{x}_i \sim \mathbf{U}(x_i \,|\, \min(x_i), \max(x_i))$ into the original data space. With this strategy, the clustering problem becomes a canonical classification problem that can be solved by the classification forests training method discussed above.

10.4.2 Prototype Discovery

Now we discuss how to achieve feature importance mining through a clustering-classification forests model. First, we describe how to achieve prototype discovery by clustering forests (Fig. 10.3a–e). In contrast to a top-down approach to specifying appearance attributes and mining features to support each attribute class independently [23], in this study we investigate bottom-up approach to discovering automatically representative clusters (prototypes) corresponding to similar constitutions of multiple classes of appearance attributes.

To that end, we first perform unsupervised clustering to group a given set of unlabelled images into several *prototypes* or clusters. Each prototype is composed of images that possess similar appearance attributes, e.g. wearing colourful shirt, with backpack, dark jacket (Fig. 10.3e). More precisely, given an input of n unlabelled images $\{I_i\}$, where $i = 1, \ldots, n$, feature extraction $f(\cdot)$ is first performed on every image to extract a D-dimensional feature vector, that is $f(I) = \mathbf{x} = (x_1, \ldots, x_D)^\top \in \mathbb{R}^D$ (Fig. 10.3b). We wish to discover a set of prototypes $c \in \mathcal{C} = \{1, \ldots, K\}$, i.e. low-dimensional manifold clusters that group images $\{I\}$ with similar appearance attributes.

We treat the prototype discovery problem as a graph partitioning problem, which requires us to first estimate the pairwise similarity between images. We adopt the clustering forests [12, 25] for pairwise similarity estimation. Formally, we construct clustering forests as an ensemble of T_{cluster} clustering trees (Fig. 10.3c). Each clustering tree t defines a partition of the input samples $\mathbf{x}$ at its leaves, $l(\mathbf{x}) : \mathbb{R}^D \rightarrow \mathcal{L} \subset \mathbb{N}$, where l represents a leaf index and $\mathcal{L}$ is the set of all leaves in a given tree. Now for each tree, we are able to compute an $n \times n$ affinity matrix A^t, with each element A^t_{ij} defined as

$$A^t_{ij} = \exp^{-\text{dist}^t(\mathbf{x}_i, \mathbf{x}_j)}, \tag{10.8}$$

where

$$\text{dist}^t\left(\mathbf{x}_i, \mathbf{x}_j\right) = \begin{cases} 0 & \text{if } l(\mathbf{x}_i) = l(\mathbf{x}_j) \\ \infty & \text{otherwise} \end{cases}. \tag{10.9}$$

Following Eq. (10.9), we assign closest affinity $= 1$ (distance $= 0$) to samples $\mathbf{x}_i$ and $\mathbf{x}_j$ if they fall into the same leaf node, and affinity $= 0$ (distance $= \infty$) otherwise. To obtain a smooth forests affinity matrix, we compute the final affinity matrix as

$$A = \frac{1}{T_{\text{cluster}}} \sum_{t=1}^{T_{\text{cluster}}} A^t. \tag{10.10}$$

Given the affinity matrix, we perform spectral clustering algorithm [31] to partition the weighted graph into K prototypes. Thus, each unlabelled probe image $\{I_i\}$ is assigned to a prototype c_i (Fig. 10.3e). In this study, K is the cluster number and pre-defined, but its value can be readily estimated automatically using alternative methods such as [32, 38].

10.4.3 Prototype-Sensitive Feature Importance

In this section, we discuss how to derive the feature importance for each prototype generated by the previous prototype discovery. As discussed in Sect. 10.1, unlike the global feature importance that is assumed to be universally good for all images, prototype-sensitive feature importance is designed to be specific to prototype characterised by different appearance characteristics. That is, each prototype c has its own prototype-sensitive weighting or feature importance (PSFI)

$$\mathbf{w}^c = \left(w_1^c, \ldots, w_D^c\right)^{\mathsf{T}}, \tag{10.11}$$

of which high value should be assigned to unique features of that prototype. For example, texture features gain higher weights than others if the images in the prototype have rich textures but less bright colours.

Based on the above consideration, we compute the importance of a feature according to its ability in discriminating different prototypes. The forests model naturally reserves a validation set or out-of-bag (oob) samples for each tree during bagging (Sect. 10.4.1). This property permits a convenient and robust way of evaluating the importance of individual features.

Specifically, we train a classification random forests [12] using $\{\mathbf{x}\}$ as inputs and treating the associated prototype labels $\{c\}$ as classification outputs (Fig. 10.3f). To compute the feature importance, we first compute the classification error $\varepsilon_d^{c,t}$ for every dth feature in prototype c. Then we randomly permute the value of the dth feature in the oob samples and compute the $\widetilde{\varepsilon}_d^{c,t}$ on the perturbed oob samples of prototype c. The importance of the dth feature of prototype c is then computed as the error gain

$$w_d^c = \frac{1}{T_{\text{class}}} \sum_{t=1}^{T_{\text{class}}} (\widetilde{\varepsilon}_d^{c,t} - \varepsilon_d^{c,t}). \tag{10.12}$$

Higher value in w_d^c indicates higher importance of the dth feature in prototype c. Intuitively, the dth feature is important if perturbing its value in the samples causes a drastic increase in classification error, therefore suggests its critical role in discriminating between different prototypes.

10.4.4 Ranking

With the method described in Sect. 10.4.3, we obtain PSFI for each prototypes. This subsequently permits us to evaluate bottom-up feature importance of an unseen probe image, $\mathbf{x}^p$ on-the-fly driven by its intrinsic appearance prototype. Specifically, following Eq. (10.4), we classify $\mathbf{x}^p$ using the learned classification forests to obtain its prototype label c_p

$$c_p = \text{argmax}_c p(c|\mathbf{x}^p), \tag{10.13}$$

and obtain accordingly its feature importance $\mathbf{w}^{c_p}$ (Fig. 10.3h). Then we compute the distance between $\mathbf{x}^p$ against a feature vector of a gallery/target image $\mathbf{x}^g$ using the following function:

$$\text{dist}(\mathbf{x}^p, \mathbf{x}^g) = \|\left(\mathbf{w}^{c_p}\right)^\mathsf{T} |\mathbf{x}^p - \mathbf{x}^g|\|_1 \tag{10.14}$$

The matching ranks of $\mathbf{x}^p$ against a gallery of images can be obtained by sorting the distances computed from Eq. (10.14). A smaller distance results in a higher rank.

10.4.5 Fusion of Different Feature Importance Strategies

Contemporary methods [33, 41] learn a weight function that captures the global environmental viewing condition changes which cannot be derived from the unsupervised method described so far. Thus we investigate the fusion between the global feature weight matrix obtained from [33, 41] and our prototype-sensitive feature importance vector $\mathbf{w}$ to gain more accurate person re-identification performance.

In general, methods [33, 41] aim to optimise a distance metric so that a true match pair lies closer than a false match pair, given a set of relevance rank annotations. The distance metric can be written as

$$d(\mathbf{x}_i^p, \mathbf{x}_j^g) = (\mathbf{x}_i^p - \mathbf{x}_j^g)^\mathsf{T} \mathbf{V} (\mathbf{x}_i^p - \mathbf{x}_j^g). \tag{10.15}$$

The optimisation process involves finding a semi-positive definite global feature weight matrix $\mathbf{V}$. There exist several global feature weighting methods, most of them differing by different constraints and optimisation schemes they use (see Sect. 10.2 for discussion).

To combine our proposed prototype-sensitive feature importance with the global feature importance, we adopt a weighted sum scheme as follows:

$$\mathrm{dist}_{\mathrm{fusion}}(\mathbf{x}^p, \mathbf{x}^g) = \alpha \| \left(\mathbf{w}^{c_p}\right)^{\mathsf{T}} |\mathbf{x}^p - \mathbf{x}^g| \|_1 + (1-\alpha) \| \mathbf{V}^{\mathsf{T}} |\mathbf{x}^p - \mathbf{x}^g| \|_1, \quad (10.16)$$

where $\mathbf{V}$ is the global weight matrix obtained from Eq. (10.15) and α is a parameter that balances global and prototype-sensitive feature importance scores. We found that setting α in the range of [0.1, 0.3] gives stable empirical performance across all the datasets we tested. We fix it to 0.1 in our experiments. Note that setting a small α implies a high emphasis on the global weight derived from supervised learning. This is reasonable since performance gain in re-identification still has to rely on the capability of capturing the global viewing condition changes, which requires supervised weight learning. We shall show in the following evaluation that this fused metric is able to benefit from both feature importance mining from individual visual appearance changes, whilst taking into account the generic global environmental viewing condition changes between camera views.

10.5 Evaluation

In Sect. 10.5.2, we first investigate the re-identification performance of using different features given individuals with different inherent appearance attributes. In Sect. 10.5.3, the qualitative results of prototype discovery are presented. Sect. 10.5.4 then compares feature importance produced by the proposed unsupervised bottom-up prototype discovery method and two top-down GFI methods, namely RankSVM [33] and PRDC [41]. Finally, we report the results on combining the bottom-up and the top-down feature importance mining strategies.

10.5.1 Settings

We first describe the experimental settings and implementation details.

Datasets Four publicly available person re-identification datasets are used for evaluation. They are VIPeR [19], i-LIDS Multiple-Camera Tracking Scenario (i-LIDS) [40], QMUL underGround Re-IDentification (GRID) [27] and Person Re-IDentification 2011 (PRID2011) [20]. Example images of these datasets are shown in Fig. 10.4. More specifically,

1. The VIPeR dataset (see Fig. 10.4a) contains 632 persons, each of which has two images captured in two different outdoor views. The dataset is challenging due to drastic appearance difference between most of the matched image pairs caused by viewpoint variations and large illumination changes at outdoor environment (see also Fig. 10.5a, b).
2. The i-LIDS dataset (see Fig. 10.4b) was captured in a busy airport arrival hall using multiple cameras. It contains 119 people with a total of 476 images, with an average of four images per person. Apart from illumination changes and pose

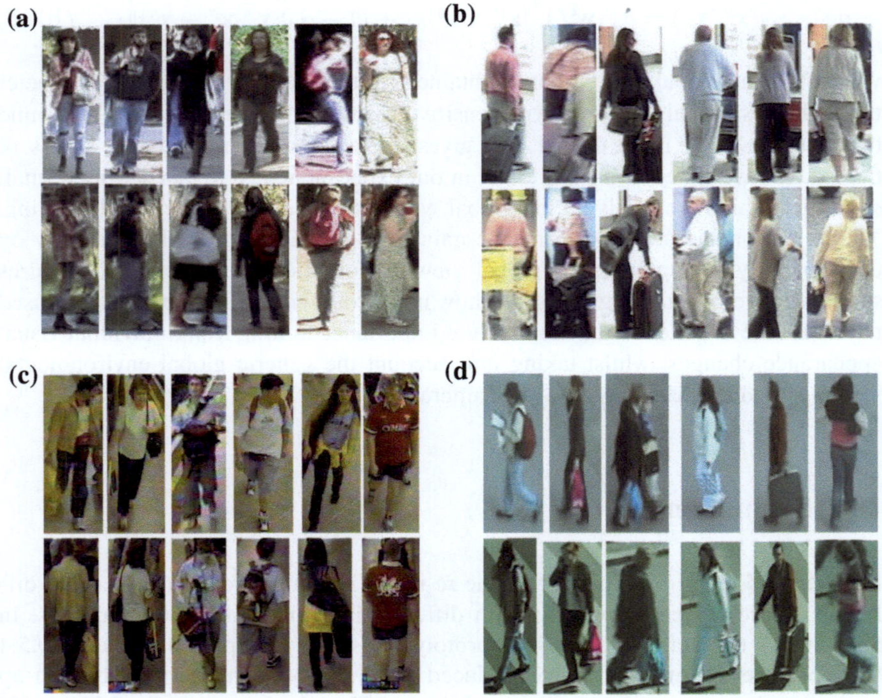

Fig. 10.4 Example images of different datasets used in our evaluation. Each column denotes an image pair of the same person. Note the large appearance variations within an image pair. In addition, note the unique appearance characteristics of different individuals, which can potentially be used to discriminate him/her from other candidates. **a** VIPeR, **b** i-LIDS, **c** GRID, **d** PRID2011

variations, many images in this dataset are also subject to severe inter-object occlusion (see also Fig. 10.5c, d).

3. The GRID dataset (see Fig. 10.4c) was captured from eight disjoint camera views installed in a busy underground station. It was divided into probe and gallery sets. The probe set contains 250 persons, whilst the gallery set contains 1,025 persons in which an additional 775 persons were collected who do not match any images in the probe set. The dataset is challenging due to severe inter-object occlusion, large viewpoint variations and poor image quality (see also Fig. 10.5e, f).
4. The PRID2011 dataset (see Fig. 10.4d) was captured from two outdoor cameras. We use the single-shot version in which each person is only associated with one picture in a camera. The two cameras contains 385 and 749 individuals separately, within which the first 200 persons have two views. The challenge lies in severe lighting changes caused by the sunlight (see also Fig. 10.5g, h).

A summary of these datasets is given in Table 10.1.

Features In Sect. 10.5.2, we employ all the feature types discussed in Sect. 10.3 for a comprehensive evaluation of their individual performance in person

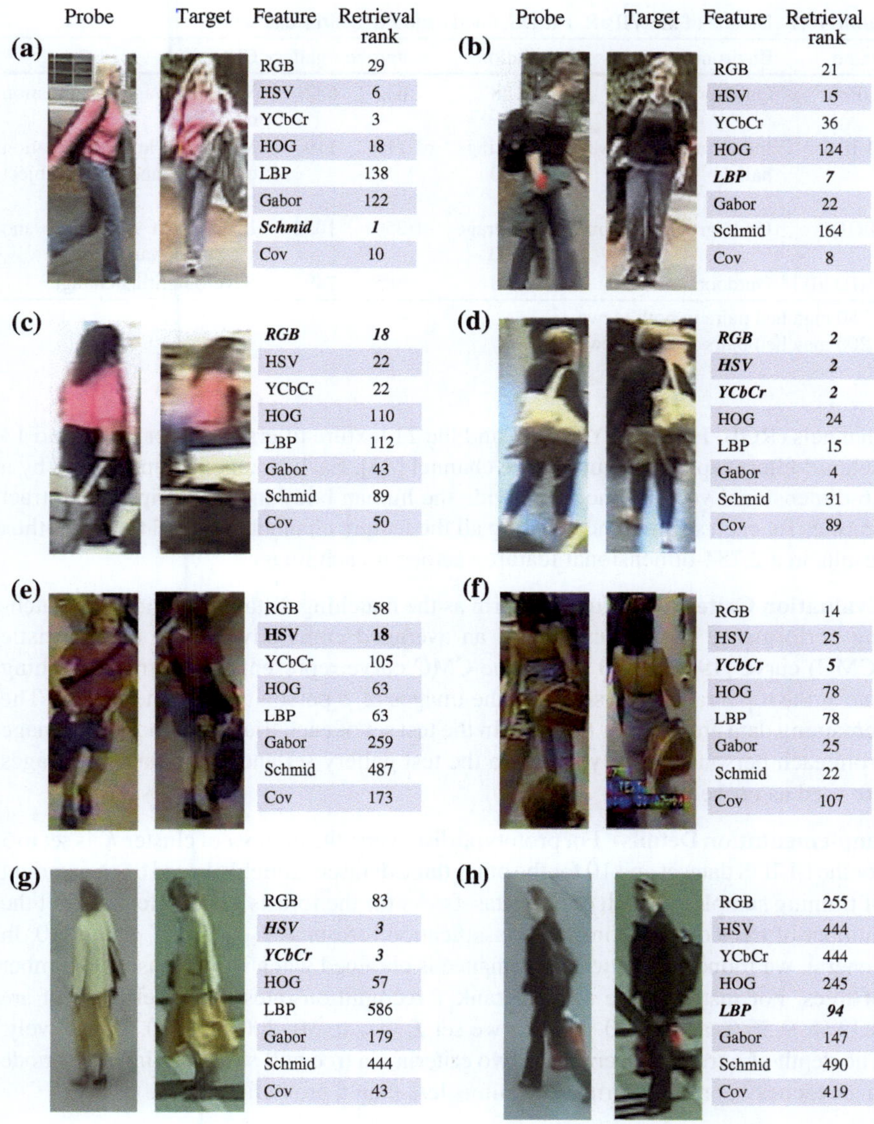

Fig. 10.5 Feature effectiveness in re-identification—in each subfigure, we show the probe image and the target image, together with the rank of correct matching by using different feature types separately

re-identification. In Sect. 10.5.3, we select from the aforementioned feature channels to form a feature subset, which is identical to those used in existing GFI mining methods [30, 33, 41]. Having the similar set of features allows a fair and comparable evaluation against the methods. Specifically, we consider 8 colour

Table 10.1 Details of the VIPeR, i-ILDS, GRID and PRID2011 datasets

Name	Environment	Resolution	#probe	#gallery	Challenges
VIPeR	Outdoor	48 × 128	632	632	Viewpoint and illumination changes
i-LIDS	Indoor airport arrival hall	An average of 60 × 150	119	119	Viewpoint and illumination changes and inter-object occlusion
GRID[a]	Underground station	An average of 70 × 180	250	1050	Inter-object occlusion and viewpoint variations
PRID2011[b]	Outdoor	64 × 128	385	749	Severe lighting change

[a] 250 matched pairs in both views
[b] 200 matched pairs in both views

channels (RGB, HSV and YCbCr)[1] and the 21 texture filters (8 Gabor filters and 13 Schmid filters) applied to luminance channel [33]. Each channel is represented by a 16-dimensional vector. Since we divide the human body into six strips and extract features for each strips, concatenating all the feature channels from all the strips thus results in a 2,784-dimensional feature vector for each image.

Evaluation Criteria We use $\ell 1$-norm as the matching distance metric. The matching performance is measured using an averaged cumulative match characteristic (CMC) curve [19] over 10 trials. The CMC curve represents the correct matching rate at the top r ranks. We select all the images of p person to build the test set. The remaining data are used for training. In the test set of each trial, we choose one image from each person randomly to set up the test gallery set and the remaining images are used as probe images.

Implementation Details For prototype discovery, the number of cluster K is set to 5 for the i-LIDS dataset and 10 for the other three datasets, roughly based on the amount of training samples in each of the datasets. As for the forests' parameters, we set the number of trees of clustering and classification forests as $T_{\text{cluster}} = T_{\text{class}} = 200$. In general, we found that better performance is obtained when we increase the number of trees. For instance, the average rank 1 recognition rates on VIPeR dataset are 8.32 %, 9.56 % and 10.00 % when we set T_{cluster} to 50, 200 and 500, respectively. The depth of a tree is governed by two criteria—a tree will stop growing if the node size reaches 1, or the information gain is less than a pre-defined value.

10.5.2 Comparing Feature Effectiveness

We assume that certain features can be more important than others in describing an individual and distinguishing him/her from other people. To validate this hypothesis,

[1] Since HSV and YCbCr share similar luminance/brightness channel, dropping one of them results in a total of 8 channels.

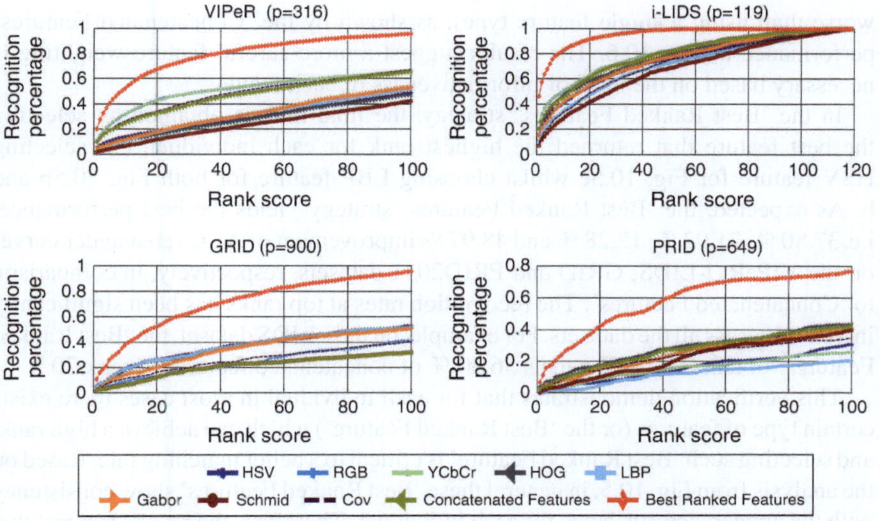

Fig. 10.6 The CMC performance comparison of using different features on various datasets. 'Concatenated Features' refers to the concatenation of all feature histograms with uniform weighting. In the 'Best Ranked Features' strategy, ranking for each individual was selected based on the best feature that returned the highest rank during matching. Its better performance suggests the importance and potential of selecting the right features specific to different individuals/groups

we analyse the matching performance of using different features individually as a proof of concept.

We first provide a few examples in Fig. 10.5 (also presented in Fig. 10.1) to compare the ranks returned by using different feature types. It is observed that no single feature type is able to constantly outperform the others. For example, for individuals wearing textureless but colourful and bright clothing, e.g. Fig. 10.5a, c and g, the colour features generally yield a higher rank. For persons wearing clothing with rich texture (on the shirt or skirt), e.g. Fig. 10.5b and d, texture features especially the Gabor features and the LBP features tend to dominate. The results suggest that certain features can be more informative than others given different appearance attributes.

The overall matching performance of using individual feature types is presented in Fig. 10.6. In general, HSV and YCbCr features exhibit very close performances, which are much superior over all other features. This observation of colours being the most informative features agreed with the past studies [19]. Among the texture and structure features, the Gabor filter banks produce the best performance across all the datasets. Note that the performance of covariance feature can be further improved when combined with a more elaborative region partitioning scheme, as shown in [5].

One may consider concatenating all the features together, with the hope that these features could complement each other leading to better performance. From our experiments, we found that a naive concatenation of all feature histograms with uniform weighting does not necessary yield better performance (sometimes even worse than using a single feature type), as shown by the 'Concatenated Features' performance in Fig. 10.6. The results suggest a more careful feature weighting is necessary based on the level of informativeness of each feature.

In the 'Best Ranked Features' strategy, the final rank is obtained by selecting the best feature that returned the highest rank for each individual, e.g. selecting HSV feature for Fig. 10.5e whilst choosing LBP feature for both Fig. 10.5b and h. As expected, the 'Best Ranked Features' strategy yields the best performance, i.e. 37.80 %, 21.92 %, 15.28 % and 48.97 % improvement of AUC (area under curve) on the VIPeR, i-LIDS, GRID and PRID2011 datasets, respectively, in comparison to 'Concatenated Features'. The recognition rates at top ranks has been significantly increased across all the datasets. For example, on the i-LIDS dataset, the 'Best Ranked Features' obtains 92.02 % versus 56.30 % of concatenated features at rank 20.

This verification demonstrates that for each individual in most cases there exists certain type of features (or the 'Best Ranked Feature') which can achieve a high rank, and selecting such 'Best Ranked Feature' is critical to a better matching rate. Based on the analysis from Fig. 10.5, in general these 'Best Ranked Features' show consistency with the appearance attributes for each individual. Therefore, the results suggest that the overall matching performance can be boosted potentially by weighting features selectively according to the inherent appearance attributes.

10.5.3 Discovered Prototypes

It is non-trivial to weigh features in accordance to their associated inherent appearance attributes. We formulate a method to first discover prototypes, i.e. low-dimensional manifold clusters that aim to correlate features contributing towards similar appearance attributes.

Some examples of prototypes discovered from the VIPeR dataset are depicted in Fig. 10.7. Each colour-coded row represents a prototype. A short list of possible attributes discovered/interpreted in each prototype is given in the caption. Note that these inherent attributes are neither pre-defined nor pre-labelled, but discovered automatically by the unsupervised clustering forests (Sect. 10.4.2).

As shown by the example members in each prototype, images with similar attributes are likely to be categorised into the same cluster. For instance, a majority of images in the second prototype can be characterised with bright and high contrast attributes. In the fourth prototype, the key attributes are 'carrying backpack' and 'side pose'. These results demonstrate that the formulated prototype discovering mechanism is capable of generating reasonably good clusters of inherent attributes, which can be employed in subsequent step for prototype-sensitive feature importance mining.

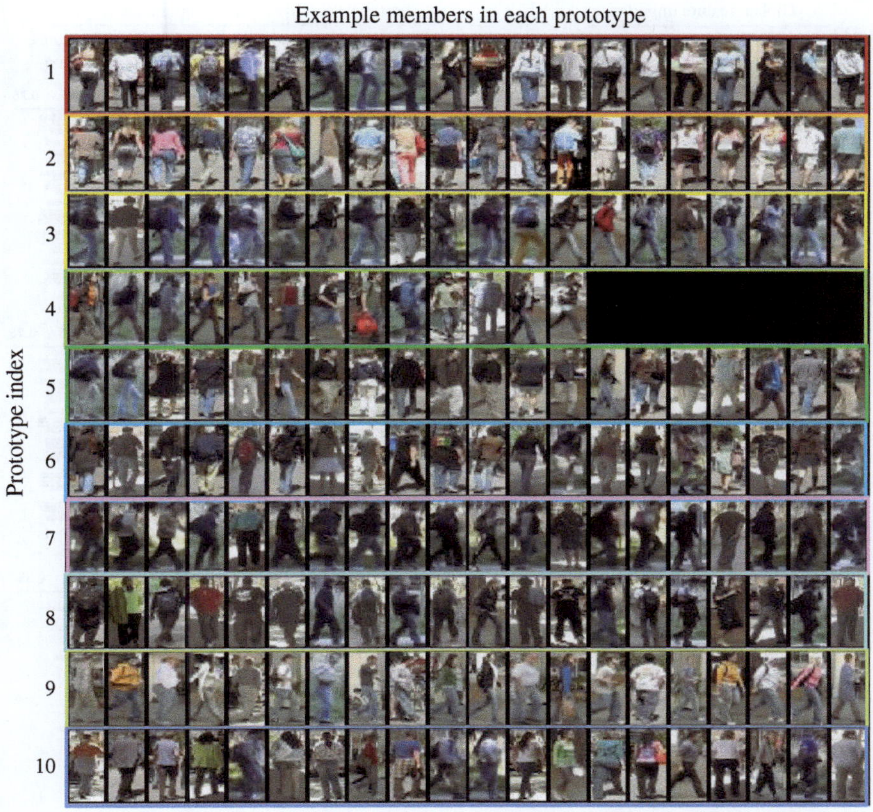

Fig. 10.7 Examples of prototypes discovered automatically from the VIPeR dataset. Each prototype represents a low-dimensional manifold cluster that models similar appearance attributes. Each image row in the figure shows a few examples of images in a particular prototype, with their interpreted unsupervised attributes listed as follows: (1) white shirt, dark trousers; (2) bright and colourful shirt; (3) dark jacket and jeans; (4) with backpack and side pose; (5) dark jacket and light colour trousers; (6) dark shirt with texture, back pose; (7) dark shirt and side pose; (8) dark shirt and trousers; (9) colourful shirt jeans; (10) colourful shirt and dark trousers

10.5.4 Prototype-Sensitive Versus Global Feature Importance

Comparing Prototype-Sensitive and Global Feature Importance The aim of this experiment is to compare different feature importance measures computed by existing GFI approaches [33, 41] and the proposed PSFI mining approach. The RankSVM [33] and PRDC [41] (see Sect. 10.1) were evaluated using the authors' original code. The global feature importance scores/weights were learned using the labelled images, and averaged over tenfold cross-validation. We set the penalty parameter C in RankSVM to 100 for all the datasets and used the default parameter values for PRDC.

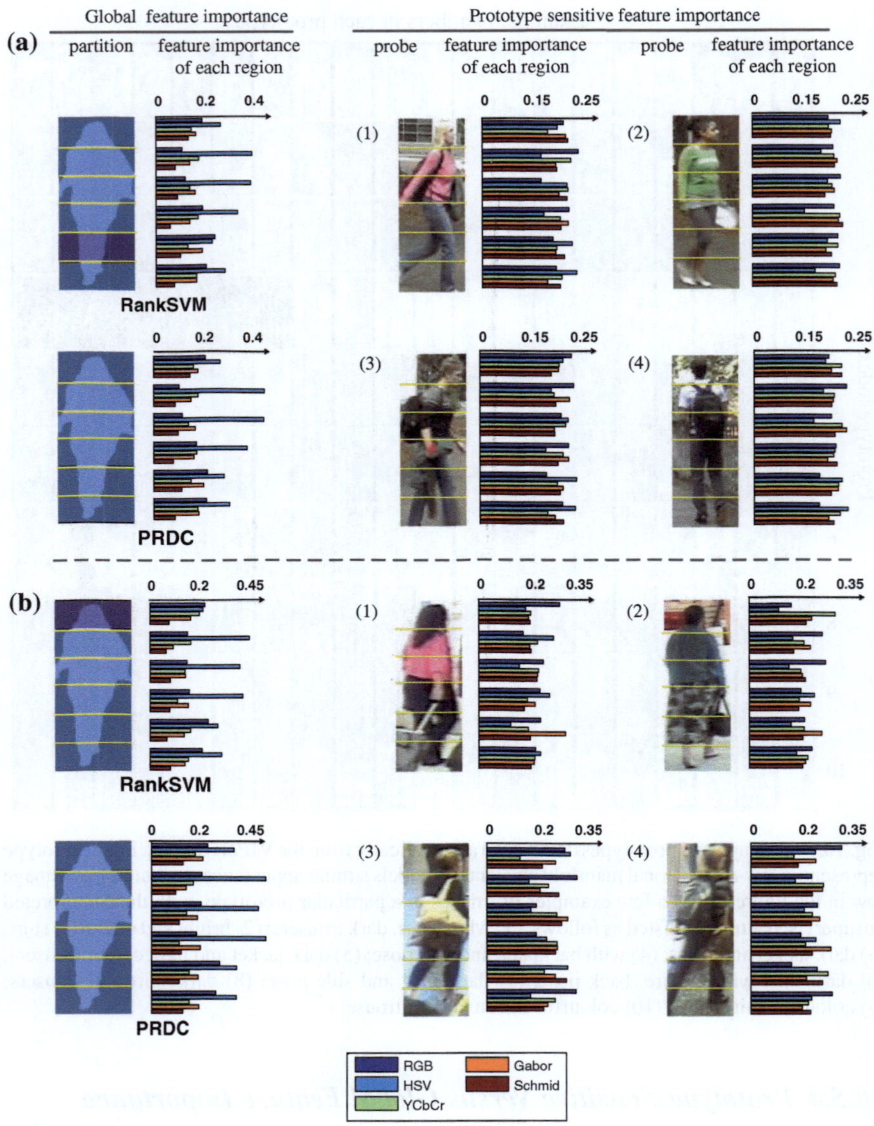

Fig. 10.8 Comparison of global feature importance weights produced by RankSVM [33] and PRDC [41] against those by prototype-sensitive feature importance. These results are obtained from the VIPeR and i-LIDS datasets

The left pane of Figs. 10.8 and 10.9 shows the feature importance discovered by both the RankSVM and PRDC. For PRDC, we only show the first learned orthogonal projection, i.e.feature importance. Each region in the partitioned silhouette images are masked with the labelling colour of the dominant feature. In the feature importance

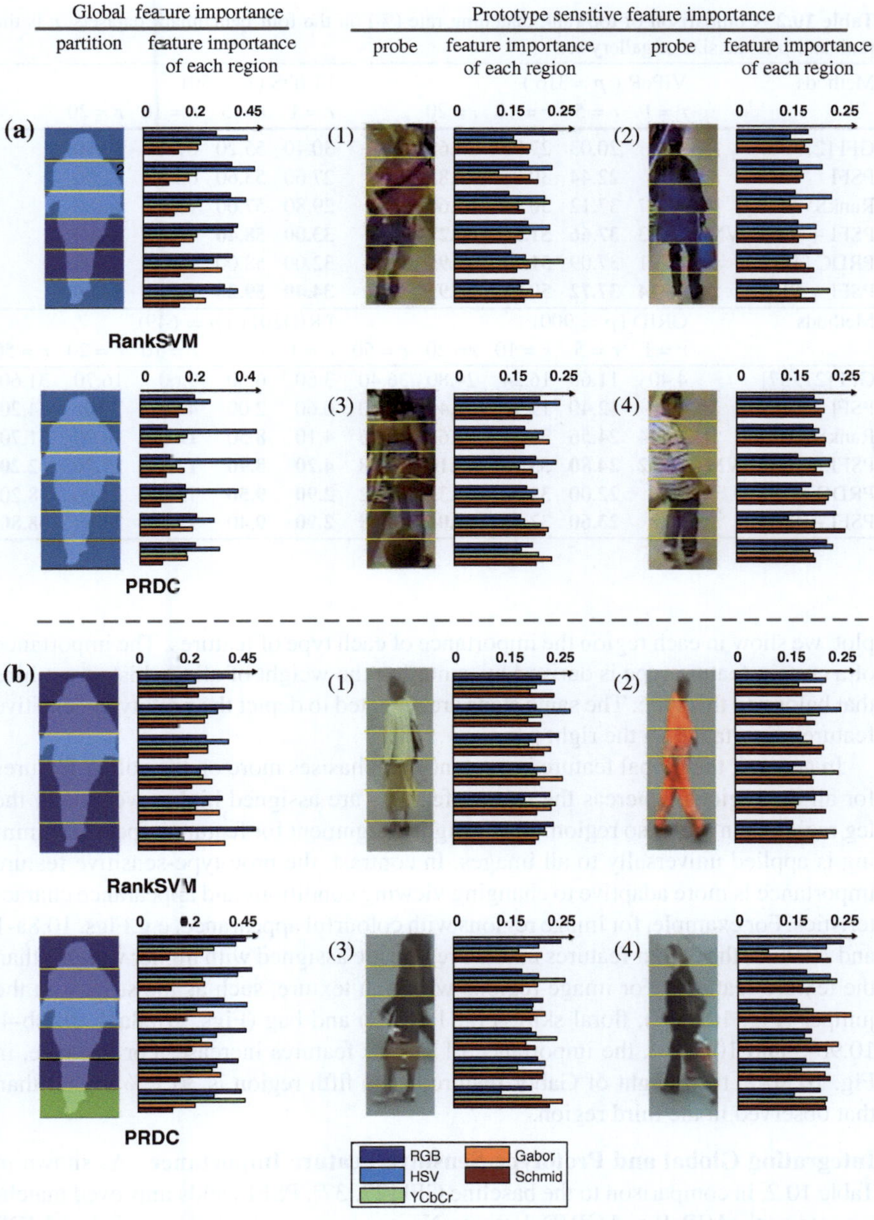

Fig. 10.9 Comparison of global feature importance weights produced by RankSVM [33] and PRDC [41] against those by prototype-sensitive feature importance. These results are obtained from the GRID and PRID2011 datasets

Table 10.2 Comparison of top rank matching rate (%) on the four benchmark datasets. r is the rank and p is the size of gallery set

Methods	VIPeR ($p = 316$)					i-LIDS (p = 50)				
	$r = 1$	$r = 5$	$r = 10$	$r = 20$		$r = 1$	$r = 5$	$r = 10$	$r = 20$	
GFI [27, 37]	9.43	20.03	27.06	34.68		30.40	55.20	67.20	80.80	
PSFI	9.56	22.44	30.85	42.82		27.60	53.60	66.60	81.00	
RankSVM [33]	14.87	37.12	50.19	65.66		29.80	57.60	73.40	84.80	
PSFI + RankSVM	**15.73**	**37.66**	**51.17**	**66.27**		**33.00**	**58.40**	**73.80**	**86.00**	
PRDC [41]	16.01	37.09	**51.27**	**65.95**		32.00	58.00	71.00	83.00	
PSFI + PRDC	**16.14**	**37.72**	50.98	**65.95**		**34.40**	**59.20**	**71.40**	**84.60**	
Methods	GRID (p = 900)					PRID2011 (p = 649)				
	$r = 1$	$r = 5$	$r = 10$	$r = 20$	$r = 50$	$r = 1$	$r = 5$	$r = 10$	$r = 20$	$r = 50$
GFI [27, 37]	4.40	11.68	16.24	24.80	36.40	3.60	6.60	9.60	16.70	31.60
PSFI	5.20	12.40	19.92	28.48	40.80	0.60	2.00	4.00	7.30	14.20
RankSVM [33]	10.24	24.56	33.28	43.68	**60.96**	4.10	8.50	**12.50**	18.90	31.70
PSFI + RankSVM	**10.32**	**24.80**	**33.76**	**44.16**	60.88	**4.20**	**8.90**	**12.50**	**19.70**	**32.20**
PRDC [41]	**9.68**	22.00	**32.96**	44.32	64.32	**2.90**	**9.50**	15.40	23.00	38.20
PSFI + PRDC	9.28	**23.60**	32.56	**45.04**	**64.48**	**2.90**	9.40	**15.50**	**23.60**	**38.80**

plot, we show in each region the importance of each type of features. The importance of a certain feature type is derived by summing the weight of all the histogram bins that belong to this type. The same steps are repeated to depict the prototype-sensitive feature importance on the right pane.

In general, the global feature importance emphasises more on the colour features for all the regions, whereas the texture features are assigned higher weights in the leg region than the torso region. This weight assignment for feature importance mining is applied universally to all images. In contrast, the prototype-sensitive feature importance is more adaptive to changing viewing conditions and appearance characteristics. For example, for image regions with colourful appearance, e.g.Figs. 10.8a-1 and 10.9b-2, the colour features in torso region are assigned with higher weights than the texture features. For image regions with rich texture, such as the stripes on the jumper (Fig. 10.8a-3), floral skirt (Fig. 10.8b-2) and bag (Figs. 10.8a-4, 10.8b-4, 10.9b-3 and 10.9b-4), the importance of texture features increase. For instance, in Fig. 10.8b-2, the weight of Gabor feature in the fifth region is 36.7 % higher than that observed in the third region.

Integrating Global and Prototype-Sensitive Feature Importance As shown in Table 10.2, in comparison to the baseline GFI [27, 37], PSFI yields improved matching rate on the VIPeR and GRID datasets. No improvement is observed on the i-LIDS and PRID2011 datasets. A possible reason is the small training size in the i-LIDS and PRID2011 dataset, which leads to suboptimal prototype discovery. This can be resolved by collecting more unannotated images for unsupervised prototype discovery. We integrate both global and prototype-sensitive feature importance following the method described in Sect. 10.4 by setting $\alpha = 0.1$. An improvement as much as 3.2 % on rank 1 matching rate can be obtained when we combine our method with RankSVM [33] and PRDC [41] on these datasets. It is not surprising to observe that the supervised learning-based approaches [33, 41] outperform our unsupervised approach. Nevertheless, the global approaches benefit from slight bias of feature weights driven by specific appearance attributes of individuals. The results suggest that these two kinds of feature importance are not exclusive, but can complement each other to improve re-identification accuracy.

10.6 Findings and Analysis

In this study, we investigated the effect of feature importance for person re-identification. We summarise our main findings as follows:

Mining Feature Importance for Person Re-Identification Our evaluation shows that certain appearance features are more important than others in describing an individual and distinguishing him/her from other people. In general, colour features are dominant, not surprisingly, for person re-identification and outperform the texture or structure features, though illumination changes may cause instability in the colour features. However, texture and structure features take greater effect when the appearances contain noticeable local statistics, caused by bag, logo and repetitive patterns.

Combining various features for robust person re-identification is non-trivial. Naively concatenating all the features and applying uniform global weighting to them does not necessarily yield better performance in re-identification. Our results show a tangible indication that instead of biasing all the weights to features that are presumably good for all individuals, distributing selectively some weights to informative feature specific to certain appearance attributes can lead to better re-identification performance.

We also find that the effectiveness of prototype-sensitive feature importance mining is dependent on the quantity and quality of training data, in terms of the available size of the training data and the diversity of underlying attributes in appearance, i.e. sufficient and non-biased sampling in the training data. First, as shown in the experiment on the i-LIDS dataset, a sufficient number of unlabelled data are desired to generate robust prototypes. Second, it would be better to prepare a training set of unlabelled images that cover a variety of different prototypes, in order to have non-biased contributions from different feature types. For example, in the PRID2011 dataset, images with rich structural and texture features are rare. Therefore, the derived feature importance scores for those features are prone to be erroneous.

Hierarchical Feature Importance for Person Re-Identification The global feature importance and prototype-sensitive feature importance can be seen organising themselves in a hierarchical structure, as shown in Fig. 10.10. Specifically, the global feature importance exploited by existing rank learning [33] or distance learning method [21, 41] learns a feature weighting function to accommodate the

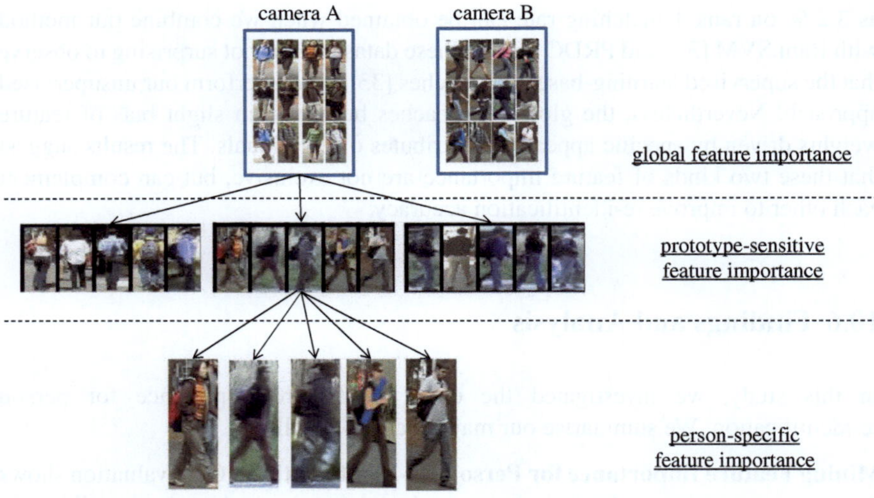

Fig. 10.10 Hierarchical structure of feature importance. Global feature importance aims at weighing more on those features that remain consistent between cameras from a statistical point of view. Prototype-sensitive feature importance emphasises more on the intrinsic features which can discriminate a given prototype from the others. Person-specific feature importance should be capable of distinguishing a given person from those who are categorised into the same prototype

feature inconsistency between different cameras, caused by illumination changes or viewpoint variations. The discovered feature weights can be treated as feature importance in the highest level of the hierarchy, without taking specific individual appearance characteristics into account. Whilst the prototype-sensitive feature importance aims to emphasise more on the intrinsic feature properties that can discriminate a given prototype from the others. Our study shows that these two kinds of feature importance in different levels of the hierarchy can be complementary to each other in improving re-identification accuracy.

Though the proposed prototype-sensitive feature importance is capable of reflecting the intrinsic/salient appearance characteristics of a given person, it still lacks the ability to differentiate the disparity between two different individuals who fall into the same prototype. Thus, it would be interesting to investigate person-specific feature importance that is unique to a specific person, which allows the manifestation of subtle differences among individuals belong to the same prototype.

References

1. Alahi, A., Vandergheynst, P., Bierlaire, M., Kunt, M.: Cascade of descriptors to detect and track objects across any network of cameras. Comput. Vis. Image Underst. **114**(6), 624–640 (2010)
2. Avraham, T., Gurvich, I., Lindenbaum, M., Markovitch, S.: Learning implicit transfer for person re-identification. In: European Conference on Computer Vision, First International Workshop on Re-Identification, pp. 381–390 (2012)

3. Bak, S., Corvee, E., Bremond, F., Thonnat, M.: Multiple-shot human re-identification by mean Riemannian covariance grid. In: IEEE International Conference on Advanced Video and Signal Based Surveillance, pp. 179–184 (2011)
4. Bak, S., Corvee, E., Brémond, F., Thonnat, M.: Person re-identification using haar-based and DCD-based signature. In: IEEE International Conference on Advanced Video and Signal Based Surveillance, pp. 1–8 (2010)
5. Bak, S., Corvee, E., Brémond, F., Thonnat, M.: Person re-identification using spatial covariance regions of human body parts. In: IEEE International Conference on Advanced Video and Signal Based Surveillance, pp. 435–440 (2010)
6. Bak. S., Charpiat, G., Corvée, E., Brémond, F., Thonnat, M.: Learning to match appearances by correlations in a covariance metric space. In: European Conference on Computer Vision, pp. 806–820 (2012)
7. Bauml, M., Stiefelhagen, R.: Evaluation of local features for person re-identification in image sequences. In: IEEE International Conference on Advanced Video and Signal Based Surveillance, pp. 291–296 (2011)
8. Bazzani, L., Cristani, M., Perina, A., Murino, V.: Multiple-shot person re-identification by chromatic and epitomic analyses. Pattern Recogn. Lett. **33**(7), 898–903 (2012)
9. Bazzani, L., Cristani, M., Murino, V.: Symmetry-driven accumulation of local features for human characterization and re-identification. Comput. Vis. Image Underst. **117**(2), 130–144 (2013)
10. Berg, T.L., Berg, A.C., Shih, J.: Automatic attribute discovery and characterization from noisy web data. In: European Conference on Computer Vision, pp. 663–676 (2010)
11. Breiman, L., Friedman, J., Stone, C., Olshen, R.: Classification and regression trees. Chapman and Hall/CRC, Boca Raton (1984)
12. Breiman, L.: Random forests. Mach. Learn. **45**(1), 5–32 (2001)
13. Caruana, R., Karampatziakis, N., Yessenalina, A.: An empirical evaluation of supervised learning in high dimensions. In: International Conference on, Machine learning, pp. 96–103 (2008)
14. Cheng, D., Cristani, M., Stoppa, M., Bazzani, L., Murino, V.: Custom pictorial structures for re-identification. In: British Machine Vision Conference, pp. 68.1–68.11 (2011)
15. Dalal, N., Triggs, B.: Histograms of oriented gradients for human detection. IEEE Comput. Vis. Pattern Recogn. **1**, 886–893 (2005)
16. Doretto, G., Sebastian, T., Tu, P., Rittscher, J.: Appearance-based person reidentification in camera networks: problem overview and current approaches. J. Ambient Intell. Humanized Comput. **2**(2), 127–151 (2011)
17. Farenzena, M., Bazzani, L., Perina, A., Murino, V., Cristani, M.: Person re-identification by symmetry-driven accumulation of local features. In: IEEE Conference Computer Vision and, Pattern Recognition, pp. 2360–2367 (2010)
18. Farhadi, A., Endres, I., Hoiem, D., Forsyth, D.: Describing objects by their attributes. In: IEEE Conference on Computer Vision and, Pattern Recognition, pp. 1778–1785 (2009)
19. Gray, D., Tao, H.: Viewpoint invariant pedestrian recognition with an ensemble of localized features. In: European Conference on Computer Vision, pp. 262–275 (2008)
20. Hirzer, M., Beleznai, C., Roth, P., Bischof, H.: Proceedings of the 17th Scandinavian Conference on Image Analysis, Springer-Verlag, 91–102 (2011)
21. Hirzer, M., Roth, P., Köstinger, M., Bischof, H.: Relaxed pairwise learned metric for person re-identification. In: European Conference on Computer Vision, pp. 780–793 (2012)
22. Javed, O., Rasheed, Z., Shafique, K., Shah, M.: Tracking across multiple cameras with disjoint views. In: International Conference on Computer Vision, pp. 952–957 (2003)
23. Layne, R., Hospedales, T., Gong, S.: Person re-identification by attributes. In: British Machine Vision Conference (2012)
24. Liu, C., Wang, G., Lin, X.: Person re-identification by spatial pyramid color representation and local region matching. IEICE Trans. Inf. Syst. **E95-D**(8), 2154–2157 (2012)
25. Liu, B., Xia, Y., Yu, P.S.: Clustering through decision tree construction. In: International Conference on Information and, Knowledge Management, pp. 20–29 (2000)

26. Loy, C.C., Liu, C., Gong, S.: Person re-identification by manifold ranking. In: IEEE International Conference on Image Processing (2013)
27. Loy, C.C., Xiang, T., Gong, S.: Time-delayed correlation analysis for multi-camera activity understanding. Int. J. Comput. Vis. **90**(1), 106–129 (2010)
28. Loy, C.C., Xiang, T., Gong, S.: Incremental activity modelling in multiple disjoint cameras. IEEE Trans. Pattern Anal. Mach. Intell. **34**(9), 1799–1813 (2012)
29. Ma, S., Sclaroff, S., Ikizler-Cinbis, N.: Unsupervised learning of discriminative relative visual attributes. In: European Conference on Computer Vision, Workshops and Demonstrations, pp. 61–70 (2012)
30. Mignon, A., Jurie, F.: PCCA: A new approach for distance learning from sparse pairwise constraints. In: IEEE Conference Computer Vision and, Pattern Recognition, pp. 2666–2672 (2012)
31. Ng, A.Y., Jordan, M.I., Weiss, Y., et al.: On spectral clustering: analysis and an algorithm. Adv. Neural Inf. Process. Syst. **2**, 849–856 (2002)
32. Perona, P., Zelnik-Manor, L.: Self-tuning spectral clustering. Adv. Neural Inf. Process. Syst. **17**, 1601–1608 (2004)
33. Prosser, B., Zheng, W., Gong, S., Xiang, T.: Person re-identification by support vector ranking. In: British Machine Vision Conference, pp. 21.1–21.11 (2010)
34. Satta, R., Fumera, G., Roli, F.: Fast person re-identification based on dissimilarity representations. Pattern Recogn. Lett. **33**(14), 1838–1848 (2012)
35. Schulter, S., Wohlhart, P., Leistner, C., Saffari, A., Roth, P.M., Bischof, H.: Alternating decision forests. In: IEEE Conference Computer Vision and Pattern Recognition (2013)
36. Schwartz, W., Davis, L.: Learning discriminative appearance-based models using partial least squares. In: Brazilian Symposium on, Computer Graphics and Image Processing, pp. 322–329 (2009)
37. Wang, X.G., Doretto, G., Sebastian, T., Rittscher, J., Tu, P.: Shape and appearance context modeling. In: International Conference on Computer Vision, pp. 1–8 (2007)
38. Xiang, T., Gong, S.: Spectral clustering with eigenvector selection. Pattern Recogn. **41**(3), 1012–1029 (2008)
39. Zhang, Y., Li, S.: Gabor-LBP based region covariance descriptor for person re-identification. In: International Conference on Image and Graphics, pp. 368–371 (2011)
40. Zheng, W., Gong, S., Xiang, T.: Associating groups of people. In: British Machine Vision Conference, pp. 23.1–23.11 (2009)
41. Zheng, W., Gong, S., Xiang, T.: Re-identification by relative distance comparison. IEEE Trans. Pattern Anal. Mach. Intell. **35**(3), 653–668 (2013)

Part II
Matching and Distance Metric

Chapter 11
Learning Appearance Transfer for Person Re-identification

Tamar Avraham and Michael Lindenbaum

Abstract In this chapter we review methods that model the transfer a person's appearance undergoes when passing between two cameras with non-overlapping fields of view. While many recent studies deal with re-identifying a person at any new location and search for universal signatures and metrics, here we focus on solutions for the natural setup of surveillance systems in which the cameras are specific and stationary, solutions which exploit the limited transfer domain associated with a specific camera pair. We compare the performance of explicit transfer modeling, implicit transfer modeling, and camera-invariant methods. Although explicit transfer modeling is advantageous over implicit transfer modeling when the inter-camera training data are poor, implicit camera transfer, which can model multi-valued mappings and better utilize negative training data, is advantageous when a larger training set is available. While camera-invariant methods have the advantage of not relying on specific inter-camera training data, they are outperformed by both camera-transfer approaches when sufficient training data are available. We therefore conclude that camera-specific information is very informative for improving re-identification in sites with static non-overlapping cameras and that it should still be considered even with the improvement of camera-invariant methods.

11.1 Introduction

The first studies to deal with person re-identification for non-overlapping cameras extended work whose goal was to continue to track people who moved between overlapping cameras. To account for the appearance changes, they modeled the transfer

T. Avraham (✉) · M. Lindenbaum
Computer Science Department,Technion Israel Institute of Technology, Haifa, Israel
e-mail: tammya@cs.technion.ac.il

M. Lindenbaum
e-mail: mic@cs.technion.ac.il

S. Gong et al. (eds.), *Person Re-Identification*,
Advances in Computer Vision and Pattern Recognition,
DOI: 10.1007/978-1-4471-6296-4_11, © Springer-Verlag London 2014

of colors associated with the two specific cameras [15, 26–29, 34, 39, 47]. Following the work of Porikli [46], designed for overlapping cameras, these early studies proposed different ways to estimate a brightness transfer function (BTF).

The BTF approach for modeling the appearance transfer between two cameras has some limitations. First, it assumes that a perfect foreground-background segmentation is available both for the training set and at real-time, when the re-identification decision should take place, and second, it is not always sufficient for modeling all the variability in the possible appearance changes.

The recently proposed Implicit Camera Transfer (ICT) method [1] introduced a novel way for modeling the camera-dependent transfer, while addressing the two shortcomings of BTF-based methods mentioned above: (1) The camera transfer is modeled by a binary relation $R = \{(x, y)|x$ and y describe the same person seen from cameras A and B respectively$\}$. This allows the camera transfer function to be a multi-valued mapping, and provides a generalized representation of the variabilities of the changes in appearance. (2) The ICT algorithm does not rely on high-level descriptors, or even on pre-processes of background subtraction. Rather, it uses the bounding-box surrounding the people as input both in the training stage and in the classification stage. The figure-ground separation is performed implicitly by automatic feature selection.

A common alternative approach for solving the person re-identification problem is the one that searches for optimal metrics under which instances belonging to the same person are more similar than instances belonging to different people. We refer to such methods as *similarity-based* (as opposed to the *transfer-based* methods mentioned above). The similarity-based methods can be divided to *direct* methods that use fixed metrics following high-level image analysis [2, 3, 8–10, 16, 18, 19, 25, 37, 38, 42, 44, 50] and to *learning-based* methods that perform feature selection [24, 49, 59] or metric selection [14, 40, 41]. The direct methods are camera-invariant while the learning-based ones, depending on a training set, can be either camera-dependent, if trained with camera-specific data, or camera-invariant, if trained from data from a wide variety of locations.

It was shown in [1] that when there is a sufficient set of inter-camera data available, the ICT algorithm, which is transfer-based, and does not assume there is a metric with distinctive capabilities, performs better than direct and learning-based similarity-based methods.

Yet, not depending on pre-annotated training data is an important advantage of direct methods, which is not shared with transfer-based methods. In order to eliminate the dependence on a camera-specific pre-annotated training set some works suggested on automatic collection of training data and on unsupervised learning methods [6, 7, 20, 21, 36, 48]. Here we present an alternative transfer-based algorithm denoted ECT (Explicit Camera Transfer) that is designed to deal with situations where the available inter-camera training data are poor. It models the camera appearance transfer by a single function, while exploiting intra-camera data (that are available easily without supervision) for modeling appearance variability. As we show below, when only a rather small set of inter-camera training data are available, the explicit transfer approach outperforms the implicit transfer approach, while for larger training sets the implicit approach performs better. This may be considered as a demonstration for the general bias-variance tradeoff [4]. We also show that the camera-invariant approach outperforms both transfer approaches when no training data are available, or when the training set size is very small, while both transfer approaches outperform camera-invariant methods when more training data are added.

We conclude that camera-dependent data are very informative for improving re-identification, and that it should still be considered even with the improvement of camera-invariant methods. We believe that future directions should combine both camera-dependent transfer-based modeling and camera-invariant methods. In addition, an effort should be made to propose better ways to automatically collect training data.

Chapter outline: Most methods that model transfer between non-overlapping cameras do not rely only on appearance change modeling, but also combine spatio-temporal cues such as traveling time between cameras, and modeling the likelihood of different trajectories (e.g., [6, 7, 12, 17, 20, 21, 27, 28, 30, 33, 43, 44, 52, 53]). Efforts have also been made to exploit cues of gait [22, 32, 54] and height [55]. In this chapter we focus only on methods for modeling the transfer in appearance. As other chapters in this book detail similarity-based methods (direct and learning-based), here we focus on camera-specific transfer-based methods. In Sect. 11.2 we review BTF-based methods, in Sect. 11.3 we review attempts to collect inter-camera training data automatically, in Sect. 11.4 we review the ICT method, and in Sect. 11.5 we present the ECT method. Experimentally we compare the performance of ECT and ICT under different conditions (Sect. 11.6.1) and compare the performance of these two methods with the performance of camera-invariant methods (Sect. 11.6.2). We show that the choice for a re-identification technique should depend on the amount of training data available. Finally, Sect. 11.7 concludes the chapter.

11.2 BTF (Brightness Transfer Function) Based Methods

The attempts to automatically re-identify people in CCTV sites with non-overlapping camera views is a natural extension of the attempts to continuously keep track of people who pass between viewpoints of overlapping cameras. Many works on automatic re-identification have thus focused on inter-camera appearance transfer, and most of them have followed the work of Porikli [46]. Porikli suggested learning the Brightness Transfer Function (BTF) that an object's colors undergo when passing between the viewpoints of two cameras and used the overlapping region that appeared in both cameras in order to learn how the colors acquired by the first camera are changed when the second camera acquires the same objects and background. These changes arise as usually two different cameras have different properties and are calibrated with different parameters (exposure time, focal length, aperture size, etc.). This transformation of colors can be inferred—regardless of camera properties and parameters—by comparing color histograms that captured the same scene region at the same time, and to learn the transformation, i.e., which function will transfer one histogram to the other. In [46], a correlation matrix between such two histograms is computed, and using dynamic programming, a minimum cost path is computed that defines the BTF.

Javed et al. [27, 28] were the first to extend this approach for non-overlapping cameras, where the color transfer is not caused only by camera parameters, but also by illumination and pose changes. In this case a training pair is not taken from a common scene region captured by the two cameras at the same time, but from samples taken at different times and locations. The training data consist of pairs of images with known correspondences, i.e., the same person appears in two corresponding images and the exact silhouette of the person for each image is assumed to be provided. A BTF is estimated for each training pair indexed i, and for each brightness level b by

$$\text{BTF}_i(b) = H_{B_i}^{-1}(H_{A_i}(b)), \tag{11.1}$$

where H_{A_i} is the (one color channel) normalized cumulative histogram associated with camera A and H_{B_i} is the normalized cumulative histogram associated with camera B. It is suggested that all transfer functions associated with a pair of cameras lie in a low dimensional feature space. A Probabilistic Principle Component Analysis (PPCA) is used for approximating the sub-space by a normal distribution,

$$\text{BTF}_j \sim N(\text{mean}_i(\text{BTF}_i), Z), \tag{11.2}$$

where BTF_i is the BTF computed using the i-th training set pair, and $Z = WW^T + \sigma I$, where W is the PCA projection matrix and σ is the variance of the information lost in the projection. During system activation a BTF is estimated for each candidate pair, and the probability that this BTF is sampled from the normal distribution described in Eq. (11.2) is calculated. The final classification decision is taken by combining this probability with the probability that spatio-temporal features associated with that pair are sampled from the distribution of location cues, which are modeled by Parzen windows.

Prosser et al. [47] suggested the CBTF method (Cumulative Brightness Transfer Function) which, instead of using separate histograms for each person, cumulates the pixels from all the training set. As a result, the BTF can be learned from denser information, and one, more robust BTF can be inferred. During activation the system measures the similarity between an instance from one camera and the instance from the other camera, after converting the latter with the estimated BTF. A few similarity measures were tested and the best results were obtained for a measure based on a bi-directional Bhattacharya distance.

D'Orazio et al. [15], Kumar et al. [34], and Ilyas et al. [26] further tested the CBTF approach. D'Orazio et al. and Ilyas et al. empirically showed comparable results for CBTF and MBTF, which uses the mean of the BTFs learned for each individual member of the training set. Kumar et al. test different shortest path algorithms for finding the optimal BTF and test re-identification performance as a function of the number of histogram bins used. Ilyas et al. suggested an improvement to CBTF, denoted MCBTF (Modified Cumulative BTF). As averaging the histograms as in CBTF caused information to be lost, they suggest cumulating in each bin only information from single examples for which a large number of pixels are associated with the bin.

Lian et al. [39] use the learned CBTF in order to transfer the appearance of a person captured by one camera to an estimated appearance in the second camera, and then use textural descriptors that separably describe the lower and upper garments, following by a chi-square goodness of fit measure used as a similarity measure. Kumar et al. [35] fused CBTF learnt information with eigen-faces based recognition.

All the above described methods compute a separate BTF for each color channel. Jeong et al. [29] suggested a variation that models dependencies between the chromatic color channels. The colors of each object in each camera are represented by a Gaussian mixture model in the (U,V) 2D-space using Expectation-Maximization (EM). Given two mixtures of Gaussians, the dissimilarity between them is defined to be the minimum out of the dissimilarities of the $m!$ possible fits between modes. The approximated minimum fit is found by sorting the modes of each mixture by their similarity to a Gaussian centered at the 2D-space origin, using an angular metric. The order of the two sorted descriptors defines the fitting of modes. Given these corresponding fits, the parameters of an affine BTF is estimated.

As discussed before, one of the drawbacks in all the methods reviewed in this section is that they require annotated example pairs. Obtaining such manually annotated examples is not simple and sometimes inapplicable. Moreover, illumination may change with time, which would make the training set unrepresentative of the changed conditions. In the next section we discuss a few attempts to automatically collect data and/or to make BTF based methods adaptive to illumination changes.

11.3 Unsupervised Methods for Collecting Training Data

A few studies suggested methods for automatic learning of spatio-temporal cues for re-identification in non-overlapping cameras (e.g., [6, 7, 17, 20, 21, 30, 43, 44, 52, 53]), or for automatically learning the topography of a camera network [11, 51, 57]. Some of these works used the BTF-based methods described above to model the inter-camera appearance changes, and some use similarity-based methods to compare appearances. We have found only a few studies that also propose automatic gathering of the appearance training data that are used for training the appearance transfer models. Gilbert and Bowden's [20, 21] completely unsupervised method interacts between color similarity and spatio-temporal cues. The color transformation is modeled by a transformation matrix T. ($V_A \times T = V_B$, where V_A and V_B are the color histograms associated with camera A and camera B, respectively.) At system initialization, T is the identity matrix, and the similarity measure is based only on color similarity. Examples are then collected using a model that quantifies the probability of exit/entry points for pairs of cameras as a function of the time interval. This probability is associated with the intersection of color histograms collected from people who appeared in both cameras in limited time intervals. As examples are collected, SVD is used for updating T. The improvement of the system as its activation time lengthens is proven empirically.

Kuo et al. [36] suggests a method that learns a discriminative appearance affinity model without manually-labeled pairs, using Multiple Instance Learning (MIL) [13]. High confident negative examples are collected by simply pairing tracks from the two cameras captured at the same time, and are used as members of 'negative bags'. 'Positive bags' are collected using spatio-temporal considerations. Each positive bag consists of a set of pairs, out of which at least one (unknown in advance) is positive. The MIL boosting process performs an iterative feature selection process that outputs a classifier for single pairs.

Chen et al. [6, 7] also suggest an unsupervised method that combines spatio-temporal and appearance information which automatically annotates a training set. Given n tracks of people in camera A and n tracks of the same people in camera B, the method selects a likely pairing out of the $n!$ possibilities. They rely on the assumption that, for the correct correspondences, the BTF subspace (Javed et al. [28]) estimated from a subset of pairs will provide high probabilities for the complementary pairs. (Markov Chain Monte Carlo) MCMC sampling by the Metropolis-Hastings algorithm is used to find a sample that improves an initial spatio-temporal based fit.

Chen et al. [7], as well as Prosser et al. [48], suggested ways to make the system tolerant to illumination changes. Prosser et al. model background illumination changes and infer an updated CBTF. Chen et al. change the BTF subspace with time, to adapt to gradual illumination changes with new arrival data, using incremental probabilistic PCA [45]. In addition, when sudden illumination changes are detected [56], the weight of the appearance cues is temporally lowered, while spatio-temporal cue weights are increased.

We believe that there is place to further develop ways for automatic collection of inter-camera training data for camera-dependent re-identification algorithms. In addition, improvements of re-identification algorithms should reduce the dependency on accurate foreground-background segmentation, and should lead to more robust models of the appearance transfer. The method described in the next section addresses the last two issues.

11.4 Implicit Camera Transfer

Most previously mentioned methods, both transfer-based and similarity-based, assume that an accurate segmentation of the people from the background is available. Although figure-ground separation of moving objects is much simpler to solve than the segmentation of static images, inaccurate background removal is still a problem, due to, for instance, shadows. Re-identification that trusts the output masks to be reliable may fail for such cases. (See, for instance, failure cases reported in [47], where it is shown that many re-identification failure cases are caused by imperfect segmentations.)

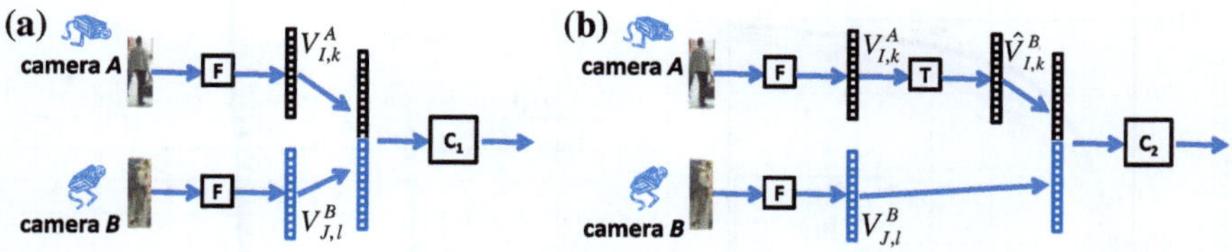

Fig. 11.1 Illustration of the classification process used by the ICT and ECT algorithms. From each of the instances captured by cameras A and B, features are extracted (F). In ICT, the concatenation of those two feature vectors, $V^A_{I,k}$ and $V^B_{J,l}$, is the input to the classifier C_1. In ECT, $V^A_{I,k}$ undergoes a learned transformation function (T) that returns the estimate $\hat{V}^B_{I,k}$. Then, the concatenation of $\hat{V}^B_{I,k}$ and $V^B_{J,l}$ is classified by C_2

Another drawback of most methods that model appearance transfer between two specific cameras is that they try to model the camera transfer by a single transfer function, or a sub-space modeled by a single Gaussian. This cannot capture all the variations associated with each of the two cameras. (See, for instance, [31], where it was observed that one global color mapping is not enough to account for the color transfers associated even with only two images captured from different viewpoints and illuminations.)

The ICT algorithm [1] models camera transfer by a binary relation R whose members are pairs (x, y) that describe the same person seen from cameras A and B respectively. This solution implies that the camera transfer function is a multi-valued mapping and not a single-valued transformation. Given a person's appearance described by a feature vector of length d, the binary relation models a (not necessarily continuous) sub-space in $\mathbb{R}^{2d}$. That is, the $\mathbb{R}^{2d}$ space is divided to 'positive' regions (belonging to the relation) and 'negative' regions (not belonging to the relation). Let $V^A_{I,k}$ describe the k'th appearance of a person with identity I captured by camera A, and let $V^B_{J,l}$ describe the l'th appearance of a person with identity J captured by camera B. Given a pair $(V^A_{I,k}, V^B_{J,l})$, the goal is to distinguish between *positive* pairs with the same identity ($I = J$), and *negative* pairs ($I \neq J$). The concatenation $[V^A_{I,k} \| V^B_{J,l}]$ of such two vectors provides a $2d$-dimensional vector in $\mathbb{R}^{2d}$. The algorithm trains a binary SVM (Support Vector Machine) classifier with an RBF (Radial Basis Function) kernel using such concatenations of both positive and negative pairs coming from training data. Then it classifies new such pairs by querying the classifier on their concatenations. The decision value output of the SVM is used for ranking different candidates. The classification stage of ICT is illustrated in Fig. 11.1a.

The algorithm uses common and simple descriptions of the bounding boxes surrounding the people: each bounding box is divided into five horizontal stripes. Each stripe is described by a histogram with 10 bins for each of the color components H, S, and V. This results in feature vectors with 150 dimensions.

The implicit transfer approach implemented by the ICT algorithm has a few unique properties compared to other re-identification methods: (1) By learning a relation and not a metric it does not assume that instances of the same person are

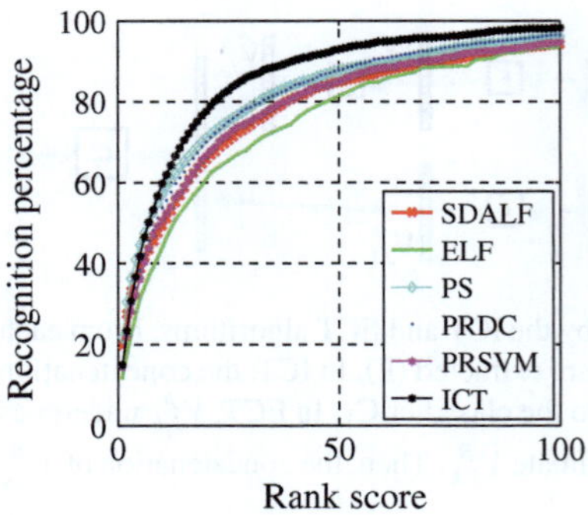

method	expectation	rank(1)	rank(10)	rank(20)	nAUC
SDALF	25.5	19.9	49.4	65.7	92.2
ELF	28.9	12	44	61	91.2
PS	21.2	**21.8**	57.2	71.2	93.6
PRDC	21.5	15.7	53.9	70.1	93.5
PRSVM	27.9	14.6	50.9	66.8	91.4
ICT	**15.9**	14.4	**59.7**	**78.3**	**95.3**

Fig. 11.2 CMC curves and acceptable performance metrics comparing ICT's results on VIPeR [23] with state-of-the-art similarity-based methods, including direct methods (SDALF [2], PS [8]) and learning-based methods (ELF [24], PRSVM [49], and PRDC [59])

necessarily more similar to one another than instances of two different people. (2) Unlike previous transfer-based methods, it exploits negative examples, which have an important role in defining the limits of the positive "clouds". (3) It does not depend on a preprocess that accurately segments the person from the background. An implicit feature selection process allows the automatic separation of foreground data that are person dependent and background data that are location dependent but not person dependent and is similar for positive and negative pairs. (4) It does not build only on a feature-by-feature comparison, but also learns dependencies between different features.

See Fig. 11.2 for results reported in [1], where it was shown that ICT outperforms state-of-the-art similarity-based methods, both direct and learning-based.

11.5 An Explicit Camera Transfer Algorithm (ECT)

In this section we present the ECT algorithm. This algorithm addresses the situation where we have only a very small set of inter-camera data. In such a case it is harder to generalize a domain of implicit transformations. We propose to compensate for the dearth of inter-camera examples by exploiting the easily available intra-camera data.

ECT computes the average inter-camera transfer function by a linear regression of the transfers associated with the inter-camera training data. As an alternative to modeling variations in the transfer, it models intra-camera variations using data from single-camera tracks.

ECT is built from two components, trained in the following way:

- **Inter-camera Regression**: Learn a regression $T : R^d \rightarrow R^d$ using pairs of instances associated with the same person (i.e, use only 'positive' inter-camera example pairs).
- **Intra-camera Concatenation**: Train a classier using concatenations of positive and negative intra-camera pairs, where both instances of each pair are associated with camera B. Note that non-annotated tracks can also be used for training this classifier.

The classification/decision stage is illustrated in Fig. 11.1b. For each pair of input descriptor vectors $(V^A_{I,k}, V^B_{J,l})$ it includes: (1) applying the learned regressions on $V^A_{I,k}$. This provides an estimation $\hat{V}^B_{I,k} = T(V^A_{I,k})$ of how person indexed I may be described when captured by camera B; (2) applying the trained SVM classifier on the concatenation $[\hat{V}^B_{I,k} \| V^B_{J,l}]$, and obtaining the decision value.

In our implementation of ECT we train d linear regressions, $T_1, ..., T_d$ (using the Support Vector Regression (SVR) implementation of LibSVM [5]). Each $T_i : R^d \rightarrow R$ is associated with the relation between a vector describing an instance captured by camera A, and component i in a vector describing an instance captured by camera B.

11.6 Experiments

11.6.1 Explicit Versus Implicit Transfer Modeling

In this section we test the dependence of the ICT and ECT algorithms on the amount of training examples. We show that when the amount of inter-camera training data available is very poor, ECT performs better than ICT. However, when more data is available, ICT—being able to model more variations in the inter-camera transfer—performs better.

We compare the performance of ICT and ECT in different multi-shot setups using the iLIDs MCTS dataset. This dataset includes hours of video captured by 5 different cameras in a busy airport, as well as annotations of sub-images describing tracks associated with 36 people. Using these annotations, we extracted 10 bounding boxes for each person at each camera. (We are aware of the set of data annotated by [58] and corresponding to 119 people appearing in the i-LIDs videos. That set includes a few instances for each person without indication of the camera's identity. It was thus unsuitable for our setup.) We performed two sets of experiments. In the first set, for each pair of cameras (10 options), we ran the algorithms using two people as the test set and the other 34 as a training set. This was repeated for all such possible choices. (This makes 630 possibilities in most cases, excluding cases involving cameras 1 and/or 4, for which annotations are missing for 5 and 7 people respectively, and for which the number of possibilities is reduced accordingly.) As there are two people in each test set, this simulates the situation in which two people walk close to each

Table 11.1 Results of the i-LIDs experiments

Cameras	1–2	1–3	1–4	1–5	2–3	2–4	2–5	3–4	3–5	4–5
a Inter-camera data available for 34 people										
ICT	**86.9**	**84.7**	**86.3**	**86.7**	87.9	**87.3**	93.3	88.4	**89.1**	**94.4**
ECT	83.7	84.1	82.3	79.4	**90.6**	85.7	**94.6**	**89.4**	89.0	91.0
b Inter-camera training data available only for 8 people										
ICT	73.7	76.3	**74.7**	72.9	76.9	**77.9**	83.6	76.0	76.6	77.5
ECT	**78.3**	**79.3**	73.6	**74.0**	**80.9**	71.7	**85.5**	**80.8**	**78.5**	**80.4**

The bold numbers are the higher values of two compared cases. **a** Results (%) for the first set of experiments in which inter-camera training data is available for 34 people. ICT is shown to be more suitable in this setup. **b** Results (%) for the second, harder, set of experiments in which inter-camera training data is available only for 8 people. ECT, which exploits also additional intra-camera training data, is shown to be more suitable in this setup

other when captured by camera A and then walk close to each other again when captured, a few minutes later, by camera B. Let $D_{i,j}$ be the decision value output of the SVM classifier associated with matching the appearance of person i in camera A with the appearance of person j in camera B. A successful match is counted if $D_{i,i} + D_{j,j} > D_{i,j} + D_{j,i}$.

For training we used a random choice of 10 positive concatenated pairs for each person, and 10 negative concatenated pairs for each two different people. Each of the 630 runs may output a 'success' or a 'failure'. The percent of 'successes' is reported in Table 11.1a. We see that in this setup, where inter-camera data are available for 34 people, ICT performs better.

In the second set of experiments we tested a harder setup. We repeated the following for each pair of cameras: we randomly chose 2 people for the test set, and 8 other people as the inter-camera training set. For the rest of the people (26) we used only the data available for camera B, i.e., only intra-camera tracks (which can be exploited only by the ECT algorithm). The two algorithms were tested for 1,000 such random divisions. The percent of 'successes' is reported in Table 11.1b. We see that in this setup ECT performs better.

As expected, ECT, which uses a more restricted transfer modeling, is a "fast learner," but more limited than ICT. In the next section we compare ECT and ICT performance also with that of camera-invariant methods.

11.6.2 Camera-Dependent Transfer-Based Versus Camera-Invariant Similarity-Based Methods

In [1] ICT's performance was compared to the performance of a few state-of-the-art similarity-based methods using the VIPeR and the CAVIAR4REID datasets. Here we report an extension of the experiment with CAVIAR4REID which includes testing ECT and shows the relative performance for ECT, ICT, and camera-invariant similarity-based methods as a function of the inter-camera training set size. As

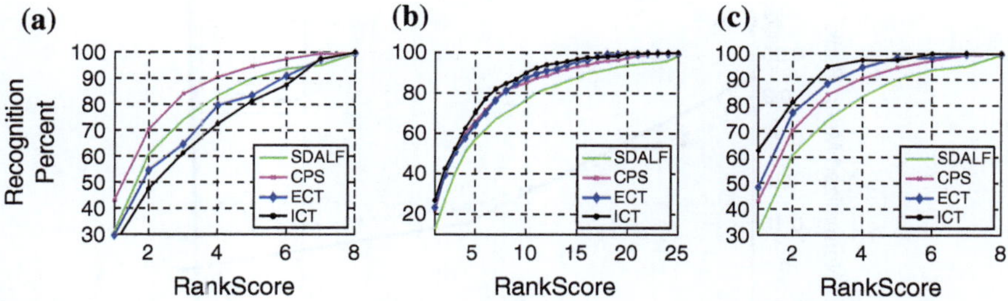

Fig. 11.3 CMC curves comparing ICT and ECT's results on CAVIAR4REID with those of SDALF [18] and CPS [8]

in [1], and as acceptable in all recent re-identification studies, Cumulative Match Characteristic (CMC) curves are used for reporting performance: for each person in the test set, each algorithm ranks the matching of his or her appearance in camera A (denoted "the probe") with the appearances of all the people in the test set in camera B (denoted "the gallery set"). The CMC curve summarizes the statistics of the ranks of the true matches by a cumulative histogram.

The CAVIAR4REID dataset includes 50 pedestrians, each captured by two different cameras, and 22 additional pedestrians captured by only one of the cameras. For each person in each camera there are 10 available appearances. We report results of ECT and ICT for three setups in Fig. 11.3, demonstrating the relative performance as a function of the size of the training data available. In the first setup (Fig. 11.3a), only 8 people are included in the inter-camera training set, and 8 other people are included in the test set. In the second setup (Fig. 11.3b) the 50 people who appear in both cameras are equally divided into a training set of 25 and a test set of 25. In the third setup (Fig. 11.3c) 42 people are included in the inter-camera training set and 8 others in the test set. In the first setup, ECT also exploits the intra-camera data associated with the remaining 56 people from one of the cameras, while in the two other setups ECT uses only the additional 22 intra-camera people. For each setup, we average results on 10 random divisions. The results obtained for ICT and ECT are compared to those of SDALF [2, 18] and CPS [8] reported in [8]. In [8] the test set consisted of all 50 inter-camera people. We estimated the performance for test sets of 25 and 8 by normalizing the CMC curves reported in [8] (i.e., if a person's true match was rated m among n people, then on the average it will be ranked $(m-1)*(k-1)/(n-1)+1$ among k people.).

If we compare the performance of ICT and ECT we again see (as in the iLIDS experiments in Sec. 11.6.1) that ICT is better if we have more annotated inter-camera people to train from, and the ECT has advantages when fewer inter-camera annotations are available. Comparing the performance of ICT and ECT to that of SDALF and CPS, we see that for training sets of 25 people or more, both camera-dependent algorithms outperform the camera-invariant methods, while for smaller training sets they do not perform as well.

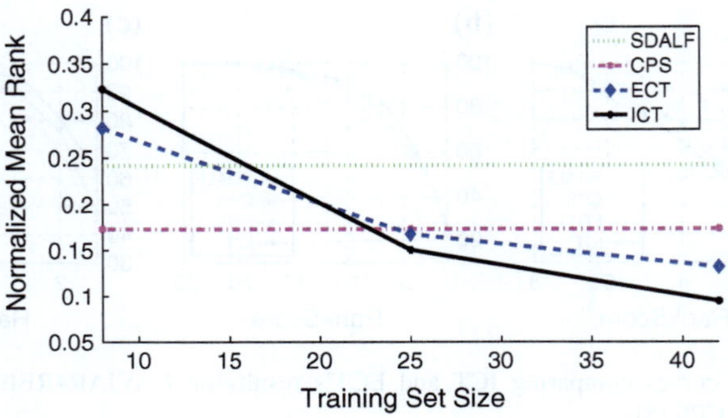

Fig. 11.4 Re-identification performance as a function of inter-camera training set size: this plot reports results of a set of experiments with the CAVIAR4REID. This demonstrates that without—or with a small—training set, similarity-based camera-invariant methods perform best, while with larger training sets the camera-dependent transfer-based methods perform better. This implies that there is place to exploit specific camera data and make an effort to collect it. The plot reports the normalized-mean-rank as a function of the inter-camera training set size (a smaller normalized-mean-rank is better)

Two common measures for comparison, extracted from the CMC curve, are *rank*(r) and *CMC-expectation*. rank(r) is the percent of people in the probe set for whom the correct match was found within the first r ranked people in the gallery set. *CMC-expectation*, or as we prefer to call it, *the mean-rank*, describes the rank of the true match in the average case, when there are n members in the gallery set, and is defined as

$$\text{CMC-expectation} = \text{mean-rank} = \sum_{r=1}^{n} r \frac{\text{rank}(r) - \text{rank}(r-1)}{100} . \qquad (11.3)$$

These measures are lacking in the sense that they do not allow comparing performance for different gallery set sizes. The *nAUC* (normalized Area under Curve) measure, used sometimes, also does not enable direct comparison between tests that differ in the gallery set size. We therefore define here the *normalized-mean-rank* performance measure,

$$\text{normalized-mean-rank} = \frac{(\text{mean-rank} - 1)}{n-1} . \qquad (11.4)$$

That is, the normalized-mean-rank is the fraction of the gallery set members ranked before the true match. The objective is, of course, obtaining a normalized-mean-rank as small as possible.

In Fig. 11.4 the normalized-mean-rank associated with the results presented in Fig. 11.3 is plotted as a function of the inter-camera training set size. This plot clearly demonstrates that without Ű— or with a small —Ű training set, camera-invariant methods perform best, while with larger training sets the camera-dependent methods perform better. This implies that there is place to exploit specific camera data and to make an effort to collect them.

11.7 Conclusions

In this chapter we focused on re-identification methods that do not ignore the camera ID, and that aim to model the transformation that the person's appearance undergoes when passing between two specific cameras. The illumination, background, resolution, and sometimes human pose, associated with a certain camera, are limited, and there is place to exploit this information. Indeed, we have shown that when an inter-camera training set is available, transfer-based camera-dependent methods outperform camera-invariant methods.

Most transfer-based methods try to fit parameters for some explicit transfer function. This approach, being restricted to one possible transformation of a specific form, has the advantage of requiring only a small training set. However, it may not be general enough and as such may not be able to model all possible transformations. As observed in [31], one global color mapping is not enough to account for the color transfers associated even with only two images captured from different viewpoints and illuminations. An algorithm that implicitly models the camera transfer by means of a more general model was recently proposed. We have shown that while explicit transfer modeling performs better when trained with a small inter-camera training sets, implicit transfer modeling has a steeper learning curve, and outperforms both explicit transfer and similarity-based methods for larger training sets.

We believe that future appearance-based re-identification techniques should combine both implicit transfer modeling and camera-invariant methods. In addition, an effort should be made to devise better ways to automatically collect training data.

References

1. Avraham, T., Gurvich, I., Lindenbaum, M., Markovitch, S.: Learning implicit transfer for person re-identification. In: The 1st International Workshop on Re-Identification (Re-Id 2012), in conjunction with ECCV, LNCS, vol. 7583, pp. 381–390 (2012)
2. Bazzani, L., Cristani, M., Murino, V.: Symmetry-driven accumulation of local features for human characterization and re-identification. Comput. Vis. Image Underst. **117**, 130–144 (2013)
3. Bazzani, L., Cristani, M., Perina, A., Murino, V.: Multiple-shot person re-identification by chromatic and epitomic analyses. Pattern Recogn. Lett. **33**(7), 898–903 (2012). (Special Issue on Awards from ICPR 2010)
4. Bishop, C.: Pattern recognition and machine learning (Information Science and Statistics). Springer-Verlag New York (2006)
5. Chang, C., Lin, C.: LIBSVM: a library for support vector machines (2001). Software available at http://www.csie.ntu.edu.tw/~cjlin/libsvm

6. Chen, K.W., Lai, C.C., Hung, Y.P., Chen, C.S.: An adaptive learning method for target tracking across multiple cameras. In: IEEE Conference on Computer Vision and Pattern Recognition (2008)
7. Chen, K.W., Lai, C.C., Lee, P., Chen, C.S., Hung, Y.P.: Adaptive learning for target tracking and true linking discovering across multiple non-overlapping cameras. IEEE Trans. Multimedia **13**(4), 625–638 (2011)
8. Cheng, D.S., Cristani, M., Stoppa, M., Bazzani, L., Murino, V.: Custom pictorial structures for re-identification. In: British Machine Vision Conference (2011)
9. Cheng, E.D., Piccardi, M.: Matching of objects moving across disjoint cameras. In: IEEE International Conferenct on Image Processing, pp. 1769–1772 (2006)
10. Cong, D., Khoudour, L., Achard, C., Meurie, C., Lezoray, O.: People re-identification by spectral classification of silhouettes. Signal Process. **90**(8), 2362–2374 (2010)
11. Detmold, H., Hengel, A., Dick, A., Cichowski, A., Hill, R., Kocadag, E., Falkner, K.E., Munro, D.S.: Topology estimation for thousand-camera surveillance networks. In: ACM/IEEE International Conference on Distributed Smart Cameras, pp. 195–202 (2007)
12. Dick, A., Brooks, M.: A stochastic approach to tracking objects across multiple cameras. In: Australian Conference on, Artificial Intelligence, pp. 160–170 (2004)
13. Dietterich, T.G., Lathrop, R.H., Lozano-Perez, T.: Solving the multiple instance problem with axis-parallel rectangles. Artif. Intell. **89**(1–2), 31–71 (1997)
14. Dikmen, M., Akbas, E., Huang, T.S., Ahuja, N.: Pedestrian recognition with a learned metric. In: Asian Conference on Computer Vision, pp. 501–512 (2010)
15. D'Orazio, T., Mazzeo, P.L., Spagnolo, P.: Color brightness transfer function evaluation for non overlapping multi camera tracking.In: ACM/IEEE International Conference on Distributed Smart Cameras (2009)
16. Doretto, G., Sebastian, T., Tu, P., Rittscher, J.: Appearance-based person reidentification in camera networks: problem overview and current approaches. J. Ambient Intell. Humanized Comput. **2**(2), 127–151 (2011)
17. Ellis, T., Makris, D., Black, J.: Learning a multi-camera topology. Joint IEEE Workshop on Visual Surveillance and Performance Evaluation of Tracking and Surveillance, pp. 165–171 (2003)
18. Farenzena, M., Bazzani, L., Perina, A., Murino, V., Cristani, M.: Person re-identification by symmetry-driven accumulation of local features. In: IEEE Conference on Computer Vision and Pattern Recognition (2010)
19. Gheissari, N., Sebastian, T., Hartley, R.: Person reidentification using spatiotemporal appearance. In: IEEE Conference on Computer Vision and Pattern Recognition (2006)
20. Gilbert, A., Bowden, R.: Tracking objects across cameras by incrementally learning inter-camera colour calibration and patterns of activity. In: European Conference on Computer Vision, pp. 125–136 (2006)
21. Gilbert, A., Bowden, R.: Incremental, scalable tracking of objects inter camera. Comput. Vis. Image Underst. **111**, 43–58 (2008)
22. Goffredo, M., Bouchrika, I., Carter, J., Nixon, M.: Self-calibrating view-invariant gait biometrics. IEEE Trans. Syst. Man Cybern. B Cybern. **4**, 997–1008 (2010)
23. Gray, D., Brennan, S., Tao, H.: Evaluating appearance models for recognition, reacquisition, and tracking. In: IEEE Workshop on Visual Surveillance and Performance Evaluation of Tracking and Surveillance (2007)
24. Gray, D., Tao, H.: Viewpoint invariant pedestrian recognition with an ensemble of localized features. In: European Conference on Computer Vision, pp. 262–275 (2008)
25. Hu, W., Hu, M., Zhou, X., Tan, T., Lou, J.: Principal axis-based correspondence between multiple cameras for people tracking. IEEE Trans. Pattern Anal. Mach. Intell. **28**(4), 663–671 (2006)
26. Ilyas, A., Scuturici, M., Miguet, S.: Inter-camera color calibration for object re-identification and tracking. In: IEEE International Conference of Soft Computing and Pattern Recognition, pp. 188–193 (2010)

27. Javed, O., Shafique, K., Rasheed, Z., Shah, M.: Modeling inter-camera space-time and appearance relationships for tracking across non-overlapping views. Comput. Vis. Image Underst. **109**, 146–162 (2008)
28. Javed, O., Shafique, K., Shah, M.: Appearance modeling for tracking in multiple non-overlapping cameras. In: IEEE Conference on Computer Vision and Pattern Recognition (2005)
29. Jeong, K., Jaynes, C.: Object matching in disjoint cameras using a color transfer approach. Mach. Vis. Appl. **19**, 443–455 (2010)
30. KaewTrakulPong, P., Bowden, R.: A real-time adaptive visual surveillance system for tracking low resolution colour targets in dynamically changing scenes. J. Image Vis. Comput. **21**(10), 913–929 (2003)
31. Kagarlitsky, S., Moses, Y., Hel-Or, Y.: Piecewise-consistent color mappings of images acquired under various conditions. In: IEEE International Conference on Computer Vision, pp. 2311–2318 (2009)
32. Kale, A., Chowdhury, A., Chellappa, R.: Towards a view invariant gait recognition algorithm. In: IEEE Conference on Advanced Video and Signal Based Surveillance, pp. 143–150 (2003)
33. Kettnaker, V., Zabih, R.: Bayesian multi-camera surveillance.In: IEEE Conference on Computer Vision and Pattern Recognition (1999)
34. Kumar, P., Dogancay, K.: Analysis of brightness transfer function for matching targets across networked cameras. In: IEEE International Conference on Digital Image Computing: Techniques and Applications, pp. 250–255 (2011)
35. Kumar, P., Dogancay, K.: Fusion of colour and facial features for person matching in a camera network. In: IEEE Seventh International Conference on Intelligent Sensors, Sensor Networks and Information Processing, pp. 490–495 (2011)
36. Kuo, C.H., Huang, C., Nevatia, R.: Inter-camera association of multi-target tracks by on-line learned appearance affinity models. In: European Conference on Computer Vision, pp. 383–396. Springer (2010)
37. Kviatkovsky, I., Adam, A., Rivlin, E.: Color invariants for person reidentification. IEEE Trans. Pattern Anal. Mach. Intell. **35**(7), 1622–1634 (2013)
38. Layne, R., Hospedales, T., Gong, S.: Towards person identification and re-identification with attributes. In: The 1st International Workshop on Re-Identification (Re-Id 2012), in conjunction with ECCV, LNCS, vol. 7583, pp. 402–412 (2012)
39. Lian, G., Lai, J., Suen, C.Y., Chen, P.: Matching of tracked pedestrians across disjoint camera views using CI-DLBP. IEEE Trans. Circuits Syst. Video Technol. **22**(7), 1087–1099 (2012)
40. Liu, C., Gong, S., Loy, C.C., Lin, X.: Person re-identification: What features are important. In: The 1st International Workshop on Re-Identification (Re-Id 2012), in conjunction with ECCV, LNCS, vol. 7583, pp. 391–401 (2012)
41. Ma, B., Su, Y., Jurie, F.: Local descriptors encoded by Fisher vectors for person re-identification. In: The 1st International Workshop on Re-Identification (Re-Id 2012), in conjunction with ECCV, LNCS, vol. 7583, pp. 413–422 (2012)
42. Madden, C., Cheng, E.D., Piccardi, M.: Tracking people across disjoint camera views by an illumination-tolerant appearance representation. Mach. Vis. Appl. **18**, 233–247 (2007)
43. Makris, D., Ellis, T., Black, J.: Bridging the gaps between cameras. In: IEEE Conference on Computer Vision and Pattern Recognition (2004)
44. Mazzon, R., Tahir, S.F., Cavallaro, A.: Person re-identification in crowd. Pattern Recogn. Lett. **33**, 1828–1837 (2012)
45. Nguyen, H., Qiang, J., Smeulders, A.: Spatio-temporal context for robust multitarget tracking. IEEE Trans. Pattern Anal. Mach. Intell. **29**(1), 52–64 (2007)
46. Porikli, F.: Inter-camera color calibration by correlation model function. In: IEEE International Conference on Image Processing, vol. 2 (2003)
47. Prosser, B., Gong, S., Xiang, T.: Multi-camera matching using bi-directional cumulative brightness transfer functions. In: British Machine Vision Conference (2008)
48. Prosser, B., Gong, S., Xiang, T.: Multi-camera matching under illumination change over time. In: Workshop on Multi-camera and Multi-modal Sensor Fusion Algorithms and Application (2008)

49. Prosser, B., Zheng, W., Gong, S., Xiang, T.: Person re-identification by support vector ranking. In: British Machine Vision Conference (2010)
50. Satta, R., Fumera, G., Roli, F.: Fast person re-identification based on dissimilarity representations. Pattern Recogn. Lett. **33**(14), 1838–1848 (2012)
51. Shafique, K., Hakeem, A., Javed, O., Haering, N.: Self calibrating visual sensor networks. In: IEEE Workshop Applications of Computer Vision (2008)
52. Stauffer, C.: Learning to track objects through unobserved regions. In: IEEE Workshop Motion and Video, Computing (2005)
53. Tieu, K., Dalley, G., Grimson, W.: Inference of non-overlapping camera network topology by measuring statistical dependence. In: IEEE International Conference on Computer Vision (2005)
54. Wang, L., Tan, T., Ning, H., Hu, W.: Silhouette analysis based gait recognition for human identification. IEEE Trans. Pattern Anal. Mach. Intell. **25**(12), 1505–1518 (2003)
55. Wang, Y., He, L., Velipasalar, S.: Real-time distributed tracking with non-overlapping cameras. In: IEEE International Conferenct on Image Processing, pp. 697–700 (2010)
56. Xie, B., Ramesh, V., Boult, T.: Sudden illumination change detection using order consistency. Image Vis. Comput. **22**(2), 117–125 (2004)
57. X.Wang, Tieu, K., Grimson, W.E.L.: Correspondence-free multicamera activity analysis and scene modeling. In: IEEE Conference on Computer Vision and Pattern Recognition (2008)
58. Zheng, W., Gong, S., Xiang, T.: Associating groups of people. In: British Machine Vision Conference (2009)
59. Zheng, W., Gong, S., Xiang, T.: Person re-identification by probabilistic relative distance comparison. In: IEEE Conference on Computer Vision and Pattern Recognition (2011)

Chapter 12
Mahalanobis Distance Learning for Person Re-identification

Peter M. Roth, Martin Hirzer, Martin Köstinger, Csaba Beleznai and Horst Bischof

Abstract Recently, Mahalanobis metric learning has gained a considerable interest for single-shot person re-identification. The main idea is to build on an existing image representation and to learn a metric that reflects the visual camera-to-camera transitions, allowing for a more powerful classification. The goal of this chapter is twofold. We first review the main ideas of Mahalanobis metric learning in general and then give a detailed study on different approaches for the task of single-shot person re-identification, also comparing to the state of the art. In particular, for our experiments, we used Linear Discriminant Metric Learning (LDML), Information Theoretic Metric Learning (ITML), Large Margin Nearest Neighbor (LMNN), Large Margin Nearest Neighbor with Rejection (LMNN-R), Efficient Impostor-based Metric Learning (EIML), and *KISSME*. For our evaluations we used four different publicly available datasets (i.e., *VIPeR*, *ETHZ*, *PRID 2011*, and *CAVIAR4REID*). Additionally, we generated the new, more realistic *PRID 450S* dataset, where we also provide detailed segmentations. For the latter one, we also evaluated the influence of using well-segmented foreground and background regions. Finally, the corresponding results are presented and discussed.

P. M. Roth (✉) · M. Hirzer · M. Köstinger · H. Bischof
Graz University of Technology, Graz, Austria
e-mail: pmroth@icg.tugraz.at

M. Hirzer
e-mail: hirzer@icg.tugraz.at

M. Köstinger
e-mail: koestinger@icg.tugraz.at

H. Bischof
e-mail: bischof@icg.tugraz.at

C. Beleznai
Austrian Institute of Technology, Vienna, Austria
e-mail: Csaba.Beleznai@ait.ac.at

S. Gong et al. (eds.), *Person Re-Identification*,
Advances in Computer Vision and Pattern Recognition,
DOI: 10.1007/978-1-4471-6296-4_12, © Springer-Verlag London 2014

12.1 Introduction

Person re-identification has become one of the major challenges in visual surveillance, showing a rather wide range of applications such as searching for criminals or tracking and analyzing individuals or crowds. In general, there are two main strategies: single-shot and multishot recognition. For the first one, an image pair is matched: one image given as input and one stored in a database. In contrast, for multishot scenarios multiple images (i.e., trajectories) are available. In this chapter, we mainly focus on the single-shot case, even though the ideas can simply be extended to the multishot scenario.

Even for humans, person re-identification is very challenging for several reasons. First, the appearance of an individual can vary extremely across a network of cameras due to changing view points, illumination, different poses, etc. Second, there is a potentially high number of "similar" persons (e.g., people wear rather dark clothes in winter). Third, in contrast to similar large-scale search problems, typically no accurate temporal and spatial constraints can be exploited to ease the task. Having these problems in mind and motivated by the high number of practical applications, there has been a significant scientific interest during the last years (e.g., [3, 8, 11, 14, 16, 22, 26, 28, 29]), and also various benchmark datasets (e.g., [13, 16]) have been published.

In general, the main idea is to find a suitable image description and then to perform a matching step using a standard distance. For describing images there exist two different strategies: (a) invariant and (b) discriminative description. The goal of invariant methods (e.g., [4, 11, 16, 27, 29]) is to extract visual features that are both, distinctive and stable under changing viewing conditions between different cameras. The large intraclass appearance variations, however, make the computation of distinctive and stable features often impossible under realistic conditions. To overcome this limitation, discriminative methods (e.g., [3, 14, 16, 28] on the other hand take advantage of class information to exploit the discriminative information to find a more distinctive representation. However, as a drawback such methods tend to overfit to the training data. Moreover, they are often based on local image descriptors, which might be a severe disadvantage. For instance, a red bag visible in one view would be very discriminative, however, if it is not visible in the other view it becomes impossible to re-identify a specific person.

An alternative to these two approaches, also incorporating label information, is to adopt metric learning for the given task (e.g., [8, 17, 18, 20, 21, 31]). Similar to the idea of intercamera color calibration (e.g., [25]), using labeled samples transitions in feature space between two camera views can be modeled. Hence, using a non-Euclidean distance even less distinctive features, which do not need to capture the visual invariances between different cameras, are sufficient for getting considerable matching results. However, to estimate such a metric, a training stage is necessary, but once learned, metric learning approaches are very efficient during evaluation, since additionally to the feature extraction and the matching only a linear projection has to be computed.

When dealing with person re-identification, we have to cope with three main problems. First, to capture all relevant information often complex, high dimensional feature representations are required. Thus, widely used metric learners such as arge Margin Nearest Neighbor (LMNN) [30], Information Theoretic Metric Learning (ITML) [7], and Logistic Discriminant Metric Learning (LDML) [15] building on complex optimization schemes run into high computational costs and memory requirements, making them infeasible in practice. Second, these methods typically assume a multiclass classification problem, which is not the case for person re-identification. In fact, we are typically given image pairs, so existing methods have to be adapted. There are only a few methods such as [1, 12] which directly intend learning a metric from data pairs. Third, we have to deal with a partially ill-posed problem. In fact, two images showing the same person might not be similar (e.g., due to camera noise, geometry, or different viewpoints: frontal vs. back). On the other hand, images not showing the same person can be very similar (e.g., in winter many people wear black/dark gray coats). Thus, for standard methods there is a high tendency to overfit to the training data yielding insufficient results during testing.

The goal of this chapter is to analyze the applicability of metric learning for the task of single-shot person re-identification from a more general point of view. Thus, we first review the main idea of Mahalanobis distance metric learning and give an overview of selected approaches targeting at the problem of discriminative metric learning via different strategies. In particular, we selected established methods applied to diverse visual classification tasks, (i.e., LDML [15], ITML [7], and LMNN [30]), as well as approaches that have been developed in particular for person re-identification (i.e., Large Margin Nearest Neighbor with Rejection (LMNN-R) [8], Efficient Impostor-based Metric Learning (EIML) [17], and *KISSME* [20]).

To show that metric learning is widely applicable, we run experiments on five different datasets showing different characteristics. Four of them, namely *VIPeR*, *ETHZ*, *PRID 2011*, and *CAVIAR4REID* are publicly available and widely used. For a more thorough evaluation and as additional contribution we created a new, more realistic dataset, *PRID 450S*, where we also provide detailed foreground/background segmentations. The results are summarized and compared to the state-of-the-art results for the specific datasets. In addition, to have a generative and discriminative baseline, the same experiments were also run using the standard Mahalanobis distance and a slightly adapted version of Linear Discriminant Analysis (LDA) [10].

The rest of the chapter is organized as follows. First, in Sect. 12.2 Mahalanobis metric learning in general is introduced and the approaches used in the study are summarized. Then, in Sect. 12.3, our specific person re-identification framework consisting of three stages is presented. In Sects. 12.4 and 12.5, we first review the five datasets used for our study and then present the obtained results. Finally, in Sect. 12.6 we summarize and conclude the chapter.

12.2 Mahalanobis Distance Metric Learning

In this section, we first introduce the general idea of Mahalanobis metric learning and then give an overview of the approaches used in this study. We selected generic methods that have shown good performance for diverse visual classification tasks as well as specific methods that have been developed for the task of person re-identification. Moreover, to give a more generic analysis, we tried to select methods tackling the same problem from different points of view: generative data analysis, statistical inference, information theoretic aspect, and discriminative learning. Additionally, we consider LDA and standard Mahalanobis metric learning, which can be considered simple baselines. For all methods the implementations are publicly available, thus allowing (a) for a fair comparison, and (b) for easily exchanging the used representation.

12.2.1 Mahalanobis Metric

Mahalanobis distance learning is a prominent and widely used approach for improving classification results by exploiting the structure of the data. Given n data points $\mathbf{x}_i \in \mathbb{R}^m$, the goal is to estimate a matrix $\mathbf{M}$ such that

$$d_{\mathbf{M}}(\mathbf{x}_i, \mathbf{x}_j) = (\mathbf{x}_i - \mathbf{x}_j)^\top \mathbf{M}(\mathbf{x}_i - \mathbf{x}_j) \tag{12.1}$$

describes a pseudometric. In fact, this is assured if $\mathbf{M}$ is positive semidefinite, i.e., $\mathbf{M} \succeq 0$. If $\mathbf{M} = \Sigma^{-1}$ (i.e., the inverse of the sample covariance matrix), the distance defined by Eq. (12.1) is referred to as the *Mahalanobis distance*. An alternative formulation for Eq. (12.1), which is more intuitive, is given via

$$d_{\mathbf{L}}(\mathbf{x}_i, \mathbf{x}_j) = ||\mathbf{L}(\mathbf{x}_i - \mathbf{x}_j)||^2 , \tag{12.2}$$

which is easily obtained from

$$(\mathbf{x}_i - \mathbf{x}_j)^\top \mathbf{M}(\mathbf{x}_i - \mathbf{x}_j) = (\mathbf{x}_i - \mathbf{x}_j)^\top \underbrace{\mathbf{L}^\top \mathbf{L}}_{\mathbf{M}}(\mathbf{x}_i - \mathbf{x}_j) = ||\mathbf{L}(\mathbf{x}_i - \mathbf{x}_j)||^2 . \tag{12.3}$$

Hence, either directly the metric matrix $\mathbf{M}$ or the factor matrix $\mathbf{L}$ can be estimated from the data. A discussion on factorization and the corresponding optimality criteria can be found in, e.g., [5, 19].

If additionally for a sample $\mathbf{x}$ its class label $y(\mathbf{x})$ is given, not only the generative structure of the data but also discriminative information can be exploited. For many problems (including person re-identification), however, we are lack class labels. Thus, given a pair of samples $(\mathbf{x}_i, \mathbf{x}_j)$m, we break down the original multiclass problem into a two-class problem in two steps. First, we transform the samples from the data space to the label agnostic difference space $\mathcal{X} = \{\mathbf{x}_{ij} = \mathbf{x}_i - \mathbf{x}_j\}$, which is inherently given by the metric definitions in Eqs. (12.1) and (12.2). Moreover, $\mathcal{X}$ is invariant to the actual locality of the samples in the feature space. Second, the original class labels are discarded and the samples are arranged using pairwise equality and inequality constraints, where we obtain the classes *same* $\mathcal{S}$ and *different* $\mathcal{D}$:

$$\mathcal{S} = \{(\mathbf{x}_i, \mathbf{x}_j) | y(\mathbf{x}_i) = y(\mathbf{x}_j)\} \tag{12.4}$$

$$\mathcal{D} = \{(\mathbf{x}_i, \mathbf{x}_j) | y(\mathbf{x}_i) \neq y(\mathbf{x}_j)\} . \tag{12.5}$$

In our particular case the pair $(\mathbf{x}_i, \mathbf{x}_j)$ consists of images showing persons in different camera views, and sharing a label means that the samples $\mathbf{x}_i$ and $\mathbf{x}_j$ describe the same person. In the following, we exemplarily discuss different approaches dealing with the problem described above.

To increase readability, we introduce the notation $\mathbf{C}_{ij} = (\mathbf{x}_i - \mathbf{x}_j)(\mathbf{x}_i - \mathbf{x}_j)^\top$ and the similarity variable

$$y_{ij} = \begin{cases} 1 & y(\mathbf{x}_i) = y(\mathbf{x}_j) \\ 0 & y(\mathbf{x}_i) \neq y(\mathbf{x}_j) . \end{cases} \tag{12.6}$$

12.2.2 Linear Discriminant Analysis

Let $\mathbf{x}_i \in \mathbb{R}^m$ be a sample and c its corresponding class label. Then, the goal of LDA [10] is to compute a classification function $g(\mathbf{x}) = \mathbf{L}^\top \mathbf{x}$ such that the Fisher criterion

$$\mathbf{L}_{\text{opt}} = \arg\max_{\mathbf{L}} \frac{|\mathbf{L}^\top \mathbf{S}_b \mathbf{L}|}{|\mathbf{L}^\top \mathbf{S}_w \mathbf{L}|}, \tag{12.7}$$

where $\mathbf{S}_w$ and $\mathbf{S}_b$ are the within-class scatter and between-class scatter matrices, is optimized. This is typically realized via solving the generalized eigenvalue problem

$$\mathbf{S}_B \mathbf{w} = \lambda \mathbf{S}_W \mathbf{w} \tag{12.8}$$

or directly by computing the eigenvectors for $\mathbf{S}_W^{-1} \mathbf{S}_B$.

However, it is known that the Fisher criterion given by Eq. (12.7) is only optimal in Bayes' sense for two classes (see, e.g., [23]). Thus, if the number of classes (image pairs in our case) is increasing LDA is going to fail. To overcome this problem, we can reformulate the original multiclass objective Eq. (12.7) to a binary formulation by using the two classes defined in Eqs. (12.4) and (12.5). In other words, Eq. (12.7) tries to minimize the distance between similar pairs and to maximize the distance between dissimilar pairs.

12.2.3 Logistic Discriminant Metric Learning

A similar idea is followed by LDML of Guillaumin et al. [15], however, from a probabilistic point of view. Thus, to estimate the Mahalanobis distance the probability p_{ij} that a pair $(\mathbf{x}_i, \mathbf{x}_j)$ is similar is modeled as

$$p_{ij} = p(y_{ij} = 1|\mathbf{x}_i, \mathbf{x}_j; \mathbf{M}, b) = \sigma(b - d_{\mathbf{M}}(\mathbf{x}_i, \mathbf{x}_j)), \tag{12.9}$$

where $\sigma(z) = (1 + \exp(-z))^{-1}$ is a sigmoid function and b is a bias term. As Eq. (12.9) is a standard linear logistic model, $\mathbf{M}$ can be optimized by maximizing the log-likelihood

$$\mathcal{L}(\mathbf{M}) = \sum_{ij} y_{ij} \ln(p_{ij}) + (1 - y_{ij}) \ln(1 - p_{ij}). \tag{12.10}$$

The optimal solution is then obtained by gradient ascent in direction

$$\frac{\partial \mathcal{L}(\mathbf{M})}{\partial \mathbf{M}} = \sum_{ij} (y_{ij} - p_{ij})\mathbf{C}_{ij}, \tag{12.11}$$

where the influence of each pair on the gradient direction is controlled over the probability. No further constraints, in particular no positive semi-definiteness on $\mathbf{M}$, are imposed on the problem!

12.2.4 Information Theoretic Metric Learning

Similarly, ITML was presented by Davis et al. [7], who regularized the estimated metric $\mathbf{M}$ by minimizing the distance to a predefined metric $\mathbf{M}_0$ via an information theoretic approach. In particular, they exploit the existence of a bijection between the set of Mahalanobis distances and the set of equal-mean multivariate Gaussian distributions. Let $d_{\mathbf{M}}$ be a Mahalanobis distance, then its corresponding multivariate Gaussian is given by

$$p(\mathbf{x}, \mathbf{M}) = \frac{1}{Z} \exp\left(-\frac{1}{2} d_{\mathbf{M}}(\mathbf{x}, \mu)\right), \tag{12.12}$$

where Z is a normalizing factor, μ is the mean, and the covariance is given by $\mathbf{M}^{-1}$.

Thus, the goal is to minimize the relative entropy between $\mathbf{M}$ and $\mathbf{M}_0$ arising the following optimization problem:

$$\min \; KL(g(\mathbf{x}, \mathbf{M}_0) || g(\mathbf{x}, \mathbf{M})) \tag{12.13}$$

$$s.t.$$

$$d_{\mathbf{M}}(\mathbf{x}_i, \mathbf{x}_j) \leq u \quad (\mathbf{x}_i, \mathbf{x}_j) \in \mathcal{S} \tag{12.14}$$

$$d_{\mathbf{M}}(\mathbf{x}_i, \mathbf{x}_j) \geq l \quad (\mathbf{x}_i, \mathbf{x}_j) \in \mathcal{D}\,, \tag{12.15}$$

where KL is the Kullback–Leibler divergence, and the constraints in Eqs. (12.14) and (12.15) enforce that the distances between similar pairs are small while they are large for dissimilar pairs.

As the optimization problem Eqs. (12.13)–(12.15) can be expressed via Bregman divergence, starting from $\mathbf{M}_0$ the Mahalanobis distance matrix $\mathbf{M}$ can be obtained by the following update rule:

$$\mathbf{M}_{t+1} = \mathbf{M}_t + \beta \mathbf{M}_t \mathbf{C}_{ij} \mathbf{M}_t, \tag{12.16}$$

where β encodes both, the pair label and the step size.

12.2.5 Large Margin Nearest Neighbor

In contrast, LMNN metric learning, introduced by Weinberger and Saul [30], additionally exploits the local structure of the data. For each instance, a local perimeter surrounding the k nearest neighbors sharing the same label (target neighbors) is established. Samples having a different label that invade this perimeter (impostors) are penalized. More technically, for a target pair $(\mathbf{x}_i, \mathbf{x}_j) \in \mathcal{S}$, i.e, $y_{ij} = 1$, any sample $\mathbf{x}_l$ with $y_{il} = 0$ is an impostor if

$$||\mathbf{L}(\mathbf{x}_i - \mathbf{x}_l)||^2 \leq ||\mathbf{L}(\mathbf{x}_i - \mathbf{x}_j)||^2 + 1. \tag{12.17}$$

Thus, the objective is to pull target pairs together and to penalize the occurrence of impostors. This is realized via the following objective function:

$$\mathcal{L}(\mathbf{M}) = \sum_{j \rightsquigarrow i} \left[d_{\mathbf{M}}(\mathbf{x}_i, \mathbf{x}_j) + \beta \sum_l (1 - y_{il}) \xi_{ijl}(\mathbf{M}) \right] \tag{12.18}$$

with

$$\xi_{ijl}(\mathbf{M}) = 1 + d_{\mathbf{M}}(\mathbf{x}_i, \mathbf{x}_j) - d_{\mathbf{M}}(\mathbf{x}_i, \mathbf{x}_l) \tag{12.19}$$

and β to be a weighting factor. The first term of Eq. (12.18) minimizes the distance between target neighbors $\mathbf{x}_i$ and $\mathbf{x}_j$, indicated by $j \rightsquigarrow i$, and the second one denotes the amount by which impostors invade the perimeter of $\mathbf{x}_i$ and $\mathbf{x}_j$. To estimate the metric $\mathbf{M}$, gradient descent is performed on the objective function Eq. (12.18):

$$\frac{\partial \mathcal{L}(\mathbf{M})}{\partial \mathbf{M}} = \sum_{j \rightsquigarrow i} \mathbf{C}_{ij} + \beta \sum_{(i,j,l) \in \mathcal{N}} (\mathbf{C}_{ij} - \mathbf{C}_{il}) \tag{12.20}$$

where $\mathcal{N}$ describes the set of triplets indices corresponding to a positive slack ξ.

LMNN was later adopted for person re-identification by Dikmen et al. [8], who introduced a rejection scheme not returning a match if all neighbors are beyond a certain threshold: LMNN-R.

12.2.6 Efficient Impostor-Based Metric Learning

Since both approaches described in Sect. 12.2.5, *LMNN* and *LMNN-R*, rely on complex optimization schemes, in [17] EIML was proposed that allows for exploiting the information provided by impostors more efficiently. In particular, Eq. (12.17) is relaxed to the original difference space. Thus, given a target pair $(\mathbf{x}_i, \mathbf{x}_j)$, a sample $\mathbf{x}_l$ is an impostor if

$$||(\mathbf{x}_i - \mathbf{x}_l)||^2 \leq ||(\mathbf{x}_i - \mathbf{x}_j)||^2. \tag{12.21}$$

To estimate the metric $\mathbf{M} = \mathbf{L}^\top \mathbf{L}$ the following objective function has to be minimized:

$$\mathcal{L}(\mathbf{L}) = \sum_{(\mathbf{x}_i, \mathbf{x}_j) \in \mathcal{S}} ||\mathbf{L}(x_i - x_j)||^2 - \sum_{(\mathbf{x}_i, \mathbf{x}_l) \in \mathcal{I}} ||\mathbf{L}\ w_{il}\ (x_i - x_l)||^2, \tag{12.22}$$

where $\mathcal{I}$ is the set of all impostor pairs and

$$w_{il} = e^{-\frac{||x_i - x_l||}{||x_i - x_j||}} \tag{12.23}$$

is a weighting factor also taking into account how much an impostor invades the perimeter of a target pair. By adding the orthogonality constraint $\mathbf{L}\mathbf{L}^\top = \mathbf{I}$, Eq. (12.22) can be reformulated to an eigenvalue problem:

$$(\Sigma_{\mathcal{S}} - \Sigma_{\mathcal{I}})\mathbf{L} = \Lambda \mathbf{L}, \tag{12.24}$$

where

$$\Sigma_{\mathcal{S}} = \frac{1}{|\mathcal{S}|} \sum_{(\mathbf{x}_i, \mathbf{x}_j) \in \mathcal{S}} \mathbf{C}_{ij} \quad \text{and} \quad \Sigma_{\mathcal{I}} = \frac{1}{|\mathcal{I}|} \sum_{(\mathbf{x}_i, \mathbf{x}_j) \in \mathcal{I}} \mathbf{C}_{ij} \tag{12.25}$$

are the covariance matrices for $\mathcal{S}$ and $\mathcal{I}$, respectively. Hence, the problem is much simpler and can be solved efficiently.

12.2.7 *KISSME*

The goal of the Keep It Simple and Straightforward MEtric (*KISSME*) [20] is to address the metric learning approach from a statistical inference point of view. Therefore, we test the hypothesis H_0 that a pair $(\mathbf{x}_i, \mathbf{x}_j)$ is dissimilar against H_1 that it is similar using a likelihood ratio test:

$$\delta(\mathbf{x}_i, \mathbf{x}_j) = \log\left(\frac{p(\mathbf{x}_i, \mathbf{x}_j | H_0)}{p(\mathbf{x}_i, \mathbf{x}_j | H_1)}\right) = \log\left(\frac{f(\mathbf{x}_i, \mathbf{x}_j, \theta_0)}{f(\mathbf{x}_i, \mathbf{x}_j, \theta_1)}\right), \tag{12.26}$$

where δ is the log-likelihood ratio, and $f(\mathbf{x}_i, \mathbf{x}_j, \theta)$ is a PDF with the parameter set θ. Assuming zero-mean Gaussian distributions Eq. (12.26) can be rewritten to

$$\delta(\mathbf{x}_i, \mathbf{x}_j) = \log\left(\frac{\frac{1}{\sqrt{2\pi|\Sigma_{\mathcal{D}}|}} \exp(-1/2\ (\mathbf{x}_i - \mathbf{x}_j)^\top\ \Sigma_{\mathcal{D}}^{-1}\ (\mathbf{x}_i - \mathbf{x}_j))}{\frac{1}{\sqrt{2\pi|\Sigma_{\mathcal{S}}|}} \exp(-1/2\ (\mathbf{x}_i - \mathbf{x}_j)^\top\ \Sigma_{\mathcal{S}}^{-1}\ (\mathbf{x}_i - \mathbf{x}_j))}\right), \tag{12.27}$$

where $\Sigma_{\mathcal{S}}$ and $\Sigma_{\mathcal{D}}$ are the covariance matrices of $\mathcal{S}$ and $\mathcal{D}$ according to Eq. (12.25).

The maximum likelihood estimate of the Gaussian is equivalent to minimizing the distances from the mean in a least squares manner. This allows *KISSME* to find respective relevant directions for $\mathcal{S}$ and $\mathcal{D}$. By taking the log and discarding the constant terms, we can simplify Eq. (12.27) to

$$\begin{aligned}\delta(\mathbf{x}_i, \mathbf{x}_i) &= (\mathbf{x}_i - \mathbf{x}_j)^\top\ \Sigma_{\mathcal{S}}^{-1}\ (\mathbf{x}_i - \mathbf{x}_j) - (\mathbf{x}_i - \mathbf{x}_j)^\top\ \Sigma_{\mathcal{D}}^{-1}\ (\mathbf{x}_i - \mathbf{x}_j)\\ &= (\mathbf{x}_i - \mathbf{x}_j)^\top (\Sigma_{\mathcal{S}}^{-1} - \Sigma_{\mathcal{D}}^{-1})(\mathbf{x}_i - \mathbf{x}_j). \end{aligned} \tag{12.28}$$

Hence, the Mahalanobis distance matrix $\mathbf{M}$ is defined by

$$\mathbf{M} = \left(\Sigma_{\mathcal{S}}^{-1} - \Sigma_{\mathcal{D}}^{-1}\right). \tag{12.29}$$

12.3 Person Re-identification System

In the following, we introduce the person re-identification system used for our study consisting of three stages: (1) feature extraction, (2) metric learning, and (3) classification. The overall system is illustrated in Fig. 12.1. During training the metric between two cameras is estimated, which is then used for calculating the distances between an unknown sample and the samples given in the database. The three steps are discussed in more detail in the next sections.

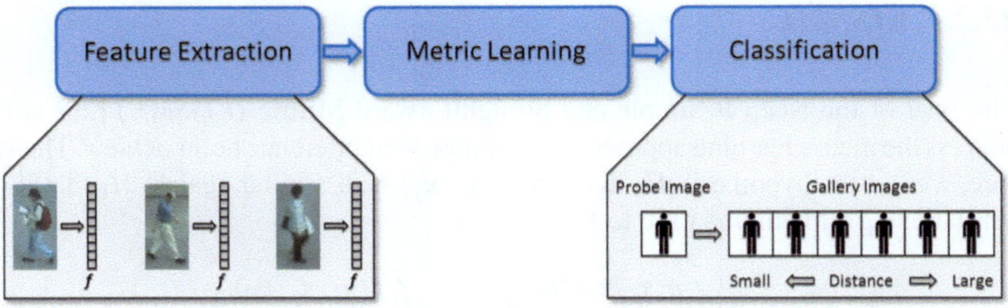

Fig. 12.1 Person re-identification system consisting of three stages: (1) feature extraction—dense sampling of color and texture features, (2) metric learning—exploiting the structure of similar and dissimilar pairs, and (3) classification—nearest neighbor search under the learned metric

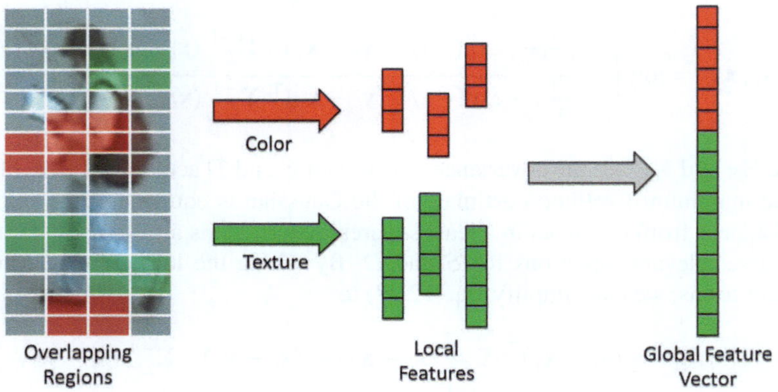

Fig. 12.2 Global image descriptor: different local features (HSV, Lab, LBP) are extracted from overlapping regions and then concatenated to a single feature vector

12.3.1 Representation

Color and texture features have proven to be successful for the task of person re-identification. We use HSV and Lab color channels as well as Local Binary Patterns to create a person image representation. The features are extracted from 8×16 rectangular regions sampled from the image with a grid of 4×8 pixels, i.e., 50 % overlap in both directions, which is illustrated in Fig. 12.2. In each rectangular patch, we calculate the mean values per color channel, which are then discretized to the range 0–40. Additionally, a histogram of LBP codes is generated from a gray value representation of the patch. These values are then put together to form a feature vector. Finally, the vectors from all regions are concatenated to generate a representation for the whole image.

(a) (b) (c) (d)

Fig. 12.3 Example image pairs from **a** the *VIPeR*, **b** the *PRID 2011*, **c** the *ETHZ*, and **d** the *CAVIAR4REID* dataset. The *upper* and *lower* rows correspond to different appearances of the same person, respectively

12.3.2 Metric Learning

First of all, we run a PCA step to reduce the dimensionality and for noise removal. In general, this step is not critical (the particular settings are given in Sect. 12.5), but we recognized that for smaller datasets a lower dimensional representation is also sufficient. During training we learn a Mahalanobis metric $\mathbf{M}$ according to Eq. (12.1). Once $\mathbf{M}$ has been estimated, during evaluation the distance between two samples $\mathbf{x}_i$ and $\mathbf{x}_j$ is calculated via Eq. (12.1). Hence, additionally to the actual classification effort only linear projections are required.

12.3.3 Classification

In person re-identification we want to recognize a certain person across different, nonoverlapping camera views. In our setup, we assume that we have already detected the persons in all camera views, i.e., we do not tackle the detection problem. The goal of person re-identification now is to find a person image that has been selected in one view (*probe image*) in all the images from another view (*gallery images*). This is achieved by calculating the distances between the probe image and all gallery images using the learned metric, and returning those gallery images with the smallest distances as potential matches.

12.4 Re-identification Datasets

In the following, we give an overview of the datasets used in our evaluations and explain the corresponding setups. In particular, these are *VIPeR* [13], *PRID 2011* [16], *ETHZ* [28], *CAVIAR4REID* [6], and *PRID 450S*. The first four (see Fig. 12.3) are publicly available and widely used for benchmarking person re-identification methods; the latter one was newly generated for this study.

Although there are other datasets like *iLIDS*, we abstained from using them in this study. "The" *iLIDS* dataset was not used since there are at least four different datasets available that arbitrarily cropped patches from the huge (publicly not available!) *iLIDS* dataset, making it difficult to give fair comparisons.

12.4.1 VIPeR Dataset

The *VIPeR* dataset contains 632 person image pairs taken from two different camera views. Changes of viewpoint, illumination and pose are the most prominent sources of appearance variation between the two images of a person. For evaluation we followed the procedure described in [14]. The set of 632 image pairs is randomly split into two sets of 316 image pairs each, one for training and one for testing. In the test case, the two images of an image pair are randomly assigned to a probe and a gallery set. A single image from the probe set is then selected and matched with all images from the gallery set. This process is repeated for all images in the probe set.

12.4.2 ETHZ Dataset

The *ETHZ* dataset [28], originally proposed for pedestrian detection [9] and later modified for benchmarking person re-identification approaches, consists of three video sequences: SEQ. #1 containing 83 persons (4.857 images), SEQ. #2 containing 35 persons (1.961 images), and SEQ. #3 containing 28 persons (1.762 images). All images have been resized to 64×32 pixels. The most challenging aspects of this dataset are illumination changes and occlusions. However, as the person images are captured from a single moving camera, the dataset does not provide a realistic scenario for person re-identification (i.e., no disjoint cameras, different viewpoints, different camera characteristics, etc.). Despite this limitation it is commonly used for person re-identification. We use a single-shot evaluation strategy, i.e., we randomly sample two images per person to build a training pair; and another pair for testing. The images of the test pairs are then assigned to the probe and the gallery set.

12.4.3 PRID 2011 Dataset

The *PRID 2011* dataset[1] consists of person images recorded from two different static cameras. Two scenarios are provided: multishot and single-shot. Since we are focusing on single-shot methods in this work, we use only the latter one. Typical

[1] The dataset is publicly available under https://lrs.icg.tugraz.at/download.php.

challenges on this dataset are viewpoint and pose changes as well as significant differences in illumination, background and camera characteristics. Camera view A contains 385 persons, camera view B contains 749 persons, with 200 of them appearing in both views. Hence, there are 200 person image pairs in the dataset. These image pairs are randomly split into a training and a test set of equal size. For evaluation on the test set, we followed the procedure described in [16], i.e., camera A is used for the probe set and camera B is used for the gallery set. Thus, each of the 100 persons in the probe set is searched in a gallery set of 649 persons (all images of camera view B except the 100 training samples).

12.4.4 CAVIAR4REID Dataset

The *CAVIAR4REID* dataset [6] contains images of 72 individuals captured from two different cameras in a shopping center, where the original images have been resized to 128×64. 50 of them appear in both camera views, the remaining 22 only in one view. Since we are interested in person re-identification in different cameras, we only use individuals appearing in both views in our experiments. Each person is represented by 10 appearances per camera view. Typical challenges on this dataset are viewpoint and pose changes, different light conditions, occlusions, and low resolution. To compare the different methods we use a multishot evaluation strategy similar to [2]. The set of 50 persons is randomly split into a training set of 42 persons, and a test set of 8 persons. Since every person is represented by 10 images per camera view, we can generate 100 different image pairs between the views of two individuals. During training, we use all possible combinations of positive pairs showing the same person, and negative pairs showing different persons. When comparing two individuals in the evaluation stage, we again use all possible combinations in order to calculate the mean distance between the two persons.

12.4.5 PRID 450S Dataset

The *PRID 450S* dataset[2] builds on *PRID 2011*, however, is arranged according to *VIPeR* by image pairs and contains more linked samples than *PRID 2011*. In particular, the dataset contains 450 single-shot image pairs depicting walking humans captured in two spatially disjoint camera views. From the original images with resolution of 720×576 pixels, person patches were annotated manually by bounding boxes with a vertical resolution of 100–150 pixels. To form the ground truth for re-identification, persons with the same identity seen in the different views were associated. In addition, for each image instance we generated binary segmentation

[2] The dataset is publicly available under https://lrs.icg.tugraz.at/download.php.

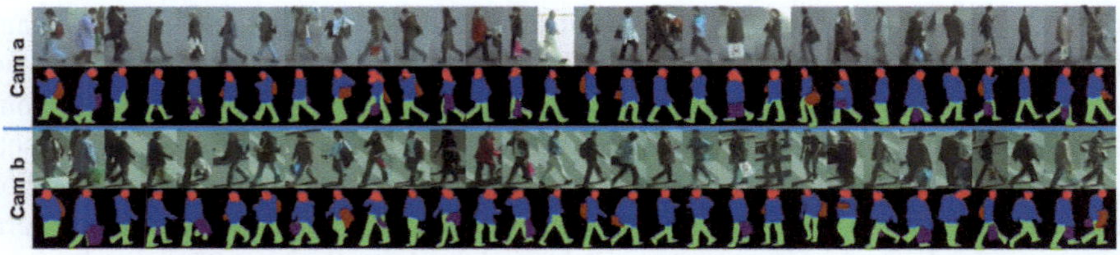

Fig. 12.4 *PRID 450S* dataset: original images (*top*) and multilabel segmentations (*bottom*) for both camera views

masks separating the foreground from the background. Moreover, we further provide a part-level segmentation[3] describing the following regions: head, torso, legs, carried object at torso level (if any), and carried object below torso (if any). The union of these part segmentations is equivalent to the foreground segment. Exemplary images and corresponding segmentations for both cameras are illustrated in Fig. 12.4.

12.5 Experiments

In the following, we give a detailed study on metric learning for person re-identification using the framework introduced in Sect. 12.4. In particular, we compare the methods discussed in Sect. 12.2 using the datasets presented in Sect. 12.4, where all methods get exactly the same data (training/test splits, representation). The results are presented in form of CMC scores [29], representing the expectation of finding the true match within the first r ranks. In particular, we plot the CMC scores for the different metric learning approaches and additionally provide tables for the first ranks, where the best scores are given in boldface, respectively. If available, comparisons to state-of-the-art methods are also given. The reported results are averaged over 10 random runs. Regarding the number of PCA dimensions, we use 100 dimensions for *VIPeR* and *CAVIAR4REID*, 40 for *PRID 2011*, *PRID 450S*, and *ETHZ* SEQ. #1, and 20 for *ETHZ* SEQ. #2 and SEQ. #3.

12.5.1 Dataset Evaluations

The first experiment was carried out on the *VIPeR* dataset, which can be considered the standard benchmark for single-shot re-identification scenarios. The CMC curves for the different metric learning approaches are shown in Fig. 12.5a. It can be seen that besides *LDA* and *LDML*, which either have too weak discriminative power or are overfitting to the training data, all approaches significantly improve the

[3] The more detailed segmentations were actually not used for this study, but as they could be beneficial for others they are also provided.

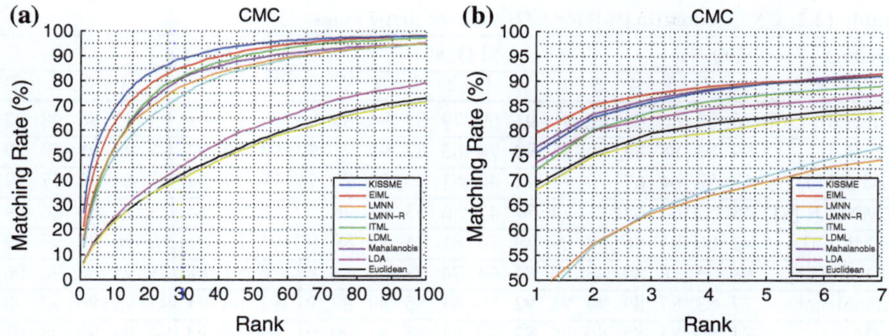

Fig. 12.5 CMC curves for **a** *VIPeR* and **b** *ETHZ* SEQ. #1

Table 12.1 CMC scores (in [%]) and average training times per trial for *VIPeR*

Method	$r = 1$	10	20	50	100	t_{train}
KISSME [20]	**27**	**70**	**83**	**95**	**99**	0.1 sec
EIML [17]	22	63	79	93	**99**	0.3 s
LMNN [30]	17	54	69	87	96	2 min
LMNN-R [8]	13	50	65	86	95	45 min
ITML [7]	13	54	73	91	98	25 s
LDML [15]	6	24	35	54	72	0.8 s
Mahalanobis	16	54	72	89	96	**0.001 s**
LDA	7	25	37	61	79	0.1 s
Euclidean	7	24	34	55	73	–
ELF [14]	12	43	60	81	93	5 h
SDALF [4]	20	50	65	85	–	–
ERSVM [26]	13	50	67	85	94	13 min
DDC [16]	19	52	65	80	91	–
PS [6]	22	57	71	87	–	–
PRDC [31]	16	54	70	87	97	15 min
PCCA [24]	19	65	80	–	–	–

classification results over all rank levels. In addition, we provide these results compared to state-of-the-art methods (i.e., *ELF* [14], *SDALF* [4], *ERSVM* [26], *DDC* [16], *PS* [6], *PRDC* [31], and *PCCA* [24]) in Table 12.1. As for many methods timings are available, these are also included in the table. The results show that metric learning boosts the performance of the originally quite simple representation and finally yields competitive results; however, at dramatically reduced computational complexity.

Next, we show results for *ETHZ*, another widely used benchmark, containing trajectories of persons captured from a single camera. Thus, the image pairs show the same characteristics and metric learning has only little influence. Nevertheless, the CMC curves in Fig. 12.5b for SEQ. #1, where metric learning has the largest

Table 12.2 CMC scores (in [%]) for *ETHZ* for the first 7 ranks

Method	SEQ. #1							SEQ. #2							SEQ. #3						
	1	2	3	4	5	6	7	1	2	3	4	5	6	7	1	2	3	4	5	6	7
KISSME [20]	76	83	86	88	**90**	90	91	69	79	83	86	89	90	91	83	91	93	95	96	98	98
EIML [17]	**80**	**85**	**88**	**89**	**90**	**91**	**92**	**74**	**83**	**87**	**90**	**91**	**92**	**93**	**90**	**94**	95	**96**	**98**	**99**	**99**
LMNN [7]	47	58	64	67	70	73	74	40	51	59	66	70	75	79	34	51	61	66	72	77	79
LMNN-R [8]	45	57	64	68	71	74	77	47	56	65	72	76	79	83	49	64	73	79	83	86	89
ITML [30]	72	80	84	86	88	89	89	70	81	85	87	89	90	91	88	93	**96**	**96**	**98**	98	**99**
LDML [15]	68	75	78	80	82	83	84	64	74	78	81	84	85	86	81	88	91	95	96	96	96
Mahalanobis	77	83	87	**89**	**90**	**91**	**92**	70	81	85	89	89	91	91	84	91	93	95	96	98	98
LDA	74	80	83	85	86	86	87	70	81	85	87	90	91	92	88	**94**	**96**	**96**	**98**	98	98
Euclidean	69	75	80	81	83	84	85	68	77	81	83	85	87	89	85	91	94	95	96	96	97

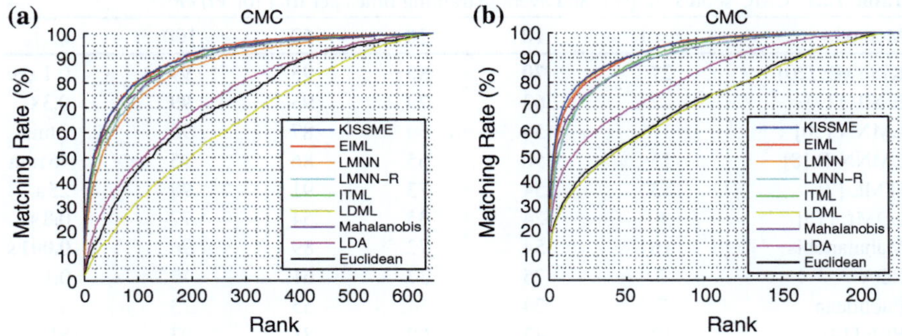

Fig. 12.6 CMC curves for **a** *PRID 2011* and **b** *PRID 450S*

impact, reveal that a performance gain of more than 5 % can be obtained over all ranks. The decrease of *LMNN* can be explained by the evaluation protocol, which generates impostors resulting in an overfitting model (Table 12.2).

In contrast, *PRID 2011* defines a more realistic setup. In fact, the images stem from multiple cameras and especially the number of gallery images is much higher. Again from the CMC curves in Fig. 12.6a it can be seen that for all methods besides *LDA* and *LDML* a significant improvement can be obtained, especially for the first ranks. The results in Table 12.3 reveal that in this case using a standard Mahalanobis distance yields competitive results. Moreover, it can be seen that the descriptive approach [16], which uses a much more complex representation, can clearly be outperformed.

As the newly created *PRID 450S* dataset builds on *PRID 2011*, it has similar characteristics, however, provides much more linked samples. In addition, we also generated detailed foreground/background masks, allowing us to analyze the effect of using an exact foreground/background segmentation. The CMC curves exploiting the given segmentations are shown in Fig. 12.6b. Again it can be seen that using *LDML* has no and using *LDA* has only a little influence on the classification

Table 12.3 CMC scores (in [%]) for *PRID 2011*

Method	$r = 1$	10	20	50	100
KISSME [20]	15	39	**52**	**68**	80
EIML [17]	**16**	39	51	**68**	**81**
LMNN [30]	10	30	42	59	73
LMNN-R [8]	9	32	43	60	76
ITML [7]	12	36	47	64	79
LDML [15]	2	6	11	19	32
Mahalanobis	**16**	**41**	51	64	76
LDA	4	14	21	35	48
Euclidean	3	10	14	28	45
Descr. M. [16]	4	24	37	56	70

Table 12.4 CMC scores (in [%]) for *PRID 450S*: (a) without segmentation and (b) with segmentation

Method	$r = 1$	10	20	50	100
KISSME [20]	33	**71**	**79**	**90**	97
EIML [17]	**35**	68	77	**90**	**98**
LMNN [30]	29	68	78	**90**	97
LMNN-R [8]	22	59	71	86	95
ITML [7]	24	59	71	87	97
LDML [15]	12	31	39	55	73
Mahalanobis	31	62	73	85	95
LDA	20	46	54	69	86
Euclidean	13	32	41	55	74

results, whereas for all other approaches a significant improvement can be obtained. The impact of segmentation is analyzed in Table 12.4, where both, the results with and without segmentation, are compared. It can be recognized that using the foreground information is beneficial for all approaches increasing the performance by up to 5 %.

Finally, we show the results for the *CAVIAR4REID* dataset for two reasons. First, to demonstrate that metric learning can also be applied if the number of training samples is small, and, second, to show that the single-shot setup can easily be extended to multishot. The corresponding CMC scores (due to the small number of samples averaged over 100 runs) are shown in Fig. 12.7 and Table 12.5, where we also compare to [2]. Again for all approaches except *LDML* an improvement can be obtained. The higher variance in performance can be explained by the smaller number of training samples, resulting in a higher overfitting tendency.

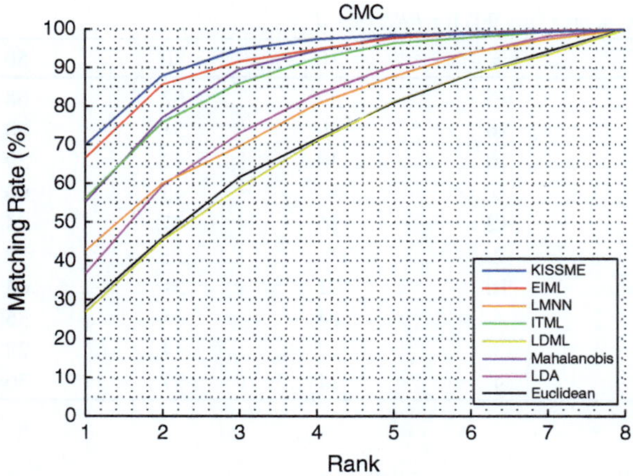

Fig. 12.7 CMC curves for *CAVIAR4REID*

Table 12.5 CMC scores (in [%]) for *CAVIAR4REID*

Method	$r = 1$	2	3	4	5	6	7	8
KISSME [20]	**70**	**88**	**95**	**98**	**99**	99	99	**100**
EIML [17]	67	86	92	95	98	99	**100**	**100**
LMNN [30]	43	60	70	81	88	94	98	**100**
ITML [7]	56	76	86	93	97	98	**100**	**100**
LDML [15]	27	46	59	71	81	88	94	**100**
Mahalanobis	55	77	90	95	98	99	**100**	**100**
LDA	37	60	73	83	91	94	98	**100**
Euclidean	28	46	62	71	81	88	94	**100**
ICT [2]	62	81	**95**	97	97	**100**	**100**	**100**

12.5.2 Discussion

The results shown above clearly indicate that metric learning, in general, can drastically boost the performance for single-shot (and even for multishot) person re-identification. In fact, by learning a metric we can describe the visual transition from one camera to the other, thus, the applied features do not have to cope with all variabilities, allowing for more meaningful feature matchings. Hence, even if rather simple features are used competitive results can be obtained.

In particular, we used only block-based color and texture descriptions for two reasons. On the one side since they are easy and fast to compute and on the other side to demonstrate that using even such simple features, state-of-the-art or better results can be obtained. However, it is clear that better features, e.g., exploiting temporal information in a multishot scenario will further improve the results.

Surprisingly, even using the standard Mahalanobis distance allows for improving the results and finally yields considerable results. Nevertheless, incorporating discriminative information yields a further performance gain. However, we have to consider the specific constraints given by the task: (a) images showing the same person might not have a similar visual description whereas (b) images not showing the same person could be very close in the original feature space. Thus, the problem is somehow ill-posed and highly prone to overfitting. This can for instance be recognized for *LDML*, *LMNN*, and *ITML*.

As *LDML* does not use any regularization, it is totally overfitting to the training data and thus yields rather weak results (comparable to the Euclidean distance). The results of *LMNN* are typically better, however, since the impostor handling is not robust against outliers, the problems described above cannot be handled sufficiently. The same applies for *ITML*, which often yields similar results as the original Mahalanobis distance, clearly showing that given somehow "ambiguously labeled" samples no additional discriminative information can be gained. In contrast, *KISSME* and *EIML*, following different strategies, provide some regularization by relaxing the original problem, which seems to be better suited for the given task. Moreover, the metric estimation is computationally much more efficient.

Results on five different datasets showing totally different characteristics clearly demonstrate that metric learning is a general purpose strategy. In fact, the same features were used, only the parameter for PCA was adjusted, which has only a little influence on the results. However, we recognized that for smaller datasets less PCA dimensions are sufficient. The results also indicate the characteristics of the datasets. For *VIPeR* and *CAVIAR4REID*, showing a larger variety in the appearance the discriminative power can fully be exploited. For *PRID 2011* and *PRID 450S* containing a larger amount of "similar" instances the improvement from generative to discriminative metric is less significant. Finally, for the *ETHZ* dataset, where the images are taken from the same camera view, metric learning has, as expected, only a little influence.

Thus, if we are given enough data to learn a meaningful metric, metric learning could be highly beneficial in the context of person re-identification. However, more important than much data is good data. Hence, it would be more meaningful to use temporal information to select good candidates for learning than just using larger amounts of data. Similarly, it was also revealed by the improved results for the *PRID 450S* dataset that using better data (i.e., estimating the metric on the foreground regions only) is beneficial.

12.6 Conclusions

The goal of this chapter was to analyze the applicability of Mahalanobis metric learning in the context of single-shot person re-identification. We first introduced the main ideas of metric learning and gave an overview on specific approaches addressing the same problem following different paradigms. These were evaluated within a fixed framework on five different benchmark datasets (where one was newly generated). If applicable, we also gave a comparison to the state of the art. Even though some approaches tend to overfit to the training data, we can conclude that metric learning can dramatically boost the classification performance and that even less complex (non-handcrafted) representations could be sufficient for the given task. Moreover, one interesting result is that even a standard Mahalanobis metric not using any discriminative information yields quite good classification results. We also showed that having a perfect segmentation further improves the classification and that it is straight forward to extend the current framework toward multishot scenarios. In a similar way temporal information or a better image representation can also be used.

References

1. Alipanahi, B., Biggs, M., Ghodsi, A.: Distance metric learning vs. fisher discriminant analysis. Proceedings of the AAAI Conference on Artificial Intelligence (2008)
2. Avraham, T., Gurvich, I., Lindenbaum, M., Markovitch, S.: Learning implicit transfer for person re-identification. In: Proceedings of the ECCV Workshop on Re-Identification (2012)
3. Bak, S., Corvee, E., Brémond, F., Thonnat, M.: Person re-idendification using Haar-based and DCD-based signature. In: Workshop on Activity Monitoring by Multi-Camera Surveillance Systems (2010)
4. Bazzani, L., Cristani, M., Murino, V.: Symmetry-driven accumulation of local features for human characterization and re-identification. Comput. Vision Image Underst. **117**(2), 130–144 (2013)
5. Burer, S., Monteiro, R.: A nonlinear programming algorithm for solving semidefinite programs via low-rank factorization. Math. Program. **95**(2), 329–357 (2003)
6. Cheng, D.S., Cristani, M., Stoppa, M., Bazzani, L., Murino, V.: Custom pictorial structures for re-identification. In: Proceedings of the British Machine Vision Conference (2011)
7. Davis, J.V., Kulis, B., Jain, P., Sra, S., Dhillon, I.S.: Information-theoretic metric learning. In: Proceedings of the Int'l Conference on Machine Learning (2007)
8. Dikmen, M., Akbas, E., Huang, T.S., Ahuja, N.: Pedestrian recognition with a learned metric. In: Proceedings of the Asian Conference on Computer Vision (2010)
9. Ess, A., Leibe, B., Gool, L.V.: Depth and appearance for mobile scene analysis. In: Proceedings of the IEEE Int'l Conference on Computer Vision (2007)
10. Fisher, R.A.: The use of multiple measurements in taxonomic problems. Ann. Eugenics **7**, 179–188 (1936)
11. Gheissari, N., Sebastian, T.B., Hartley, R.: Person reidentification using spatiotemporal appearance. In: Proceedings of the IEEE Conference on Computer Vision and, Pattern Recognition (2006)
12. Ghodsi, A., Wilkinson, D.F., Southey, F.: Improving embeddings by flexible exploitation of side information. In: Proceedings of the Int'l Joint Conference on, Artificial Intelligence (2007)
13. Gray, D., Brennan, S., Tao, H.: Evaluating appearance models for recognition, reacquisition, and tracking. In: Proceedings of the IEEE Workshop on Performance Evaluation of Tracking and Surveillance (2007)
14. Gray, D., Tao, H.: Viewpoint invariant pedestrian recognition with an ensemble of localized features. In: Proceedings of the European Conference on Computer Vision (2008)
15. Guillaumin, M., Verbeek, J., Schmid, C.: Is that you? Metric learning approaches for face identification. In: Proceedings of the IEEE Int'l Conference on Computer Vision (2009)

16. Hirzer, M., Beleznai, C., Roth, P.M., Bischof, H.: Person re-identification by descriptive and discriminative classification. In: Proceedings of the Scandinavian Conference on Image, Analysis (2011)
17. Hirzer, M., Roth, P.M., Bischof, H.: Person re-identification by efficient imposter-based metric learning. In: Proceedings of the IEEE Int'l Conference on Advanced Video and Signal-Based Surveillance (2012)
18. Hirzer, M., Roth, P.M., Köstinger, M., Bischof, H.: Relaxed pairwise learned metric for person re-identification. In: Proceedings of the European Conference on Computer Vision (2012)
19. Journée, M., Bach, F., Absil, P.A., Sepulchre, R.: Low-rank optimization of the cone of positive semidefinite matrices. SIAM J. Optim. **20**(5), 2327–2351 (2010)
20. Köstinger, M., Hirzer, M., Wohlhart, P., Roth, P.M., Bischof, H.: Large scale metric learning from equivalence constraints. In: Proceedings of the IEEE Conference on Computer Vision and, Pattern Recognition (2012)
21. Li, W., Zhao, R., Wang, X.: Human reidentification with transferred metric learning. In: Proceedings of the Asian Conference on Computer Vision (2012)
22. Lin, Z., Davis, L.S.: Learning pairwise dissimilarity profiles for appearance recognition in visual surveillance. In: Advances Int'l Visual Computing, Symposium (2008)
23. Loog, M., Duin, R.P.W., Haeb-Umbach, R.: Multiclass linear dimension reduction by weighted pairwise fisher criteria. IEEE Trans. Pattern Anal. Mach. Intell. **23**(7), 762–766 (2001)
24. Mignon, A., Jurie, F.: PCCA: A new approach for distance learning from sparse pairwise constraints. In: Proceedings of the IEEE Conference on Computer Vision and Pattern Recognition (2012)
25. Porikli, F.: Inter-camera color calibration by correlation model function. In: Proceedings of the Int'l Conference on Image Processing (2003)
26. Prosser, B., Zheng, W.S., Gong, S., Xiang, T.: Person re-identification by support vector ranking. In: Proceedings of the British Machine Vision Conference (2010)
27. Rahimi, A., Dunagan, B., Darrell, T.: Simultaneous calibration and tracking with a network of non-overlapping sensors. In: Proceedings of the IEEE Conference on Computer Vision and Pattern Recognition (2004)
28. Schwartz, W.R., Davis, L.S.: Learning discriminative appearance-based models using partial least squares. In: Proceedings of the Brazilian Symposium on, Computer Graphics and Image Processing (2009)
29. Wang, X., Doretto, G., Sebastian, T.B., Rittscher, J., Tu, P.H.: Shape and appearance context modeling. In: Proceedings of the IEEE Int'l Conference on Computer Vision (2007)
30. Weinberger, K.Q., Saul, L.K.: Fast solvers and efficient implementations for distance metric learning. In: Proceedings of the Int'l Conference on, Machine Learning (2008)
31. Zheng, W.S., Gong, S., Xiang, T.: Reidentification by relative distance comparison. IEEE Trans Pattern Anal. Mach. Intell. **35**(3), 653–668 (2013)

Chapter 13
Dictionary-Based Domain Adaptation Methods for the Re-identification of Faces

Qiang Qiu, Jie Ni and Rama Chellappa

Abstract Re-identification refers to the problem of recognizing a person at a different location after one has been captured by a camera at a previous location. We discuss re-identification of faces using the domain adaptation approach which tackles the problem where data in the target domain (different location) are drawn from a different distribution as the source domain (previous location), due to different view points, illumination conditions, resolutions, etc. In particular, we discuss the adaptation of dictionary-based methods for re-identification of faces. We first present a domain adaptive dictionary learning (DADL) framework for the task of transforming a dictionary learned from one visual domain to the other, while maintaining a domain-invariant sparse representation of a signal. Domain dictionaries are modeled by a linear or nonlinear parametric function. The dictionary function parameters and domain-invariant sparse codes are then jointly learned by solving an optimization problem. We then discuss an unsupervised domain adaptive dictionary learning (UDADL) method where labeled data are only available in the source domain. We propose to interpolate subspaces through dictionary learning to link the source and target domains. These subspaces are able to capture the intrinsic domain shift and form a shared feature representation for cross-domain identification.

Q. Qiu (✉)
Department of Electrical and Computer Engineering, Duke University, Durham, NC 27708, USA
e-mail: qiang.qiu@duke.edu

J. Ni · R. Chellappa
Department of Electrical and Computer Engineering, University of Maryland,
College Park, MD 20742, USA
e-mail: jni@umiacs.umd.edu

J. Ni · R. Chellappa
Center for Automation Research, UMIACS, University of Maryland,
College Park, MD 20742, USA
e-mail: rama@umiacs.umd.edu

S. Gong et al. (eds.), *Person Re-Identification*,
Advances in Computer Vision and Pattern Recognition,
DOI: 10.1007/978-1-4471-6296-4_13, © Springer-Verlag London 2014

13.1 Introduction

Re-identification refers to identify a subject initialized at one location with a feasible set of candidates at other locations and over time. We are interested in face re-identification as face is an important biometric signature to determine the identity of a person. Re-identification is a fundamentally challenging problem due to the large visual appearance changes caused by variations in view angle, lighting, background clutter, and occlusion [37]. It is well known that traditional face recognition techniques perform well when constrained face images are acquired at close range, with controlled studio lights and cooperative subjects. Yet these ideal assumptions are usually violated in the scenario of re-identification, which poses serious challenges to standard face recognition algorithms [5]. As it is very difficult to address the large appearance changes through physical models of individual degradations, we formulate the face re-identification problem as a domain adaptation problem to handle the distribution shift between query and candidate images.

Domain Adaptation (DA) aims to utilize a *source domain* (early location) with plenty of labeled data to learn a classifier for a *target domain* (different location) which belongs to a different distribution. It has drawn much attention in the computer vision community [12, 13, 16, 28]. Based on the availability of labeled data in the target domain, DA methods can be classified into two categories: *semi-supervised* and *unsupervised* DA. Semi-supervised DA leverages the few labels in the target data or correspondence between the source and target data to reduce the divergence between two domains. Unsupervised DA is inherently a more challenging problem without any labeled target data to build associations between two domains.

In this chapter, we investigate the DA problem using dictionary learning and sparse representation approaches. Sparse and redundant modeling of signals has received a lot of attention from the vision community [33]. This is mainly due to the fact that signals or images of interest are sparse or compressible in some dictionary. In other words, they can be well approximated by a linear combination of a few atoms of a redundant dictionary. It has been observed that dictionaries learned directly from data achieved state-of-the-art results in a variety of tasks in image restoration [9, 19] and classification [34, 36].

When designing dictionaries for image classification tasks, we are often confronted with situations where conditions in the training set are different from those present during testing. For example, in the case of face re-identification, more than one familiar view may be available for training. Such training faces may be obtained from a live or recorded video sequence, where a range of views are observed. However, the test images can contain conditions that are not necessarily presented in the training images such as a face in a different pose. For such cases where the same set of signals are observed in several visual domains with correspondence information available, we discuss the proposed domain adaptive dictionary learning (DADL) method in [26] to learn a dictionary for a new domain associated with no observations. We formulate this problem of dictionary transformation in a function learning framework, i.e., dictionaries across different domains are modeled by a parametric

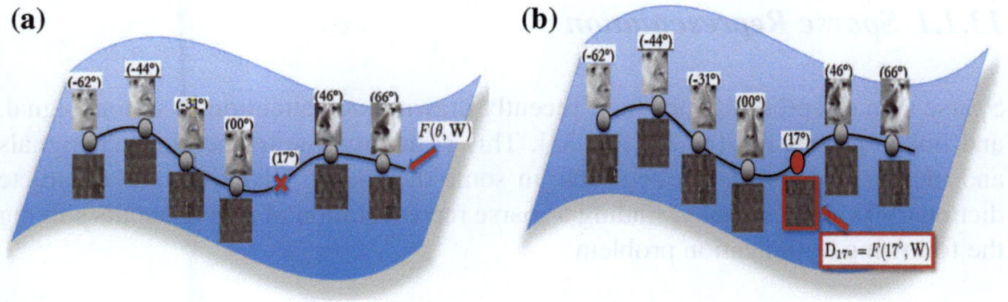

Fig. 13.1 Overview of DADL. Consider example dictionaries corresponding to faces at different azimuths. **a** shows a depiction of example dictionaries over a curve on a dictionary manifold which will be discussed later. Given example dictionaries, our approach learns the underlying dictionary function $F(\boldsymbol{\theta}, \mathbf{W})$. In **b**, the dictionary corresponding to a domain associated with observations is obtained by evaluating the learned dictionary function at the corresponding domain parameters [26]

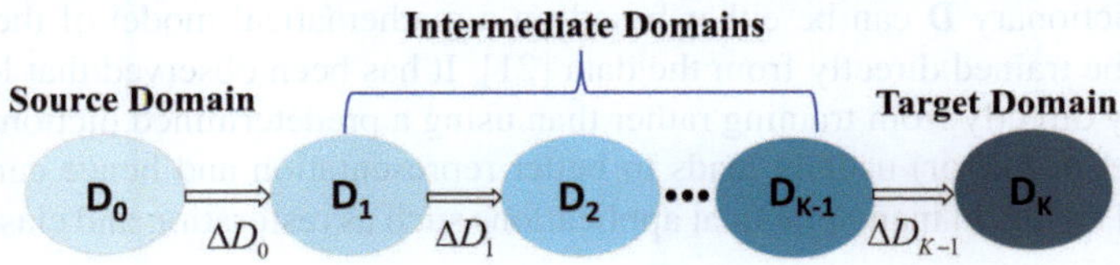

Fig. 13.2 Given labeled data in the source domain and unlabeled data in the target domain, our DA procedure learns a set of intermediate domains (represented by dictionaries $\{\mathbf{D}_k\}_{k=1}^{K-1}$) and the target domain (represented by dictionary $\mathbf{D}_K$) to capture the intrinsic domain shift between two domains. $\{\Delta\mathbf{D}_k\}_{k=0}^{K-1}$ characterize the gradual transition between these subspaces

function. The dictionary function parameters and domain-invariant sparse codes are then jointly learned by solving an optimization problem. As shown in Fig. 13.1, given a learned dictionary function, a dictionary adapted to a new domain is obtained by evaluating such a dictionary function at the corresponding domain parameters, e.g., pose angles. The domain invariant sparse representations are used here as shared feature representation for cross-domain face re-identification.

We further discuss the unsupervised DA with no correspondence information or labeled data in the target domain. Unsupervised DA is more representative of real-world scenarios for re-identification. In addition to individual degradation factors due to view points, lighting, resolution, etc, sometimes the coupling effect among these different factors give rise to more variations in the target domain. As it is very costly to obtain labels for target images under all kinds of acquisition condition, it is more desirable that our identification system can adapt in an unsupervised fashion. We discuss an unsupervised domain adaptive dictionary learning (UDADL) method to learn a set of intermediate domain dictionaries between the source and target domains, as shown in Fig. 13.2. We then apply invariant sparse codes across the source, intermediate, and target domains to render intermediate representations, which provide a shared feature space for face re-identification. A more detailed discussion of UDADL can be found in [20].

13.1.1 Sparse Representation

Sparse signal representations have recently drawn much attention in vision, signal, and image processing [1, 25, 27, 33]. This is mainly due to the fact that signals and images of interest can be sparse in some dictionary. Given an over-complete dictionary $\mathbf{D}$ and a signal $\mathbf{y}$, finding a sparse representation of $\mathbf{y}$ in $\mathbf{D}$ entails solving the following optimization problem

$$\hat{\mathbf{x}} = \arg\min_{\mathbf{x}} \|\mathbf{x}\|_0 \text{ subject to } \mathbf{y} = \mathbf{D}\mathbf{x}, \tag{13.1}$$

where the ℓ_0 sparsity measure $\|\mathbf{x}\|_0$ counts the number of nonzero elements in the vector $\mathbf{x}$. Problem (13.1) is NP-hard and cannot be solved in a polynomial time. Hence, approximate solutions are usually sought [1, 6, 24, 30].

The dictionary $\mathbf{D}$ can be either based on a mathematical model of the data [1] or it can be trained directly from the data [21]. It has been observed that learning a dictionary directly from training rather than using a predetermined dictionary (such as wavelet or Gabor) usually leads to better representation and hence can provide improved results in many practical applications such as restoration and classification [27, 33].

Various algorithms have been developed for the task of training a dictionary from examples. One of the most commonly used algorithms is the K-SVD algorithm [1]. Let $\mathbf{Y}$ be a set of N input signals in a n-dimensional feature space $\mathbf{Y} = [\mathbf{y_1}...\mathbf{y_N}]$, $\mathbf{y_i} \in \mathbb{R}^n$. In K-SVD, a dictionary with a fixed number of K items is learned by finding a solution iteratively to the following problem:

$$\arg\min_{\mathbf{D},\mathbf{X}} \|\mathbf{Y} - \mathbf{D}\mathbf{X}\|_F^2 \quad s.t. \ \forall i, \|\mathbf{x_i}\|_0 \le T \tag{13.2}$$

where $\mathbf{D} = [\mathbf{d_1}...\mathbf{d_K}]$, $\mathbf{d_i} \in \mathbb{R}^n$ is the learned dictionary, $\mathbf{X} = [\mathbf{x_1}, ..., \mathbf{x_N}]$, $\mathbf{x_i} \in \mathbb{R}^K$ are the sparse codes of input signals $\mathbf{Y}$, and T specifies the sparsity that each signal has fewer than T items in its decomposition. Each dictionary atom $\mathbf{d_i}$ is l_2-normalized.

Organization of the chapter: The structure of the rest of the chapter is as follows: in Sect. 13.2, we relate our work to existing work on domain adaptation. In Sect. 13.3, we discuss the domain adaptive dictionary learning framework for domain adaptation with correspondence available. In Sect. 13.4, we present the details of our unsupervised domain adaptive dictionary learning method. We report experimental results on face pose alignment and face re-identification in Sect. 13.5. The chapter is summarized in Sect. 13.6.

13.2 Related Work

Several DA methods have been discussed in the literature. We briefly review relevant work below. Semi-supervised DA methods rely on labeled target data or correspondence between two domains to perform cross-domain classification. Daume [7] proposes a feature augmentation technique such that data points from the same domain are more similar than those from different domains. The Adaptive-SVM introduced in [35] selects the most effective auxiliary classifiers to adapt to the target dataset. The method in [8] designed an adaptive classifier based on multiple base kernels. Metric learning approaches were also proposed [16, 28] to learn a cross-domain transformation to link two domains. Recently, Jhuo et al. [15] utilized low-rank reconstructions to learn a transformation, so that the transformed source samples can be linearly reconstructed by the target samples.

Given no labels in the target domain to learn the similarity measure between data instances across domains, unsupervised DA is more difficult to tackle. It usually enforces certain prior assumption to relate the source and target data. Structural correspondence learning [4] induces correspondence among features from the two domains by modeling their relations with *pivot* features, which appear frequently in both domains. Manifold-alignment based DA [32] computes similarity between data points in different domains through the local geometry of data points within each domain. The techniques in [22, 23] learn a latent feature space where domain similarity is measured using maximum mean discrepancy. Two recent approaches [12, 13] in the computer vision community are more relevant to our methodology of UDADL, where the source and target domains are linked by sampling finite or infinite number of intermediate subspaces on the Grassmannian manifold. These intermediate subspaces are able to capture the intrinsic domain shift. Compared to their abstract manifold walking strategies, our UDADL approach emphasizes on synthesizing intermediate subspaces in a manner which gradually reduces the reconstruction error of the target data.

13.3 Domain Adaptive Dictionary Learning

We denote the same set of P signals observed in N different domains as $\{\mathbf{Y_1}, ..., \mathbf{Y_N}\}$, where $\mathbf{Y_i} = [\mathbf{y_{i1}}, ..., \mathbf{y_{iP}}]$, $\mathbf{y_{ip}} \in \mathbb{R}^n$. Thus, $\mathbf{y_{ip}}$ denotes the pth signal observed in the ith domain. In the following, we will use $\mathbf{D_i}$ as the vector-space embedded dictionary. Let $\mathbf{D_i}$ denote the dictionary for the ith domain, where $\mathbf{D_i} = [\mathbf{d_{i1}}...\mathbf{d_{iK}}]$, $\mathbf{d_{ik}} \in \mathbb{R}^n$. We define a *vector transpose* (VT) operation over dictionaries as shown in Fig. 13.3. The VT operator treats each individual dictionary atom as a value and then perform the typical matrix transpose operation. Let $\mathbf{D}$ denote the stack dictionary shown in Fig. 13.3b over all N domains. It is noted that $\mathbf{D} = [\mathbf{D^{VT}}]^{\mathbf{VT}}$.

The domain dictionary learning problem can be formulated as (13.3). Let $\mathbf{X} = [\mathbf{x_1}, ..., \mathbf{x_P}]$, $\mathbf{x_p} \in \mathbb{R}^K$, be the sparse code matrix. The set of domain dictionary $\{\mathbf{D_i}\}_i^N$

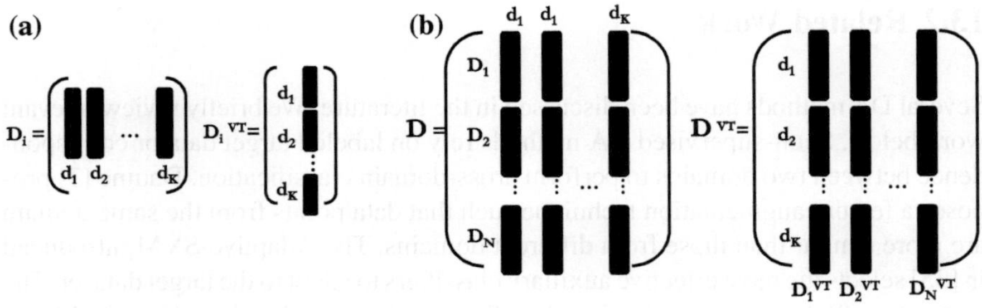

Fig. 13.3 The vector transpose (VT) operator over dictionaries

learned through (13.3) enable the same sparse codes $\mathbf{x_p}$ for a signal $\mathbf{y_p}$ observed across N different domains to achieve domain adaptation.

$$\underset{\{\mathbf{D_i}\}_i^N, \mathbf{X}}{\arg\ \min} \sum_i^N \|\mathbf{Y_i} - \mathbf{D_i X}\|_F^2 \ \text{ s.t. } \forall p \ \|\mathbf{x}_p\|_o \le T, \tag{13.3}$$

where $\|\mathbf{x}\|_o$ counts the number of nonzero values in $\mathbf{x}$. T is a sparsity constant.

We propose to model domain dictionaries $\mathbf{D_i}$ through a parametric function in (13.4), where $\boldsymbol{\theta}_i$ denotes a vector of domain parameters, e.g., view point angles, illumination conditions, etc., and $\mathbf{W}$ denotes the dictionary function parameters.

$$\mathbf{D_i} = F(\boldsymbol{\theta}_i, \mathbf{W}) \tag{13.4}$$

Applying (13.4) to (13.3), we formulate the domain dictionary function learning as (13.5).

$$\underset{\mathbf{W}, \mathbf{X}}{\arg\min} \sum_i^N \|\mathbf{Y_i} - F(\boldsymbol{\theta}_i, \mathbf{W})\mathbf{X}\|_F^2 \ \text{ s.t. } \forall p \ \|\mathbf{x}_p\|_o \le T. \tag{13.5}$$

We adopt power polynomials to model $\mathbf{D_i^{VT}}$ in Fig. 13.3a through the following dictionary function $F(\boldsymbol{\theta}_i, \mathbf{W})$,

$$F(\theta_i, \mathbf{W}) = w_0 + \sum_{s=1}^S w_{1s}\theta_{is} + ... + \sum_{s=1}^S w_{ms}\theta_{is}^m \tag{13.6}$$

where we assume S-dimensional domain parameter vectors and an mth-degree polynomial model. For example, given $\boldsymbol{\theta}_i$ a 2-dimensional domain parameter vector, a quadratic dictionary function is defined as,

$$F(\theta_i, \mathbf{W}) = w_0 + w_{11}\theta_{i1} + w_{12}\theta_{i2} + w_{21}\theta_{i1}^2 + w_{22}\theta_{i2}^2$$

Given $\mathbf{D_i}$ contains K atoms and each dictionary atom is in the $\mathbb{R}^n$ space, as $\mathbf{D_i^{VT}} = F(\boldsymbol{\theta}_i, \mathbf{W})$, it can be noted from Fig. 13.3 that w_{ms} is a nK-sized vector. We define the function parameter matrix $\mathbf{W}$ and the domain parameter matrix $\boldsymbol{\Theta}$ as

$$\mathbf{W} = \begin{bmatrix} w_0^{(1)} \; w_0^{(2)} \; w_0^{(3)} \ldots w_0^{(nK)} \\ w_{11}^{(1)} \; w_{11}^{(2)} \; w_{11}^{(3)} \ldots w_{11}^{(nK)} \\ \cdot \\ \cdot \\ \cdot \\ w_{mS}^{(1)} \; w_{mS}^{(2)} \; w_{mS}^{(3)} \ldots w_{mS}^{(nK)} \end{bmatrix} \qquad \boldsymbol{\Theta} = \begin{bmatrix} 1 \; 1 \; 1 \ldots 1 \\ \theta_{11} \; \theta_{21} \; \theta_{31} \ldots \theta_{N1} \\ \cdot \\ \cdot \\ \cdot \\ \theta_{1S}^m \; \theta_{2S}^m \; \theta_{3S}^m \ldots \theta_{NS}^m \end{bmatrix}$$

Each row of $\mathbf{W}$ corresponds to the nK-sized w_{ms}^T, and $\mathbf{W} \in \mathbb{R}^{(mS+1)\times nK}$. N different domains are assumed and $\boldsymbol{\Theta} \in \mathbb{R}^{(mS+1)\times N}$. With the matrix $\mathbf{W}$ and $\boldsymbol{\Theta}$, (13.6) can be written as,

$$\mathbf{D^{VT}} = \mathbf{W^T}\boldsymbol{\Theta} \tag{13.7}$$

where $\mathbf{D^{VT}}$ is defined in Fig. 13.3b. Now dictionary function learning formulated in (13.5) can be written as,

$$\underset{\mathbf{W},\mathbf{X}}{\arg\min} \|\mathbf{Y} - [\mathbf{W^T}\boldsymbol{\Theta}]^{\mathbf{VT}}\mathbf{X}\|_F^2 \;\; s.t. \, \forall p \; \|\mathbf{x}_p\|_o \leq T \tag{13.8}$$

where $\mathbf{Y}$ is the stacked training signals observed in different domains. With the objective function defined in (13.8), the dictionary function learning can be performed as described below.

Step 1: Obtain the sparse coefficients X and $[\mathbf{W^T}\boldsymbol{\Theta}]^{\mathbf{VT}}$ via any dictionary learning method, e.g., K-SVD [1].

Step 2: Given the domain parameter matrix $\boldsymbol{\Theta}$, the optimal dictionary function can be obtained as [18],

$$\mathbf{W} = [\boldsymbol{\Theta}\boldsymbol{\Theta}^{\mathbf{T}}]^{-1}\boldsymbol{\Theta}[[[\mathbf{W^T}\boldsymbol{\Theta}]^{\mathbf{VT}}]^{\mathbf{VT}}]^{\mathbf{T}}. \tag{13.9}$$

13.4 Unsupervised Domain Adaptive Dictionary Learning

In this section, we present the UDADL method for face re-identification. We first describe some notations to facilitate subsequent discussions.

Let $\mathbf{Y}_s \in \mathbb{R}^{n*N_s}$, $\mathbf{Y}_t \in \mathbb{R}^{n*N_t}$ be the data instances from the source and target domain respectively, where n is the dimension of the data instance, N_s and N_t denote the number of samples in the source and target domains. Let $\mathbf{D}_0 \in \mathbb{R}^{n*m}$ be the dictionary learned from $\mathbf{Y}_s$ using standard dictionary learning methods, e.g., K-SVD [1], where m denotes the number of atoms in the dictionary.

We hypothesize there is a virtual path which smoothly connects the source and target domain. Imagine the source domain consists of face images in the frontal view while the target domain contains those in the profile view. Intuitively, face images which gradually transform from the frontal to profile view will form a smooth transition path. Our approach samples several intermediate domains along this virtual path, and associate each intermediate domain with a dictionary $\mathbf{D}_k$, $k \in [1, K]$, where K is the number of intermediate domains.

13.4.1 Learning Intermediate Domain Dictionaries

Starting from the source domain dictionary $\mathbf{D}_0$, we learn the intermediate domain dictionaries $\{\mathbf{D}_k\}_{k=1}^K$ sequentially to gradually adapt to the target data. This is also conceptually similar to incremental learning. The final dictionary $\mathbf{D}_K$ which best represents the target data in terms of reconstruction error is taken as the target domain dictionary. Given the k-th domain dictionary $\mathbf{D}_k$, $k \in [0, K-1]$, we learn the next domain dictionary $\mathbf{D}_{k+1}$ based on its coherence with $\mathbf{D}_k$ and the remaining residue of the target data. Specifically, we decompose the target data $\mathbf{Y}_t$ with $\mathbf{D}_k$ and get the reconstruction residue $\mathbf{J}_k$:

$$\boldsymbol{\Gamma}_k = \arg\min_{\boldsymbol{\Gamma}} \|\mathbf{Y}_t - \mathbf{D}_k \boldsymbol{\Gamma}\|_F^2, \; s.t. \forall i, \; \|\alpha_i\|_0 \leq T \tag{13.10}$$

$$\mathbf{J}_k = \mathbf{Y}_t - \mathbf{D}_k \boldsymbol{\Gamma}_k \tag{13.11}$$

where $\boldsymbol{\Gamma}_k = [\alpha_1, ..., \alpha_{N_t}] \in \mathbb{R}^{m*N_t}$ denote the sparse coefficients of $\mathbf{Y}_t$ decomposed with $\mathbf{D}_k$, and T is the sparsity level. We then obtain $\mathbf{D}_{k+1}$ by estimating $\Delta\mathbf{D}_k$, which is the adjustment in the dictionary atoms between $\mathbf{D}_{k+1}$ and $\mathbf{D}_k$:

$$\min_{\Delta\mathbf{D}_k} \|\mathbf{J}_k - \Delta\mathbf{D}_k \boldsymbol{\Gamma}_k\|_F^2 + \lambda \|\Delta\mathbf{D}_k\|_F^2 \tag{13.12}$$

This formulation consists of two terms. The first term ensures that the adjustments in the atoms of $\mathbf{D}_k$ will further decrease the current reconstruction error $\mathbf{J}_k$. The second term penalizes abrupt changes between adjacent intermediate domains, so as to obtain a smooth path. The parameter λ controls the balance between these two terms. This is a ridge regression problem. By setting the first-order derivatives to be zeros, we obtain the following closed form solution:

$$\Delta\mathbf{D}_k = \mathbf{J}_k \boldsymbol{\Gamma}_k^T (\lambda \mathbf{I} + \boldsymbol{\Gamma}_k \boldsymbol{\Gamma}_k^T)^{-1} \tag{13.13}$$

where $\mathbf{I}$ is the identity matrix. The next intermediate domain dictionary $\mathbf{D}_{k+1}$ is then obtained as:

$$\mathbf{D}_{k+1} = \mathbf{D}_k + \Delta\mathbf{D}_k \tag{13.14}$$

Starting from the source domain dictionary $\mathbf{D}_0$, we apply the above adaptation framework iteratively, and terminate the procedure when the magnitude of $\|\Delta\mathbf{D}_k\|_F$ is below certain threshold, so that the gap between the two domains is absorbed into the learned intermediate domain dictionaries. This stopping criterion also automatically gives the number of intermediate domains to sample from the transition path. We summarize our approach in Algorithm 1.

Algorithm 1 Algorithm to generate intermediate subspaces between source and target domains.

1: Input:Dictionary $\mathbf{D}_0$ trained from the source data, target data $\mathbf{Y}_t$, sparsity level T, stopping threshold δ, parameter λ, $k = 0$.
2: Output: Dictionaries $\{\mathbf{D}_k\}_{k=1}^{K-1}$ for the intermediate domains, dictionary $\mathbf{D}_K$ for the target domain.
3: **while** stopping criteria is not reached **do**
4: Decompose the target data with the current intermediate domain dictionary $\mathbf{D}_k$, get the reconstruction residue $\mathbf{J}_k$ using equation (12.10),(12.11)
5: Get an estimate of the adjustment in dictionary atoms $\Delta\mathbf{D}_k$ and the next intermediate domain dictionary $\mathbf{D}_{k+1}$ using equation (12.13),(12.14). Normalize the atoms in $\mathbf{D}_{k+1}$ to have unit norm.
6: $k \leftarrow k + 1$
7: check the stopping criteria $\|\Delta\mathbf{D}_k\|_F \leq \delta$
8: **end while**

13.4.2 Recognition Under Domain Shift

To this end, we have learned a transition path which is encoded with the underlying domain shift. This provides us with rich information to obtain new representations to associate source and target data. Here, we simply apply invariant sparse codes across source, intermediate, target domain dictionaries $\{\mathbf{D}_k\}_{k=0}^{K}$. The new augmented feature representation is obtained as follows:

$$[(\mathbf{D}_0\mathbf{x})^T, (\mathbf{D}_1\mathbf{x})^T, ..., (\mathbf{D}_K\mathbf{x})^T]^T$$

where $\mathbf{x} \in \mathbb{R}^m$ is the sparse code of a source data signal decomposed with $\mathbf{D}_0$, or a target data signal decomposed with $\mathbf{D}_K$. This new representation incorporates the smooth domain transition recovered in the intermediate dictionaries into the signal space. It brings source and target data into a shared space where the data distribution shift is mitigated. Therefore, it can serve as a more robust characteristic across different domains. Given the new feature vectors, we apply PCA for dimension reduction, and then employ an SVM classifier for cross-domain recognition.

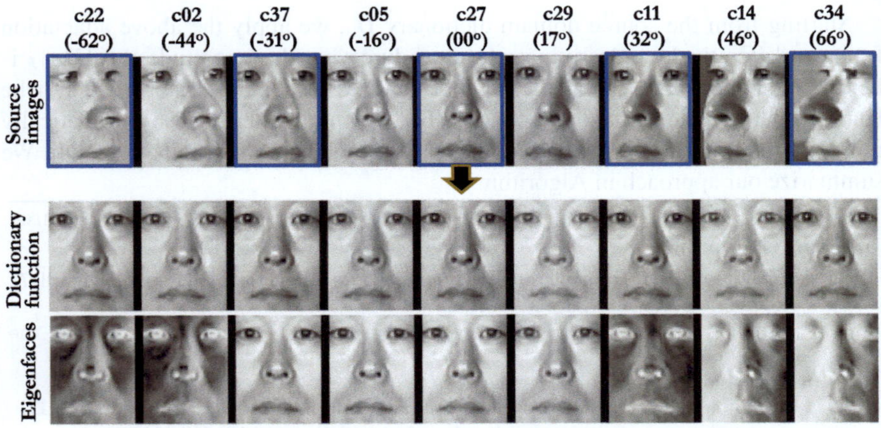

Fig. 13.4 Frontal face alignment. For the first row of source images, pose azimuths are shown below the camera numbers. Poses highlighted in blue are known poses to learn a linear dictionary function (m=4), and the remaining are unknown poses. The second and third rows show the aligned face to each corresponding source image using the linear dictionary function and Eigenfaces, respectively

13.5 Experiments

We present the results of experiments using two public face datasets: the CMU-PIE dataset [29] and the Extended YaleB dataset [11]. The CMU-PIE dataset consists of 68 subjects in 13 poses and 21 lighting conditions. In our experiments we use 9 poses which have approximately the same camera altitude, as shown in the first row of Fig. 13.4. The Extended YaleB dataset consists of 38 subjects in 64 lighting conditions. All images are in 64×48 size. We will first evaluate the basic behaviors of DADL through pose alignment. Then we will demonstrate the effectiveness of both DADL and UDADL in face re-identification across domain.

13.5.1 DADL for Pose Alignment

Frontal Face Alignment In Fig. 13.4, we align different face poses to the frontal view. We learn for each subject in the PIE dataset a linear dictionary function $F(\boldsymbol{\theta}, \mathbf{W})$ (m=4) using 5 out of 9 poses. The training poses are highlighted in blue in the first row of Fig. 13.4. Given a source image $\mathbf{y_s}$, we first estimate the domain parameters $\boldsymbol{\theta}_s$, i.e., the pose azimuth here, as discussed in [26]. We then obtain the sparse representation $\mathbf{x_s}$ of the source image as $\min_{\mathbf{x_s}} \|\mathbf{y_s} - F(\boldsymbol{\theta}_s, \mathbf{W})\mathbf{x_s}\|_2^2,\ s.t.\ \|\mathbf{x_s}\|_o \leq T$ (sparsity level) using any pursuit methods such as OMP [10]. We specify the frontal pose azimuth (00°) as the parameter for the target domain $\boldsymbol{\theta}_t$, and obtain the frontal view image $\mathbf{y_t}$ as $\mathbf{y_t} = F(\boldsymbol{\theta}_t, \mathbf{W})\mathbf{x_s}$. The second row of Fig. 13.4 shows the aligned frontal view images to the respective poses in the first row. These aligned frontal faces are close

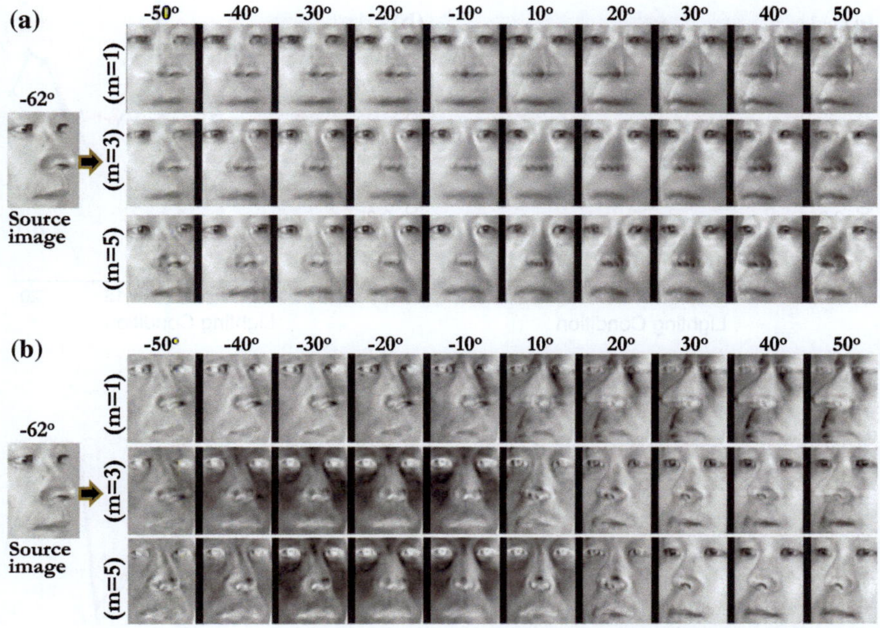

Fig. 13.5 Pose synthesis using various degrees of dictionary polynomials. All the synthesized poses are unknown to learned dictionary functions and associated with no actual observations. m is the degree of a dictionary polynomial in (13.6)

to the actual image, i.e., c27 in the first row. It is noted that images with poses c02, c05, c29, and c14 are unknown poses to the learned dictionary function.

For comparison purposes, we learn Eigenfaces for each of the 5 training poses and obtain adapted Eigenfaces at 4 unknown poses using the same function fitting method in our framework. We then project each source image (mean-subtracted) on the respective Eigenfaces and use frontal Eigenfaces to reconstruct the aligned image shown in the third row of Fig. 13.4. The proposed method of jointly learning the dictionary function parameters and domain-invariant sparse codes in (13.8) significantly outperforms the Eigenfaces approach, which fails for large pose variations.

Pose Synthesis In Fig. 13.5, we synthesize new poses at any given pose azimuth. We learn for each subject in the PIE dataset a linear dictionary function $F(\boldsymbol{\theta}, \mathbf{W})$ using all 9 poses. In Fig. 13.5a, given a source image $\mathbf{y_s}$ in a profile pose (-62°), we first estimate the domain parameters $\boldsymbol{\theta}_s$ for the source image, and sparsely decompose it over $F(\boldsymbol{\theta}_s, \mathbf{W})$ for its sparse representation $\mathbf{x_s}$. We specify every 10° pose azimuth in $[-50^o, 50^o]$ as parameters for the target domain $\boldsymbol{\theta}_t$, and obtain a synthesized pose image $\mathbf{y_t}$ as $\mathbf{y_t} = F(\boldsymbol{\theta}_t, \mathbf{W})\mathbf{x_s}$. It is noted that none of the target poses are associated with actual observations. As shown in Fig. 13.5a, we obtain reasonable synthesized images at poses with no observations. We observe improved synthesis performance by increasing the value of m, i.e., the degree of a dictionary polynomial.

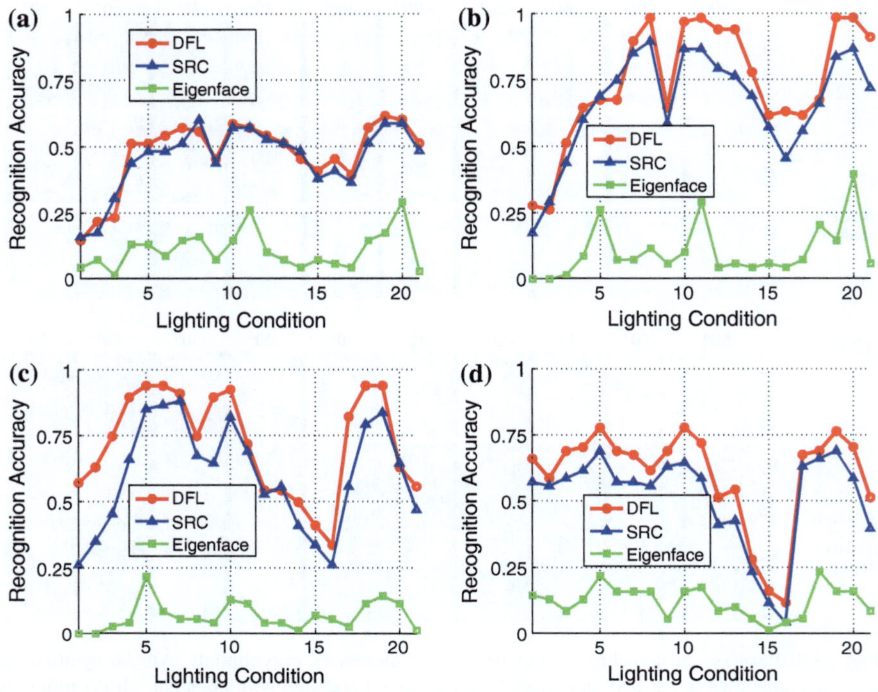

Fig. 13.6 Face recognition accuracy on the CMU-PIE dataset. The proposed method is denoted as DFL in red color

In Fig. 13.5b, we perform curve fitting over Eigenfaces as discussed. The proposed dictionary function learning framework exhibits better synthesis performance.

13.5.2 DADL for Face Re-identification

Two face recognition methods are adopted for comparisons: Eigenfaces [31] and SRC [34]. SRC is a state-of-the-art method to use sparse representation for face recognition. We denote our method as the Dictionary Function Learning (DFL) method. For a fair comparison, we adopt exactly the same configurations for all the three methods, i.e., we use 68 subjects in 5 poses c22, c37, c27, c11, and c34 in the PIE dataset for training, and the remaining 4 poses for testing.

For the SRC method, we form a dictionary from the training data for each pose of a subject. For the proposed DFL method, we learn from the training data a dictionary function across pose for each subject. In SRC and DFL, a test image is classified using the subject label associated with the dictionary or the dictionary function, respectively, that gives the minimal reconstruction error. In Eigenfaces, a nearest neighbor classifier is used. In Fig. 13.6, we present the face recognition accuracy on

Table 13.1 Face recognition under pose variation on CMU-PIE dataset [29]

	c11	c29	c05	c37	average
Ours	76.5	**98.5**	**98.5**	88.2	**90.4**
GFK [12]	63.2	92.7	92.7	76.5	81.3
SGF [13]	51.5	82.4	82.4	67.7	71.0
Eigen light-field [14]	**78.0**	91.0	93.0	**89.0**	87.8
K-SVD [1]	48.5	76.5	80.9	57.4	65.8

the PIE dataset for different testing poses under each lighting condition. The proposed DFL method outperforms both Eigenfaces and SRC methods for all testing poses.

13.5.3 Unsupervised DADL for Face Re-identification

Across pose variation: We present the results of face recognition across pose variation using the CMU-PIE dataset [29]. This experiment includes 68 subjects under 5 different poses. Each subject has 21 images at each pose, with variations in lightings. We select the frontal face images as the source domain, with a total of 1428 images. The target domain contains images at different poses, which are denoted as $c05$ and $c29$ (yawning about $\pm 22.5^o$), $c37$ and $c11$ (yawning bout $\pm 45^o$), respectively. We choose the front-illuminated source images to be the labeled data in the source domain. The task is to determine the identity of faces in the target domain with the same illumination condition. The classification results are in Table 13.1. We compare our method with the following methods. 1) Baseline K-SVD [1], where the target data are represented using the dictionary learned from the source domain, and the resulting sparse codes are compared using nearest neighbor classifier. 2) GFK [12] and SGF [13], which perform subspace interpolation via infinite or finite sampling on the Grassmann manifold. 3) Eigen light-field [14] method, which is specifically designed to handle face recognition across pose variations. We observe that the baseline is heavily biased under domain shift, and all the DA methods improve upon it. Our method has advantages over other two DA methods when the pose variation is large. Further, our average performance is competitive with [14], which relies on a generic training set to build pose specific models, while DA methods do not make such an assumption. We also show some of the synthesized intermediate images in Fig.13.7 for an illustration. As our DA approach gradually updates the dictionary learned from frontal face images using non-frontal images, these transformed representations thus convey the transition process in this scenario. These transformations could also provide additional information for certain applications, e.g., face reconstruction across different poses.

Across blur and illumination variations: Next, we performed a face recognition experiment across combined blur and illumination variations. All frontal images of the first 34 subjects under 21 lighting conditions from the CMU-PIE dataset [29] are included in this experiment. We randomly select images under 11 different

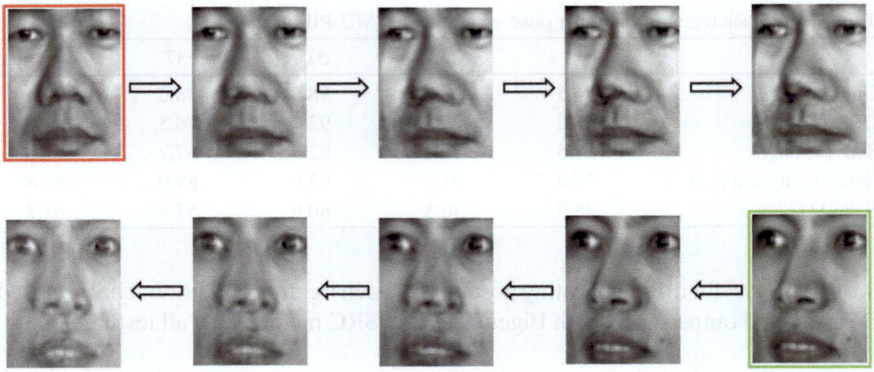

Fig. 13.7 Synthesized intermediate representations between frontal face images and face images at pose $c11$. The first row shows the transformed images from a source image (*in red box*) to the target domain. The second row shows the transformed images from a target image (*in green box*) to the source domain

Table 13.2 Face recognition across illumination and blur variations on CMU-PIE dataset [29]

	$\sigma = 3$	$\sigma = 4$	$L = 9$	$L = 11$
Ours	**80.29**	**77.94**	**85.88**	**81.18**
GFK [12]	78.53	77.65	82.35	77.65
SGF [13]	63.82	52.06	70.29	57.06
LPQ [2]	66.47	32.94	73.82	62.06
Albedo [3]	50.88	36.76	60.88	45.88
K-SVD [1]	40.29	25.59	42.35	30.59

illumination conditions to form the source domain. The remaining images with the other 10 illumination conditions are convolved with a blur kernel to form the target domain. Experiments are performed with the Gaussian kernels with standard deviations of 3 and 4, and motion blurs with lengths of 9 (angle $\theta = 135^o$) and 11 (angle $\theta = 45^o$), respectively. We compare our results with those of K-SVD [1], GFK [12], and SGF [13]. Besides, we also compare with the Local Phase Quantization [2] method, which is a blur insensitive descriptor, and the method based in [3], which estimates an albedo map (Albedo) as an illumination robust signature for matching. We report the results in Table 13.2. Our method is competitive with [12], and outperforms all other algorithms by a large margin. Since the domain shift in this experiment consists of both illumination and blur variation, traditional methods which are only illumination insensitive or robust to blur are not able to fully handle both variations. DA methods are useful in this scenario as they do not rely on the knowledge of physical domain shift. We also show transformed intermediate representations along the transition path of our approach in Fig. 13.8, which clearly captures the transition from

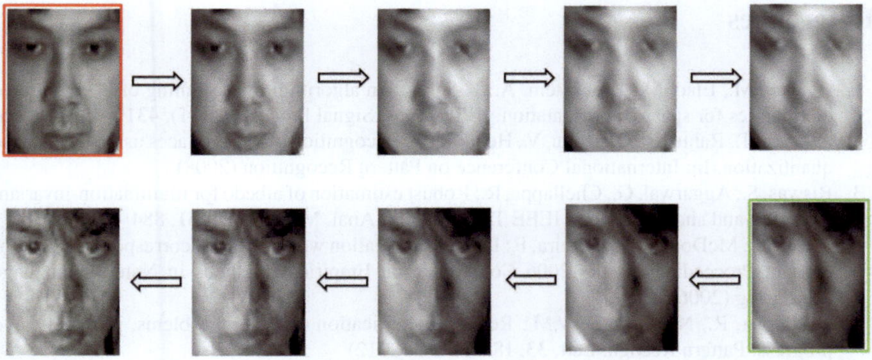

Fig. 13.8 Synthesized intermediate representations from the experiment on face recognition across illumination and blur variations (motion blur with length of nine). The first row demonstrates the transformed images from a source image (*in red box*) to the target domain. The second row demonstrates the transformed images from a target image (*in green box*) to the source domain

clear to blur images and vice versa. Particularly, we believe that the transformation from blur to clear conditions is useful for blind deconvolution, which is a highly under-constrained and costly problem [17].

13.6 Conclusions

In this chapter, we presented two different methods for the face re-identification problem using the domain adaptive dictionary learning approach. We first presented a general dictionary function learning framework to transform a dictionary learned from one domain to the other. Domain dictionaries are modeled by a parametric function. The dictionary function parameters and domain-invariant sparse codes are then jointly learned by solving an optimization problem with a sparsity constraint. We then discussed a fully unsupervised domain adaptive dictionary learning method with no prior knowledge of the underlying domain shift. This unsupervised DA method learns a set of intermediate domain dictionaries between the source and target domains, and renders intermediate domain representations to form a shared feature space for re-identification of faces. Extensive experiments on real datasets demonstrate the effectiveness of these methods on applications such as face pose alignment and face re-identification across domains.

Acknowledgments The work reported here is partially supported by a MURI Grant N00014-08-1-0638 from the Office of Naval Research

References

1. Aharon, M., Elad, M., Bruckstein, A.: K-SVD : An algorithm for designing of overcomplete dictionaries for sparse representation. IEEE Trans. Signal Process. **54**(11), 4311–4322 (2006)
2. Ahonen, T., Rahtu, E., Ojansivu, V., Heikkilä, J.: Recognition of blurred faces using local phase quantization. In: International Conference on Pattern Recognition (2008)
3. Biswas, S., Aggarwal, G., Chellappa, R.: Robust estimation of albedo for illumination-invariant matching and shape recovery. IEEE Trans. Pattern Anal. Mach. Intell. **31**, 884–899 (2009)
4. Blitzer, J., McDonald, R., Pereira, F.: Domain adaptation with structural correspondence learning. In: Proceedings of the 2006 Conference on Empirical Methods in Natural Language Processing (2006)
5. Chellappa, R., Ni, J., Patel, V.M.: Remote identification of faces: Problems, prospects, and progress. Pattern Recogn. Lett. **33**, 1849–1859 (2012)
6. Chen, S., Donoho, D., Saunders, M.: Atomic decomposition by basis pursuit. SIAM J. Sci. Comp. **20**, 33–61 (1998)
7. Daume III, H.: Frustratingly easy domain adaptation. In: Proceedings of the 45th Annual Meeting of the Association of, Computational Linguistics (2007)
8. Duan, L., Xu, D., Tsang, I.W.H., Luo, J.: Visual event recognition in videos by learning from web data. IEEE Trans. Pattern Anal. Mach. Intell. **99**, 1785–1792 (2011)
9. Elad, M., Aharon, M.: Image denoising via sparse and redundant representations over learned dictionaries. IEEE Trans. Imag. Process. **15**(12), 3736–3745 (2006)
10. Engan, K., Aase, S.O., Hakon Husoy, J.: Method of optimal directions for frame design. In: International Conference on Acoustics, Speech, and, Signal Processing (1999)
11. Georghiades, A.S., Belhumeur, P.N., Kriegman, D.J.: From few to many: Illumination cone models for face recognition under variable lighting and pose. IEEE Trans. Pattern Anal. Mach. Intell. **23**, 643–660 (2001)
12. Gong, B., Shi, Y., Sha, F., Grauman, K.: Geodesic flow kernel for unsupervised domain adaptation. In: Proceedings of the IEEE Computer Society Conference on Computer Vision and Pattern Recognition (2012)
13. Gopalan, R., Li, R., Chellappa, R.: Domain adaptation for object recognition: An unsupervised approach. In: International Conference on Computer Vision (2011)
14. Gross, R., Matthews, I., Baker, S.: Appearance-based face recognition and light-fields. IEEE Trans. Pattern Anal. Mach. Intell. **26**, 449–465 (2004)
15. Jhuo, I.H., Liu, D., Lee, D.T., Chang, S.F.: Robust visual domain adaptation with low-rank reconstruction. In: Proceedings of IEEE Computer Society Conference on Computer Vision and Pattern Recognition (2012)
16. Kulis, B., Saenko, K., Darrell, T.: What you saw is not what you get: Domain adaptation using asymmetric kernel transforms. In: Proceedings of IEEE Computer Society Conference on Computer Vision and Pattern Recognition (2011)
17. Levin, A., Weiss, Y., Durand, F., Freeman, W.T.: Understanding and evaluating blind deconvolution algorithms. In: Proceedings of IEEE Computer Society Conference on Computer Vision and Pattern Recognition (2009)
18. Machado, L., Leite, F.S.: Fitting smooth paths on riemannian manifolds. Int. J. Appl. Math. Stat. **4**, 25–53 (2006)
19. Mairal, J., Elad, M., Sapiro, G.: Sparse representation for color image restoration. IEEE Trans. Imag. Process. **17**(1), 53–69 (2008)
20. Ni, J., Qiu, Q., Chellappa, R.: Subspace interpolation via dictionary learning for unsupervised domain adaptation. In: Proceedings of IEEE Computer Society Conference on Computer Vision and Pattern Recognition (2013)
21. Olshausen, B.A., Field, D.J.: Emergence of simple-cell receptive field properties by learning a sparse code for natural images. Nature **381**(6583), 607–609 (1996)
22. Pan, S.J., Kwok, J.T., Yang, Q.: Transfer learning via dimensionality reduction. In: Proceedings of the 23rd National Conference on Artificial Intelligence (2008)

23. Pan, S.J., Tsang, I.W., Kwok, J.T., Yang, Q.: Domain adaptation via transfer component analysis. In: International Joint Conferences on Artificial Intelligence (2009)
24. Pati, Y.C., Rezaiifar, R., Krishnaprasad, P.S.: Orthogonal matching pursuit: Recursive function approximation with applications to wavelet decomposition. In: Proceedings of 27th Asilomar Conference on Signals, Systems and Computers, pp. 40–44 Pacific Grove, CA (1993)
25. Qiu, Q., Jiang, Z., Chellappa, R.: Sparse dictionary-based representation and recognition of action attributes. In: International Conference on Computer Vision, pp. 707–714 (2011)
26. Qiu, Q., Patel, V., Turaga, P., Chellappa, R.: Domain adaptive dictionary learning. In: Proceedings of European Conference on Computer Vision (2012)
27. Rubinstein, R., Bruckstein, A., Elad, M.: Dictionaries for sparse representation modeling. Proc. IEEE **98**(6), 1045–1057 (2010)
28. Saenko, K., Kulis, B., Fritz, M., Darrell, T.: Adapting visual category models to new domains. In: Proceedings of European Conference on Computer Vision (2010)
29. Sim, T., Baker, S., Bsat, M.: The CMU pose, illumination, and expression (PIE) database. IEEE Trans. Pattern Anal. Mach. Intell. **25**(12), 1615–1618 (2003)
30. Tropp, J.: Greed is good: Algorithmic results for sparse approximation. IEEE Trans. Inf. Theor. **50**, 2231–2242 (2004)
31. Turk, M., Pentland, A.: Face recognition using eigenfaces. In: Proceedings of IEEE Computer Society Conference on Computer Vision and Pattern Recognition (1991)
32. Wang, C., Mahadevan, S.: Manifold alignment without correspondence. In: International Joint Conferences on, Artificial Intelligence, pp. 1273–1278 (2009)
33. Wright, J., Ma, Y., Mairal, J., Sapiro, G., Huang, T., Yan, S.: Sparse representation for computer vision and pattern recognition. Proc. IEEE **98**(6), 1031–1044 (2010)
34. Wright, J., Yang, A.Y., Ganesh, A., Sastry, S.S., Ma, Y.: Robust face recognition via sparse representation. IEEE Trans. Pattern Anal. Mach. Intell. **31**, 210–227 (2009)
35. Yang, J., Yan, R., Hauptmann, A.G.: Cross-domain video concept detection using adaptive svms. In: ACM Multimedia, pp. 188–197. ACM (2007)
36. Zhang, Q., Li, B.: Discriminative K-SVD for dictionary learning in face recognition. In: Proceedings of IEEE Computer Society Conference on Computer Vision and Pattern Recognition (2010)
37. Zheng, W.S., Gong, S., Xiang, T.: Reidentification by relative distance comparison. IEEE Trans. Pattern Anal. Mach. Intell. **35**, 653–668 (2013)

Chapter 14
From Re-identification to Identity Inference: Labeling Consistency by Local Similarity Constraints

Svebor Karaman, Giuseppe Lisanti, Andrew D. Bagdanov and Alberto DelBimbo

Abstract In this chapter, we introduce the problem of identity inference as a generalization of person re-identification. It is most appropriate to distinguish identity inference from re-identification in situations where a large number of observations must be identified without knowing a priori that groups of test images represent the same individual. The standard single- and multishot person re-identification common in the literature are special cases of our formulation. We present an approach to solving identity inference by modeling it as a labeling problem in a Conditional Random Field (CRF). The CRF model ensures that the final labeling gives similar labels to detections that are similar in feature space. Experimental results are given on the ETHZ, i-LIDS and CAVIAR datasets. Our approach yields state-of-the-art performance for multishot re-identification, and our results on the more general identity inference problem demonstrate that we are able to infer the identity of very many examples even with very few labeled images in the gallery.

14.1 Introduction

Person re-identification is traditionally defined as the recognition of an individual at different times, possibly imaged from different camera views and/or locations, and

S. Karaman (✉) · G. Lisanti · A. D. Bagdanov · A. Del Bimbo
Media Integration and Communication Center, University of Florence,
Viale Morgagni 65, Florence, Italy
e-mail: svebor.karaman@unifi.it

G. Lisanti
e-mail: lisanti@dsi.unifi.it

A. D. Bagdanov
e-mail: bagdanov@dsi.unifi.it

A. Del Bimbo
e-mail: delbimbo@dsi.unifi.it

S. Gong et al. (eds.), *Person Re-Identification*,
Advances in Computer Vision and Pattern Recognition,
DOI: 10.1007/978-1-4471-6296-4_14,

considering a large number of candidate individuals in a known gallery. It is a standard component of multicamera surveillance systems as it is a way to associate multiple observations of the same individual over time. Particularly in scenarios in which the long-term behavior of persons must be characterized, accurate re-identification is essential. In realistic, wide-area surveillance scenarios such as airports, metro, and train stations, re-identification systems should be capable of robustly associating a unique identity with hundreds, if not thousands, of individual observations collected from a distributed network of many sensors.

Re-identification performance has traditionally been evaluated as a retrieval problem. Given a gallery consisting of a number images of known individuals, for each test image or group of test images of an unknown person, the goal of re-identification is to return a ranked list of individuals from the gallery. Configurations of the re-identification problem are generally categorized according to how much group structure is available in the gallery and test image sets. In a *single-shot image set* there is no grouping information available. Though there might be multiple images of an individual, there is no knowledge of which images correspond to that person. In a *multishot image set*, on the other hand, there is explicit grouping information available. That is, it is known which images correspond to the same individual, though of course the identities corresponding to each group are not known and the re-identification problem is to determine them.

The categorization of re-identification scenarios into multi- and single-shot configurations is useful for establishing benchmarks and standardized datasets for experimentation on the discriminative power of descriptors for person re-identification. However, these scenarios are not particularly realistic with respect to many real-world application scenarios. In video surveillance scenarios, for example, it is more common to have a few individuals of interest and to desire that all occurrences of them be labeled. In this case, the number of unlabeled test images to re-identify is typically much larger than the number of gallery images available. Another unrealistic aspect of traditional person re-identification is its formulation as a retrieval problem. In most video surveillance applications, the accuracy of re-identification at rank-1 is the most critical metric and higher ranks are of much less interest.

Based on these observations, in this chapter we describe a generalization of person re-identification which we call *identity inference*. The identity inference formulation is expressive enough to represent existing single- and multishot scenarios, while at the same time also modeling a larger class of problems not considered in the literature. In particular, we demonstrate how identity inference models problems where only a few labeled examples are available, but where identities must be inferred for a large number of probe images. In addition to describing identity inference problems, our formalism is also useful for precisely specifying the various multi- and single-shot re-identification modalities in the literature. We show how a Conditional Random Field (CRF) can then be used to efficiently and accurately solve a broad range of identity inference problems, including existing person re-identification scenarios as well as more difficult tasks involving a lot of test images.

In the next section, we review the literature on person re-identification. In Sect. 14.3 we introduce our formulation of the identity inference problem and in Sect. 14.4 propose a solution based on label inference in a CRF. Section 14.5 contains a number of experiments illustrating the effectiveness of our approach for both the re-identification and identity inference problems. We conclude in Sect. 14.6 with a discussion of our results.

14.2 Related Work

Person re-identification has applications in tracking, target reacquisition, verification, and long-term activity modeling. The most popular approaches to person re-identification are *appearance-based* techniques which must overcome problems such as varying illumination conditions, poses changes, and target occlusion. Within the broad class of appearance-based approaches to person re-identification, we distinguish *learning-based* methods, which generally require a training stage in which statistics of multiple images of each person is used to build a discriminative models of persons to be re-identified, from *direct* methods which require no initial training phase.

The majority of existing research on the person re-identification problem has concentrated on the development of sophisticated features for describing the visual appearance of targets. In [20] were introduced discriminative appearance-based models using Partial Least Squares (PLS) over texture, gradients, and color features. The authors of [13] use an ensemble of local features learned using a boosting procedure, while in [1] the authors use a covariance matrix of features computed in a grid of overlapping cells. The SDALF descriptor introduced in [11] exploits axis symmetry and asymmetry and represents each part of a person by a weighted color histogram, maximally stable color regions (MSCR), and texture information from recurrent highly structured patches. In [8] the authors fit a Custom Pictorial Structure (CPS) model consisting of head, chest, thighs, and legs part descriptors using color histograms and MSCR. The Global Color Context (GCC) of [7] uses a quantization of color measurements into color words and then builds a color context modeling the self-similarity for each word using a polar grid. The Asymmetry-based Histogram Plus Epitome (AHPE) approach in [4] represents a person by a global mean color histogram and recurrent local patterns through epitomic analysis. A common feature of most appearance-based approaches is that they compute an aggregate or mean appearance model over multiple observations of the same individual (for multishot modalities).

The approaches mentioned above concentrate on feature representation and not specifically on the classification or ranking technique. An approach which does concentrate specifically on the ranking approach is the Ensemble RankSVM technique of [18], which learns a ranking SVM model to solve single-shot re-identification problems. The Probabilistic Distance Comparison (PRDC) approach [25] introduced a comparison model which maximizes the probability of a pair of correctly matched images having a smaller distance than that of an incorrectly matched pair. The same authors in [26] then model person re-identification as a transfer ranking problem where the goal is to transfer similarity observations from a small gallery to a larger, unlabeled probe set. Metric learning approaches in which the metric space is adapted to the gallery data have also been successfully applied recently to the re-identification problem [10, 16].

We believe that in realistic scenarios many unlabeled images will be available while only few detections with known identities will be given, which is a scenario not covered by the standard classification of single- and multishot cases. We propose a CRF model that is able to encode a "soft grouping" property of unlabeled images. Our application of CRFs to identity inference is similar in spirit to semi-supervised techniques based on the graph Laplacian-like manifold ranking [17, 27]. These techniques, however, do not immediately generalize to multishot modalities and it is unclear how to use them for batch re-identification of more than one probe image at a time.

14.3 Identity Inference as Generalization of Re-identification

In this section, we give a formal definition of the re-identification and identity inference problems. The literature on person re-identification considers several different configurations of gallery and test images. The modality of a specific re-identification problem depends on whether the gallery and/or test subsets contain single or multiple instances of each individual. Here we consider each modality in turn and show how each can be represented as an instance of our definition of re-identification. A summary of the different protocols is given in Fig. 14.1. Despite the importance of the problem, there is much confusion about how each of the classical re-identification modalities are defined. One of our goals in this chapter is to formally define how, given a set of images of people extracted from video sequences, each type of re-identification problem is determined.

Let $\mathcal{L} = \{1, \ldots N\}$ be a label set for a re-identification scenario, where each element represents a unique individual appearing in a video sequence or collection of sequences. Given a number of instances (images) of individuals from $\mathcal{L}$ detected in a video collection:

$$\mathcal{I} = \{x_i \mid i = 1 \ldots D\},$$

we assume that each image x_i of an individual is represented by a feature vector $\mathbf{x}_i \equiv \mathbf{x}(x_i)$ and that the label corresponding to instance x_i is given by $y_i \equiv y(x_i)$. Note that we interchangeably use the implicit notation y_i and $\mathbf{x}_i$ for the label and feature vector corresponding to image x_i, or the explicit functional notation $y(x_i)$ and $\mathbf{x}(x_i)$, as appropriate.

An instance of a re-identification problem, represented as a tuple $\mathcal{R} = (\mathcal{G}, \mathcal{T})$, is completely characterized by its gallery and test image sets ($\mathcal{G}$ and $\mathcal{T}$, respectively). Formally, the gallery images are defined as:

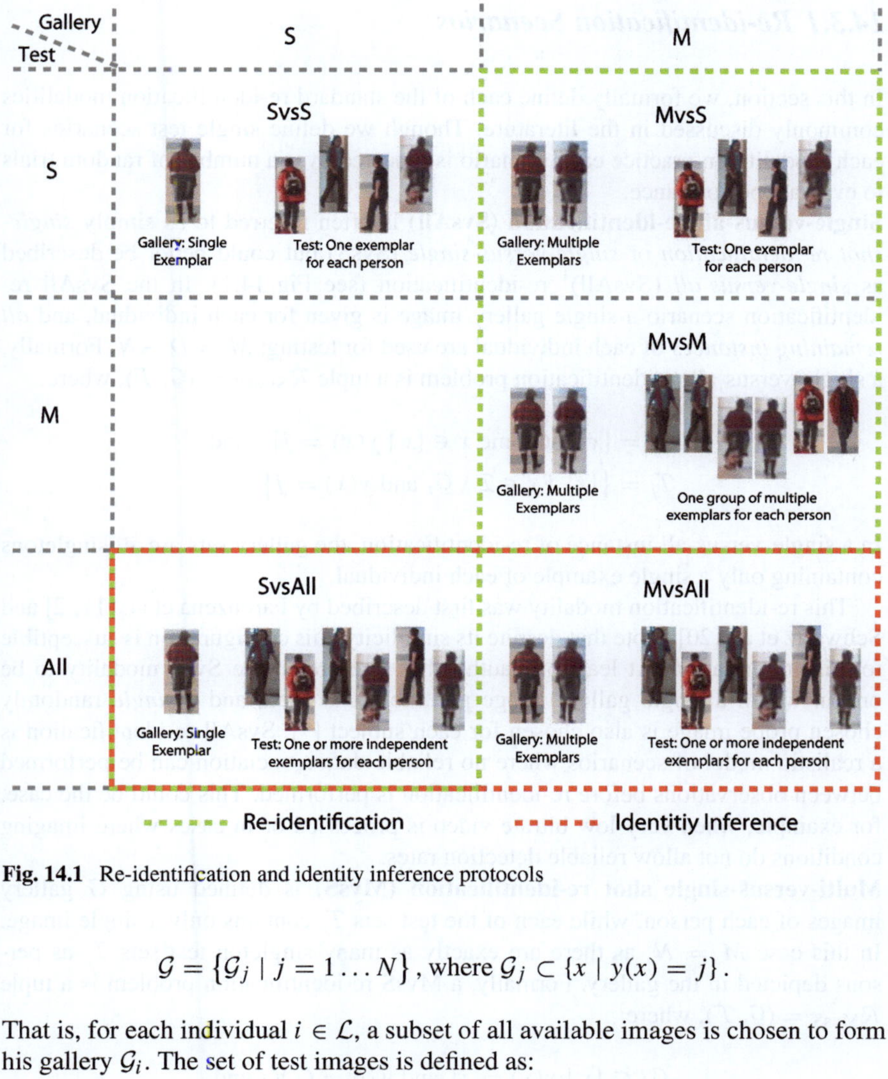

Fig. 14.1 Re-identification and identity inference protocols

$$\mathcal{G} = \{\mathcal{G}_j \mid j = 1 \dots N\}, \text{ where } \mathcal{G}_j \subset \{x \mid y(x) = j\}.$$

That is, for each individual $i \in \mathcal{L}$, a subset of all available images is chosen to form his gallery $\mathcal{G}_i$. The set of test images is defined as:

$$\mathcal{T} = \{\mathcal{T}_j \mid j = 1 \dots M\} \subset \mathcal{P}(\mathcal{I}),$$

where $\mathcal{P}$ is the powerset operator (i.e. $\mathcal{P}(I)$ is the set of all subsets of $\mathcal{I}$). We further require for all $\mathcal{T}_j \in \mathcal{T}$ that $x, x' \in \mathcal{T}_j \Rightarrow y(x) = y(x')$ (sets in $\mathcal{T}$ have homogeneous labels), and $\mathcal{T}_j \in \mathcal{T} \Rightarrow \mathcal{T}_j \cap \mathcal{G}_i = \emptyset, \forall i \in \{1 \dots N\}$ (the test and gallery sets are disjoint). A solution to an instance of a re-identification problem is a mapping from the test images $\mathcal{T}$ to the set of all permutations of $\mathcal{L}$.

14.3.1 Re-identification Scenarios

In this section, we formally define each of the standard re-identification modalities commonly discussed in the literature. Though we define single test scenarios for each modality, in practice each scenario is repeated over a number of random trials to evaluate performance.

Single-versus-all re-identification (SvsAll) is often referred to as simply *single-shot re-identification* or *single-versus-single* (SvsS) but could better be described as *single-versus-all* (SvsAll)[1] re-identification (see Fig. 14.1). In the SvsAll re-identification scenario a single gallery image is given for each individual, and *all remaining instances* of each individual are used for testing: $M = D - N$. Formally, a single-versus-all re-identification problem is a tuple $\mathcal{R}_{\text{SvsAll}} = (\mathcal{G}, \mathcal{T})$, where:

$$\begin{aligned}\mathcal{G}_j &= \{x\} \text{ for some } x \in \{x \mid y(x) = j\}, \text{ and}\\ \mathcal{T}_j &= \big\{\{x\} \mid x \in \mathcal{I} \setminus \mathcal{G}_j \text{ and } y(x) = j\big\}.\end{aligned}$$

In a single-versus-all instance of re-identification, the gallery sets are all singletons containing only a single example of each individual.

This re-identification modality was first described by Farenzena et al. [11, 2] and Schwartz et al. [20]. Note that despite its simplicity, this configuration is susceptible to misinterpretation. At least one author has interpreted the SvsS modality to be one in which a single gallery image per subject is used, and a *single* randomly chosen probe image is also chosen for each subject [7]. SvsAll re-identification is a realistic model of scenarios where no reliable data association can be performed between observations before re-identification is performed. This could be the case, for example, when very low bitrate video is processed or in cases where imaging conditions do not allow reliable detection rates.

Multi-versus-single shot re-identification (MvsS) is defined using G gallery images of each person, while each of the test sets $\mathcal{T}_j$ contains only a single image. In this case $M = N$, as there are exactly as many singleton test sets $\mathcal{T}_j$ as persons depicted in the gallery. Formally, a MvsS re-identification problem is a tuple $\mathcal{R}_{\text{MvsS}} = (\mathcal{G}, \mathcal{T})$, where:

$$\begin{aligned}\mathcal{G}_j &\subset \{x \mid y(x) = j\} \text{ and } |\mathcal{G}_j| = G \ \forall j \text{ and}\\ \mathcal{T}_j &= \{x\} \text{ for some } x \notin \mathcal{G}_j \text{ s.t. } y(x) = j.\end{aligned}$$

The MvsS configuration is not precisely a generalization of the SvsAll person re-identification problem in that, after selecting G gallery images for each individual, only a *single* test image is selected to form the test sets $\mathcal{T}_j$.

The MvsS re-identification scenario has been used in only a few works in the literature [7, 11]. We do not consider it to be an important modality, though it might be

[1] We prefer the SvsAll terminology as the SvsS terminology has been misinterpreted at least once in the literature.

an appropriate model of verification scenarios where a fixed set of gallery individuals are enrolled and then must be unambiguously re-identified on the basis of a single image.

Multi-versus-multi shot re-identification (MvsM) is the case in which the gallery and test sets of each person both have G images. In this case $M = N$, there is again as many gallery sets as test sets. After selecting the G gallery images for each of the N individuals, only a fraction of the remaining images of each person are used to form the test set. Formally, a MvsM re-identification problem is a tuple $\mathcal{R}_{\text{MvsM}} = (\mathcal{G}, \mathcal{T})$, where:

$$\mathcal{G}_j \subset \{x \mid y(x) = j\} \text{ and } |\mathcal{G}_j| = G \; \forall j \text{ and}$$
$$\mathcal{T}_j \subset \left\{x \mid y(x) = j \text{ and } x \notin \mathcal{G}_j\right\} \text{ and } |\mathcal{T}_j| = G \; \forall j.$$

Note that the MvsM configuration is *not* a generalization of the SvsAll case in which all of the available imagery for each target is used as test imagery. The goal in MvsM re-identification is to re-identify each *group* of test images, leveraging the knowledge that images in each group are all of the same individual.

The MvsM re-identification modality is the most commonly reported one in the literature [1, 4, 8, 11]. It is representative of scenarios in which some amount of reliable data association can be performed before re-identification. However, it is not a completely realistic formulation since data association is never completely correct and there will always be uncertainty about group structure in probe observations.

14.3.2 Identity Inference

Identity inference addresses the problem of having *few* labeled images while desiring to label *many* unknown images without explicit knowledge that groups of images represent the same individual. The formulation of the *single-versus-all* re-identification falls within the scope of identity inference, but neither the multi-versus-single nor the multi-versus-multi formulations are a generalization of this case to multiple gallery images. In the MvsS and MvsM cases, the test set is either a singleton for each person (MvsS) or a group of images (MvsM) of the same size as the gallery image set for each person. Identity inference could be described as a *multi-versus-all* configuration. Formally, it is a tuple $\mathcal{R}_{\text{MvsAll}} = (\mathcal{G}, \mathcal{T})$, where:

$$\mathcal{G}_j \subset \{x \mid y(x) = j\} \text{ and } |\mathcal{G}_j| = G \text{ and}$$
$$\mathcal{T}_j = \left\{\{x\} \mid x \in \mathcal{I} \setminus \mathcal{G}_j \text{ and } y(x) = j\right\}.$$

In instances of identity inference a set of G gallery images is chosen for each individual. All remaining images of each individual are then used as an element of the test set without any identity grouping information. As in the SvsAll case, the test images sets are all singletons.

Identity inference as a generalization of person re-identification was first introduced in [14]. It encompasses both SvsAll and MvsAll person re-identification and represents, in our opinion, one of the most realistic scenarios in practice. The re-identification modalities accurately model situations where an operator is interested in inferring the identity of past, unlabeled observations on the basis of very few labeled examples of each person. In practice, the number of labeled images available is significantly less than the number of images for which labels are desired.

14.4 A CRF Model for Identity Inference

Conditional Random Fields (CRFs) have been used to model the statistical structure of problems such as semantic image segmentation [5] and stereo matching [19]. In this section, we show how we model the identity inference problem as a minimum energy labelling problem in a CRF.

A CRF is defined in general case by a graph $\mathcal{G} = (\mathcal{V}, \mathcal{E})$, a set of random variables $\mathcal{Y} = \{Y_j \mid j = 1 \ldots |V|\}$ which represents the statistical structure of the problem being modeled, and a set of possible labels $\mathcal{L}$. The vertices $\mathcal{V}$ index the random variables in $\mathcal{Y}$ and the edges $\mathcal{E}$ encode the statistical dependence relations between the random variables. The labeling problem is then to find an assignment $\tilde{\mathbf{y}}$ of labels to nodes that minimizes an energy function E over possible labellings $\mathbf{y}^* = (y_i^*)_{i=1}^{|V|}$: $\tilde{\mathbf{y}} = \arg\min_{\mathbf{y}^*} E(\mathbf{y}^*)$.

The energy function $E(\mathbf{y}^*)$ is defined as:

$$E(\mathbf{y}^*) = \sum_{i \in \mathcal{V}} \phi_i(y_i^*) + \lambda \sum_{(i,j) \in \mathcal{E}} \psi_{ij}(y_i^*, y_j^*), \tag{14.1}$$

where $\phi_i(y_i^*)$ is a unary data cost encoding the penalty of assigning label y_i^* to vertex i and $\psi_{ij}(y_i^*, y_j^*)$ is a binary smoothness cost representing the conditional penalty of assigning labels y_i^* and y_j^*, respectively, to vertices i and j. The parameter λ in Eq. (14.1) controls the trade-off between data and smoothness costs.

To minimize properly defined energy functions [15] and find an optimal labeling $\tilde{\mathbf{y}}$ in a CRF, the graph cut approach has been shown to be competitive [21] against other methods proposed in the literature such as Max-Product Loopy Belief Propagation [12] and Tree-Reweighted Message Passing [23]. The multilabel problem is solved by iterating the α-expansion move [6] where the binary labeling is expressed as each node either keeping its current label or taking the label α selected for the iteration. In all experiments, we use the graph cut approach to minimizing the energy in Eq. (14.1). If higher ranks are desired, an inference algorithm like loopy belief propagation that returns the marginal distributions at each node can be used [12]. We feel that rank-1 precision is the most significant performance measure, and found loopy belief propagation to be many times slower than graph cuts on our inference problem.

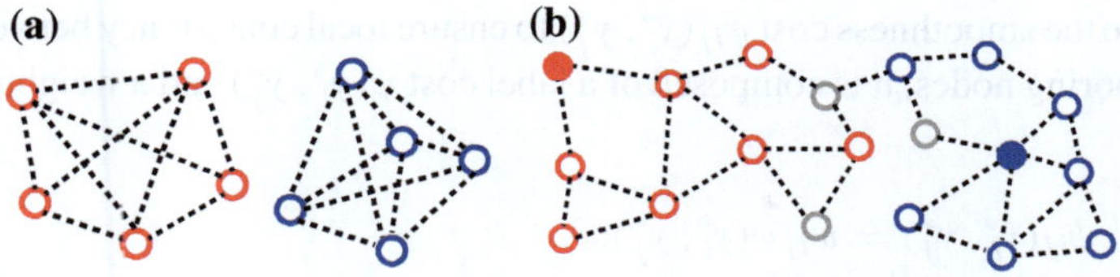

Fig. 14.2 Illustrations of the CRF topology for the MvsM (**a**) and SvsAll (**b**) modalities. *Filled circles* represent gallery images, *unfilled circles* probes. *Color* indicates the ground truth label

CRF topology: We can map an identity inference problem $\mathcal{R} = (\mathcal{G}, \mathcal{T})$ onto a CRF by defining the vertex and edge sets $\mathcal{V}$ and $\mathcal{E}$ in terms of the gallery and test image sets defined by $\mathcal{G}$ and $\mathcal{T}$. We have found two configurations of vertices and edges to be useful for solving identity inference problems. The first uses vertices to represent groups of images in the test set $\mathcal{T}$ and is particularly useful for modeling MvsM re-identification problems:

$$\mathcal{V} = \bigcup_{i=1}^{N} \mathcal{T}_i \text{ and } \mathcal{E} = \left\{(x_i, x_j) \mid x_i, x_j \in \mathcal{T}_l \text{ for some } l\right\}.$$

The edge topology in this CRF is completely determined by the group structure as expressed by the $\mathcal{T}_j$.

When no identity grouping information is available for the test set, as in the general identity inference case as well as in SvsAll re-identification, we instead use the following formulation of the CRF:

$$\mathcal{V} = I \text{ and } \mathcal{E} = \bigcup_{x_i \in \mathcal{V}} \left\{(x_i, x_j) \mid x_j \in \text{kNN}(x_i)\right\},$$

where the kNN(x_i) maps an image to its k most similar images in feature space. Note that we hence treat equally training and test images when building $\mathcal{E}$. The topology of this CRF formulation, in the absence of explicit group information, uses feature similarity to form connections between nodes. Illustrations of the topology for the MvsM and SvsAll scenarios are given in Fig. 14.2.

Data and smoothness costs: The unary data cost determines the penalty of assigning label y_i^* to vertex i given $\mathbf{x}(x_i)$, the observed feature representation of image x_i. We define it as:

$$\phi_i(y_i^*) = \min_{x \in \mathcal{G}_{y_i^*}} ||\mathbf{x}(x) - \mathbf{x}(x_i)||_2. \quad (14.2)$$

That is, the cost of assigning label y_i^* is proportional to the minimum L2-distance between the feature representation of image x_i and any gallery image of individual y_i^*. The data cost is L1-normalized for each vertex i, and hence is a cost distribution over the labels. The data cost can be seen as the individual assignment cost.

We use the smoothness cost $\psi_{ij}(y_i^*, y_j^*)$ to ensure local consistency between labels in neighboring nodes, it is composed of a label cost $\psi(y_i^*, y_j^*)$ and a weighting factor w_{ij}:

$$\psi_{ij}(y_i^*, y_j^*) = w_{ij}\psi(y_i^*, y_j^*), \tag{14.3}$$

$$\psi(y_i^*, y_j^*) = \begin{cases} 0 & \text{if } y_i^* = y_j^* \\ \frac{1}{|\mathcal{G}_{y_i^*}||\mathcal{G}_{y_j^*}|} \sum\limits_{\substack{x \in \mathcal{G}_{y_i^*} \\ x' \in \mathcal{G}_{y_j^*}}} ||\mathbf{x}(x) - \mathbf{x}(x')||_2 & \text{otherwise.} \end{cases} \tag{14.4}$$

The label cost $\psi(y_i^*, y_j^*)$ depends only on the labels. The more similar two labels are in terms of the available gallery images for them, the lower the cost for them will be to coexist in a neighborhood of the CRF. The label cost are L1-normalized, and thus is a cost distribution over all labels. Note that the label cost is fixed to 0 if $y_i^* = y_j^*$ (see Eq. (14.4)). The weighting factors w_{ij} allow the smoothness cost between nodes i and j to be flexibly controlled according to the problem at hand. In the experiments presented in this chapter, we define the weights w_{ij} from Eq. (14.3) between vertices i and j in the CRF in terms of feature similarity:

$$w_{ij} = exp(-||\mathbf{x}(x_i) - \mathbf{x}(x_j)||_2). \tag{14.5}$$

This definition gives a higher cost to a potential labeling $\mathbf{y}^*$ that labels similar images differently. But as the similarity between nodes decreases, so does the cost of keeping these two different labels. Hence, our method would still allow connected nodes to share different labels but will tend to discourage this situation especially for very similar images and/or very different identities.

14.5 Experiments

In this section, we describe a series of experiments we performed to evaluate the effectiveness of the approach described in Sect. 14.4 for solving identity inference problems. In the next section, we describe the three datasets we use in all experiments and the feature descriptor we use to represent appearance. In Sect. 14.5.2 we report on the re-identification experiments performed on these three datasets, and in Sect. 14.5.3 we report results for identity inference.

Note that in all experiments we only evaluate performance at rank-1 and not across all ranks (as is done in some re-identification works). We believe that the rank-1 performance, that is *classification* performance, is the most important metric for person re-identification since it is indicative of how well fully automatic re-identification performs. Consequently, all plots in this section are not CMC curves, but rather plots of rank-1 re-identification accuracies for various parameter settings.

Table 14.1 Re-identification dataset characteristics

	ETHZ1	ETHZ2	ETHZ3	CAVIAR	i-LIDS
Number of cameras	1	1	1	2	3
Environment	Outdoor	Outdoor	Outdoor	Indoor	Indoor
Number of identities	83	35	28	72	119
Minimum number of images/person	7	6	5	10	2
Average number of images/person	58	56	62	17	4
Maximum number of images/person	226	206	356	20	8
Average detection size	60×132	63×135	66×148	34×81	68×175

14.5.1 Datasets and Feature Representation

We evaluate the performance of our CRF approach on a variety of commonly used datasets for re-identification. To describe the visual appearance of images in gallery and probe sets we use a simple descriptor that captures color and shape information.

Datasets

For evaluating identity inference performance we are particularly interested in test scenarios where there are many images of each test subject. However, most publicly available datasets for re-identification possess exactly the opposite property in that they contain very few images per person. The VIPER [13] dataset only provides a pair of images for each identity and thus no meaningful structure can be defined for our approach. Another popular dataset for re-identification is i-LIDS [24] which has on average four images per person. Although this is a rather small number of images per person we want to demonstrate the robustness of our approach also on this dataset. The most interesting publicly available datasets for our approach are CAVIAR [8], which contains between 10 and 20 images for each person extracted from two different view of an indoor environment, and the ETHZ [20] dataset, which consists of three video sequences, where on average each person appears in more than 50 images. The characteristics of the selected datasets are summarized in Table 14.1 and some details are given below:

- **ETHZ.** The ETHZ Zurich dataset [20] consists of detections of persons extracted from three sequences acquired outdoors. This dataset is divided into distinct datasets, corresponding to three different sequences in which different persons appear.

  1. The ETHZ1 sequence contains images of 83 persons. The number of detections per person ranges from 7 to 226, the average being 58. Detections have an average resolution of 60×132.
  2. ETHZ2 contains images of 35 persons. The number of detections per person ranges from 6 to 206, with an average of 56. This sequence seems to have been

recorded on a bright, sunny day and the strong illumination tends to partially wash out differences in appearance, making this sequence one of the most difficult.

3. ETHZ3 contains images of 28 persons, with the number of detections per person ranging between 5 and 356 (62 on average). The resolution of each detection is quite high (66 × 148), facilitating better description of each one. The small number of persons, the high resolution and the large number of images per person make this sequence the easiest of the three ETHZ datasets.

- **CAVIAR.** The CAVIAR dataset consists of several sequences recorded in a shopping center. It contains 72 persons imaged from two different views and was designed to maximize variability with respect to resolution changes, illumination conditions, and pose changes. As detailed in Table 14.1, the number of images per person is either 10 or 20 with an average of 17. While the number of persons, cameras, and detections make this dataset interesting, the very small average resolution of 34 × 81 makes it difficult to extract discriminative features.
- **i-LIDS.** The i-LIDS dataset consists of multiple camera views from a busy airport arrival hall. It contains 119 people imaged in different lighting conditions and most of the time with baggage that in part occlude the person. The number of images per person is low, with a minimum of two, a maximum of eight, and an average of four. The average resolution of the detections (68 × 175) is rather high, especially with respect to the other datasets.

A Descriptor for Re-identification

In our experiments we use a descriptor based on both color and shape information that requires no foreground/background segmentation and does not rely on body-part localization. Given an input image of a target, it is resized to a canonical size of 64 × 128 pixels with coordinates between $[-1, 1]$ and origin $(0, 0)$ at the center of the image. Then we divide it into overlapping horizontal stripes of 16 pixels in height and from each stripe we extract an RGB histogram. The use of horizontal stripes allows us to capture the vertical color distribution in the image, while overlapping stripes allow us to maintain color correlation information between adjacent stripes in the final descriptor. We equalize all RGB color channels before extracting the histogram. Histograms are quantized to 4 × 4 × 4 bins.

Descriptors of visual appearance for person recognition can be highly susceptible to background clutter, and many approaches to person re-identification use sophisticated background modeling techniques to separate foreground from background signals [3, 4, 11]. We use a more straightforward approach that weights the contribution of each pixel to its corresponding histogram bin according to an Epanechnikov kernel centered on the target image:

$$K(x, y) = \begin{cases} \frac{3}{4}(1 - (\frac{x}{W})^2 - (\frac{y}{H})^2) & \text{if } |(\frac{x}{W})^2 + (\frac{y}{H})^2| \leq 1 \\ 0 & \text{otherwise} \end{cases} \quad (14.6)$$

where W and H are, respectively, the width and height of the target image. This discards (or diminishes the influence of) background information and avoids the need to learn a background model for each scenario. To the weighted RGB histograms, we concatenate a set of Histogram of Oriented Gradients (HOG) descriptors computed on a grid over the image as described in [9]. The HOG descriptor captures local structure and texture in the image that are not captured by the color histograms.

The use of the Hellinger kernel, which is a simple application of the square root to all descriptor bins, is well known in the image classification community [22] and helps control the influence of dimensions in the descriptor that tend to have disproportionately high values with respect to the others. In preliminary experiments we found this to improve robustness of Euclidean distances between descriptors and we therefore take the square root of all histogram bins (both RGB and HOG) to form our final descriptor.

14.5.2 Multishot Re-identification Results

To evaluate our approach in comparison with other state-of-the-art methods [4, 8, 11], we performed experiments on each dataset described above for the MvsM re-identification scenario. We evaluate performance for galleries varying in size: 2, 5, and 10 images per person for ETHZ; 2, 3, and 5 images per person for CAVIAR; and 2 and 3 images per person for i-LIDS.

Note that grouping information in the test set is explicitly encoded in the CRF. Edges only link test images that correspond to the same individual, and one test image is connected to all other test images of that individual. In these experiments we fix $\lambda = 1$ in the energy function of Eq. (14.1), and the weight on the edges is defined according to the features similarity as detailed in Eq. (14.5).

Results for MvsM person re-identification are presented in Fig. 14.3a, c and e for ETHZ, Fig. 14.4a for CAVIAR, and Fig. 14.5a for i-LIDS. The NN curve in these figures corresponds to labeling each test image with the nearest gallery image label without exploiting group knowledge, while the GroupNN approach exploits group knowledge by assigning each group of test images the label for which the average distance between test images of that group and gallery individuals of that label is minimal. We refer to our approach as "CRF" in all plots, and for each configuration we randomly select the gallery and test images and average performance over ten trials.

Multishot Re-identification Performance on ETHZ

For the MvsM scenarios on ETHZ we tested $M \in \{2, 5, 10\}$. We now detail results on each sequence and compare with the state-of-the-art when available.

ETHZ1: Performance on ETHZ1 (Fig. 14.3a) starts at 84 % accuracy at rank-1 for the simple NN classification approach and at 91 % for both the GroupNN and our CRF

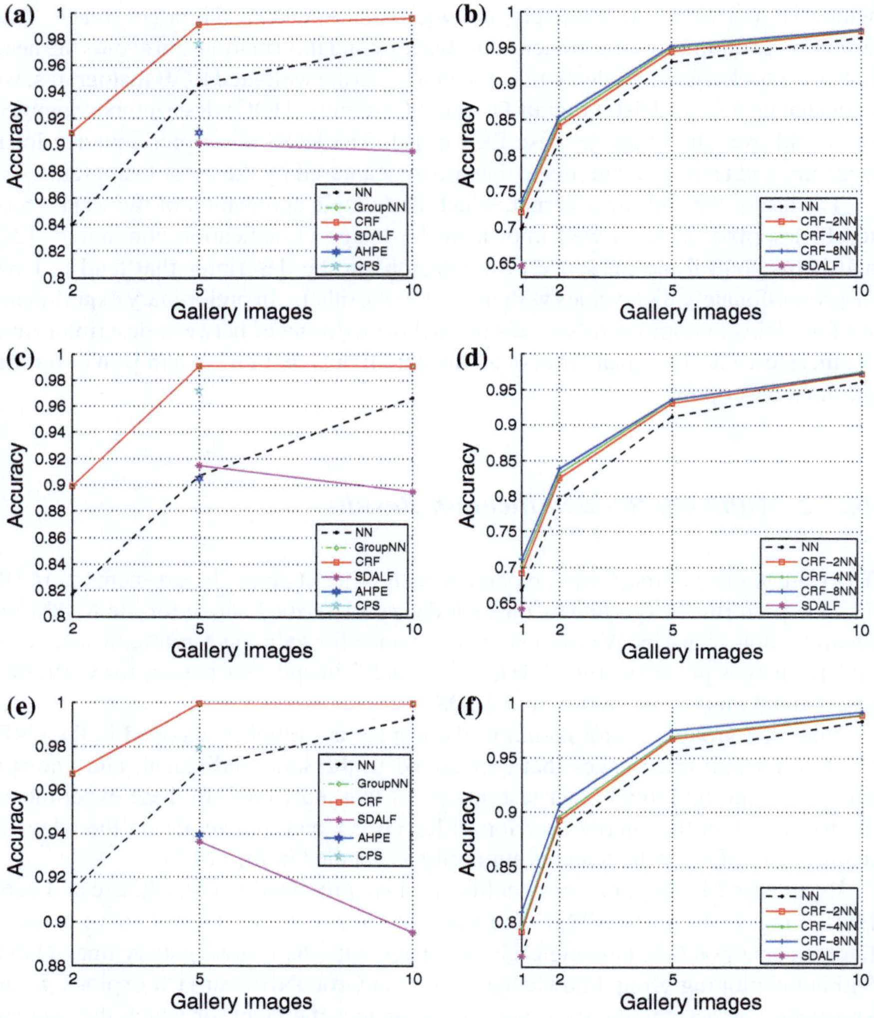

Fig. 14.3 MvsM (*left column*) and SvsAll and MvsAll (*right column*) re-identification accuracy on ETHZ. Note that these are *not CMC curves*, but are rank-1 *classification* accuracies over varying gallery and test set sizes. **a** ETHZ1 MvsM, **b** ETHZ1 M,SvsAll, **c** ETHZ2 MvsM, **d** ETHZ2 M,SvsAll, **e** ETHZ3 MvsM, **f** ETHZ3 M,SvsAll

approach for $M = 2$. Using five images, GroupNN and the CRF reach an accuracy of about 99.2 %. The state-of-the-art on ETHZ1 for $M = 5$ is CPS at 97.7 %. With 10 gallery and test images per subject, the CRF approach reaches 99.6 % accuracy while the NN classification peaks at 97.7 %. The SDALF approach obtains 89.6 % on this scenario.

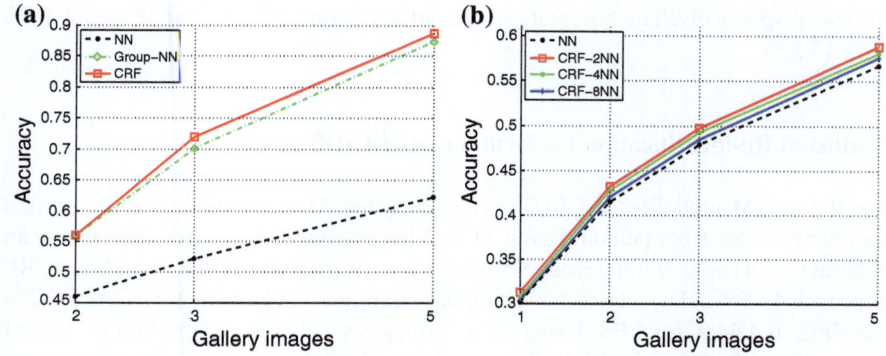

Fig. 14.4 MvsM (*left*) and SvsAll and MvsAll (*right*) re-identification accuracy on CAVIAR **a** CAVIAR MvsM **b** CAVIAR M,SvsAll

ETHZ2: On ETHZ2 (Fig. 14.3c), which is the most difficult of the ETHZ datasets, performance at $M = 2$ starts at 81.7 % for the simple NN baseline and 90 % for both the GroupNN and our CRF approach. Using five images, methods exploiting group knowledge reach 99.1 %. The state-of-the-art on this dataset is SDALF at 91.6 %, AHPE at 90.6 %, and CPS which reaches 97.3 %. Finally, when using 10 images for the gallery and test sets, methods using grouping knowledge stays at 99.1 %. Note that, as with ETHZ1, SDALF performance at $M = 10$ is less than at $M = 5$ with 89.6 % rank-1 accuracy.

ETHZ3: On ETHZ3 (Fig. 14.3e), which is the "easiest" of the ETHZ datasets, performance start at 91.4 % rank-1 accuracy for the simple NN baseline and at 96.8 % for both the groupNN and our CRF approach with $M = 2$. The NN classification reaches 97 % using 5 images and 99.3 % using 10 images. Methods using group knowledge saturate the performance on this dataset for both 5 and 10 images. The SDALF approach obtains 93.7 and 89.6 % accuracy using 5 and 10 images, respectively. For $M = 5$, the AHPE approach obtains 94 % while the CPS method arrives at 98 % rank-1 accuracy.

Multishot Re-identification Performance on CAVIAR

For MvsM re-identification on CAVIAR we performed experiments with $M \in \{2, 3, 5\}$ (see Fig. 14.4a). Performance begins at 46 % accuracy for the NN classification and 55 % for approaches that use group structure in the probe image set. Using three images, the NN baseline reaches an accuracy of 52 %, the GroupNN approach reaches 70.6 %, while the CRF reaches 72 %. These results can be compared with SDALF performance at 8.5 %, HPE at 7.5 %, and the CPS performance at 13 % for $M = 3$. Finally, with $M = 5$ the difference between methods exploiting group structure and those that do not becomes even more prominent. Nearest neighbor achieves 62.7 %, while the GroupNN and CRF approaches reach 86.9 and

88.4 %, respectively. The best state-of-the-art result on CAVIAR for $M = 5$ is CPS with 17.5 %.

Multishot Re-identification Performance on i-LIDS

For the MvsM modality on i-LIDS we only tested $M \in \{2, 3\}$ due to the limited number of images per person. Using $M = 2$ images, the NN classification yields an accuracy of 41.8 %, while GroupNN yields a performance of 48.4 % and our CRF approach 47.5 %. The state-of-the-art on this configuration is: 39 % for SDALF, 32 % for HPE, and 44 % for CPS. Using $M = 3$ images yields only a small improvement as for many identities in i-LIDS there are fewer than six images. In these cases, the gallery and probe image sets are limited to two images. GroupNN outperforms our CRF approach on this dataset due to the limited number of images per person.

Summary of Multishot Re-identification Results

From these experiments on multishot person re-identification it is evident that significant improvement is obtained by exploiting group structure in the probe image set. The simple GroupNN rule and our CRF approach yield similar performance on MvsM re-identification scenarios on the ETHZ datasets where many images of each person are available. In combination with the discriminative power of our descriptor, our approach outperforms the state-of-the-art on the three ETHZ datasets. On the other hand, performance gains on the i-LIDS dataset are limited by the low number of images available for each person in the dataset.

Group structure in the probe image set yields a large boost in multishot re-identification performance also on the CAVIAR dataset, with an improvement of almost 30 % between the simple NN classification and our CRF formulation exploiting group knowledge (note also the large improvement with respect to state-of-the-art methods). This is likely due to the fact that our approach does not compute mean or aggregate representations of groups and that our descriptor does not fit complex background or part models to the resolution-limited images in the CAVIAR dataset.

14.5.3 Identity Inference Results

To evaluate the performance of our approach in comparison with other state-of-the-art methods [4, 8, 11], we performed experiments on all datasets using the SvsAll and MvsAll modalities described above. For the general identity inference case, unlike MvsM person re-identification, we have no information about relationships between test images. In the CRF model proposed in Sect. 14.4 for identity inference, the local neighborhood structure is determined by the K nearest neighbors to each image in feature space. For all experiments we tested $K \in \{2, 4, 8\}$. We set $\lambda = \frac{|\mathcal{V}|}{|\mathcal{E}|}$ in Eq. (14.1) for the SvsAll and the MvsAll scenarios. Since there may be up to four times more smoothness terms than unary data cost terms in Eq. (14.1), setting λ in this way prevents smoothness from dominating the energy function. In identity inference, gallery images are randomly selected and *all remaining images* define the test set. All reported results are averages over 10 trials as before. Results for identity inference are presented in Figs. 14.3b, d and f for ETHZ, Fig. 14.4b for CAVIAR and 14.5b for i-LIDS. We will now analyze the results for each dataset and then draw general conclusions.

Identity Inference Results on ETHZ

For the MvsAll configuration on ETHZ we tested $M \in \{2, 5, 10\}$.

ETHZ1: On ETHZ1 (Fig. 14.3b) we can observe that on the SvsAll modality the NN baseline using our descriptor yields an accuracy of 69.7 %, while SDALF obtains 64.8 %. The CRF approach improves this performance to 72 % using two neighbors and 73.7 % using eight neighbors. Using two gallery images per person increases CRF results to a rank-1 accuracy ranging from 84.2 to 85.6 %. Adding more gallery images yields continued improvement, reaching 97.7 % accuracy with 10 gallery images per person and our CRF approach with eight neighbors. Performance of the CRF with different neighborhood sizes seems to converge at this point.
ETHZ2: On ETHZ2 (Fig. 14.3d), which is the most challenging of the ETHZ datasets, we can see that performance is slightly lower. The NN classification on the SvsAll modality obtains an accuracy of 66.9 % compared to the SDALF performance of 64.4 %, while our approach yields 69.3 and 71.3 % accuracy using, respectively, two and eight neighbors. With two gallery images, the gap between the NN baseline (79.5 %) and the CRF (84 %) slightly widens. Using 10 model images, the performances stabilizes at 97.7 % and we observe the same convergence as on ETHZ1.
ETHZ3: Finally, on ETHZ3 (Fig. 14.3f) the NN baseline and SDALF obtain the same accuracy of 77 %, while the performance of our CRF approach ranges from 79.2 to 81 % depending on the neighborhood size. The performance quickly saturates with a maximum accuracy of 97.7 % using 5 training images and 99 % with 10 images.

Identity Inference on CAVIAR

Identity inference on the CAVIAR dataset is significantly more challenging than on ETHZ. We evaluate performance on the SvsAll modality and on MvsAll modalities for $M \in \{2, 3, 5\}$ (see Fig. 14.4b). With only one gallery images per person, both the nearest neighbor and CRF approaches yield a rank-1 accuracy of about 30 %. This is significantly higher than the state-of-the-art of about 8 % for SDALF, AHPE,

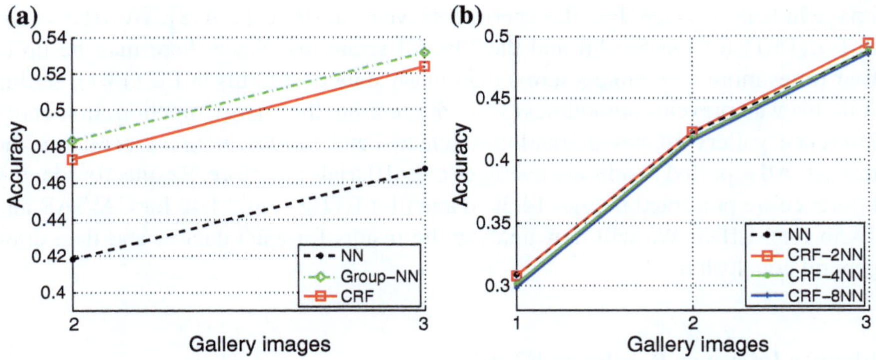

Fig. 14.5 MvsM (*left*) and SvsAll and MvsAll (*right*) re-identification accuracy on i-LIDS. **a** i-LIDS MvsM **b** i-LIDS M,SvsAll

and CPS, which is likely due to the simplicity of our descriptor and its robustness to occlusion and illumination changes. For all MvsAll modalities, we note a significant gain in accuracy when adding more gallery images per person, with performance peaking at about 59 % for the $M = 10$ case. This demonstrates that our CRF approach is able to effectively exploit multiple gallery examples.

Identity Inference on i-LIDS

Due to the relatively small average number of images per person in the i-LIDS dataset, we only test the MvsAll modality for $M \in \{2, 3\}$. Results are summarized in Fig. 14.5b. For only one training example per gallery individual, our approach yields a rank-1 accuracy of about 31 %, which is comparable to the state-of-the-art result of 28 % reported for SDALF. Adding more examples, as with the CAVIAR dataset, consistently improves rank-1 performance.

Summary of Identity Inference Results

Using the CRF framework proposed in Sect. 14.4 clearly improves accuracy over the simple NN re-identification rule. With our approach it is possible to label a very large number of probe images using very few gallery images for each person. For example, on the ETHZ3 dataset, we are able to correctly label 1,553 out of 1,706 test images *using only two model images per person*. The robustness of our method with respect to occlusions and illumination changes is shown in the qualitative results in Fig. 14.6. The CRF approach yields correct labels even in strongly occluded cases thanks to the neighborhood edges connecting it to less occluded, yet similar, images. This property of our descriptor pays off particularly well for the resolution-limited

Fig. 14.6 Identity inference results (SvsAll). *First row* test image, *second row* incorrect NN result, *third row* correct result given by our CRF approach

CAVIAR dataset, for which we outperform the state-of-the-art already for the SvsAll case.

14.6 Conclusions

In this chapter, we introduced the identity inference problem which we propose as a generalization of the standard person re-identification scenarios described in the literature. Identity inference can be thought of as a generalization of the single-versus-all person re-identification modality, and at the same time as a relaxation of the multi-versus-multi shot case. Instances of identity inference problems do not require hard knowledge about relationships between test images (e.g., that they correspond to the same individual). We have also attempted to formalize the specification of person re-identification and identity inference modalities through the introduction of a set-theoretic notation for precise definition of scenarios. This notation is useful in that it establishes a common, unambiguous language for talking about person re-identification problems.

We also proposed a CRF-based approach to solving identity inference problems. Using feature space similarity to define the neighborhood topology in the CRF, our approach is able to exploit the soft-grouping structure present in feature space rather than requiring explicit group information as in classical MvsM person re-identification. Our experimental results show that the CRF approach can efficiently solve standard re-identification tasks, achieving classification performance beyond the state-of-the-art rank-1 results in the literature. The CRF model can also be used to solve more general identity inference problems in which no hard grouping information and very many test images are present in the probe set.

It is our opinion that in practice it is almost always more common to have many more unlabeled images than labeled ones, and thus that the standard MvsM formulation is unrealistic for most application scenarios. Further exploration of identity inference requires datasets containing many images of many persons imaged from many cameras. Most standard datasets like CAVIAR and i-LIDS are very limited in this regard.

References

1. Bak, S., Corvee, E., Bremond, F., Thonnat, M.: Multiple-shot human re-identification by mean riemannian covariance grid. In: Proceedings of AVSS, pp. 179–184 (2011)
2. Bazzani, L., Cristani, M., Murino, V.: Symmetry-driven accumulation of local features for human characterization and re-identification. Comput. Vis. Image Underst. **117**(2), 130–144 (2013)
3. Bazzani, L., Cristani, M., Perina, A., Farenzena, M., Murino, V.: Multiple-shot person re-identification by hpe signature. In: 20th International Conference on Pattern Recognition, pp. 1413–1416 (2010)
4. Bazzani, L., Cristani, M., Perina, A., Murino, V.: Multiple-shot person re-identification by chromatic and epitomic analyses. Pattern Recogn. Lett. **33**(7), 898–903 (2012)
5. Boix, X., Gonfaus, J.M., van de Weijer, J., Bagdanov, A.D., Serrat, J., Gonzàlez, J.: Harmony potentials. Int. J. Comput. Vision **96**(1), 83–102 (2012)
6. Boykov, Y., Veksler, O., Zabih, R.: Fast approximate energy minimization via graph cuts. IEEE Trans. Pattern Anal. Mach. Intell. **23**(11), 1222–1239 (2001)
7. Cai, Y., Pietikäinen, M.: Person re-identification based on global color context. In: Proceedings of the Asian Conference on Computer Vision Workshops, pp. 205–215 (2011)
8. Cheng, D.S., Cristani, M., Stoppa, M., Bazzani, L., Murino, V.: Custom pictorial structures for re-identification. In: Proceedings of the British Machine Vision Conference, vol. 2, p. 6 (2011)
9. Dalal, N., Triggs, B.: Histograms of oriented gradients for human detection. In: Proceedings of the IEEE Conference on Computer Vision and, Pattern Recognition, vol. 1, pp. 886–893, 2005
10. Dikmen, M., Akbas, E., Huang, T.S., Ahuja, N.: Pedestrian recognition with a learned metric. In: Proceedings of the Asian conference on Computer Vision, pp. 501–512 (2011)
11. Farenzena, M., Bazzani, L., Perina, A., Murino, V., Cristani, M.: Person re-identification by symmetry-driven accumulation of local features. In: Proceedings of the IEEE Conference on Computer Vision and, Pattern Recognition, pp. 2360–2367 (2010)
12. Felzenszwalb, P., Huttenlocher, D.: Efficient belief propagation for early vision. Int. J. Comput. Vision **70**(1), 41–54 (2006)
13. Gray, D., Tao, H.: Viewpoint invariant pedestrian recognition with an ensemble of localized features. In: Proceedings of the European Conference on Computer Vision, pp. 262–275 (2008)
14. Karaman, S., Bagdanov, A.D.: Identity inference: generalizing person re-identification scenarios. In: Computer Vision. Workshops and Demonstrations, pp. 443–452. Springer, Heidelberg (2012)
15. Kolmogorov, V., Zabin, R.: What energy functions can be minimized via graph cuts? IEEE Trans. Pattern Anal. Mach. Intell. **26**(2), 147–159 (2004)
16. Köstinger, M., Hirzer, M., Wohlhart, P., Roth, P.M., Bischof, H.: Large scale metric learning from equivalence constraints. In: Proceedings of the IEEE Conference on Computer Vision and Pattern Recognition (2012)
17. Loy, C.C., Liu, C., Gong, S.: Person re-identification by manifold ranking. In: Proceedings of IEEE International Conference on Image Processing (2013)
18. Prosser, B., Zheng, W., Gong, S., Xiang, T.: Person re-identification by support vector ranking. In: Proceedings of the British Machine Vision Conference (2010)
19. Scharstein, D., Szeliski, R.: A taxonomy and evaluation of dense two-frame stereo correspondence algorithms. Int. J. Comput. Vision **47**(1), 7–42 (2002)

20. Schwartz, W.R., Davis, L.S.: Learning discriminative appearance-based models using partial least squares. In: Brazilian Symposium on Computer Graphics and Image Processing (SIBGRAPI), pp. 322–329. IEEE, New York (2009)
21. Szeliski, R., Zabih, R., Scharstein, D., Veksler, O., Kolmogorov, V., Agarwala, A., Tappen, M., Rother, C.: A comparative study of energy minimization methods for markov random fields with smoothness-based priors. IEEE Trans. Pattern Anal. Mach. Intell. **30**(6), 1068–1080 (2008)
22. Vedaldi, A., Zisserman, A.: Efficient additive kernels via explicit feature maps. IEEE Trans. Pattern Anal. Mach. Intell. **34**(3), 480–492 (2012)
23. Wainwright, M., Jaakkola, T., Willsky, A.: Map estimation via agreement on trees: message-passing and linear programming. IEEE Trans. Inf. Theory **51**(11), 3697–3717 (2005)
24. Zheng, W., Gong, S., Xiang, T.: Associating groups of people. In: Proceedings of British Machine Vision Conference (2009)
25. Zheng, W., Gong, S., Xiang, T.: Re-identification by relative distance comparison.IEEE Trans. Pattern Anal. Mach. Intell. PP(99), 1 (2012)
26. Zheng, W., Gong, S., Xiang, T.: Transfer re-identification: From person to set-based verification. In: Proceedings of the IEEE Conference on Computer Vision and Pattern Recognition (2012)
27. Zhou, D., Weston, J., Gretton, A., Bousquet, O., Schölkopf, B.: Ranking on data manifolds. Adv. Neural Inf. Proc. Syst. **16**, 169–176 (2003)

Chapter 15
Re-identification for Improved People Tracking

François Fleuret, Horesh Ben Shitrit and Pascal Fua

Abstract Re-identification is usually defined as the problem of deciding whether a person currently in the field of view of a camera has been seen earlier either by that camera or another. However, a different version of the problem arises even when people are seen by multiple cameras with overlapping fields of view. Current tracking algorithms can easily get confused when people come close to each other and merge trajectory fragments into trajectories that include erroneous identity switches. Preventing this means re-identifying people across trajectory fragments. In this chapter, we show that this can be done very effectively by formulating the problem as a minimum-cost maximum-flow linear program. This version of the re-identification problem can be solved in real-time and produces trajectories without identity switches. We demonstrate the power of our approach both in single- and multicamera setups to track pedestrians, soccer players, and basketball players.

15.1 Introduction

Person re-identification is often understood as determining whether the same person has been seen at different locations in nonoverlapping camera views and other chapters in this book deal with this issue. However, a different version of the problem

This work was funded in part by the Swiss National Science Foundation.

F. Fleuret (✉)
IDIAP, Martigny CH-1920, Switzerland
e-mail: francois.fleuret@idiap.ch

H. Ben Shitrit · P. Fua
EPFL, Lausanne CH-1015, Switzerland
e-mail: horesh.benshitrit@epfl.ch

P. Fua
e-mail: pascal.fua@epfl.ch

S. Gong et al. (eds.), *Person Re-Identification*,
Advances in Computer Vision and Pattern Recognition,
DOI: 10.1007/978-1-4471-6296-4_15, © Springer-Verlag London 2014

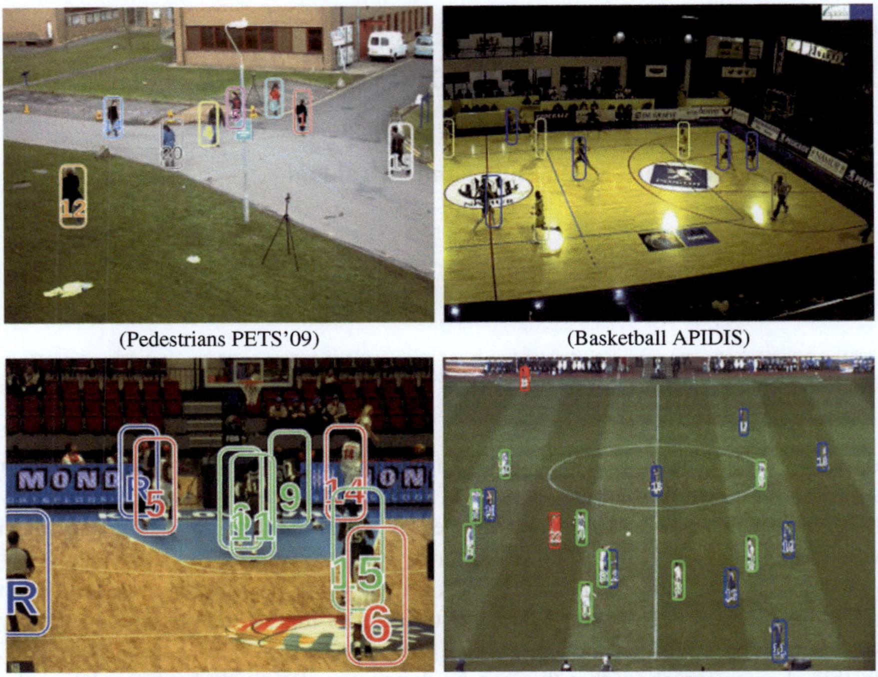

(Pedestrians PETS'09) (Basketball APIDIS)

(Basketball FIBA) (Soccer ISSIA)

Fig. 15.1 Representative detection results on four different datasets. The pedestrian results were obtained using a single camera while the others were obtained with multiple cameras

arises when attempting to track people over long periods of time to provide long-lived and persistent characterizations. Even though the problem may seem easier than the traditional re-identification one, state-of-the-art algorithms [4, 6, 23, 26, 28, 29, 36] are still prone to produce trajectories with identity switches, that is, combining trajectory fragments of several individuals into a single path. Preventing this and guaranteeing that the resulting trajectories are those of a single person can therefore be understood as a re-identification problem since the algorithm must understand which trajectory fragments correspond to the same individual.

This is the re-identification problem we address in this chapter. We will show that by formulating the multiobject tracking as a minimum-cost maximum-flow linear program, we can make appearance-free tracking robust enough so that relatively simple appearance cues, such as using color histograms or simple face-recognition technology, yield real-time solutions that produce trajectories free from the above-mentioned identity switches.

More specifically, we have demonstrated in earlier work [12] that, given probabilities of presence of people at various locations in individual time frames, finding the most likely set of trajectories is a global optimization problem whose objective function is convex and depends on very few parameters. Furthermore, it can be efficiently solved using the K-Shortest Paths algorithm (KSP) [30]. However, this formulation completely ignores appearance, which can result in unwarranted identity switches in complex scenes. We therefore later extended it [10] to allow the exploitation of *sparse* appearance information to keep track of people's identities, even when their paths come close to each other or intersect. By sparse, we mean that the appearance needs only be discriminative in a very limited number of frames. For example, in the basketball and soccer sequences of Fig. 15.1, all teammates wear the same uniform and the numbers on the back of their shirts can only be read once in a long while. Furthermore, the appearance models are most needed when the players are bunched together, and it is precisely then that they are the least reliable [25]. Our algorithm can disambiguate such situations using the information from temporally distant frames. This is in contrast with many state-of-the-art approaches that depend on associating appearance models across *successive* frames [3, 5, 22, 23].

In this chapter, we first introduce our formulation of the multitarget tracking problem as a Linear Program. We then discuss our approach to estimate the required probabilities, and present our results, first without re-identification and then with it.

15.2 Tracking as Linear Programming

In this section, we formulate the multitarget tracking as an integer program (IP), which can be relaxed to a Linear Program (LP) and efficiently solved. We begin with the case where appearance can be ignored and then extend our approach to take it into account.

15.2.1 Tracking without Using Appearance

We represent the ground plane by a discrete grid and, at each time step over a potentially long period of time, we compute a *Probability Occupancy Map* (POM) that associates to each grid cell a probability of presence of people, as will be discussed in Sect. 15.3. We then formulate the inference of trajectories from these often noisy POMs as a LP [12], which can be solved very efficiently using the KSP [30]. In this section, we first introduce our LP and then our KSP approach to solving it.

Linear Program Formulation

We model people's trajectories as continuous flows going through an area of interest. More specifically, we first discretize the said area into K grid locations, and the time interval into T instants. Let $\mathbf{I}$ stand for the images we are processing. For any $(i, t) \in \{1, \ldots, K\} \times \{1, \ldots, T\}$ let $X_i(t)$ be a Boolean random variable standing for the presence of someone at location i at time t, and

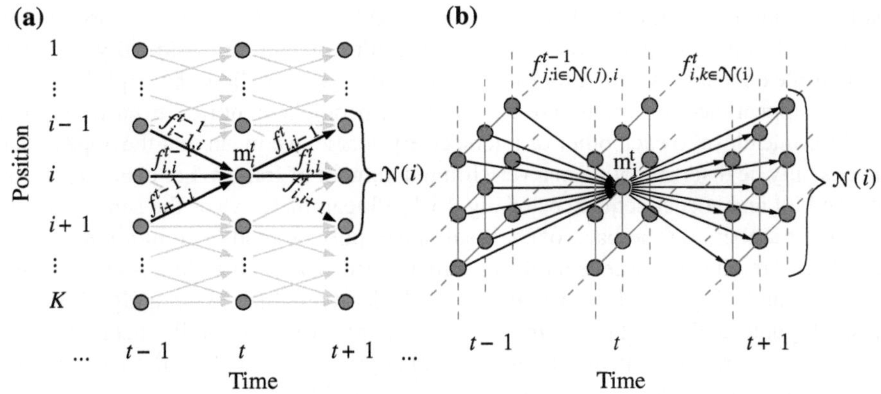

Fig. 15.2 Directed Acyclic Graph and corresponding flows. **a** Positions are arranged on one dimension and edges created between vertices corresponding to neighboring locations at consecutive time instants. **b** Basic flow model used for tracking people moving on a 2D grid. For the sake of readability, only the flows to and from location i at time t are printed

$$\rho_i(t) = P(X_i(t) = 1 \mid \mathbf{I}) \tag{15.1}$$

be the posterior probability that someone stands at location k at time t, given the images.

For any location i, let $\mathcal{N}(i) \subset \{1, \ldots, K\}$ denote its neighborhood, that is, the locations a person located at i at time t can reach at time $t+1$. To model occupancy over time, let us consider a labeled directed acyclic graph with $K \times T$ vertices such as the one depicted by Fig. 15.2a, which represents every location at every instant. As shown in Fig. 15.2b, these locations represent spatial positions on successive grids, one for every instant t. The edges connecting locations correspond to admissible motions, which means that there is one edge $e^t_{i,j}$ from (t, i) to $(t+1, j)$ if, and only if, $j \in \mathcal{N}(i)$. Note that to allow people to remain static, we have $\forall i, i \in \mathcal{N}(i)$. Hence, there is always an edge from a location at time t to itself at time $t+1$.

As shown in Fig. 15.2b, each vertex is labeled with a discrete variable m^t_i standing for the number of people located at i at time t. Each edge is labeled with a discrete variable $f^t_{i,j}$ standing for the number of people moving from location i at time t to location j at time $t+1$. For instance, the fact that a person remains at location i between times t and $t+1$ is represented by $f^t_{i,i} = 1$. These notations and those we will introduce later are summarized in Table 15.1.

In general, the number of people being tracked may vary over time, meaning some may appear inside the tracking area and others may leave. Thus, we introduce two additional nodes υ_{source} and υ_{sink} into our graph. They are linked to all the nodes representing positions through which people can respectively enter or exit the area, such as doors and borders of the cameras' fields of view. In addition, edges connect υ_{source} to all the nodes of the first frame to allow the presence of people anywhere in that frame, and reciprocally edges connect all the nodes of the last frame to υ_{sink},

Table 15.1 Notations used in this chapter. When appearance is ignored as in Sect. 15.2.1, L the number of groups is equal to one and the ^{l} superscripts are omitted

T	:	Number of time steps
$\mathbf{I} = (\mathbf{I}^1, \ldots, \mathbf{I}^T)$	:	Captured images
K	:	Number of locations on the ground plane
L	:	Number of *labeled* groups of people
N_l	:	Maximum number of people in group l
$\mathcal{N}(i) \subset \{1, \ldots, K\}$	:	Neighborhood of location i, all locations which can be reached in one time step
m_i^t	:	Number of people at location i at time t
$e_{i,j}^t$	:	Directed edge in the graph
$f_{i,j}^t$	:	Number of people moving from location i to location j at time t in group l
$Q_i(t)$	:	R. V. standing for the true identity group of a person in location i, at time t
$X_i(t)$	:	R. V. standing for the true occupancy of location i at time t
$\varphi_i^l(t)$	:	Estimated probability of a location i to be occupied by a person from group l according to the appearance model
$\rho_i(t)$	:	Estimated probability of location i to be occupied by an unidentified person according to the pedestrian detector

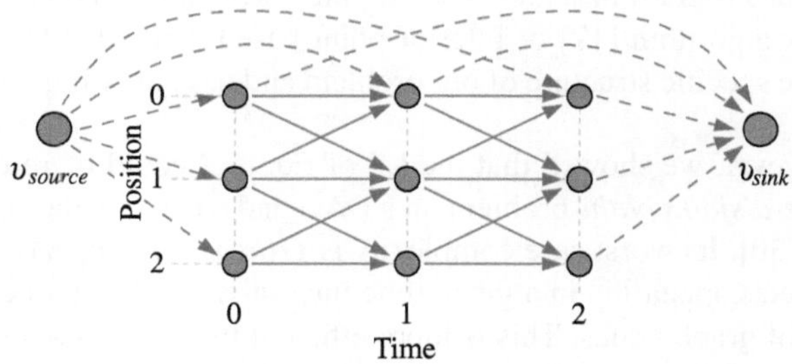

Fig. 15.3 Complete graph for a small area of interest consisting only of three positions and three time frames. Here, we assume that position 0 is connected to the virtual positions and therefore a possible entrance and exit point. Flows to and from the virtual positions are shown as dashed lines while flows between physical positions are shown as solid lines

to allow for people to still be present in that frame. As an illustration, consider the case of a small area of interest that can be modeled using only three locations, one of which is both entrance and exit, over three time steps. This yields the directed acyclic graph (DAG) depicted by Fig. 15.3. υ_{source} and υ_{sink} are *virtual locations*, because, unlike the other nodes of the graph, they do not represent any physical place.

Under the constraints that people may not enter or leave the area of interest by any other location than those connected to υ_{sink} or υ_{source} and that there can never be more than one single person at each location, we showed in [12] that the flow with the maximum a posteriori probability is the solution of the IP

$$
\begin{aligned}
\text{Maximize} \; & \sum_{t,i} \log \left(\frac{\rho_i(t)}{1-\rho_i(t)} \right) \sum_{j \in \mathcal{N}(i)} f_{i,j}(t), \\
\text{subject to} \; & \forall t, i, j, \;\; f_{i,j}(t) \geq 0 \;, \\
& \forall t, i, \;\; \sum_{j \in \mathcal{N}(i)} f_{i,j}(t) \leq 1 \;, \\
& \forall t, i, \;\; \sum_{j \in \mathcal{N}(i)} f_{i,j}(t) - \sum_{k: i \in \mathcal{N}(k)} f_{k,i}(t-1) \leq 0 \;, \\
& \sum_{j \in \mathcal{N}(\upsilon_{\text{source}})} f_{\upsilon_{\text{source}},j} - \sum_{k: \upsilon_{\text{sink}} \in \mathcal{N}(k)} f_{k,\upsilon_{\text{sink}}} \leq 0 \;,
\end{aligned}
\tag{15.2}
$$

where $\rho_i(t)$ is the probability that someone is present at location i at time t computed from either one or multiple-images, as will be discussed in Sect. 15.3.

Using the K-Shortest Path Algorithm

The constraint matrix of the IP of Eq. 15.2 can be shown to be *totally unimodular*, which means it could be solved exactly by relaxing the integer assumption and solving a Linear Program instead. However, most available solvers rely on variants of the Simplex algorithm [17] or interior point based methods [24], which do not make use of the specific structure of our problem and have very high worst case time complexities.

In [12] however, we showed that the LP of Eq. 15.2 could be reformulated as a *k shortest node-disjoint paths* problem on a DAG and solved by the computationally efficient KSP [30]. Its worst case complexity is $O(k(m + n \cdot \log n))$, where k is the number of objects appearing in a given time interval, m is the number of edges and n the number of graph nodes. This is more efficient than the min-cost flow method of [36], which exhibits a worst case complexity of $O(kn^2 m \log n)$. Furthermore, due to the acyclic nature of our graph, the average complexity is almost linear with the number of nodes, and we have observed 1,000-fold speed gains over general LP solvers.

As a result, we have been able to demonstrate real-time performance on realistic scenarios by splitting sequences in overlapping batches of 100 frames. This results in a constant 4-s delay between input and output, which is acceptable for many applications.

15.2.2 Tracking with Sparse Appearance Cues

The KSP algorithm of "Using the K-Shortest Path Algorithm" completely ignores appearance, which can result in unwarranted identity switches when people come close and separate again. If reliable appearance cues were available, this would be easy to avoid but such cues are often undependable, especially when people are in close proximity. For example, in the case of the basketball players, the appearance of

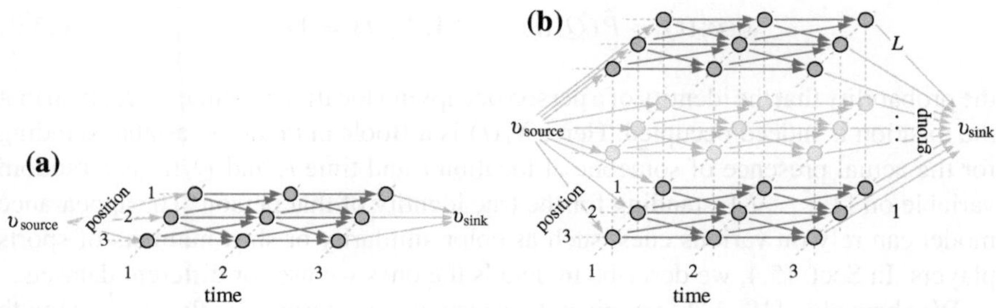

Fig. 15.4 Our tracking algorithm involves computing flows on a Directed Acyclic Graph (DAG). **a** The DAG of "Using the K-Shortest Path Algorithm" includes source and sink nodes that allow people to enter and exit at selected locations, such as the boundaries of the playing field. This can be interpreted as a single-commodity network flow. **b** To take image-appearance into account, we formulate the tracking problem as a multicommodity network flow problem which can be illustrated as a duplication of the graph for each appearance-group

teammates is very similar and they can only reliably be distinguished by reading the numbers on the back of their jerseys. In practice, this can only be done at infrequent intervals.

Multicommodity Network Flow Formulation

To take advantage of this kind of sparse appearance information, we extend the framework of "Linear Program Formulation" by computing flows on the expanded DAG of Fig. 15.4b. It is obtained by starting from the graph of Fig. 15.4a, which is the one we used before, and duplicating it for each possible appearance group.

More precisely, we partition the total number of tracked people into L groups and assign a separate appearance model to each. In a constrained scene, such as a ball game, we can restrict each group l to include at most N_l people, but in general cases, N_l is left unbounded. The groups can be made of individual people, in which case $N_l = 1$. They can also be composed of several people that share a common appearance, such as members of the same team or referees, in sports games.

The resulting expanded DAG has $|\mathcal{V}| = K \times T \times L$ nodes. Each one represents a location i at time t occupied by a member of identity group l. Edges represent admissible motions between locations at consecutive times. Since individuals cannot change their identity, there are no edges linking groups, and therefore no vertical edge in Fig. 15.4b. The resulting graph is made of disconnected layers, one per identity group. This is in contrast to the approach of "Linear Program Formulation", which relies on a single-layer graph such as the one of Fig. 15.4a.

As before, let us assume that we have access to a person detector that estimates the probability of presence $\rho_i(t)$ of someone at every location i and time t. Let us further assume that we can compute an appearance model that we use to estimate

$$\varphi_i^l(t) = \hat{P}(Q_i(t) = l \mid \mathbf{I}, X_i(t) = 1)\,, \tag{15.3}$$

the probability that the identity of a person occupying location i at time t is l, given that the location is indeed occupied. Here, $X_i(t)$ is a Boolean random variable standing for the actual presence of someone at location i and time t, and $Q_i(t)$ is a random variable on $\{1, \ldots, L\}$, standing for the true identity of that person. The appearance model can rely on various cues, such as color similarity or shirt numbers of sports players. In Sect. 15.4, we describe in details the ones we use for different datasets.

We showed in [10, 11] that, given these appearance terms, the flows $f_{i,j}^l(t)$ with the maximum a posteriori probability are the solution of the IP

$$\begin{aligned}
\text{Maximize } & \sum_{t,i,l} \log\left(\frac{\rho_i(t)\varphi_i^l(t)L}{1-\rho_i(t)}\right) \sum_{j\in\mathcal{N}(i)} f_{i,j}^l(t) \\
\text{subject to } & \forall t, l, i, j,\ \ f_{i,j}^l(t) \geq 0\,. \\
& \forall t, i,\ \ \sum_{j\in\mathcal{N}(i)} \sum_{l=1}^{L} f_{i,j}^l(t) \leq 1\,, \\
& \forall t, l, i,\ \ \sum_{j\in\mathcal{N}(i)} f_{i,j}^l(t) - \sum_{k:i\in\mathcal{N}(k)} f_{k,i}^l(t-1) \leq 0\,, \\
& \sum_{j\in\mathcal{N}(\upsilon_{\text{source}})} f_{\upsilon_{\text{source}},j} - \sum_{k:\upsilon_{\text{sink}}\in\mathcal{N}(k)} f_{k,\upsilon_{\text{sink}}} \leq 0\,, \\
& \forall t, l,\ \ \sum_{i=1}^{K} \sum_{j\in\mathcal{N}(i)} f_{i,j}^l(t) \leq N_l\,.
\end{aligned} \tag{15.4}$$

Since Integer Programming is NP-complete, we relax the problem of Eq. 15.4 into a multicommodity network flow (MCNF) problem of polynomial complexity as in "Using the K-Shortest Path Algorithm" by making the variables real numbers between zero and one. However unlike the one of Eq. 15.2, this new problem is not totally unimodular. As a result, the LP solution is not guaranteed to be integral and real values that are far from either zero or one may occur [9]. In practice this only happens rarely, and typically when two or more targets are moving so close to each other that appearance information is unable to disambiguate their respective identities. These noninteger results can be interpreted as an uncertainty about identity assignment by our algorithm. This represents valuable information that could be used. However, as this happens rarely, we simply round off noninteger results in our experiments.

Making the Problem Computationally Tractable

A more severe problem is that the graphs that we have to deal with are much larger than those of "Using the K-Shortest Path Algorithm". The massive number of variables and constraints involved usually results in too large a problem to be directly handled by regular solvers for real-life cases. Furthermore, the problem cannot be solved anymore using the efficient KSP [30].

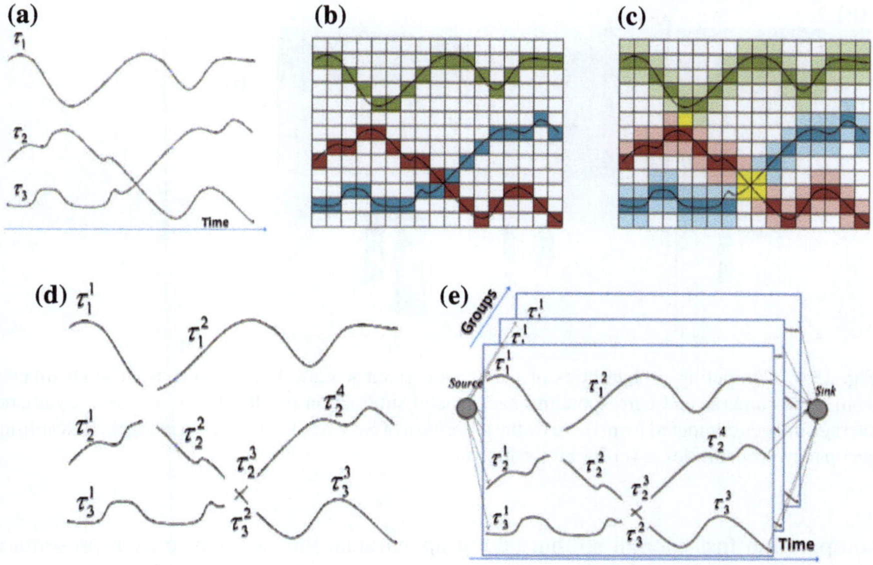

Fig. 15.5 Pruning the graph and splitting trajectories into tracklets. **a** For simplicity, we represent the trajectories as being one-dimensional and assume that we have three of them. **b** Each trajectory is a set of vertices from successive time instants. We assigned a different color to each. **c** The neighborhoods of the trajectories within a distance of 1 are shown in a color similar to that of the trajectory, but less saturated. The vertices that are included in more than one neighborhood appear in yellow and are used along with those on the trajectories themselves to build the expanded graph. **d** The yellow vertices are also used as trajectory splitting points to produce tracklets. Note that two trajectories do not necessarily have to cross to be split; it is enough that they come close to each other. **e** The tracklet-based multicommodity network flow algorithm can be interpreted as finding paths from the source node to the sink node, on a multiple layer graph whose nodes are the tracklets

In practice we address this problem by removing unnecessary nodes from the graph of Fig. 15.4b. To this end, we first ignore appearance and run the KSP on the DAG of Fig. 15.4a. The algorithm tracks all the people in the scene very efficiently but is prone to identity switches. We account for this by eliminating all graph nodes except those that belong to trajectories found by the algorithm plus those that could be used to connect one trajectory to the other, such as the yellow vertices of Fig. 15.5c. We then turn the pruned graph into a multilayer one and solve the multicommodity network flow problem of Eq. 15.4 on this expanded graph, which is now small enough to be handled by standard solvers.

The computational complexity can be further reduced by not only removing obviously empty nodes from the graph but, in addition, by grouping obviously connected ones into *tracklets*, such as those of Fig. 15.5d. The Linear Program of Eq. 15.4 can then be solved on a reduced graph such as the one of Fig. 15.5e whose nodes are the tracklets instead of individual locations. It is equivalent to the one of Fig. 15.4b, but with a much reduced number of vertices and edges [11]. In practice, this makes the

Fig. 15.6 Computing probabilities of occupancy given a static background. **a** Original images from three cameras and corresponding background subtraction results shown in green. Synthetic average images computed from them by the algorithm of Sect. 15.3.1 are shown in black. **b** Resulting occupancy probabilities $\rho_i(t)$ for all locations i

computation fast enough so that taking appearance into account only represents a small overhead over not using and we can still achieve real-time performance.

15.3 Computing the Probabilities of Presence

The LP programs of Eqs. 15.2 and 15.4 both depend on the estimated probabilities $\rho_i(t)$ that someone is present at location i at time t. In this section, we explain how we compute these probabilities. We will discuss the appearance-based probabilities $\varphi_i^l(t)$ that the person belongs to group $l \in L$ that the program of Eq. 15.4 also requires in the following section.

We describe here two alternatives to estimating the probabilities of presence $\rho_i(t)$ depending on whether the background is static or not. In the first case, we can rely on background subtraction and in the second on people detectors to compute the Probability Occupancy Maps (POMs) introduced in Sect. 15.2.1, that is, values of $\rho_i(t)$ for all locations i.

15.3.1 Detecting People Against a Static Background

When the background is static, a background subtraction algorithm can be used to find the people moving about the scene. As shown in Fig. 15.6a, this results in very rough binary masks $\mathbf{B}_c$, one per image, where the pixels corresponding to the moving people are labeled as ones and the others as zeros. Our goal then is to infer a POM such as the one of Fig. 15.6b from these. A key challenge is to account for the fact that people often occlude each other.

To this end, we introduced a generative model-based approach [22] that has been shown to be competitive against start-of-the-art ones [19]. We represent humans as cylinders that project to rectangles in individual images, as depicted by the black rectangles in Fig. 15.6a. If we knew the true state of occupancy $X_i(t)$ at location i and time t for all locations, this model could be used to generate synthetic images such as those in the bottom row of Fig. 15.6a. Given probability estimates $\rho_i(t)$ for the $X_i(t)$, we consider the *average* synthetic image these probabilities imply. We select them to minimize the distance between the average synthetic image and the background subtraction results in all images simultaneously. In [22], we showed that under a mean-field assumption this amounts to minimizing the Kullback-Leibler divergence between the resulting product law, and the "true" conditional posterior distribution of occupancy given the background subtraction output under our generative model.

In practice, given the binary masks $\mathbf{B}_1, \ldots, \mathbf{B}_C$ from the one or more images acquired at time t and omitting the time indices in the remainder of this section, this allows us to compute the corresponding ρ_i as the fixed point of a large system of equations of the form

$$\rho_i = \frac{1}{1 + \exp\left(\lambda_i + \sum_c \Psi\left(\mathbf{B}_c, \mathbf{S}_c^{X_i=1}\right) - \Psi\left(\mathbf{B}_c, \mathbf{S}_c^{X_i=0}\right)\right)}, \tag{15.5}$$

with λ_k a small constant that accounts for the *a priori* probability of presence in individual grid cells, $\mathbf{S}_c^{X_k=b}$ is the average synthetic image in view c given all the ρ_k for $k \neq i$ and assuming that $X_k = b$. Ψ measures the dissimilarity between images and is defined as

$$\Psi(B, S) = \frac{1}{\sigma} \frac{\|B \otimes (1 - S) + (1 - B) \otimes S\|}{\|S\|}, \tag{15.6}$$

where $\otimes$ denote the pixel-wise product of two images and σ accounts for the expected quality of the background subtraction.

Equation 15.5 is one of a large system of equations whose unknowns are the ρ_i values. To compute their values, we iteratively update them in parallel until we find a fixed point of the system, which typically happens in 100 iterations given a uniform initialization of the ρ_i. Computationally, the dominant term is the estimation of the synthetic images, which can be done very fast using integral images. As a result, computing POMs in this manner is computationally inexpensive and using them to instantiate the LPs of Eqs. 15.2 and 15.4 is the key to a real-time people-tracking pipe-line.

Vanila DPM Trained DPM (ours)

Fig. 15.7 Detection results of the DPM trained using only the INRIA pedestrian database (*left*) versus our retrained DPM (*right*). In both cases, we use the same parameters at run-time and obtain clearly better results with the retrained DPM

15.3.2 Detecting People Against a Dynamic Background

If the environment changes or if the camera moves, we replace the background subtraction based estimation of the marginal probabilities of presence $\rho_i(t)$ of Sect. 15.3.1 by the output of a modified Deformable Part Model object detector (DPM) [21]. We chose it because it has consistently been found to be competitive against other state-of-the-art approaches but we could equally well have used another one, such as [8, 16, 27].

Given a set of high-scoring detections, we assign a large occupancy probability value to the corresponding ground locations. In practice, when using the DPM detector, the top of the head tends to be the most accurately detected body part. We therefore estimate ground locations by projecting the center of the top of the bounding boxes, assumed to be at a pre-specified height above ground. The occupancy probability at locations where no one has been detected are set to a low value to account for the fact that the detector could have failed to detect somebody who was actually there. Note that these probabilities could also be learned in an automated fashion given sufficient amounts of training data.

Re-training the DPM Model. We performed most of our experiments with dynamic backgrounds on tracking basketball players and found that in such a context the performance of the original DPM model [21] is insufficient for our purposes. This is due in large part to the fact that it is trained using videos and images of pedestrians whose range of motion is very limited. By contrast and as shown in Fig. 15.7, the basketball players tend to perform large amplitude motions.

To overcome this difficulty, we used our multi camera setup [10] to acquire additional training data from two basketball matches for which we have multiple synchronized views, which we add to the standard INRIA pedestrian database [16]. We use the bounding boxes corresponding to unoccluded players as positive examples and images of empty courts as negative ones.

Geometric Constraints and Non-Maximum Suppression. It is well known that imposing geometric consistency constraints on the output of a people detector significantly improves detection accuracy [14, 34]. In the specific case of basketball, we can use the court markings to accurately compute the camera intrinsic and extrinsic parameters [31]. This allows us to reject all detections that are clearly out of our area of interest, the court in this case.

Non-Maximum Suppression (NMS) is widely used to post-process the output of object detectors that rely on a sliding window search. This is necessary because their responses for windows translated by a few pixels are virtually identical, which usually results in multiple detections for a single person. In the specific case of the DPM we use, the head usually is the most accurately detected part and, in the presence of occlusions, it is not uncommon for detection responses to correspond to the same head but different bodies. In our NMS procedure, we therefore first sort the detections based on their score. We then eliminate all those whose head overlaps by more than a fraction with that of a higher scoring one or whose body overlaps by more than a similar fraction.

15.3.3 Appearance-Free Experimental Results

By using the approaches described above at every time-frame independently, we obtain the $\rho_i(t)$ probabilities the KSP algorithm of "Using the K-Shortest Path Algorithm" requires. We tested them on two very different basketball datasets:

- The FIBA dataset comprises several multiview basketball sequences captured during matches at the 2010 women's world championship. We manually annotated the court locations of the players and the referees on 1000 frames of the Mali versus Senegal match and 6000 frames of the Czech Republic versus Belarus match. Individual frames from these matches are shown in Figs. 15.1 and 15.7. They were acquired either by one of six stationary synchronized cameras or by a single moving broadcast camera.
- The APIDIS dataset [7] is a publicly available set of video sequences of a basketball match captured by seven stationary unsynchronized cameras placed above and around the court. It features challenging lighting conditions produced by the many direct light sources that are reflected on the court while other regions are shaded. We present results using either all seven cameras or only Camera #6, which captures half of the court as shown at the top right of Fig. 15.1.

Figure. 15.8 depicts our results. They are expressed in terms of the standard MODA CLEAR metric [13], which stands for *Multiple Object Detection Accuracy* and is defined as

$$\text{MODA} = 1 - \frac{\sum_t (m_t + fp_t)}{\sum_t g_t}, \tag{15.7}$$

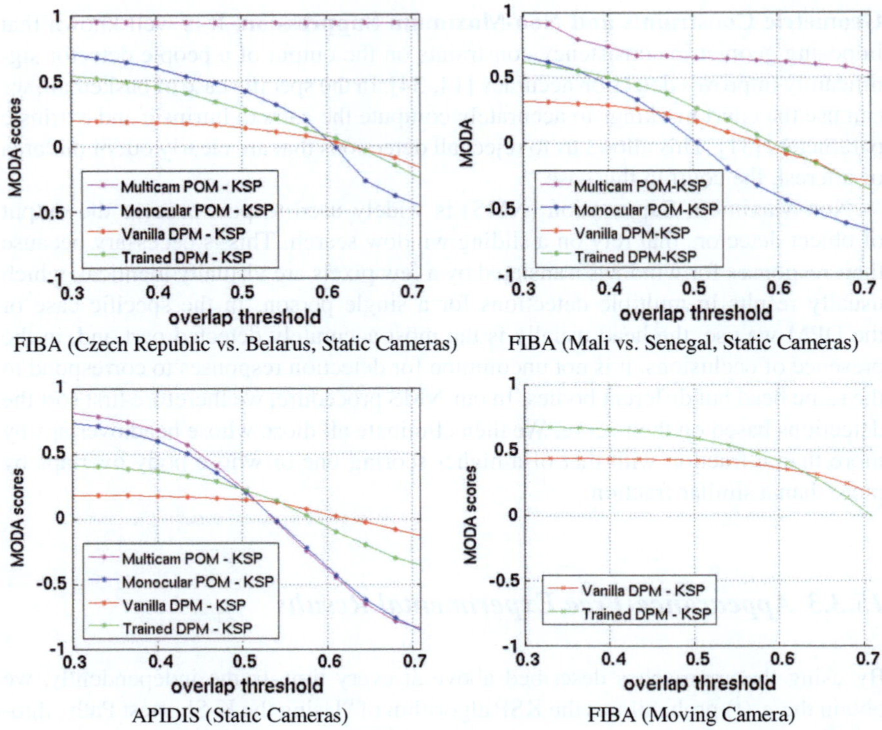

Fig. 15.8 MODA scores obtained for two different FIBA matches and one APIDIS using one or more static cameras and the different approaches to people detection of Sect. 15.3. The corresponding curves are labeled as **Multicam POM**, multicamera generative-model; **Monocular POM**, single-camera generative-model; **Vanilla DPM**, DPM trained only with INRIA pedestrian dataset; **Trained DPM**, DPM trained using both pedestrian and basketball datasets. The MODA scores were calculated as functions of the bounding-box overlap value used to decide whether two detections correspond to the same person

where g_t is the number of ground truth detections at time t, m_t the number of misdetections, fp_t the false positive count.

Following standard Computer Vision practice, we decide whether two detections correspond to the same person on the basis of whether the overlap of the corresponding bounding boxes is greater or smaller than a fraction of their area, which is usually taken to be between 0.3 and 0.7 [19]. In Fig. 15.8, we therefore plot our results as functions of this threshold.

When background-subtraction can be used, the generative-model approach of Sect. 15.3.1 yields excellent results with multiple cameras. Even when using a single camera, it outperforms the people detector-based approach of Sect. 15.3.2, in part because the generative-model explicitly handles occlusion. However, when the camera moves it becomes impractical whereas people detectors remain effective.

15.4 Using Appearance-Based Clues for Re-identification Purposes

The KSP approach of "Using the K-Shortest Path Algorithm", which has been used to obtain the results of Sect. 15.3.3, does not take appearance cues into account. Thus, it does not preclude identity switches. In other words, trajectory segments corresponding to different people can be mistakenly joined into a single long trajectory. This typically happens when two people come close to each other and then separate again.

In this section, we show how we can use MCNF approach of "Using the K-Shortest Path Algorithm" to take appearance cues into account and re-identify people from one tracklet to the next. This involves first computing the appearance-based probabilities $\varphi_i^l(t)$ of Eq. 15.4 that a person belongs to group $l \in L$. Note that, even though values of $\varphi_i^l(t)$ have to be provided for all locations and all times, they do not have to be informative in every single frame. If they are in a few frames and uniform in the rest, this suffices to reliably assign identities to these because we reason in terms of whole trajectories.

In other words, we only have to guarantee that the algorithms we use to process appearance return usable results once in a while, which is much easier than doing it in every frame. In the remainder of this section, we introduce three different ways of doing so and present the corresponding results.

15.4.1 Color Histograms

Since our sequences feature groups of individuals, such as players of the same team or referees whose appearance is similar, the simplest is to use color distribution as a signature [10]. We use a few temporal frames at the beginning of the sequence to generate representative templates for each group by manually selecting a few bounding boxes such as the black rectangles of Fig. 15.6 that correspond to members of that group, converting the foreground pixels within each box to the CIE-LAB color space, and generating a color histogram for each view.

Extracting color information from closely spaced people is unreliable because it is often difficult to correctly segment them. Thus, at run time, for each camera and at each time frame, we first compute an occlusion map based on the raw probability occupancy map. If a specific location is occluded with high probability in a given camera view, we do not use it to compute color similarity. Within a detection bounding box, we use the background subtraction result to segment the person. The segmented pixels are inserted into a color histogram, in the same way as for template generation. Finally, the similarity between this observed color histogram and the templates is computed using the Kullback-Leibler divergence, and normalized to get a value between 0 and 1 to be used as a probability. If no appearance cue is available, for example because of occlusions, $\varphi_i^l(t)$ is set to $\frac{1}{L}$.

Fig. 15.9 Reading Numbers. **a** Color image. **b** *Gray-scale* image. **c, d** Distances to color prototypes for the *green* and *white* team respectively

15.4.2 Number Recognition

In team-sports, the numbers on the back of players are unique identifiers that can be used to unambiguously recognize them. However, performing number recognition at every position of an image would be much too expensive at run-time.

Instead, we manually extract templates for each player's number early in the matches when all players are standing still while the national anthem is played. Since within a team the printed numbers usually share a unique color, which is well separated from the shirt color, we create distinct shirt and number color prototypes by grouping color patches on the shirts into two separate clusters. For each prototype and each pixel in the images we want to process, we then compute distances to that prototype as shown in Fig. 15.9, and binarize the resulting distance image.

At run-time, we only attempt to read numbers at locations where the probability of presence is sufficiently high. For each one, we trim the upper and lower $1/5$ of the corresponding bounding box to crop out the head and legs. We then binarize the corresponding image window as described above and search for number candidates within it by XORing the templates with image-patches of the same size. We select the patches that maximize the number of ones and take $\varphi_i^l(t)$ to be the normalized matching score. For reliability, we only retain high-scoring detections. In all other frames, we assume a uniform prior and set $\varphi_i^l(t)$ to $\frac{1}{L}$.

15.4.3 Face Recognition

Our third approach relies on face-detection and recognition. After estimating the probability of occupancy at every location, we run a face detector in each camera view, but only at locations whose corresponding probability of occupancy $\rho_i(t)$ is large. The face detector relies on Binary Brightness Features [1] and a cascade of strong classifiers built using a variant of AdaBoost, which has proved to be faster than the standard Viola-Jones detector [32] with comparable detection performance. For each detected face, we then extract a vector of histograms of Local Binary Pattern (LBP) features [2].

In some cases, such as when a limited number of people are known to be present, L can be assumed to be given *a priori* and representative feature vectors, or prototypes, learned offline for each person. However, in more general surveillance settings, both L and the representative feature vectors must be estimated online. We have therefore implemented two different scenarios.

- **Face Identification.** When the number of people that can appear is known *a priori*, our run-time system estimates the $\varphi_i^l(t)$ probabilities by comparing the feature vectors it extracts from the images to prototypes. These are created by acquiring sequences of the L people we expect our system to recognize and run our face-detection procedure. We then label each resulting feature vector as corresponding to one of the L people and train a multiclass RBF SVM [15] to produce a L-dimensional response vector [35]. At run-time, at each location i and time t where a face is detected, the same L-dimensional vector is computed and converted into probability $\varphi_i^l(t)$ for $1 \leq l \leq L$ [33]. In the absence of a face detection, we set $\varphi_i^l(t)$ to $\frac{1}{L}$ for all l.
- **Face Re-Identification.** When the number of people can be arbitrary, the system creates the prototypes and estimates L at run-time by first clustering the feature vectors [20, 35] and only then estimating the probabilities and finally computing the probabilities as described above.

In the face identification case, people's identities are known while in face re-identification all that can be known is that different tracklets correspond to the same person. The second case is of course more challenging than the first.

We deployed a real-time version of our algorithm in one room of our laboratory. The video feed is processed in 50-frame batches at a framerate of 15 Hz on a quad-core 3.2 GHz PC [35]. In practice, this means that the result is produced with a constant 3.4 s delay, making it completely acceptable for many broadcasting or even surveillance applications.

15.4.4 Appearance-Based Experimental Results

In this section, we demonstrate that using the appearance-based information does improve our results by significantly reducing the number of identity switches. To this end, we present here results on the FIBA and APIDIS datasets introduced in Sect. 15.3.3 as well as three additional ones.

- The ISSIA soccer dataset [18] is a publicly available set of 3,000-frame sequences captured by six stationary cameras placed at two sides of the stadium. They feature 25 people, 3 referees, and 11 players per team, including the goal keepers whose uniforms are different from those of their teammates. Due to the low image resolution, the shirt numbers are unreadable. Hence, we consider $L = 5$ appearance groups and only use color-based cues.
- The PETS'09 pedestrian dataset features 10 people filmed by 7 cameras at 7 fps and has been used in a computer vision challenge to compare tracking algorithms. Even though it does not use appearance cues, the KSP approach of "Using the K-Shortest Path Algorithm" was shown to outperform the other approaches on this data [19] and constitutes therefore a very good baseline for testing the influence of the appearance terms, which we did on the 800-frame sequence `S2/L1`. Most of the pedestrians wear similar dark clothes, which makes appearance-based identification very challenging. We therefore used only $L = 2$ appearance groups, one for people wearing dark clothes and the other for those wearing reddish ones.
- We designed the CVLab dataset to explore the use of face recognition in the context of people tracking. We used 6 synchronized cameras filming a 7×8 m room at 30 fps to acquire a training set of $L = 30$ sequences, each featuring a single person looking towards the 6 cameras, and a 7,400-frame test set featuring 9 of the thirty people we trained the system for entering and leaving the room. In all these frames, 2,379 instances of faces were recognized and used to compute the appearance-based probabilities.

Our results are depicted by Figs. 15.10 and 15.11 and expressed in terms of a slightly modified version of the MOTA CLEAR metric [13], which, unlike the MODA metric we used in Sect. 15.3.3, is designed to evaluate performance in terms of identity preservation. MOTA stands for *Multiple Object Tracking Accuracy* and is defined as

$$\text{MOTA} = 1 - \frac{\sum_t (m_t + fp_t + mme_t)}{\sum_t g_t} \quad , \tag{15.8}$$

where g_t is the number of ground truth detections, m_t the number of misdetections, fp_t the false positive count and mme_t the number of *instantaneous* identity switches. In all our experiments, both KSP and MCNF algorithms yield similarly high scores [10] because this metric is not discriminative enough. To see why, consider a case where the identities of two subjects are switched in the middle of a sequence. The MOTA score is decreased because mme_t is one instead of zero, but not by much even though the identities are wrong half of the time. To remedy this, we define the metric GMOTA as

$$\text{GMOTA} = 1 - \frac{\sum_t (m_t + fp_t + gmme_t)}{\sum_t g_t} \quad , \tag{15.9}$$

where $gmme_t$ now is the number of times in the sequence where the identity is wrong. The GMOTA values are those we plot in Figs. 15.10 and 15.11 as function of the

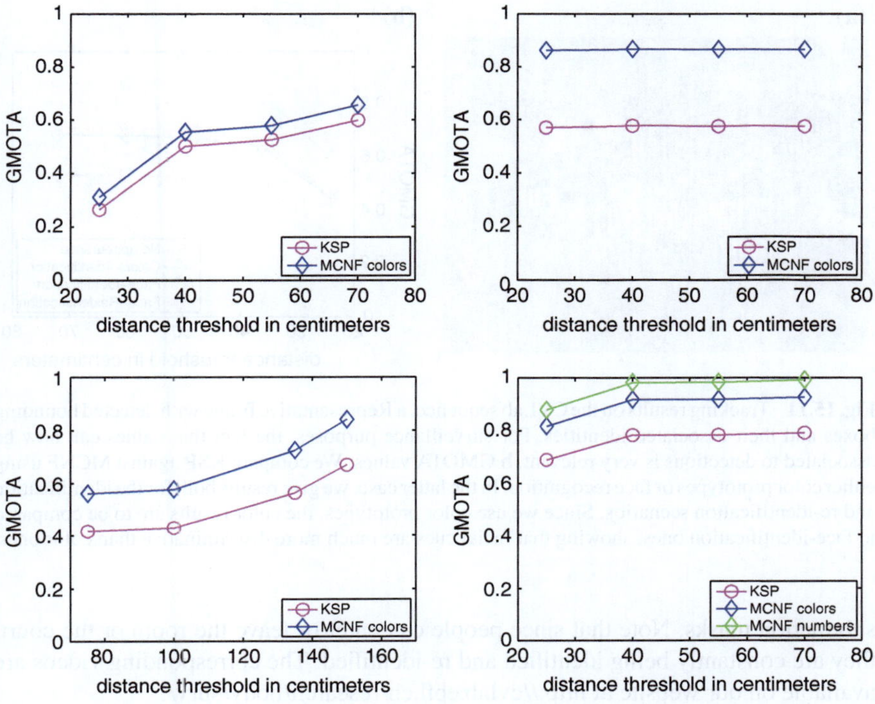

Fig. 15.10 Tracking results of four datasets depicted by Fig. 15.1 expressed in terms of GMOTA values. We compare KSP, which does not use appearance, against MCNF using color cues. For the FIBA dataset, we also show results using number recognition. The appearance information significantly improves performance in all cases.

ground-plane distance threshold we use to assess whether a detection corresponds to a ground-truth person.

In all our experiments, computing the appearance probabilities on the basis of color improves tracking performance. Moreover, for the FIBA and CVLab dataset, we show that incorporating unique identifiers such as numbers or faces is even more effective. The sequences vary in their difficulty, and this is reflected in the results. In the PETS'09 dataset, most of the pedestrians wear similar natural colors and the MCNF algorithm only delivers a small gain over KSP. The APIDIS dataset is very challenging due to strong specular reflections and poor lighting. As a result, KSP performs relatively poorly in terms of identity preservation but using color helps greatly. In the ISSIA dataset, the soccer field is big and we use large grid cells to keep the computational complexity low. Thus, localization accuracy is less and we need to use bigger distance thresholds to achieve good scores. The best tracking results are obtained on the FIBA dataset when simultaneously using color-information and the numbers, and on the CVLab sequence when using face recognition. This is largely because the images are of a much higher-resolution, yielding better background

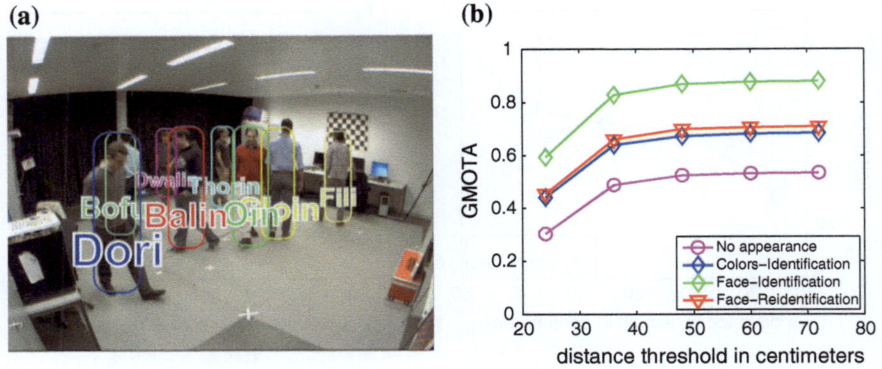

Fig. 15.11 Tracking results on the CVLab sequence. **a** Representative frame with detected bounding boxes and their associated identities. For surveillance purposes, the fact that names can now be associated to detections is very relevant. **b** GMOTA values. We compare KSP against MCNF using either color prototypes or face recognition. In the latter case, we give results both for the identification and re-identification scenarios. Since we use color prototypes, the color results are to be compared to face-identification ones, showing that facial cues are much more discriminative than color ones

subtraction masks. Note that since people can enter or leave the room or the court, they are constantly being identified and re-identified. The corresponding videos are available on our website at http://cvlab.epfl.ch/research/body/surv/.

15.5 Conclusions

In this chapter, we have described a global optimization framework for multipeople tracking that takes image-appearance cues into account, even if they are only available at infrequent intervals. We have shown that by formalizing people's displacements as flows along the edges of a graph of spatio-temporal locations and appearance groups, we can reduce this difficult estimation problem to a standard Linear Programming one.

As a result, our algorithm can identify and re-identify people reliably enough to preserve identity over very long sequences, while properly handling entrances and exits. This only requires using simple appearance-cues that can be computed easily and fast. Furthermore, by grouping spatio-temporal locations intro tracklets, we can substantially reduce the size of the Linear Program. This allows real-time processing on an ordinary computer and opens the door for practical applications, such as producing statistics of team-sport players' performance during matches.

In future work, we will focus on using these statistics for behavioral analysis and automated understanding of tactics.

References

1. Abramson, Y., Steux, B., Ghorayeb, H.: YEF Real-time object detection. In: International Workshop on Automatic Learning and Real-Time (ALaRT) (2005)
2. Ahonen, T., Hadid, A., Pietikïinen, M.: Face description with local binary patterns: application to face recognition. IEEE Trans. Pattern Anal. Mach. Intell. **28**(12), 2037–2041 (2006)
3. Andriluka, M., Roth, S., Schiele, B.: People-tracking-by-detection and people-detection-by-tracking. In: Conference on Computer Vision and Pattern Recognition (2008)
4. Andriluka, M., Roth, S., Schiele, B.: Monocular 3D pose estimation and tracking by detection. In: Conference on Computer Vision and Pattern Recognition (2010)
5. Andriyenko, A., Schindler, K.: Globally optimal multi-target tracking on a hexagonal lattice. In: European Conference on Computer Vision (2010)
6. Andriyenko, A., Schindler, K., Roth, S.: Discrete-continuous optimization for multi-target tracking. In: Conference on Computer Vision and Pattern Recognition (2012)
7. APIDIS European Project FP7-ICT-216023: (2008–2010). www.apidis.org
8. Barinova, O., Lempitsky, V., Kohli, P.: On detection of multiple object instances using hough transforms. IEEE Trans. Pattern Anal. Mach. Intell. **34**(9), 1773–1784 (2012)
9. Bazaraa, M.S., Jarvis, J.J., Sherali, H.D.: Linear programming and network flows. Wiley, Hiedelberg (2010)
10. BenShitrit, H., Berclaz, J., Fleuret, F., Fua, P.: Tracking multiple people under global apperance constraints. In: International Conference on Computer Vision (2011)
11. BenShitrit, H., Berclaz, J., Fleuret, F., Fua, P.: Multi-commodity network flow for tracking multiple people. IEEE Transactions on Pattern Analysis and Machine Intelligence (2013). Submitted for publication. Available as technical report EPFL-ARTICLE-181551.
12. Berclaz, J., Fleuret, F., Türetken, E., Fua, P.: Multiple object tracking using K-shortest paths optimization. IEEE Trans. Pattern Anal. Mach. Intell. **33**, 1806–1819 (2011). http://cvlab.epfl.ch/software/ksp
13. Bernardin, K., Stiefelhagen, R.: Evaluating multiple object tracking performance: the clear mot metrics. EURASIP J. Image Video Process. (2008)
14. Bimbo, A.D., Lisanti, G., Masi, I., Pernici, F.: Person detection using temporal and geometric context with a Pan Tilt zoom camera. In: International Conference on Pattern Recognition (2010)
15. Chang, C.C., Lin, C.J.: LIBSVM: A library for support vector machines. ACM Trans. Intell. Syst. Technol. (2011). http://www.csie.ntu.edu.tw/~cjlin/libsvm
16. Dalal, N., Triggs, B.: Histograms of oriented gradients for human detection. In: Conference on Computer Vision and Pattern Recognition (2005)
17. Dantzig, G.B.: Linear programming and extensions. Princeton University Press, Princeton (1963)
18. D'Orazio, T., Leo, M., Mosca, N., Spagnolo, P., Mazzeo, P.L.: A semi-automatic system for ground truth generation of soccer video sequences. In: International Conference on Advanced Video and Signal Based Surveillance (2009)
19. Ellis, A., Shahrokni, A., Ferryman, J.: Pets 2009 and winter pets 2009 results, a combined evaluation. In: IEEE International Workshop on Performance Evaluation of Tracking and Surveillance (2009)
20. Ester, M., Kriegel, H., Sander, J., Xu, X.: A density-based algorithm for discovering clusters in large spatial databases with noise. In: Knowledge Discovery and Data Mining (1996)
21. Felzenszwalb, P., Girshick, R., McAllester, D., Ramanan, D.: Object detection with discriminatively trained part based models. IEEE Trans. Pattern Anal. Mach. Intell. **32**(9), 1627–1645 (2010)
22. Fleuret, F., Berclaz, J., Lengagne, R., Fua, P.: Multi-camera people tracking with a probabilistic occupancy map. IEEE Trans. Pattern Anal. Mach. Intell. **30**(2), 267–282 (2008). http://cvlab.epfl.ch/software/pom
23. Jiang, H., Fels, S., Little, J.: A linear programming approach for multiple object tracking. In: Conference on Computer Vision and Pattern Recognition (2007)

24. Karmarkar, N.: A new polynomial time algorithm for linear programming. Combinatorica **4**(4), 373–395 (1984)
25. Misu, T., Matsui, A., Clippingdale, S., Fujii, M., Yagi, N.: Probabilistic integration of tracking and recognition of soccer players. In: Advances in Multimedia Modeling (2009)
26. Perera, A., Srinivas, C., Hoogs, A., Brooksby, G., Wensheng, H.: Multi-object tracking through simultaneous long occlusions and split-merge conditions. In: Conference on Computer Vision and Pattern Recognition (2006)
27. Pirsiavash, H., Ramanan, D.: Steerable part models. In: Conference on Computer Vision and Pattern Recognition (2012)
28. Pirsiavash, H., Ramanan, D., Fowlkes, C.: Globally-optimal greedy algorithms for tracking a variable number of objects. In: Conference on Computer Vision and Pattern Recognition (2011). http://www.ics.uci.edu/%7edramanan/
29. Storms, P., Spieksma, F.: An LP-based algorithm for the data association problem in multitarget tracking. Computers and Operations Research **30**(7), 1067–1085 (2003)
30. Suurballe, J.W.: Disjoint paths in a network. Networks **4**, 125–145 (1974)
31. Tsai, R.: A versatile cameras calibration technique for high accuracy 3D machine vision metrology using off-the-shelf TV cameras and lenses. Int. J. Robot. Autom. **3**(4), 323–344 (1987)
32. Viola, P., Jones, M.: Rapid object detection using a boosted cascade of simple features. In: Conference on Computer Vision and Pattern Recognition (2001)
33. Wu, T., Lin, C.: Weng, R. C.: Probability estimates for multi-class classification by pairwise coupling. J. Mach. Learn. Res. **5**, 975–1005 (2004)
34. Yuan, L., Bo, W., Nevatia, R.: Human detection by searching in 3D space using camera and scene knowledge. In: International Conference on Pattern Recognition (2008)
35. Zervos, M., BenShitrit, H., Fleuret, F., Fua, P.: Facial descriptors for identity-preserving multiple people tracking. Tech. Rep. EPFL-REPORT-187534, EPFL (2013)
36. Zhang, L., Li, Y., Nevatia, R.: Global data association for multi-object tracking using network flows. In: Conference on Computer Vision and Pattern Recognition (2008)

Part III
Evaluation and Application

Chapter 16
Benchmarking for Person Re-identification

Roberto Vezzani and Rita Cucchiara

Abstract The evaluation of computer vision and pattern recognition systems is usually a burdensome and time-consuming activity. In this chapter all the benchmarks publicly available for re-identification will be reviewed and compared, starting from the ancestors VIPeR and Caviar to the most recent datasets for 3D modeling such as SARC3d (with calibrated cameras) and RGBD-ID (with range sensors). Specific requirements and constraints are highlighted and reported for each of the described collections. In addition, details on the metrics that are mostly used to test and evaluate the re-identification systems are provided.

16.1 Introduction

Evaluation is a fundamental problem in research. We should capitalize on the lessons learned by the decades of studies in computer architecture performance evaluation, where different benchmarks are designed, such as *benchmark suites* of real programs, *kernel benchmarks* for distinct feature testing, and *synthetic benchmarks*. Similarly, in computer vision and multimedia, benchmark datasets are defined to test the efficacy and efficiency of code and algorithms. The purposes are manifold.

For assessed research and deeply explored problems, there is the need to compare new technical solutions, vendor promises, requirements, and limitations in real working conditions; typical examples are in biometrics where, although research is in continuous evolution, the market is interested in giving validation and standardization: see, as an example, the long story in the evaluation of face recognition

---

R. Vezzani (✉) · R. Cucchiara
Dipartimento di Ingegneria "Enzo Ferrari", Softech-ICT, University of Modena and Reggio Emilia, Modena, Italy
e-mail: roberto.vezzani@unimore.it

R. Cucchiara
e-mail: rita.cucchiara@unimore.it

S. Gong et al. (eds.), *Person Re-Identification*,
Advances in Computer Vision and Pattern Recognition,
DOI: 10.1007/978-1-4471-6296-4_16, © Springer-Verlag London 2014

techniques, which started with the FERET [1] contests more than 10 years ago. In some cases, when data is not easily available, some synthetic datasets have also been proposed and largely adopted (e.g. FVC2000 [2]).

For emerging activities and open problems, the need instead is to fix some common limits to the discussion and to have an acceptable starting base to compare solutions.

Often kernel benchmark datasets are defined to stress specific algorithms such as datasets for shadow detection, pedestrian detection, or other common tasks in surveillance. Among them, few datasets have been proposed to test re-identification in surveillance and forensics systems, specially for 3D/multiview approaches.

In this chapter, a review of the public datasets available for evaluating re-identification algorithms is provided in Sect. 16.2. The image resolution, the number of images for each person, the availability of entire video sequences, camera calibration parameters, and the recording conditions are some of the main features which characterize each database. In particular, new datasets specially conceived for 3D/multiview approaches are reported. In addition, metrics and performance measures used in re-identification are described in Sect. 16.3.

16.2 Datasets for Person Re-identification

The main benchmarks for re-identification are summarized in Tables 16.1 and 16.2. Other datasets proposed by a single author and not available to the community are not described here, while for additional references to generic surveillance datasets please refer to [16] or the Cantata Project repository [17].

16.2.1 VIPeR

Currently, one of the most popular and challenging datasets to test people re-identification for image retrieval is VIPeR [3], which contains 632 pedestrian image pairs taken from arbitrary viewpoints under varying illumination conditions (see Fig. 16.1). The dataset was collected in an academic setting over the course of several months. Each image is scaled to 128×48 pixels. Due to its complexity and the low resolution of the images, few researchers only have published their quantitative results on VIPeR; actually, some matches are hard to identify even by a human, such as the third couple in Fig. 16.1. A label describing the person orientation is also available. It cannot be fully employed for evaluating methods exploiting multiple shots, video frames, or 3D models, since only one pair of bounding boxes of the same person is collected. The performance of several approaches on this reference dataset are summarized in Table 16.3.

Table 16.1 Datasets available for people re-identification—PART I

Name and Ref	Image/Video	People	Additional info
VIPeR [3]	Still images	632	Scenario: Outdoor Place: Outdoor surveillance People size: 128 × 48 http://vision.soe.ucsc.edu
i-LIDS [4]	Video [fps = 25] 5 cameras PAL	1,000	Scenario: Outdoor/Indoor Place: Collection from different scenarios People size: 21 × 53 to 176 × 326 http://www.ilids.co.uk
i-LIDS-MA [4]	Still images PAL	40	Scenario: Indoor Place: Airport People size: 21 × 53 to 176 × 326 http://www.ilids.co.uk
i-LIDS-AA [4]	Still images PAL	119	Scenario: Indoor Place: Airport People size: 21 × 53 to 176 × 326 http://www.ilids.co.uk
CAVIAR4REID [5]	Still images 384 × 288	72	Scenario: Indoor Place: Shopping center People size: 17 × 39 to 72 × 144 http://www.lorisbazzani.info
ETHZ [6]	Video [fps = 15] 1 cameras 640 × 480	146	Scenario: Outdoor Place: Moving cameras on city street People size: 13 × 30 to 158 × 432 http://homepages.dcc.ufmg.br/~william/
SARC3D [7]	Still images 704 × 576	50	Scenario: Outdoor Place: University campus People size: 54 × 187 to 149 × 306 http://www.openvisor.org

16.2.2 i-LIDS

The i-LIDS Multiple-Camera Tracking Scenario (MCTS) [4] was captured indoor at a busy airport arrival hall. It contains 119 people with a total of 476 shots captured by multiple nonoverlapping cameras with an average of four images for each person. Many of these images undergo large illumination changes and are subject to occlusions (see Fig. 16.2). For instance, the i-LIDS dataset has been exploited by [18–20] for the performance evaluation of their proposals. Most of the people are carrying bags or suitcases. These accessories and carried objects can be profitably used to match their owners, but they introduce a lot of occlusions which usually act against the matching. In addition, images have been taken with different qualities (in terms

Table 16.2 Datasets available for people re-identification—PART II

Name and Ref	Image/Video	People	Additional info
3DPeS dataset [8]	Video [fps = 15] 8 cameras 704 × 576	200	Scenario: Outdoor Place: University campus People Size: 31 × 100 to 176 × 267 http://www.openvisor.org
TRECVid 2008 [9]	Video [fps = 25] 5 cameras PAL	300	Scenario: Indoor Place: Gatwick International Airport—London People Size: 21 × 53 to 176 × 326 http://www-nlpir.nist.gov/projects/tv2008/
PETS2009 [10]	Video [fps = 7] 8 cameras 768 × 576	40	Scenario: Outdoor Place: Outdoor surveillance People Size: 26 × 67 to 57 × 112 http://www.cvg.rdg.ac.uk/PETS2009/
PRID2011 [11]	Still images 2 cameras	200	Scenario: Outdoor Place: Outdoor surveillance People Size: 64 × 128 http://lrs.icg.tugraz.at/datasets/prid/
CUHK Dataset [12]	Still images 2 cameras	971	Scenario: Outdoor Place: pedestrian walkway People Size: 60 × 100 http://mmlab.ie.cuhk.edu.hk/datasets.html
GRID Dataset [13]	Still images 8 cameras	250	Scenario: Indoor Place: Underground surveillance People Size: 100 × 200 http://www.eecs.qmul.ac.uk/~ccloy
SAIVT-SOFTBIO [14]	Video [fps = 25] 8 cameras 704 × 576	150	Scenario: Indoor Place: Building surveillance People Size: 100 × 200 http://wiki.qut.edu.au/display/saivt/SAIVT-SoftBio+Database
RGBD-ID [15]	RGBD Microsoft Kinect	79	Scenario: Indoor Place: n/d People Size: 200 × 400 http://www.iit.it/en/datasets/rgbdid.html

of resolution, zoom level, noise), making very challenging the re-identification over this dataset.

16.2.3 CAVIAR4REID

It is a small dataset specifically created for evaluating person re-identification algorithms by [5]. It derives from the original CAVIAR dataset, which was initially

Fig. 16.1 Some examples from the VIPeR dataset [3]

Table 16.3 Quantitative comparison of some methods on the VIPeR dataset

Method	Rank-1	Rank-5	Rank-10	Rank-20
RGB histogram	0.04	0.11	0.20	0.27
ELF [28]	0.08	0.24	0.36	0.52
Shape and color covariance matrix [35]	0.11	0.32	0.48	0.70
Color-SIFT [35]	0.05	0.18	0.32	0.52
SDALF [36]	0.20	0.39	0.49	0.65
Ensemble-RankSVM [19]	0.16	0.38	0.53	0.69
PRDC [20]	0.15	0.38	0.53	0.70
MCC [20]	0.15	0.41	0.57	0.73
PCCA [29]	0.19	0.49	0.65	0.80
RPLM [37]	0.27	–	0.69	0.83
IML [38]				
No metric learning	0.07	0.11	0.14	0.21
using human interaction—1 iteration	0.42	0.42	0.43	0.50
using human interaction—5 iterations	0.74	0.74	0.74	0.74
using human interaction—10 iterations	0.81	0.81	0.81	0.81

created to evaluate people tracking and detection algorithms. A total of 72 pedestrians (50 of them with two camera views and the remaining 22 with one camera only) from the shopping center scenario are contained in the dataset. The ground truth has been used to extract the bounding box of each pedestrian. For each pedestrian, a set of images for each camera view (where available) is provided in order to maximize the variance with respect to resolution changes, light conditions, occlusions, and pose changes so as to make challenging the re-identification task.

16.2.4 ETHZ

The ETHZ dataset [6] for appearance-based modeling was generated from the original ETHZ video dataset [21]. The original ETHZ dataset is used for human detection and it is composed of four video sequences. Samples of testing sequence frames

Fig. 16.2 Samples from the i-LIDS dataset [4]

Fig. 16.3 Shot examples from ETZH dataset [6]

are shown in Fig. 16.3. The ETHZ dataset presents the additional challenge of being captured from moving cameras. This camera setup provides a range of variations in people's appearances, with strong changes in pose and illumination.

16.2.5 SARC3D

This dataset has been introduced in order to effectively test multiple shots methods. The dataset contains shots from 50 people, consisting of short video clips captured with a calibrated camera. To simplify the model to image alignment, four frames for each clip corresponding to predefined positions and postures of the people were

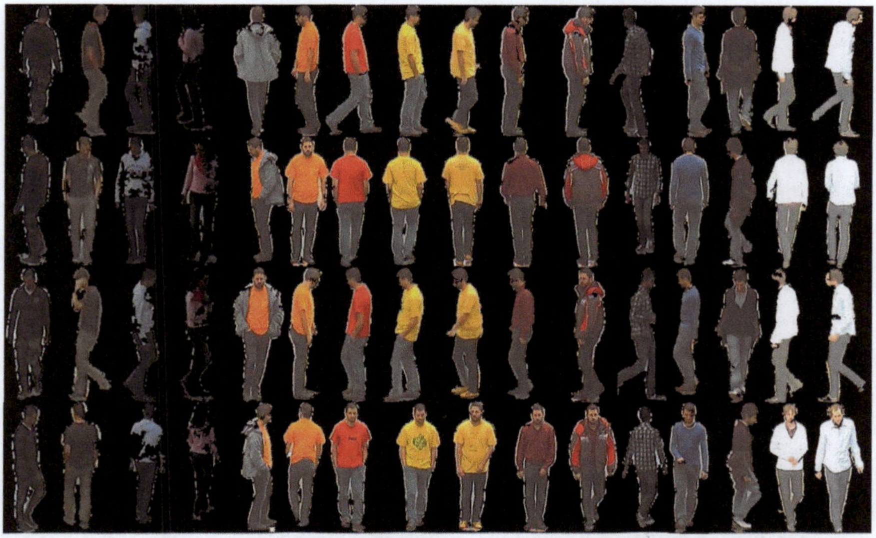

Fig. 16.4 Sample silhouettes from SARC3D re-identification dataset [7]

manually selected. Thus, the annotated dataset is composed by four views for each person, 200 snapshots in total. In addition, a reference silhouette is provided for each frame (some examples are shown in Fig. 16.4).

16.2.6 3DPeS

3DPeS dataset [8] provides a large amount of data to test all the usual steps in video surveillance (segmentation, tracking, and so on).

The dataset is captured from a real surveillance setup, composed by eight different surveillance cameras (see Fig. 16.5), monitoring a section of the campus of the University of Modena and Reggio Emilia (UNIMORE) campus. Data were collected over the course of several days. The illumination between cameras is almost constant, but people were recorded multiple times during the course of the day, in clear light and in shadowy areas, resulting in strong variations of light conditions in some cases. All cameras have been partially calibrated (position, orientation, pixel aspect ratio, and focal length are provided for each one of them). The quality of the images is mostly constant, uncompressed images with a resolution of 704×576 pixels. Depending on the camera position and orientation, people were recorded at different zoom levels. Multiple sequences of 200 individuals are available, together with reference background images, the person bounding box at key frames, and the reference silhouettes for more than 150 people. Annotation comprises: camera parameters, person IDs, and correspondences across the dataset, bounding box of the

Fig. 16.5 Sample frames from 3DPeS [8]

Fig. 16.6 Coarse 3D reconstruction of the surveilled area for the 3DPES dataset

target person in the first frame of the sequence, preselected snapshots of people appearances, silhouette, orientation and bounding box for each image, and a coarse 3D reconstruction of the surveilled area (Fig. 16.6). Each video sequence contains only the target person or a very limited number of people.

16.2.7 TRECVid 2008

The TRECVid competition has released in 2008 a dataset for Surveillance applications inside an airport. About 100 h of video surveillance data have been collected by the UK Home Office at the London Gatwick International Airport (10 days * 2 h/day * 5 cameras). About 44 individuals can be detected and matched through the 5 cameras.

Fig. 16.7 Sample images from the Person re-ID 2011 dataset [11]

16.2.8 Person re-ID 2011

The Person re-ID 2011 dataset [11] consists of images extracted from multiple person trajectories recorded from two different, static surveillance cameras. Images from these cameras contain a viewpoint change and a stark difference in illumination, background, and camera characteristics. Since images are extracted from trajectories, several different poses per person are available in each camera view. The dataset contains 385 person trajectories from one view and 749 from the other one, with 200 people appearing in both views.

Two versions of the dataset are provided, one representing the single-shot scenario and one representing the multi-shot scenario. The multi-shot version contains multiple images per person (at least five per camera view). The exact number depends on a person's walking path and speed as well as occlusions. The single-shot version contains just one (randomly selected) image per person trajectory, i.e., one image from view A and one image from view B. Sample images from the dataset are depicted in Fig. 16.7.

16.2.9 CUHK Person Re-identification Dataset

The CUHK dataset [22] contains 971 identities from two disjoint camera views. Each identity has two samples per camera view. This dataset is only for academic purpose and available by request.

16.2.10 QMUL Underground Re-identification (GRID) Dataset

The QMUL under Ground Re-IDentification (GRID) dataset [13] contains 250 pedestrian image pairs. Each pair contains two images of the same individual seen from different camera views. All images are captured from 8 disjoint camera views installed in a busy underground station. Figure 16.8 shows a snapshot of each of the camera views of the station and sample images in the dataset. This setup is

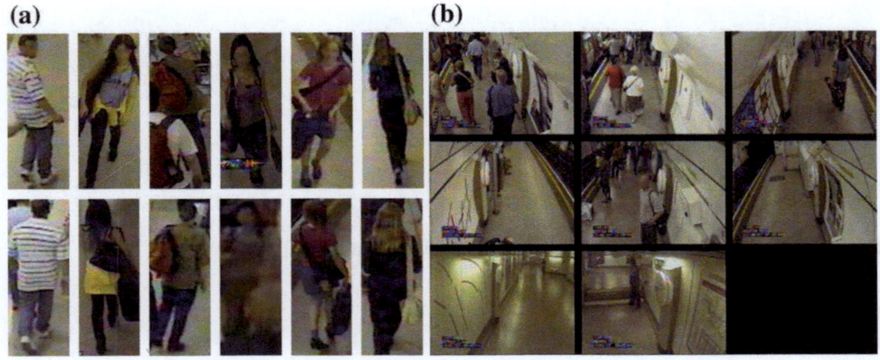

Fig. 16.8 Samples of the original videos and the cropped imaged from the QMUL dataset [13]

challenging due to variations of pose, colors, lighting changes; as well as poor image quality caused by low spatial resolution. Noisy artifacts due to video transmission and compression make the dataset more challenging.

16.2.11 SAIVT-SOFTBIO Database

SAIVT-SOFTBIO database [14] is a multi-camera surveillance database designed for the task of person re-identification. This database consists of 150 unscripted sequences of subjects traveling in a building environment though up to eight camera views, appearing from various angles and in varying illumination conditions. A flexible XML-based evaluation protocol is provided to allow a highly configurable evaluation setup, enabling a variety of scenarios relating to pose and lighting conditions to be evaluated.

16.2.12 RGBD-ID

RGBD-ID [15] is a dataset for person re-identification using depth information. The main motivation claimed by the authors is that standard techniques fail when the individuals change their clothing. Depth information can be used for long-term video surveillance since contains soft-biometric features, which do not depend on the appearance. For example, calibrated depth maps allow to recover the real height of a person, the length of his arms and legs, the ratios among body part measures, and so on. These information can be exploited in addition to the common appearance descriptors computed on the RGB images. RGBD-ID is the first color and depth dataset for re-identification which is publicly available.

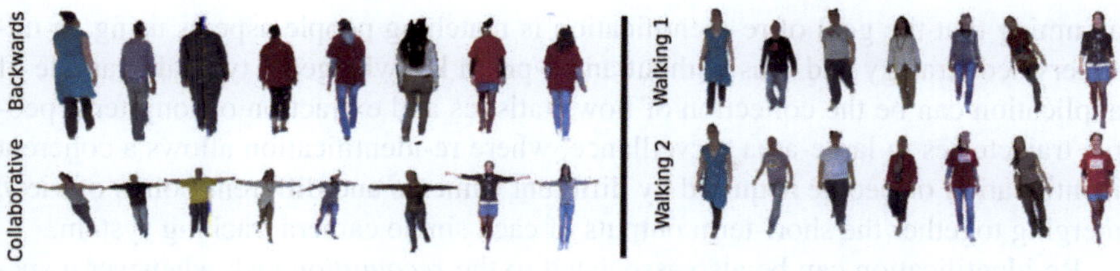

Fig. 16.9 Sample images from the RGBD-ID dataset [15]

The dataset is composed by four different groups of data collected using the Microsoft Kinect. The first group of data has been obtained by recording 79 people with a frontal view, walking slowly, avoiding occlusions and with stretched arms ("Collaborative"). This happened in an indoor scenario, where the people were at least 2 m away from the camera. The second ("Walking1") and third ("Walking2") groups of data are composed by frontal recordings of the same 79 people walking normally while entering the lab where they normally work. The fourth group ("Backwards") is a back view recording of the people walking away from the lab. Since all the acquisitions have been performed in different days, there is no guarantee that visual aspects like clothing or accessories will be kept constant. Moreover, some people dress the same t-shirt in "Walking2." This is useful to highlight the power of RGBD re-identification compared with standard appearance-based methods. Five synchronized information are available for each person: (a) a set of 5 RGB images; (b) the foreground masks; (c) the skeletons; (d) the 3d mesh; (e) the estimated floor. A MATLAB script is also provided to read the data. Sample images from the dataset are depicted in Fig. 16.9.

16.3 Evaluation Metrics for Person Re-identification

In order to correctly understand the role of re-identification, let us report the definition of the terms "detect, classify, identify, recognize, verify" as provided by the European Commission in EUROSUR-2011 [23] for surveillance:

- *Detect*: to establish the presence of an object and its geographic location, but not necessarily its nature.
- *Classify*: to establish the type (class) of object (car, van, trailer, cargo ship, tanker, fishing boat).
- *Identify*: to establish the unique identity of the object (name, number), as a rule without prior knowledge.
- *Recognize*: to establish that a detected object is a specific predefined unique object.
- *Verify*: Given prior knowledge on the object, can its presence/position be confirmed.

In agreement with the EUROSUR definition, re-identification falls in the middle between identification and recognition. It can be embraced in the *identification* task, assuming that the goal of re-identification is matching people aspects using an unsupervised strategy and thus without any a-priori knowledge. A typical example of application can be the collection of flow statistics and extraction of long-term people trajectories in large-area surveillance, where re-identification allows a coherent identification of people acquired by different cameras and different points of view, merging together the short-term outputs of each single camera tracking system.

Re-identification can be also associated to the *recognition* task whenever a specific query with a target person is provided and all the corresponding instances are searched in large datasets; the search for a suspect within the stored videos of a crime neighborhood in multimedia forensics or a visual query in an employer database are typical examples of application as a soft-biometric tool, suitable in case of low-resolution images with noncollaborative targets and when biometric recognition is not feasible.

Thus, people re-identification by visual aspect is emerging as a very interesting field and future solutions could be fruitfully exploited as a soft-biometric technology, a tool for long-term surveillance, or a support for searching in security-related databases.

In addition to the selection of the testing data, performance evaluation requires suitable metrics, which depend on the specific goal of the application. According to the above definitions, different metrics are available, also related to the specific implementation of re-identification as identification or recognition.

16.3.1 Re-identification as Identification

Since the goal is to find all the correspondences among the set of people instances without any *a-priori* knowledge, the problem is very similar to data clustering. Each expected cluster is related to one person only. Different from content-based retrieval problems, where there are relatively few clusters and very large amount of data for each cluster, here the number of desired clusters is very high with respect to the number of elements in each one. However, the same metrics adopted for clustering evaluation could be introduced [24].

Being C the set of clusters (different labels or identities associated to the people) to be evaluated, L the set of categories (reference people identities) and N the number of clustered items, $Purity$ is computed by taking the weighted average of maximal precision values:

$$Purity = \sum_i \frac{\|C_i\|}{N} \cdot \max_j Precision(C_i, L_j) \tag{16.1}$$

where

$$Precision(C_i, L_j) = \frac{\|C_i \cap L_j\|}{\|C_i\|} \tag{16.2}$$

Purity penalizes the noise in a cluster, i.e., instances of a person wrongly assigned to another person, but it does not reward grouping different items from the same category together.

Inverse Purity, instead, focuses on the cluster with maximum recall—the fraction of relevant instances that are matched—for each category, i.e., aims at verifying if all the instances of the same person are matched together and correctly re-identified.

$$Inverse\ Purity = \sum_i \frac{\|L_i\|}{N} \cdot \max_j Precision(L_i, C_j) \qquad (16.3)$$

The performance evaluation of re-identification algorithms is usually simplified taking into account each couple of items at a time. In fact, the system should state if the two items belong or not to the same person (similarly to the verification problem). In this case, *Precision* and *Recall* metrics applied to the number of hit or miss matches have been adopted [25].

Tasks of re-identification as long-term tracking also fall in this category, especially in surveillance with network of overlapped or disjoint cameras. As a tracking system, the re-identification algorithm should generate tracks as long as possible, avoiding errors such as identity switch, erroneous split, and merge of tracks, over and under segmentation of traces. For detection and tracking purposes, the ETISEO project [26] proposed some metrics which could be adopted in re-identification too. ETISEO was a project devoted to performance evaluation for video surveillance systems, studying the dependency between algorithms and the video characteristics. Sophisticated scores have been specifically proposed such as the *Tracking Time* and the *Object ID Persistence*. The first one corresponds to the percentage of time during which a reference data are detected and tracked. This metric gives us a global overview of the performance of the multi-camera tracking algorithm. Yet, it suffers from the issue that the evaluation results depend not only on the re-identification algorithms but also on the detection and single camera tracking. The second metric qualifies the re-identification precision, evaluating how many identities have been assigned to the same real person. Finally, let us cite the work by Leung et al. [27] about performance evaluation of reacquisition methods specifically conceived for public transport surveillance, which takes into account the a-priori knowledge of the scene and the normal people behaviors to estimate how the re-identification system can reduce the entropy of the surveillance framework.

16.3.2 Re-identification as Recognition

In this category the re-identification tasks aim at providing a set of ranked items given a query target, with the main hypothesis that one and only one item of the gallery could correspond to the query. This is typical of forensics problems in support of investigations, where large datasets of image and video footage must be evaluated. The overall re-identification process could be considered as a ranking problem [3] and, in this case, the *Cumulative Matching Characteristic* (CMC) curve is the proper performance evaluation metric, showing how performance improves as the allowed number of returned images increases. The CMC curve represents the expectation of finding the correct match in the top n matches. Computer Vision research community usually adopts this metric to evaluate their systems [3, 20, 28, 29].

While this performance metric is designed to evaluate recognition problems, by making some simple assumptions about the distribution of appearances in a camera network, a CMC curve can be converted into a synthetic disambiguation (SDR) or reacquisition rate (SRR) for multiple objects or multiple-camera tracking respectively [3]. Assume that a set of M pedestrians that enter a camera network are i.i.d. samples from some large testing dataset of size N. If the CMC curve for the matching function is given, the probability that any of the M best matches is correct can be computed as follows:

$$SDR(M) = SRR(M) = CMC(N/M) \tag{16.4}$$

where $CMC(k)$ is the rank k recognition rate.

Since the adopted definition of re-identification as recognition given by the Frontex committee [23] recalls the definition of identification for biometrics, the evaluation metrics defined in biometrics could be taken into account. Two biometric elements are associated with the same source if their similarity score exceeds a given threshold. Accordingly, the measures of false-acceptance rate (FAR), false-rejection rate (FRR) [30], and the decision-error trade-off (DET) curve by varying the threshold can be exploited for evaluating if two snapshots are associated to the same person or not.

16.3.3 Re-identification in Forensics

The precision/recall, FAR, FRR, and DET metrics are now standard and clearly accepted in academic world, in industrial settings for content-base image retrieval and biometrics, but they are still far to be acceptable in legal court. This is a problem very common to occidental legal courts, where image analysis is widely adopted during investigation while the final legal judgment is still devoted to the traditional use of an expert's opinion as evidence only.

Actually, a strong effort is being made to improve this practice by adding an objective, quantitative measure of evidential value [31]. To this aim the *likelihood ratio* has been introduced and suggested for different forensics problems such as speaker identification [32], DNA analysis [33], face recognition [34]. The likelihood ratio is the ratio of two probabilities of the same event under different hypotheses. Thus for events A and B, the probability of A given that B is true, divided by the probability of event A given that B is false gives a likelihood ratio.

In forensic biology, for instance, likelihood ratios are usually constructed with the numerator being the probability of the evidence if the identified person is supposed to be the source of the evidence itself, and the denominator being the probability of the evidence if an unidentified person is supposed to be the source. A similar discussions have been introduced in a survey for face recognition in forensics [34].

16.4 Conclusions

Re-identification is a very challenging task in surveillance. The introduction of several new datasets specially conceived for this problem proves the strong interest of the scientific community. However, well-assessed methodologies and reference procedures are still lacking in this field. Specific contests and competitions may probably help a global alignment of performance evaluation. Re-identification can really benefit from wide-area activities similarly to other research fields such as ImageCLEF for image retrieval [39]or TRECVID [40] for video annotation.

References

1. Phillips, P., Moon, H., Rauss, P., Rizvi, S.: The feret evaluation methodology for face-recognition algorithms. In: Proceedings of IEEE International Conference on Computer Vision and, Pattern Recognition, pp. 137–143 (1997)
2. Maio, D., Maltoni, D., Cappelli, R., Wayman, J., Jain, A.: Fvc 2000: fingerprint verification competition. IEEE Trans. Pattern Anal. Mach. Intell. **24**(3), 402–412 (2002)
3. Gray, D., Brennan, S., Tao, H.: Evaluating appearance models for recognition, reacquisition, and tracking. In: Proceedings of 10th IEEE International Workshop on Performance Evaluation of Tracking and Surveillance (PETS) (2007)
4. Nilski, A.: Evaluating multiple camera tracking systems—the i-lids 5th scenario. In: Security Technology, 2008. ICCST 2008. 42nd Annual IEEE International Carnahan Conference on, pp. 277–279 (2008)
5. Cheng, D.S., Cristani, M., Stoppa, M., Bazzani, L., Murino, V.: Custom pictorial structures for re-identification. In: British Machine Vision Conference (BMVC 2011), pp. 68.1–68.11. BMVA Press (2011)
6. Schwartz, W., Davis, L.: Learning discriminative appearance-based models using partial least squares. In: Proceedings of the XXII Brazilian Symposium on Computer Graphics and Image Processing (2009)
7. Baltieri, D., Vezzani, R., Cucchiara, R.: Sarc3d: a new 3d body model for people tracking and re-identification. In: Proceedings of IEEE International Conference on Image Analysis and Processing, pp. 197–206. Ravenna (2011)
8. Baltieri, D., Vezzani, R., Cucchiara, R.: 3dpes: 3d people dataset for surveillance and forensics. In: Proceedings of the 1st International ACM Workshop on Multimedia Access to 3D Human Objects. Scottsdale (2011)
9. Smeaton, A.F., Over, P., Kraaij, W.: Evaluation campaigns and trecvid. In: MIR '06: Proceedings of the 8th ACM International Workshop on Multimedia Information Retrieval, pp. 321–330. New York (2006)
10. Pets: performance evaluation of tracking and surveillance (2000–2009). http://www.cvg.cs.rdg.ac.uk/slides/pets.html
11. Hirzer, M., Beleznai, C., Roth, P.M., Bischof, H.: Person re-identification by descriptive and discriminative classification. In: Proceedings Scandinavian Conference on Image Analysis (2011). www.springerlink.com

12. Li, W., Wu, Y., Mukunoki, M., Minoh, M.: Common-near-neighbor analysis for person re-identification. In: Internationl Conference on Image Processing, pp. 1621–1624 (2012)
13. Loy, C.C., Xiang, T., Gong, S.: Multi-camera activity correlation analysis. In: IEEE Conference on Computer Vision and, Pattern Recognition, pp. 1988–1995 (2009)
14. Bialkowski, A., Denman, S., Sridharan, S., Fookes, C., Lucey, P.: A database for person re-identification in multi-camera surveillance networks. In: DICTA, pp. 1–8. IEEE (2012)
15. Barbosa, I.B., Cristani, M., Bue, A.D., Bazzani, L., Murino, V.: Re-identification with rgb-d sensors. In: Fusiello, A., Murino, V., Cucchiara, R. (eds.) First International ECCV Workshop on Re-Identification, Lecture Notes in Computer Science, vol. 7583, pp. 433–442. Springer (2012)
16. Vezzani, R., Cucchiara, R.: Video surveillance online repository (visor): an integrated framework. Multimedia Tools Appl. **50**(2), 359–380 (2010)
17. project, C.: Video and image datasets index. online (2008). http://www.multitel.be/cantata/
18. Bak, S., Corvee, E., Bremond, F., Thonnat, M.: Multiple-shot human re-identification by mean riemannian covariance grid. In: Proceedings of IEEE Conference on Advanced Video and Signal-Based Surveillance, pp. 179–184 (2011)
19. Brosser, B., Zheng, W., Gong, S., Xiang, T.: Person re-identification by support vector ranking. In: Proceedings of British Machine Vision Conference, pp. 21.1–2111 (2010)
20. Zheng, W.S., Gong, S., Xiang, T.: Person re-identification by probabilistic relative distance comparison. In: Proceedings of IEEE International Conference on Computer Vision and, Pattern Recognition, pp. 649–656 (2011)
21. Ess, A., Leibe, B., Gool, L.V.: Depth and appearance for mobile scene analysis. In: Proceedings IEEE International Conference on Computer Vision (2007)
22. Li, W., Zhao, R., Wang, X.: Human reidentification with transferred metric learning. In: Proceedings of Asian Conference on Computer Vision 2012 (2012)
23. Frontex: Application of surveillance tools to border surveillance - concept of operations. online (2011).
24. AmigÃş, E., Gonzalo, J., Artiles, J., Verdejo, F.: A comparison of extrinsic clustering evaluation metrics based on formal constraints. Inf. Retrieval **12**, 461–486 (2009)
25. Hamdoun, O., Moutarde, F., Stanciulescu, B., Steux, B.: Person re-identification in multi-camera system by signature based on interest point descriptors collected on short video sequences. In: Proceedings of International Conference on Distributed Smart Cameras, pp. 1–6. IEEE (2008)
26. Nghiem, A., Bremond, F., Thonnat, M., Valentin, V.: Etiseo, performance evaluation for video surveillance systems. In: Proceedings of IEEE Conference on Advanced Video and Signal-Based Surveillance, pp. 476–481 (2007)
27. Leung, V., Orwell, J., Velastin, S.A.: Performance evaluation of re-acquisition methods for public transport surveillance. In: Proceedings of International Conference on Control, Automation, Robotics and Vision, pp. 705–712. IEEE (2008)
28. Gray, D., Tao, H.: Viewpoint invariant pedestrian recognition with an ensemble of localized features. In: Proceedings of European Conference on Computer Vision, p. 262 (2008)
29. Mignon, A., Jurie, F.: Pcca: a new approach for distance learning from sparse pairwise constraints. In: Proceedings of IEEE International Conference on Computer Vision and, Pattern Recognition, pp. 2666–2672 (2012)
30. Jungling, K., Arens, M.: Local feature based person reidentification in infrared image sequences. In: Proceedings of IEEE Conference on Advanced Video and Signal-Based Surveillance, pp. 448–455 (2010)
31. Meuwly, D.: Forensic individualization from biometric data. Sci. Justice **46**(4), 205–213 (2006)
32. Gonzalez-rodriguez, J., Fierrez-aguilar, J., Ortega-Garcia, J.: Forensic identification reporting using automatic speaker recognition systems. In: Proceedings of the IEEE International Conference on Acoustics, Speech and Signal Processing, ICASSP, pp. 93–96 (2003)
33. Balding, D.: Weight-of-Evidence for Forensic DNA Profiles. Wiley, Chichester (2005)
34. Ali, T., Veldhuis, R., Spreeuwers, L.: Forensic face recognition: a survey (2010)

35. Metternich, M., Worring, M., Smeulders, A.: Color Based Tracing in Real-Life Surveillance Data. Trans. Data Hiding Multimedia Secur. **V 6010**, 18–33 (2010)
36. Bazzani, L., Cristani, M., Murino, V.: Symmetry-driven accumulation of local features for human characterization and re-identification. Comput. Vis. Image Underst. **117**(2), 130–144 (2013)
37. Hirzer, M., Roth, P.M., Köstinger, M., Bischof, H.: Relaxed pairwise learned metric for person re-identification. In: Fitzgibbon, A., Lazebnik, S., Perona, P., Sato, Y., Schmid, C. (eds.) ECCV 2012, Lecture Notes in Computer Science, vol. 7577, pp. 780–793. Springer, Berlin Heidelberg (2012)
38. Ali, S., Javed, O., Haering, N., Kanade, T.: Interactive retrieval of targets for wide area surveillance. In: Proceedings of the ACM International Conference on Multimedia, pp. 895–898. ACM, New York (2010)
39. Imageclef. http://www.imageclef.org/ (2013)
40. Over, P., Awad, G., Michel, M., Fiscus, J., Kraaij, W., Smeaton, A.F.: Trecvid 2011—an overview of the goals, tasks, data, evaluation mechanisms and metrics. In: Proceedings of TRECVID 2011. NIST, USA (2011)

Chapter 17
Person Re-identification: System Design and Evaluation Overview

Xiaogang Wang and Rui Zhao

Abstract Person re-identification has important applications in video surveillance. It is particularly challenging because observed pedestrians undergo significant variations across camera views, and there are a large number of pedestrians to be distinguished given small pedestrian images from surveillance videos. This chapter discusses different approaches of improving the key components of a person re-identification system, including feature design, feature learning, and metric learning, as well as their strength and weakness. It provides an overview of various person re-identification systems and their evaluation on benchmark datasets. Multiple benchmark datasets for person re-identification are summarized and discussed. The performance of some state-of-the-art person identification approaches on benchmark datasets is compared and analyzed. It also discusses a few future research directions on improving benchmark datasets, evaluation methodology, and system design.

17.1 Introduction

Person re-identification is to match pedestrian images observed in different camera views with visual features. The task is to match one or one set of query images with images of a large number of candidate persons in the gallery in order to recognize the identity of the query image (set). It has important applications in video surveillance including pedestrian search, multi-camera tracking, and behavior analysis. Under the settings of multi-camera object tracking, matching of visual features can be integrated with spatial and temporal reasoning [9, 29, 32]. This chapter focuses

X. Wang (✉) · R. Zhao
Department of Electronic Engineering, The Chinese University of Hong Kong, Shatin, Hong Kong
e-mail: xgwang@ee.cuhk.edu.hk

R. Zhao
e-mail: rzhao@ee.cuhk.edu.hk

S. Gong et al. (eds.), *Person Re-Identification*,
Advances in Computer Vision and Pattern Recognition,
DOI: 10.1007/978-1-4471-6296-4_17, © Springer-Verlag London 2014

Fig. 17.1 The same 12 pedestrians captured in two different camera views. Examples are from the VIPeR dataset [21]

on visual feature matching. A detailed survey on spatial and temporal reasoning in object tracking can be found in [70]. People working on the problem of person re-identification usually assume that observations of pedestrians are captured in relatively short periods, such that clothes and body shapes do not change much and can be used as cues to recognize identity. In video surveillance, the captured pedestrians are often small in size, facial components are indistinguishable in images, and face recognition techniques are not applicable. Therefore, person re-identification techniques become important. However, it is a very challenging task. Surveillance cameras may observe tens of thousands of pedestrians in a public area in one day and many of them look similar in appearance. Another big challenge comes from large variations of lightings, poses, viewpoints, blurring effects, image resolutions, camera settings, and background across camera views. Some examples are shown in Fig. 17.1. The appearance of same pedestrians observed in different camera views changes a lot.

This book chapter provides an overview of designing a person re-identification system, including feature design, feature learning, and metric learning. The strength and weakness of different person re-identification algorithms are analyzed. It also reviews the performance of state-of-the-art algorithms on benchmark datasets. Some future research directions are discussed.

17.2 System Design

17.2.1 System Diagram

The diagram of a person re-identification system is shown in Fig. 17.2. It starts with automatic pedestrian detection. Many existing works [47, 73] detect pedestrians

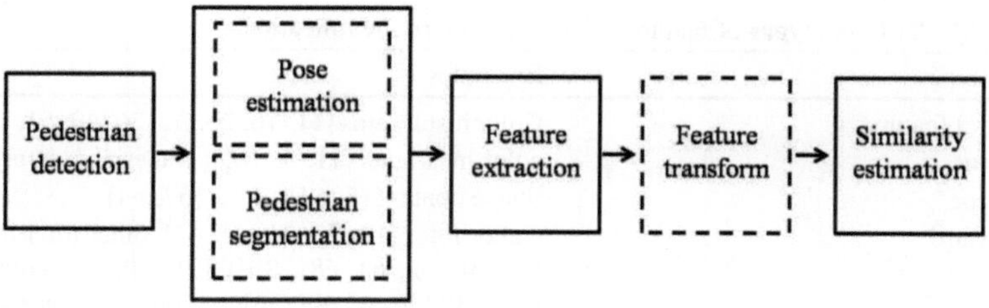

Fig. 17.2 Diagram of a person re-identification system. Dashed windows indicate steps which can be skipped in some person re-identification systems

from videos captured by static cameras with background subtraction. However, background subtraction is sensitive to lighting variations and scene clutters. It is also hard to separate pedestrians appearing in groups. In recent years, appearance-based pedestrian detectors [12, 19, 50, 63] learned from training samples become popular. There is a huge literature on this topic. The details will be skipped in this chapter. All the existing person re-identification chapters have ignored this step and assume perfect pedestrian detection by using manually cropped pedestrian images. However, perfect detection is impossible in real applications and misalignment can seriously reduce the person re-identification performance. Therefore, this factor needs to be carefully studied in the future work.

The performance of person re-identification is largely affected by variations of poses and viewpoints, which can be normalized with pose estimation [60, 69]. The change of background also has negative effect on estimating the similarity of two pedestrians. Background can be removed through pedestrian segmentation [8, 17, 55]. Although significant research works have been done on these two topics, they are still not mature enough to work robustly in most surveillance scenes. The errors of pose estimation and segmentation may lead to re-identification failures. Some person re-identification systems skip the two steps and directly extract features from detection results.

Same pedestrians may undergo significant photometric and geometric transforms across camera views. Such transforms can be estimated through a learning process. However, it is also possible to overcome such transforms by learning proper similarity metrics without the feature transform step.

Person re-identification approaches generally fall into two categories: unsupervised [4, 10, 16, 26, 39, 43–45, 64, 72] and supervised [22, 36, 37, 54, 75]. Unsupervised methods mainly focus on feature design and feature extraction. Since they do not require manually labeling training samples, they can be well generalized to new camera views without additional human labeling efforts. Supervised methods generally have better performance with the assistance of manually labeled training samples. Most existing works [4, 10, 16, 22, 24, 25, 35, 37, 39, 43, 44, 54, 64, 72, 75] choose training and test samples from the same camera views and it is uncertain about their generalization capability on new camera settings. Only very recently, people started to investigate the cases when training and test samples are from

Table 17.1 Different types of features used in person re-identification.

Feature type	Examples
Low-level features	Color histograms [11, 16, 26, 31, 36, 49, 51, 64, 75],
Color	color invariants [11, 59, 66], Gaussian mixtures [46]
Shape	Shape context [5, 64], HOG [57, 64]
Texture	Gabor filter [13, 22, 36, 43, 44], other filter banks [22, 54, 62, 64, 68], SIFT [31, 40, 72], color SIFT [1], LBP [26, 45, 48, 75], region covariance [43, 44, 61]
Global features	[11, 46, 49, 51, 57, 64]
Regional features	Shape and appearance context [64], custom pictorial structure [10], fitting articulated model [20]
Patch-based features	[4, 16, 22, 26, 31, 36, 37, 39, 43, 44, 54, 72, 75]
Semantic features	Exemplar-based representations [23, 58], attribute features [35, 39]

different camera views [36]. In surveillance applications, when the number of cameras is large, it is impractical to label training samples for every pair of camera views. Therefore, the generalization capability is important. The overview of designing each module of the person re-identification system is given below.

17.2.2 Low-Level Features

Feature design is the foundation of the person re-identification system. A summary of different types of features used in person re-identification is shown in Table 17.1. Effective low-level features usually have good generalization capability to new camera views because their design does not rely on training. Most low-level features can be integrated with the learning approaches developed in the later steps. Good features are expected to discriminate a large number of pedestrians in the gallery and to be robust to various inter- and intra-camera view variations, such as background, poses, lighting, viewpoints, and self-occlusions.

17.2.2.1 Color, Shape, and Texture Features

Like most object recognition tasks, the appearance of pedestrians from static images can be characterized from three aspects: color, shape, and texture. Color histograms of the whole images are widely used to characterize color distributions [11, 49, 51]. In order to be robust to lighting variations and the changes in photometric settings of cameras, various color spaces have been studied when computing color histograms [64]. Some components in the color spaces sensitive to photometric transformations are removed or normalized. Instead of uniformly quantizing the color spaces, Mittal and Davis [46] softly assigned pixels to color modes with a Gaussian mixture model, and estimated the correspondences of color modes across camera views. Other color invariants [11, 59, 66] can also be used as features for person re-identification.

Color distributions alone are not enough to distinguish a large number of pedestrians since the clothes of some pedestrians could be similar. Therefore, it needs to be combined with shape and texture features. Shape context [5] is widely used to characterize both global and local shape structures. Its computation is based on edge or contour detection. Histogram of Oriented Gradients (HOG) has been widely used for object detection [12], and is also effective for person re-identification [57, 64]. It characterizes local shapes by computing the histograms of gradient orientations within cells over a dense grid. In order to be robust to lighting variations, local photometric normalization is applied to histograms. Shape features are subject to the variations in viewpoints and poses.

Many texture filters and descriptors have been proposed in object recognition literature, such as Gabor filer [13, 22, 36, 43, 44] and other filter banks [22, 54, 62, 64, 68], SIFT [40, 45, 72], color SIFT [1], LBP [26, 45, 48, 75], and region covariance [43, 44, 61]. Many of them can also be used in person re-identification [24]. A typical approach is to apply these filters and descriptors to sparse interest points or on a dense grid, and then quantize their responses into visual words. The histograms of visual words can be used as features to characterize texture distributions. However, these features cannot encode spatial information. It is also possible to directly compare the responses on a fixed dense grid. But it is sensitive to misalignment, pose variation, and viewpoint variation. Therefore, correlograms [28, 64] and correlatons [56, 64] are proposed to capture the co-occurrence of visual words over spatial kernels. They balance the two extreme cases.

17.2.2.2 Global, Regional, and Patch-Based Features

Most of the visual features described above are global. They have some invariance to misalignment, pose variation, and the change in viewpoint. However, their discriminative power is not high because of losing spatial information. In order to increase the discriminative power, patch-based features are used [4, 16, 22, 26, 36, 37, 39, 43–45, 54, 72, 75]. A pedestrian image is evenly divided into multiple local patches. Visual features are computed as each patch. When computing the similarity of two images, visual features of two corresponding patches are compared. The biggest challenge of patch-based methods is to find correspondences of patches when tackling the misalignment problem. Zhao et al. [72] divided an image into horizontal stripes and find the dense correspondence of patches along each stripe with some spatial constraints.

Some patches are more distinctive and reliable when matching two persons. Some examples are shown in Fig. 17.3. In this dataset, it is easy for the human eyes to match pedestrian pairs because they have distinct patches. Person (a) carries a backpack with tilted blue stripes. Person (b) holds a red folder. Person (c) has a red bottle

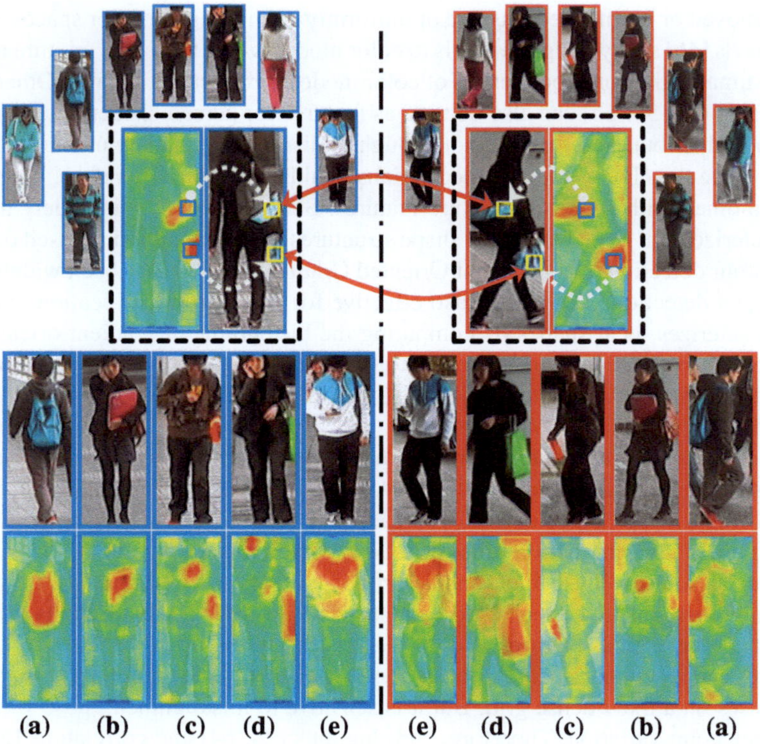

Fig. 17.3 Illustration of patch-based person re-identification with salience estimation. The dashed line in the middle divides the images observed in two different camera views. The salience maps of exemplar images are also shown

in hand. These features can well separate one person from others and they can be reliable detected across camera views. If a body part is salient in one camera view, it should also be salient in another camera view. However, most existing approaches only consider clothes and trousers as the most important regions for person re-identification. Such distinct features may be considered as outliers to be removed, since some of them do not belong to body parts. Also, these features may only take small regions in the body parts, and have little effect on computing global features. Zhao et al. [72] estimated the salience of patches through unsupervised learning and incorporate it into person matching. A patch with higher salience value gains more weight in the matching.

Pedestrians have fixed structures. If different body constitutes well detected with pose estimation and human parsing are available, region-based features can be developed and employed in person re-identification [10, 20]. Visual features are computed from each body part. Body alignment is naturally established. Cheng et al. [10] employed Custom Pictorial Structure to localize body parts, and matched their visual descriptors. Wang et al. [64] proposed shape and appearance context. The body parts

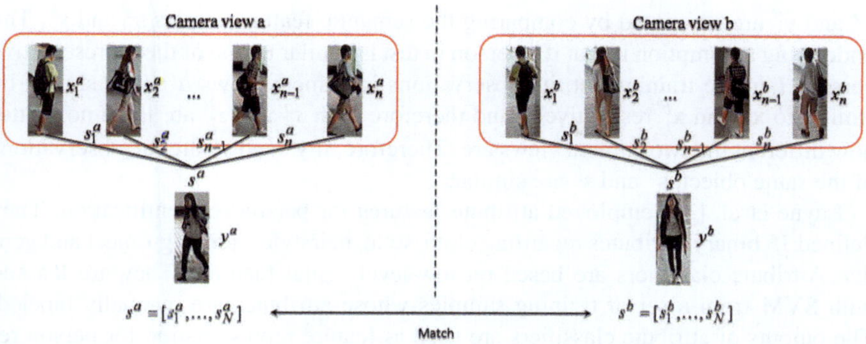

Fig. 17.4 Illustration of exemplar-based representation for person re-identification

are automatically obtained through clustering shape context. The shape and appearance context models the spatial distributions of appearance relative to body parts.

17.2.3 Semantic Features

In order to effectively reduce the cross-view variations, some high-level semantic features could be used for person re-identification besides the low-level visual features discussed above. The design of semantic features is inspired by the process of human beings recognizing person identities. For example, humans describe a person by saying "he or she looks similar to someone I know" or "he or she is tall and slim, has short hair, wears a white shirt, and carries a baggage." Such high-level descriptions are independent of camera views and have good robustness. In the computer vision field, semantic features have also been widely used in face recognition [71], general object recognition [18], and image search [65].

Shan et al. [23, 58] proposed exemplar-based representations. An illustration is shown in Fig. 17.4. The similarities of an image sample with selected representative persons in the training set are used as the semantic feature representation of the image. Suppose a and b are the two camera views to be matched. n representative pairs $\{(\mathbf{x}_1^a, \mathbf{x}_1^b), \ldots, (\mathbf{x}_n^a, \mathbf{x}_n^b)\}$ are selected as exemplars in the training set. $\mathbf{x}_i^a$ and $\mathbf{x}_i^b$ are the low-level feature vectors of the same person identity i, but are observed in different camera views a and b. If the low-level feature vector of a sample image $\mathbf{y}^a$ is observed in camera view a, it is compared against the n representative persons also observed in a, and its semantic features are represented with a $n-$dimensional vector $\mathbf{s}^a = (s_1^a, \ldots, s_n^a)$, where s_i^a is the similarity between $\mathbf{y}^a$ and $\mathbf{x}_i^a$ by matching their low-level visual features. If a sample $\mathbf{y}^b$ is observed in camera view b, its semantic feature vector $\mathbf{s}^b$ can be computed in the same way. When computing $\mathbf{s}^a$ and $\mathbf{s}^b$, the low-level visual features are only compared under the same camera view, and therefore large cross-view variations are avoided. Eventually, the similarity between

$\mathbf{y}^a$ and $\mathbf{y}^b$ are computed by comparing the semantic feature vectors $\mathbf{s}^a$ and $\mathbf{s}^b$. The underlying assumption is that if a person in test is similar to one of the representative persons i in the training set, its observations in camera views a and b should be similar to $\mathbf{x}_i^a$ and $\mathbf{x}_i^b$ respectively, and therefore both s_i^a and s_i^b are large no matter how different the two camera views are. Therefore, if $\mathbf{y}^a$ and $\mathbf{y}^b$ are the observations of the same object, $\mathbf{s}^a$ and $\mathbf{s}^b$ are similar.

Layne et al. [35] employed attribute features for person re-identification. They defined 15 binary attributes regarding cloth-style, hairstyle, carrying-object and gender. Attribute classifiers are based on low-level visual features. They are learned with SVM from a set of training samples whose attributes are manually labeled. The outputs of attribute classifiers are used as feature representation for person re-identification. They can also be combined with low-level visual features for matching. Since the training samples with the same attribute may come from different camera views, the learned attribute classifiers may have view invariance to some extent. Liu et al. [39] weighted attributes according to their importance in person re-identification. Attribute-based approaches require more labeling effort for training attribute classifiers. While in other approaches each training sample only needs one identity label, it requires all the M attributes to be labeled for a training sample.

17.2.4 Learning Feature Transforms Across Camera Views

In order to learn the feature transforms across camera views, one could first assume the photometric or geometric transform models and then learn the model parameters from training samples [30, 52, 53]. For example, Prosser et al. [53] assumed the photometric transform to be bi-directional Cumulative Brightness Transfer Functions, which map color observed in one camera view to another. Porikli and Divakaran [52] learned the color distortion function between camera views with correlation matrix analysis. Geometric transforms can also be learned from the correspondences of interest points.

However, in many cases, the assumed transform functions cannot capture the complex cross-camera transforms which could be multi-model. Even if all the pedestrian images are captured by a fixed pair of camera views, their cross-view transforms may have different configurations because of many different possible combinations of poses, resolutions, lightings, and background. Li and Wang [36] proposed a gating network to project visual features from different camera views into common feature spaces for matching without assuming any transform functions. As shown in Fig. 17.5, it automatically partitions the image spaces of two camera views into subregions, corresponding to different transform configurations. Different feature transforms are learned for different configurations. A pair of images to be matched are softly assigned to one of the configurations and their visual features are projected on a common feature space. Each common feature space has a local expert learned for matching images. The features optimal for configuration estimation and identity

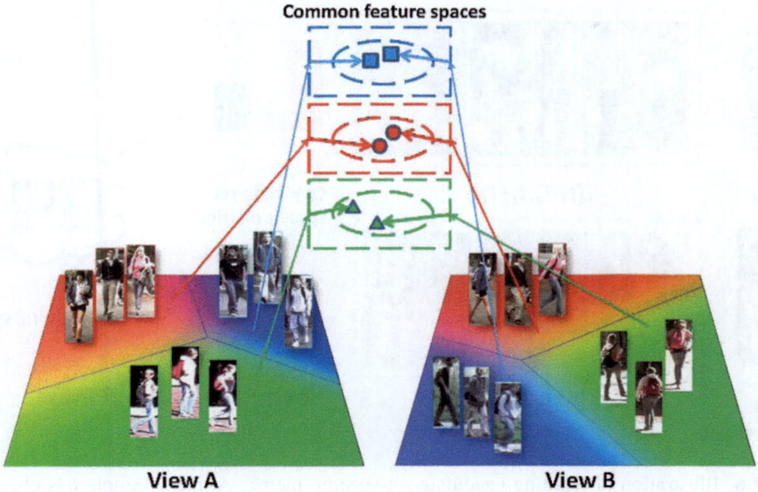

Fig. 17.5 Person re-identification in locally aligned feature transformations. The image spaces of two camera views are jointly partitioned based on the similarity of cross-view transforms. Sample pairs with similar transforms are projected to a common feature space for matching

matching are different and can be jointly learned. Experiments in [36] show that this approach not only can handle the multi-model problem but also have good generalization capability on new camera views. Given a large diversified training set, multiple cross-view transforms can be learned. The gating network can automatically choose a proper feature space to match test images from new camera views.

17.2.5 Metric Learning and Feature Selection

Given visual features, it is also important to learn a proper distance/similarity metric to further depress cross-view variations and well distinguish a large number of pedestrians. A set of reliable and discriminative features are to be selected through a learning process. Some approaches [38, 57] require that all the persons to be identified must have training samples. But this constraint largely limits their applications. In many scenarios, it is impossible to collect the training samples of pedestrians in test beforehand. Schwartz and Davis [57] learned discriminative features with Partial Least Square Reduction. The features are weighted according to the discriminative power based on one-against-all comparisons. Lin and Davis [38] learned the dissimilarity profiles under a pairwise scheme. More learned-based approaches [22, 31, 33, 54, 75] were proposed to identify persons outside the training set. Zheng et al. [75] learned a distance metric which maximizes the probability that a pair of true match has a smaller distance than a wrong match. Gray and Tao [22] employed boosting to select viewpoint invariant and discriminative features for person re-identification.

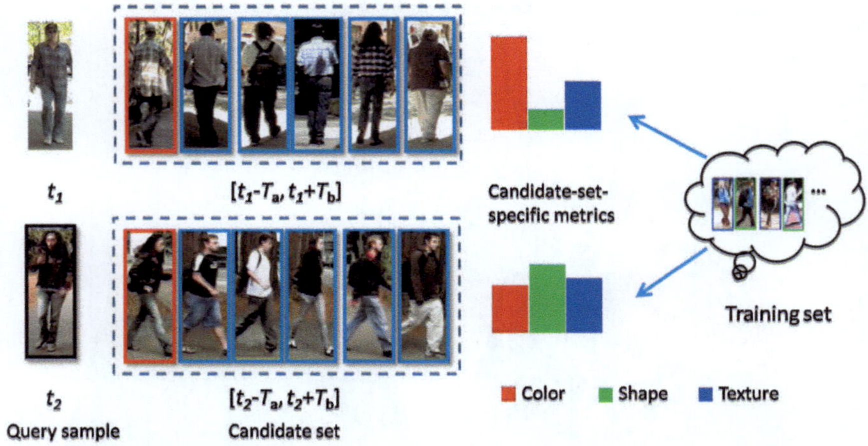

Fig. 17.6 Illustration of learning candidate-set-specific metric. A query sample i is observed at a camera view at time t_i. By reasoning the transition time, only the samples observed in another camera view during time window $[t_i - T_a, t_i + T_b]$ are considered at candidates. To distinguish persons in the first candidate set, color features are more effective. For the second candidate set, shape, and texture could be more useful. Persons in the candidate sets do not have training samples. Candidate-set-specific metrics could be learned from a large training set through transfer learning

Prosser et al. [54] formulated person re-identification as a ranking problem and used RankSVM to learn an optimal subspace.

The difficulty of person re-identification increases with the number of candidates to be matched. In cross-camera tracking, given a query image observed in one camera view, the transition time across two camera views can be roughly estimated. This simple temporal reasoning can simplify the person re-identification problem by pruning the candidate set to be matched in another camera view. All the approaches discussed above adopt a fixed metric to match a query image with any candidate. However, if the goal is to distinguish a small number of subjects in a particular candidate set, candidate-set-specific distance metrics should be preferred. An illustration is shown in Fig. 17.6. For example, the persons in one candidate set can be well distinguished with color features, while persons in another candidate set may be better distinguished with shape and texture. A better solution should tackle this problem through optimally learning different distance metrics for different candidate sets. Unfortunately, during online tracking, the correspondence of samples across camera views cannot be manually labeled for each person in the candidate set. Therefore, directly learning a candidate-set-specific metric is infeasible, since metric learning requires pairs of samples across camera views with correspondence information. Li et al. [37] tackled this problem by proposing a transfer learning approach. It assumes a large training set with pair training samples across camera views. This training set has no overlap with candidate sets on person identities. As shown in Fig. 17.7, each sample in the candidate set finds its nearest neighbors in the training set based on visual similarities. Since the training set has ground truth labels, the corresponding

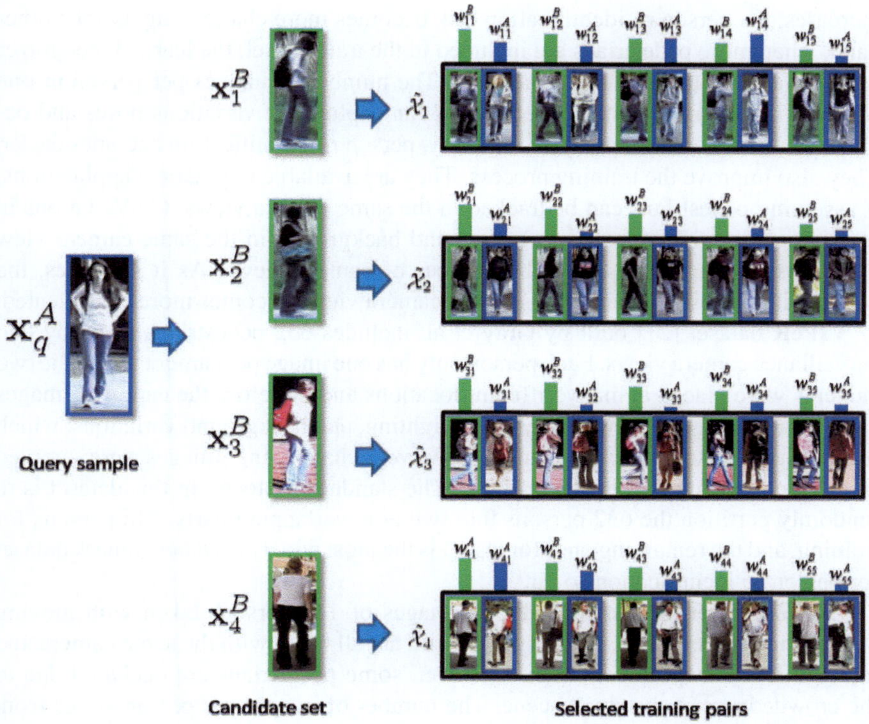

Fig. 17.7 Illustration of transfer learning for person re-identification proposed in [37]. *Blue* and *green* windows indicate samples observed in camera views A and B. $\mathbf{x}_q^A$ is a query sample observed in camera view A. $\mathbf{x}_1^B, \ldots, \mathbf{x}_4^B$ are samples of four candidate persons observed in camera view B. Each $\mathbf{x}_i^B$ finds five nearest neighbors in the same camera view B from the training set. Since the corresponds of training samples in camera views A and B are known, the paired samples of the nearest neighbors can be found to training candidate-set-specific metric. w_{ij}^A and w_{ij}^B are the weights assigned to each pair of training samples according to their visual similarities to the candidates and the query sample

training samples of the found nearest neighbors in another camera view are known. The selected training pairs are weighted according to their visual similarities to the samples in the candidate set and the query sample. Finally, the candidate-set-specific distance metric is learned from the selected and weighted training pairs.

17.3 Benchmark Datasets

Multiple benchmark datasets for person re-identification have been published in recent years. There are multiple factors to be considered when creating a benchmark dataset. (1) The number of pedestrians. As the number of pedestrians in the gallery increases, the person re-identification task becomes more challenging. On the other hand, when more pedestrians are included in the training set, the learned recognizer will be more robust at the test stage. (2) The number of images per person in one camera view. Multiple images per person can capture the variations poses and occlusions. If they are available in the gallery, person re-identification becomes easier. They also improve the training process. They are available in practical applications, if assuming pedestrians can be tracked in the same camera views. (3) Variations in resolutions, lightings, poses, occlusion, and background in the same camera view and across camera views. (4) The number of camera views. As it increases, the complexity of possible transforms across camera views becomes more complicated.

VIPeR dataset [21] built by Gray et al. includes 632 pedestrians taken by two surveillance camera views. Each person only has one image per camera view. The two cameras were placed at many different locations and therefore the captured images cover a large range of viewpoints, poses, lighting, and background variations, which makes image matching across camera views very challenging. Images were sampled from videos with compression artifacts. The standard protocol on this dataset is to randomly partition the 632 persons into two nonoverlapping parts, 316 persons for training, and the remaining ones for test. It is the most widely used benchmark dataset for person re-identification so far.

ETHZ dataset [57] includes 8,580 images of 146 persons taken with moving cameras in a street scene. Images of a person are all taken with the same camera and undergo less viewpoint variation. However, some pedestrians are occluded due to the crowdedness of the street scene. The number of images per person varies from 10 to 80.

i-LIDS MCTS dataset created by Zheng et al. [74] was collected from an airport arrival hall. It includes 476 images of 119 pedestrians. Most persons have four images captured by the same camera views or nonoverlapping different camera views.

CAVIAR4REID created by Cheng et al. [10] collected 1,220 images of 72 pedestrians from a shopping center. 50 pedestrians were captured with two camera views and the remaining ones by one camera view. Compared with other datasets, its images have large variation on resolutions.

Person Re-ID 2011 Dataset created by Hirzer et al. [25] have 931 persons captured with two static surveillance cameras. 200 of them appear in both camera views. The remaining ones only appear in one of the camera views.

RGB-D Person Re-identification Dataset created by Barbosa et al. [3] has depth information of 79 pedestrians captured in an indoor environment. For each person, the synchronized RGB images, foreground masks, skeletons, 3D meshes, and estimated floor are provided. The motivation is to evaluate the person re-identification performance for long-term video surveillance where the clothes can be changed.

QMUL underGround Re-IDentification (GRID) Dataset created by Loy et al. [41, 42] contains 250 pedestrian image pairs captured from a crowded underground train station. Each pair of images have the same identity and were captured by two nonoverlapping camera views. All the images were captured by 8 camera views.

Besides the dataset discussed above, there are also some other datasets published recently such as the **CUHK Person Re-identification Dataset** [36, 37], the **3DPes**

Dataset [2], the **Multi-Camera Surveillance Database** [7]. Baltieri et al. [2] and [7] also provided video sequences besides snapshots. The emergence of all these benchmark datasets clearly advanced the state-of-the-art on person re-identification. However, they also have several important drawbacks to be addressed in the future work.

First of all, the images in all the benchmark datasets are manually cropped. Most of the datasets even did not provide the original image frames. It means the assumption that images are perfectly aligned. Thereafter, all the developed algorithms and training process are based on this assumption. However, in practical surveillance applications, perfect alignment is impossible and pedestrian images need to be automatically cropped with pedestrian detectors [12, 19]. It is expected that the performance of existing person re-identification algorithms should drop significantly with the existence of misalignment. However, such effect has been ignored by almost all the existing publications. When building new benchmark datasets, automatically cropped image with state-of-the-art pedestrian detectors should be provided.

Secondly, the numbers of camera views in the existing datasets are small (the maximum number is 8). Moreover, in existing evaluation protocols, training and testing images are from the same camera views. The biggest challenge of person re-identification is to learn and depress cross-camera-view transforms. Given the fact that tens of thousands of surveillance cameras are available in large cities, in most surveillance applications, it is impossible to manually label training samples for every pair of camera views. Therefore, it is uncertain about the generalization capability of existing algorithms given a pair new camera views in test without extra training samples from which.

Thirdly, the numbers of persons (<1,000) and the numbers of images (<10,000) in person re-identification datasets are still much smaller than the scales of existing benchmark datasets for other computer vision problems such as object detection, object recognition, and face recognition. For example, ImageNet [15] for object classification has more than 14 million images. LFW [27] for face recognition has more than 5, 000 people. Some powerful machine learning tools such as deep models [6] have shown superior performance on computer vision challenges [34] based on large-scale training data. Therefore, it is desirable to build very large-scale datasets covering a large set of diversified camera views for person re-identification, which could not only significantly boost the performance but also enhance the generalization capability.

17.4 Evaluation

The accumulative matching characteristic (CMC) is the most widely used to evaluate the performance of person re-identification. It treats person re-identification as a ranking problem. Given one or one set of query images, the candidate images in the gallery are ranked according to their similarities to the query. CMC(k) measures the probability that the correct match has a rank equal or higher than k. As the gallery size

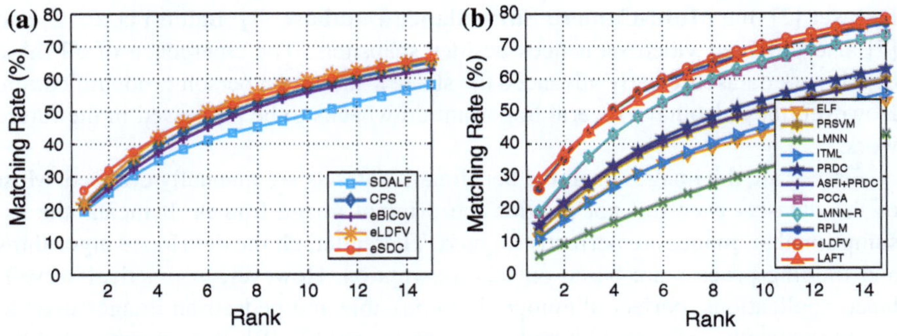

Fig. 17.8 CMC results of single-shot person re-identification on the VIPeR dataset. **a** Unsupervised methods. **b** Supervised methods

increases, it becomes more difficult to find the correct match and CMC(k) becomes lower.

Single-shot person re-identification only analyzes a single image for each person assuming no tracking information is available. Therefore, the query or any person in the gallery only has one image. Multi-shot person re-identification assume multiple images are available for each person through tracking. Therefore, a query is a set of images and the images in the gallery are also grouped into sets according to the identity information.

In Fig. 17.8, we summarize the results of single-shot person re-identification on the VIPeR dataset. VIPeR has 632 persons. They are randomly partitioned, half of the persons for training and the remaining half for test. The existing approaches are divided into two groups (unsupervised and supervised methods) for comparison. Unsupervised methods include:

- symmetry-driven accumulation of local features (SDALF) [4];
- custom pictorial structures (CPS) [10];
- biologically inspired features and covariance descriptors (BiCov) [43];
- local descriptors encoded by Fisher vectors combined with other features (eLDFV) [44];
- and salient dense correspondence combined with other features (eSDC) [72].

Supervised methods are:

- ensemble of localized features learned with AdaBoost (ELF) [22];
- person re-identification by support vector ranking (PRSVM) [54];
- distance metric learning for large margin nearest neighbor classification (LMNN) [67];
- information theoretic metric learning (ITML) [14];
- probabilistic relative distance comparison (PRDC) [75];
- attribute sensitive feature importance combined with PRDC (ASFI+PRDC) [39];
- Pairwise Constrained Component Analysis (PCCA) [45];

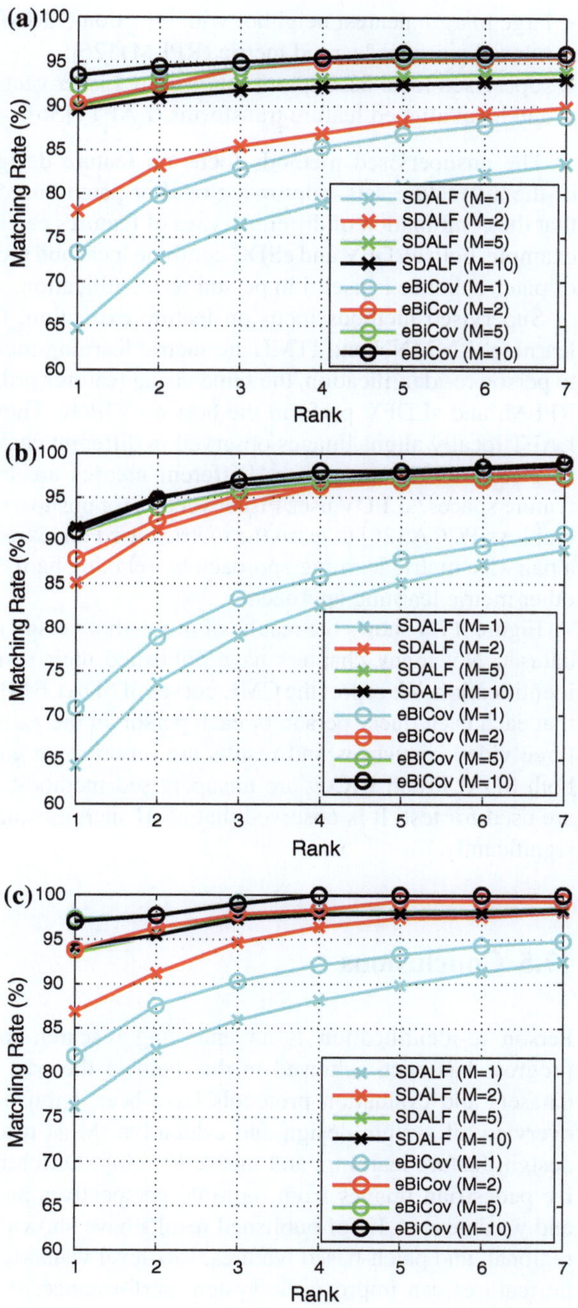

Fig. 17.9 CMC results of multi-shot person re-identification on the ETHZ dataset

- large margin nearest neighbor with rejection (LMNN-R) [16];
- relaxed pairwise learned metric (RPLM) [26];
- supervised local descriptors encoded by Fisher vectors (sLDFV) [44];
- and local aligned feature transforms (LAFT) [36].

The unsupervised methods focus on feature design. All the top ranked state-of-the-art feature sets employ regional or patch-based features. It is also observed that the combination of different types of features can improve the performance. For example, both eLDFV and eSDC combine local and global featúres. The information of patch salience is useful in person re-identification.

Supervised methods focus on feature extraction, feature transform, and metric learning. LMMNN and ITML are metric learning methods. When they are applied to person re-identification, the same visual features proposed in [75] are used. LAFT, RPLM, and sLDFV perform the best on VIPeR. They all employ metric learning. LAFT locally aligns images observed in different camera views by projecting them to a common feature space. Different metrics are learned for different common feature spaces. sLFDV uses Fisher vectors for unsupervised feature learning and then employs PCCA [26] to learn the metric based on extracted features. RPLM employs a pairwise metric learning approach by relaxing hard constraints commonly used in other metric learning approaches.

Figure 17.9 shows the results of multi-shot person re-identification on the ETHZ dataset. Not many chapters have published their results on multi-shot person re-identification. We report the CMC curves of SDALF [4] and BiCov [43]. It is assumed that each each query person or each person in the gallery has M images. ETHZ has three video sequences and results are reported on each of them separately. Since both SDALF and BiCov are unsupervised methods, all the images in the dataset are used for test. It is observed that as M increases, the CMC curves get improved significantly.

17.5 Conclusions

Person re-identification is an emerging research topic and significant research progress has been achieved in this field in the past 5 years. Multiple benchmark datasets and evaluation protocols have been published. This chapter provides an overview of system design and evaluation. Most research works focus on feature design, feature learning, and metric learning. Different types of features characterize pedestrian images from multiple perspectives and all have their own strength and weakness. A lot of published results have shown that the integration of global, regional, and patch-based features, low-level visual features, and high-level semantic features can improve the system performance. People start to pay attention to high-level semantic features, importance of different attributes, and salience of local regions, not only because they can improve the matching accuracy, but also because they are interpretable by humans and can get human feedback involved in the recognition loop. The major challenge of person re-identification is the large variations across camera views. It is tackled by learning feature transforms and distance metrics. On a complex camera network, or even just between two camera views, the cross-view transforms are multi-modal, which cannot be handled with a single feature transform or a single distance metric. Mixture models are needed.

Person re-identification is still a very challenging problem and not well solved yet. On the VIPeR dataset, the rank-1 accuracy is still below 30 %. There are multiple directions to be explored in the future. Existing works match manually cropped images. Automatic pedestrian detection should be included in the person re-identification pipeline, and the effect of misalignment caused by detection should be considered in the future research. It requires the development of new methodology as well as new benchmark datasets. When the camera network is large, it is impractical to manually label training samples for every pair of camera views. It is important to study the generalization capability of person re-identification algorithms to unseen camera views. Existing benchmark datasets are relatively small in the numbers of samples, pedestrians and camera views. The diversity of their scene coverage is also limited. In recent years, large-scale machine learning has achieved great success in many fields. It would be interesting to see its application to person re-identification. Besides online multi-camera tracking, person re-identification can be also applied to pedestrian retrieval over camera networks. Like general image retrieval systems, user feedback and linguistic descriptions can get involved in the search loop. This would be another interesting direction to be explored.

References

1. Abdel-Hakim, A.E., Farag, A.A.: Csift: A sift descriptor with color invariant characteristics. In: Proceedings of European Conference Computer Vision, (2006)
2. Baltieri, D., Vezzani, R., Cucchiara, R.: 3dpes: 3d people dataset for surveillance and forensics. In: Proceedings of the 1st International ACM Workshop on Multimedia Access to 3D Human Objects (2011)
3. Barbosa, B.I., Cristani, M., Del Bue, A., Bazzani, L., Murino, V.: Re-identification with rgb-d sensors. In: First International Workshop on Re-Identification, (2012)
4. Bazzani, L., Cristani, M., Murino, V.: Symmetry-driven accumulation of local features for human characterization and re-identification. Comput. Vis. Image Underst. **117**(2), 130–144 (2013)
5. Belongie, S., Malik, J., Puzicha, J.: Shape matching and object recognition using shape contexts. IEEE Trans. Pattern Anal. Mach. Intell. **24**, 509–512 (2002)
6. Bengio, Y.: Learning Deep Architectures for AI. Now Publishers, Hanover (2009)
7. Bialkowski, A., Denman, S., Sridharan, S., Fookes, C., Lucey, P.: A database for person re-identification in multi-camera surveillance networks. In Proceedings of International Conference on Digital Image Computing-Techniques and Applications, (2012)
8. Bo, Y., Fowlkes, C.C.: Shape-based pedestrian parsing. In: Proceedings of the IEEE International Conference on Computer Vision and, Pattern Recognition, (2011)
9. Cai, Y., Chen, W., Huang, K., Tan, T.: Continuously tracking objects across multiple widely separated cameras. In: Asian Conference on Computer Vision, (2007)
10. Cheng, D., Cristani, M., Stoppa, M., Bazzani, L., Murino, V.: Custom pictorial structures for re-identification. In: Proceedings of the British Machine Vision Conference, (2011)

11. Cheng, E.D., Piccardi, M.: Matching of objects moving across disjoint cameras. In: Proceedings of the IEEE International Conference on Image Processing (2006)
12. Dalal, N., Triggs, B.: Histograms of oriented gradients for human detection. In: Proceedings of the IEEE International Conference on Computer Vision and Pattern Recognition, (2005)
13. Daugman, J.G.: Uncertainty relation for resolution in space, spatial frequency, and orientation optimized by two-dimensional visual cortical filters. J. Opt. Soc. Am. A **2**, 1160–1169 (1985)
14. Davis, J., Kulis, B., Jain, P., Sra, S., Dhillon, I.: Information theoretic metric learning. In: Proceedings of the International Conference on Machine Learning, (2007)
15. Deng, J., Dong, W., Socher, R., Li, L.J., Li, K., Fei-Fei, L.: Imagenet: A large-scale hierarchical image database. In: Proceedings of IEEE International Conference on Computer Vision and, Pattern Recognition, (2009)
16. Dikmen, M., Akbas, E., Huang, T.S., Ahuja, N.: Pedestrian recognition with a learned metric. In: Asian Conference on Computer Vision, (2010)
17. Eslami, S.M.A., Williams, C.K.I.: A generative model for parts-based object segmentation. In: Proceedings of the Neural Information Processing Systems, (2012)
18. Farhadi, A., Endres, I., Hoiem, D., Forsyth, D.: Describing objects by their attributes. In: Proceedings of IEEE International Conference on Computer Vision and Pattern Recognition, (2009)
19. Felzenszwalb, P., McAllester, D., Ramanan, D.: A discriminatively trained, multiscale, deformable part model. In: Proceedings of the IEEE International Conference on Computer Vision and Pattern Recognition, (2008)
20. Gheissari, N., Sebastian, T.B., Rittscher, J., Hartley, R.: Person reidentification using spatiotemporal appearance. In: Proceedings of IEEE International Conference on Computer Vision and, Pattern Recognition, (2006)
21. Gray, D., Brennan, S., Tao, H.: Evaluating appearance models for recognition, reacquisition, and tracking. In: Proceedings of the IEEE International Workshop on Performance Evaluation of Tracking and Surveillance, (2007)
22. Gray, D., Tao, H.: Viewpoint invariant pedestrian recognition with an ensemble of localized features. In: Proceedings of the European Conference on Computer Vision, (2008)
23. Guo, Y., Rao, C., Samarasekera, S., Kim, J., Kumar, R., Sawhney, H.: Matching vehicles under large pose transformations using approximate 3d models and piecewise mrf model. In: Proceedings of the IEEE International Conference on Computer Vision and Pattern Recognition, (2008)
24. Hamdoun, O., Moutarde, F., Stanciulescu, B., Steux, B.: Person re-identification in multi-camera system by signature based on interest point descriptors collected on short video sequences. In: Proceedings of IEEE Conference on Distributed Smart Cameras, (2008)
25. Hirzer, M., Beleznai, C., Roth, P.M., Bischof, H.: Person re-identification by descriptive and discriminative classification. In: Proceedings of the Scandinavian Conference on Image, Analysis, (2011)
26. Hirzer, M., M., R.P., Kostinger, M., Bischof: Relaxed pairwise learned metric for person re-identification. In: Proceedings of the European Conference on Computer Vision, (2012)
27. Huang, G., Ramesh, M., Berg, T., Learned-Miller, E.: Labeled faces in the wild: A database for studying face recognition in unconstrained environments. University of Massachusetts, Amherst, Tech. rep. (2007)
28. Huang, J., Kumar, S.R., Mitra, M., Zhu, M., Zabih, R.: Image indexing using color correlograms. In: Proceedings of the IEEE International Conference on Computer Vision and, Pattern Recognition, (1997)
29. Javed, O., Rasheed, Z., Shafique, K., Shah, M.: Tracking across multiple cameras with disjoint views. In: Proceedings of the IEEE International Conference on Computer Vision and Pattern Recognition, (2003)
30. Javed, O., Shafique, K., Shah, M.: Appearance modeling for tracking in multiple non-overlapping cameras. In: Proceedings of the IEEE International Conference on Computer Vision and Pattern Recognition, (2005)

31. Jurie, F., Mignon, A.: Pcca: a new approach for distance learning from sparse pairwise constraints. In: Proceedings of IEEE International Conference on Computer Vision and Pattern Recognition, (2012)
32. Khan, S., Shah, M.: Consistent labeling of tracked objects in multiple cameras with overlapping fields of view. IIEEE Trans. Pattern Anal. Mach. Intell. **25**, 1355–1360 (2003)
33. Kostinger, M., Hirzer, M., Wohlhart, P., Roth, P., Bischof, H.: Large scale metric learning from equivalence constraints. In: Proceedings of the IEEE International Conference on Computer Vision and Pattern Recognition, (2011)
34. Krizhevsky, A., Sutskever, I., Hinton, G.: Imagenet classification with deep convolutional neural networks. In: Proceedings of the Neural Information Processing Systems, (2012)
35. Layne, R., Hospedales, T., Gong, S.: Person re-identification by attributes. In: Proceedings of the British Machine Vision Conference, (2012)
36. Li, W., Wang, X.: Locally aligned feature transforms across views. In: Proceedings of the IEEE International Conference on Computer Vision and Pattern Recognition, (2013)
37. Li, W., Zhao, R., Wang, X.: Human reidentification with transferred metric learning. In: Asian Conference on Computer Vision, (2012)
38. Lin, Z., Davis, L.: Learning pairwise dissimilarity profiles for appearance recognition in visual surveillance. In: Proceedings of the International Symposium on Advances in Visual Computing, (2008)
39. Liu, C., Gong, S., Loy, C.C., Lin, X.: Person re-identification: What features are important? In: Proceedings of the First International Workshop on Re-Identification (2012)
40. Lowe, D.: Distinctive image features from scale-invariant keypoints. Int. J. Comput. Vision **60**(2), 91–110 (2004)
41. Loy, C.C., Xiang, T., Gong, S.: Multi-camera activity correlation analysis. In: Proceedings of the IEEE International Conference on Computer Vision and Pattern Recognition, (2009)
42. Loy, C.C., Xiang, T., Gong, S.: Multi-camera activity correlation analysis. Int. J. Comput. Vision **90**, 106–129 (2010)
43. Ma, B., Su, Y., Jurie, F.: Bicov: a novel image representation for person re-identification and face verification. In: Proceedings of the British Machine Vision Conference, (2012)
44. Ma, B., Su, Y., Jurie, F.: Local descriptors encoded by fisher vectors for person re-identification. In: Proceedings of the First International Workshop on Re-identification, (2012)
45. Mignon, A., Jurie, F.: Pcca: A new approach for distance learning from sparse pairwise constraints. In: Proceedings of the IEEE Internatonal Conference on Computer Vision and Pattern Recognition, (2012)
46. Mittal, A., Davis, L.S.: M2tracker: a multi-view approach to segmenting and tracking people in a cluttered scene. Int. J. Comput. Vision **51**, 189–203 (2003)
47. Nakajima, C., Pontil, M., Heisele, B., Poggio, T.: Full-body recognition system. Pattern Recognit. **36**, 1977–2006 (2003)
48. Ojala, T., Pietikäinen, M., Mäenpää, T.: Multiresolution gray-scale and rotation invariant texture classification with local binary patterns, IEEE Trans. Pattern Anal. Mach. Intell. **24**(7):971–987 (2002)
49. Orwell, J., Remagnino, P., Jones, G.A.: Multiple camera color tracking. In: Proceedings of the IEEE Workshop on Visual Surveillance, (1999)
50. Ouyang, W., Wang, X.: A discriminative deep model for pedestrian detection with occlusion handling. In: Proceedings of the IEEE International Conference on Computer Vision and Pattern Recognition, (2012)
51. Park, U., Jain, A., Kitahara, I., Kogure, K., Hagita, N.: Vise: Visual search engine using multiple networked cameras. In: Proceedings of the IEEE International Conference on Pattern Recognition, (2006)
52. Porikli, F.: Inter-camera color calibration by correlation model function. In: Proceedings of the IEEE International Conference on Image Processing, (2003)
53. Prosser, B., Gong, S., Xiang, T.: Multi-camera matching using bi-directional cumulative brightness transfer function. In: Proceedings of the British Machine Vision Conference, (2008)

54. Prosser, B., Zheng, W., Gong, S., Xiang, T.: Person re-identification by support vector ranking. In: Proceedings of the British Machine Vision Confernce, (2010)
55. Rauschert, I., Collins, R.T.: A generative model for simultaneous estimation of human body shape and pixel-level segmentation. In: Proceedings of the European Conference on Computer Vision (2012)
56. Savarese, S., Winn, J., Criminisi, A.: Discriminative object class models of appearance and shape by correlatons. In: Proceedings of the IEEE International Conference on Computer Vision and Pattern Recognition, (2006)
57. Schwartz, W., Davis, L.: Learning discriminative appearance-based models using partial least sqaures. In: Proceedings of the XXII SIBGRAPI, (2009)
58. Shan, Y., Sawhney, H., Kumar, R.: Vehicle identification between non-overlapping cameras without direct feature matching. In: Proceedings of the IEEE International Conference on Computer Vision, (2005)
59. Slater, D., Healey, G.: The illumination-invariant recognition of 3d objects using local color invariants. IEEE Trans. Pattern Anal. Mach. Intell. **18**:206–210 (1996)
60. Tian, Y., Zitnick, C.L., Narasimhan, S.G.: Exploring the spatial hierarchy of mixture models for human pose estimation. In: Proceedings of the European Conference on Computer Vision, (2012)
61. Tuzel, O., Porikli, F., Meer, P.: Region covariance: a fast descriptor for detection and classification. In: Proceedings of the European Conference on Computer Vision, (2006)
62. Varma, M., Zisserman, A.: A statistical approach to texture classification from single images. Int. J. Comput. Vision **62**, 61–81 (2005)
63. Wang, M., Li, W., Wang, X.: Transferring a generic pedestrian detector towards specific scenes. In: Proceedings of the IEEE International Conference on Computer Vision and, Pattern Recognition, (2012)
64. Wang, X., Doretto, G., Sebastian, T., Rittscher, J., Tu, P.: Shape and appearance context modeling. In: Proceedings of the IEEE International Conference on Computer Vision, (2007)
65. Wang, X., Qiu, S., Liu, K., Tang, X.: Web image re-ranking using query-specific semantic signatures. IEEE Trans. Pattern Anal. Mach. Intell. **34**(3):436–450 (2013)
66. Weijer, J., Schmid, C.: Coloring local feature extraction. In: Proceedings of the European Conference on Computer Vision, (2006)
67. Weinberger, K., Blitzer, J., Saul, L.: Distance metric learning for large margin nearest neighbor classification. In: Proceedings of the Neural Information Processing Systems, (2006)
68. Winn, J., Criminisi, A., Minka, T.: Object categorization by learned universal visual dictionary. In: Proceedings of the IEEE International Conference on Computer Vision, (2005)
69. Yang, Y., Ramanan, D.: Articulated pose estimation using flexible mixtures of parts. In: Proceedings of the IEEE International Conference on Computer Vision and Pattern Recognition, (2011)
70. Yilmaz, A., Javed, O., Shah, M.: Object tracking: a survey. ACM Comput. Surv. **38**, 1–45 (2006)
71. Yin, Q., Tang, X., Sun, J.: An associate-predict model for face recognition. In: Proceedings of the IEEE International Conference on Computer Vision and Pattern Recognition, (2011)
72. Zhao, R., Ouyang, W., Wang, X.: Unsupervised salience learning for person re-identification. In: Proceedings of the IEEE International Conference on Computer Vision and Pattern Recognition, (2013)
73. Zhao, T., Nevatia, R., Wu, B.: Segmentation and tracking of multiple humans in crowded environments. IEEE Trans. Pattern Anal. Mach. Intell. **30**:1198–1211 (2008)
74. Zheng, W., Gong, S., Xiang, T.: Associating groups of people. In: Proceedings of the British Machine Vision Conference, (2009)
75. Zheng, W., Gong, S., Xiang, T.: Person re-identification by probabilistic relative distance comparison. In: Proceedings of the IEEE International Conference on Computer Vision and Pattern Recognition, (2011)

Chapter 18
People Search with Textual Queries About Clothing Appearance Attributes

Riccardo Satta, Federico Pala, Giorgio Fumera and Fabio Roli

Abstract Person re-identification consists of searching for an individual of interest in video sequences acquired by a camera network, using an *image* of that individual as a query. Here we consider a related task, named *people search with textual queries*, which consists of searching images of individuals that match a *textual description* of clothing appearance, given by a Boolean combination of predefined attributes. People search can be useful in applications like forensic video analysis, where the query can be obtained from a eyewitness report. We propose a general method for implementing people search as an extension of any given re-identification system that uses any *multiple part-multiple component* appearance descriptor. In our method, the same descriptor of the re-identification system at hand is used, and attributes are chosen by taking into account the information it provides. The original descriptor is then transformed into a dissimilarity one. Attribute detectors are finally constructed as supervised classifiers, using dissimilarity descriptors as the input feature vectors. We experimentally evaluate our method on a benchmark re-identification dataset.

R. Satta (✉)
European Commission—Joint Research Centre (JRC),
Institute for the Protection and Security of the Citizen,
Via E. Fermi 2749, 21027 Ispra, VA, Italy
e-mail: Riccardo.Satta@jrc.ec.europa.eu

F. Pala · G. Fumera · F. Roli
Department of Electrical and Electronic Engineering,
University of Cagliari, Piazza d'Armi SNC, 09123 Cagliari, Italy
e-mail: federico.pala@diee.unica.it

G. Fumera
e-mail: fumera@diee.unica.it

F. Roli
e-mail: roli@diee.unica.it

S. Gong et al. (eds.), *Person Re-Identification*,
Advances in Computer Vision and Pattern Recognition,
DOI: 10.1007/978-1-4471-6296-4_18, © Springer-Verlag London 2014

18.1 Introduction

Person re-identification consists of recognising an individual appearing in different video sequences acquired by a camera network, with possibly non-overlapping views. Its main application field is video-surveillance; for instance, it can enable the off-line retrieval of all video sequences containing an individual of interest. In person re-identification systems, an image or a sequence of frames containing the individual of interest is used as a query. The query image and the ones to which it has to be matched are represented using suitable descriptors. Most of the existing person re-identification methods use descriptors of clothing appearance. Other cues, like face and gait, cannot be exploited in typical video-surveillance scenarios, which are characterised by strong pose variations, partial occlusions, low resolution and unconstrained environment [6].

In this chapter we focus on a related task, which we name "people search with textual queries". It consists of retrieving images or video sequences of individuals who match a query given in terms of a *textual* description of clothing appearance, instead of an *image*. Such a functionality can be very useful in applications like forensic video analysis, where the query can be obtained from the description of the suspect author of a crime, made by an eyewitness. A preliminary version of this work appeared in [15].

A similar task was considered in [10, 11, 17, 18]. In [17, 18] it was named "person attribute search" or "attribute-based people search". The focus of [18] was on face attributes, like the presence of beard and eyeglasses, while only the dominant colours of torso and legs were considered as clothing appearance attributes. A specific detector was then developed for the above attributes. In [17] the following attributes were considered: gender, hair/hat colour, the position and colour of the bag (if any) carried by an individual, and, as in [18], the colours of torso and legs. A generative model was proposed to build the corresponding descriptors. The use of attributes was proposed instead in [10, 11] as a mean for improving the recognition performance of low-level features in person re-identification systems, based on the argument that human experts perform person re-identification by looking also at mid-level attributes like hair style, shoe type and clothing style. Fifteen specific attributes were identified in [10, 11]: shorts, skirt, sandals, backpack, jeans, logo, v-neck, open-outerwear, stripes, sunglasses, headphones, long-hair, short-hair, gender and carrying-object. The corresponding detectors were implemented as binary classifiers, using ad hoc features, and a score-level fusion with a re-identification method based on low-level features (in particular, the one of [2]) was performed. It is worth noting that the use of "attributes" as mid or high-level features has recently been proposed for more general computer vision tasks, like object recognition and image retrieval [7, 19].

Differently from [10, 11, 17, 18], our goal is to develop a *general* method for implementing people search with textual queries, as an extension of *any* given appearance-based person re-identification system. To this aim: (i) we focus on clothing appearance attributes only, disregarding attributes that require high-resolution images or rely on constrained poses or settings (e.g. face attributes like beard and sunglasses); (ii) we propose to use the *same* clothing appearance descriptor of the re-identification system at hand, instead of ad hoc features, and develop a method that can be implemented using most of the existing descriptors and (iii) we do not focus on a predetermined set of attributes, but propose to choose them by taking into account the characteristics of the appearance descriptor at hand.

We exploit the fact that most of the existing re-identification methods use *multiple part-multiple component* descriptors, i.e. they subdivide the body into parts, and represent each of them as a set of different components. Our Multiple Component Dissimilarity (MCD) framework [16] can then be used to convert any descriptor of this kind into a dissimilarity-based one, i.e. a fixed-length vector made up of dissimilarity values between a given body part and a set of visual prototypes that encode specific characteristics of that part. This allows us to devise a simple implementation of people search with textual queries. First, a set of basic attributes related to clothing appearance is identified, which can be detected by the original descriptor at hand (e.g. the colour and the texture of upper and lower garments, the presence of short or long sleeves, etc.); then, a detector for each attribute is built as a binary classifier, whose input features consist of a dissimilarity vector. In our method, each attribute corresponds to what we call a textual "basic query". We also propose a method for processing complex queries, obtained by combining basic ones through Boolean operators.

We summarise the MCD framework in Sect. 18.2, describe our approach for implementing people search with textual queries in Sect. 18.3, and experimentally evaluate it in Sect. 18.4 on a benchmark dataset for person re-identification.

18.2 Dissimilarity-Based Appearance Descriptors

Person re-identification is basically a *matching* problem: it consists of ranking a set of template images with respect to their similarity to the query image, computed as a match score. In [14] we pointed out that the descriptors used by most appearance-based re-identification methods share the high-level characteristic of adopting a *multiple part-multiple component* representation: they subdivide the human body into parts, and represent each part as a bag (set) of low-level local features, like random patches and SIFT points. This suggested us an analogy with the kind of descriptor proposed in the Multiple Component Learning (MCL) framework of [5] for the different task of object *detection*. In MCL, an object $\mathbf{X}$ (e.g. a pedestrian) is represented as an ordered sequence of m parts:

$$\mathbf{X} = [X_1, \dots, X_m] . \tag{18.1}$$

Each part X_i is defined as an unordered set of components, which are represented as local feature vectors in a feature space $\mathcal{X}_i$:

$$X_i = \{x_{i1}, x_{i2}, \dots\}, \; x_{ij} \in \mathcal{X}_i, \; i = 1, \dots, m . \tag{18.2}$$

An object detector is then constructed by combining the part detectors, which are implemented as classifiers constructed using the Multiple Instance Learning (MIL) paradigm [4], according to the corresponding multiple component representation.

Such a high-level analogy with MCL descriptors inspired us the Multiple Component Matching (MCM) framework, for constructing *multiple part-multiple component* descriptors for re-identification (matching) problems. MCM is based on the same representation as MCL, summarised above. It is also capable to frame the existing descriptors [14], including the ones that do not subdivide the body into parts, which correspond to $m = 1$ in Eqs. (18.1) and (18.2), and/or use global features instead of local ones, which amounts to defining each part in Eq. (18.2) as a singleton, $X_i = \{x_{i1}\}$.

We then developed the Multiple Component Dissimilarity (MCD) framework, which allows one to convert any MCM descriptor into a dissimilarity-based one [16]. The MCD framework, which is summarised in the rest of this section, was originally proposed for speeding up the matching step in re-identification systems, enabling real-time applications [16]. Existing descriptors are indeed rather complex, and require a relatively high matching time, while MCD descriptors are simple, fixed-length vectors of real numbers. We subsequently found that MCD exhibits other advantages; among them, it enables a general implementation of people search with textual queries.

The main idea underlying MCD stems from the dissimilarity representation for pattern recognition [13]. It was originally proposed to deal with problems in which a feature vector representation is not available or is not easy to obtain, while it is possible to define a dissimilarity measure between pairs of samples (e.g. object images). A sample can thus be represented as a fixed-size vector $\mathbf{X} = [x_1, x_2, \ldots]$, whose components $x_i \in \mathbb{R}$ are dissimilarity values to a predefined set of "prototype" objects $\{\mathbf{P}_1, \mathbf{P}_2, \ldots\}$. Prototypes are chosen depending on the task at hand, e.g. by clustering, by techniques derived from feature selection approaches, or even randomly [13]. In MCD we adapted the dissimilarity paradigm to *multiple part-multiple component* descriptors, as explained below. The main difference with the original dissimilarity representation is that the visual prototypes of MCD refer to *local* characteristics of each body part, instead of the whole individual. In particular, each prototype is defined as a set of components, according to the underlying MCM representation.

From any MCM descriptor, summarised by Eqs. (18.1) and (18.2), an MCD descriptor can be obtained as follows (we refer the reader to [16] for further details):

1. For each body part, a set of visual prototypes that are representative of low-level, local visual characteristics (e.g. a certain distribution of colours in the torso) is constructed.
2. A dissimilarity measure between a body part and any prototype of the same part is defined.
3. Given the image of an individual, each of its parts is represented as a vector of dissimilarity values to the corresponding prototypes, and the resulting vectors are concatenated into an ordered dissimilarity vector representing the whole individual.

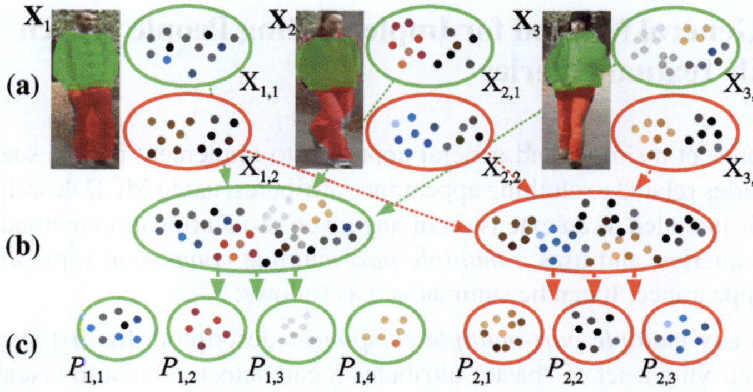

Fig. 18.1 Outline of the procedure for prototype construction in the MCD framework (taken from [16]). In this example, a body subdivision into two parts is considered: upper body (*green*) and legs (*red*). **a** A gallery of three individuals, each one represented as two sets of components, shown here as coloured dots. **b** All the components of the same part are merged. **c** The obtained sets of component are clustered, and each cluster is taken as a prototype of the corresponding body part

Step 1. The visual prototypes are obtained from a given gallery of images of n individuals, $\mathcal{G} = \{\mathbf{X}_1, \ldots, \mathbf{X}_n\}$, as follows. For each body part $i = 1, \ldots, m$:

1. The feature vectors $X_{j,i}$ of the i-th part of each image $\mathbf{X}_j \in \mathcal{G}$ are merged into a set $X'_i = \bigcup_{j=1}^{n} X_{j,i}$;
2. X'_i is clustered; we denote the resulting set of clusters as $\mathbf{P}_i = [P_{i1}, P_{i2}, \ldots]$.
3. Each cluster P_{ik} is finally taken as a distinct prototype.

This leads to an ordered sequence of a predefined number of prototypes for each part, where each prototype P_{ik} is an unordered set of components (feature vectors) itself:

$$P_{ik} = \{p_{ik1}, p_{ik2}, \ldots\}, \ p_{ikl} \in \mathcal{X}_i \ . \tag{18.3}$$

Each prototype is thus a set of visually similar image components, which can belong to different individuals of the considered gallery. As a particular case, a prototype P_{ik} can be made up of a single component (e.g. when the corresponding cluster contains a single sample, or if a single component is used in place of the whole cluster for reducing the computational complexity). Figure 18.1 summarises the prototype construction procedure.

Step 2. A dissimilarity measure $D_i(X_i, P_{ik})$ between a body part X_i and any prototype of the same part P_{ik} has to be defined. Note that it has to be a distance measure between sets, and thus it is in turn defined in terms of some distance measure between components, $d_i(x_{ij}, p_{ikl})$, $x_{ij}, p_{ikl} \in \mathcal{X}_i$. For instance, in [16] we used the K-th Hausdorff Distance, due to its robustness to outliers.

Step 3. A dissimilarity descriptor $\mathbf{X}^{\mathrm{D}}$ can be computed from any given MCM descriptor $\mathbf{X}$, by concatenating the m dissimilarity vectors obtained from each part:

$$\mathbf{X}^{\mathrm{D}} = [\mathbf{X}_1^{\mathrm{D}}, \ldots, \mathbf{X}_m^{\mathrm{D}}], \quad \mathbf{X}_i^{\mathrm{D}} = [D_i(X_i, P_{i,1}), D_i(X_i, P_{i,2}), \ldots], \ i = 1, \ldots, m \ . \tag{18.4}$$

18.3 A General Method for Implementing People Search with Textual Queries

Here we present a simple and general approach to implement people search using textual queries related to clothing appearance attributes, using MCD descriptors. Our approach is intended as an extension of any given re-identification method that uses images as queries, and uses a *multiple part-multiple component* representation of clothing appearance. It can be summarised as follows:

1. Given any *multiple part-multiple component* descriptor, the first step consists of identifying a set of "basic" attributes it can detect, related to visual clothing characteristics. We denote the chosen attributes as $\{a_1, a_2, \ldots\}$.
2. A detector f_i is built for each attribute a_i, using as input the MCD version $\mathbf{X}^{\mathrm{D}}$ (Eq. 18.4) of the descriptor at hand.
3. Each attribute a_i is associated to an "atomic" textual query q_i related to the corresponding visual characteristic. We call them *basic queries*. Complex queries can then be constructed by combining basic ones with Boolean operators, and have to be processed by suitably fusing the outputs of individual detectors.

We now discuss the three points above.

Identifying the set of attributes The attributes have to be defined taking into account the basic characteristics of clothing appearance that the descriptor at hand should be capable to detect. To this aim, each predefined body part (if more than one) can also be considered separately. For instance, if the descriptor subdivides the body into upper (torso and arms) and lower (legs) parts, and uses only colour features (e.g. the HSV colour histogram), it could enable the detection of attributes related to the colours of torso/arms, and to the colours of legs, like "blue trousers or skirt", "red upper garment with long sleeves", "upper garment with short sleeves" and "shorts trousers or skirt". Note that the two latter attributes could be detected by the presence of skin colour. More refined body part subdivisions can enable the definition of less coarse attributes. For instance, if the descriptor separates the arms from the torso, distinct attributes related to their colours can be considered. This may allow one to retrieve, e.g. images of individuals wearing a black jacket over a white shirt, by combining the corresponding basic queries (see below).

Note that the definition of attributes cannot be an automated process, but should involve human judgement. It is indeed necessary to identify which characteristics can be reasonably detected using the descriptor at hand, taking also into account the kind of images/videos of the target application scenario. For instance, the presence of eyeglasses is not likely to be detected by descriptors that do not consider the head as a distinct body part, or if the image resolution is too low. Note finally that, if a supervised procedure is used for implementing detectors (see below), a set of images of individuals exhibiting each attribute of interest must be collected and labelled.

Implementation of detectors In principle, the implementation of the detector for a given attribute depends also on the kind of the input descriptor. In fact, ad-hoc detectors were considered in [17, 18]. The MCD framework suggests us an approach

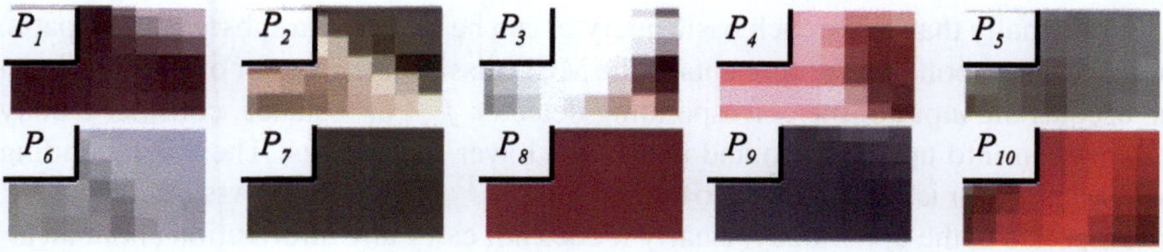

Fig. 18.2 Example of the image patches (components of a MCD descriptor) corresponding to different prototypes, obtained from the upper body parts of images of individuals taken from the VIPER dataset (see Sec.18.4)

independent on the specific descriptor, instead. Our intuition is that the clothing characteristics that can be detected by using a given appearance descriptor, according to its low-level features and its body subdivision into parts, can be encoded by one or more visual prototypes. For instance, Fig. 18.2 shows image patches extracted from the upper body parts of individuals taken from the dataset described in Sect. 18.4, using the MCD implementation of [16]. Each patch is taken from a different prototype, and corresponds to the component closest to the corresponding cluster centroid. If components are represented using colour features, images of individuals wearing a red upper garment can be expected to exhibit a high similarity to prototypes P_8, P_{10}, and perhaps P_4, and a lower similarity to the other ones. Similarly, images of individuals wearing a white upper garment should exhibit a high similarity to P_3, and, if some parts of the garment are in shade, also to P_6 and P_5. This suggests that the MCD descriptor $\mathbf{X}^{\mathrm{D}}$ of Eq. (18.4) can be conveniently exploited as the input feature vector of attribute detectors, independently on the underlying appearance descriptor.

Since people search with textual queries is a retrieval/ranking problem, as well as re-identification with image queries, we want the result of a query to be an ordered sequence of images, ranked with respect to their relevance, i.e. to the likelihood that the corresponding attribute (or combination of attributes) is present. Accordingly, detectors should output a real-valued score rather than a crisp decision, i.e. they should be implemented as functions $f_i : \mathbf{X}^{\mathrm{D}} \mapsto \mathbb{R}$. If a crisp decision is needed, a threshold can be set according to a suitable, application-dependent criterion; e.g. one can set a threshold the detector score $f_i(\mathbf{X}^{\mathrm{D}})$, or choose to retrieve only the top-N images according to their scores, for a given N.

The problem of implementing a detector for a given attribute can be seen as a supervised, binary classification problem, which consists of recognising whether the corresponding attribute is present or not in an input image, similarly to [10, 11]. The training set can be obtained from a gallery of images of individuals, labelled according to the presence or absence of the considered attribute. Any classification algorithm that outputs real-values scores can be used in principle, like support vector machines and neural networks. In particular, since the MCD descriptor $\mathbf{X}^{\mathrm{D}}$ is a fixed-size vector of scalars, it can be easily used as the feature vector of such a classification problem. This amounts to performing classification in a dissimilarity space, which allows one to exploit specific techniques developed for this purpose [12, 13].

Note finally that, since each basic query q_i can be related to a subset of body parts, only the corresponding components of the MCD dissimilarity vector of Eq. (18.4) can be used as the input of the corresponding detector f_i. For instance, consider a body subdivision into upper (torso and arms) and lower (legs) parts. The corresponding MCD descriptor is made up of two components, $\mathbf{X}^{\mathrm{D}} = [\mathbf{X}_1^{\mathrm{D}}, \mathbf{X}_2^{\mathrm{D}}]$. Assuming that $\mathbf{X}_1^{\mathrm{D}}$ corresponds to the upper body, clearly it does not carry any information about an attribute like "red trousers/skirt". The corresponding detector can thus be implemented using $\mathbf{X}_2^{\mathrm{D}}$ only as input features.

Processing basic and complex queries The answer to each basic query q_i is obtained by running the corresponding detector f_i through all the images at hand, and returning the list of images ranked according to the detectors' scores. Complex queries can be formulated by combining any subset of basic ones with Boolean operators. For instance, if q_1, q_2 and q_3 denote respectively the attributes "red upper garment", "blue trousers/skirt" and "black trousers/skirt", then the query "individual wearing blue or black trousers or skirt, and a red upper garment", is encoded by

$$Q = q_1 \wedge (q_2 \vee q_3). \tag{18.5}$$

Processing a complex query is not straightforward, instead. Two main approaches can be followed, based on interpreting the semantic of queries either using classical logic or fuzzy logic.

Classical logic approach For any given input image, the answer to each basic query, and thus of any complex query Q, is defined to be a crisp, Boolean value (either True or False). To this aim, the score of soft detectors of each basic query q_i in Q must be converted into Boolean values (see below). Let us denote as I_i the set of images of individuals for which the answer to q_i is "True". The answer to Q is then obtained by combining the sets of images I_i retrieved by each q_i in Q, through the set operators corresponding to the logic operators in Q. For instance, in the above example the answer to the complex query (18.5) would be obtained as

$$I_1 \cap (I_2 \cup I_3). \tag{18.6}$$

To convert the score $f_i(\cdot)$ of a soft detector into a Boolean value, one could set a threshold $t_i \in \mathbb{R}$, so that the attribute a_i is deemed present in the input image, if the score exceeds the threshold (i.e. q_i=True, if $f_i(\cdot) \geq t_i$), and is deemed absent otherwise. The value of the threshold t_i can be set, for instance, according to a desired precision-recall trade-off. Alternatively, one can choose to retrieve only the top-N ranked images according to their scores, for a given N. The most suitable criterion is obviously application-dependent.

Fuzzy logic approach The score $f_i(\cdot)$ produced for each basic query q_i can be considered as a fuzzy truth value (provided it is properly rescaled into [0, 1]), corresponding to the statement that the individual in the input image exhibits the attribute a_i. In other words, the detector can be seen as implementing the fuzzy membership function for such a statement. Accordingly, the answer to a complex query is obtained by combining the score values of basic queries with fuzzy logic operators, and by returning the list of input images ranked according to the resulting combined score. For instance, if the fuzzy-AND and fuzzy-OR are implemented respectively with the minimum and maximum operators, the score of an input image for the complex query (18.5) is obtained as

$$\min\{f_1(\cdot), \max\{f_2(\cdot), f_3(\cdot)\}\} . \tag{18.7}$$

18.4 Experiments

In this section we propose a possible implementation of our people search approach, and empirically evaluate it on a benchmark dataset for person re-identification.

18.4.1 Implementation

Appearance descriptors We considered two different descriptors previously proposed for person re-identification tasks. The first one is SDALF [2]. It subdivides body into torso and legs, and represents each part with three local features: an HSV colour histogram, the "Maximally Stable Colour Regions", and the "Recurrent Highly Structured Patches". We used the first two features, which are related to colour information. The second one is the descriptor we proposed in [14]. It subdivides the body as in SDALF, and represents each part with the HSV colour histograms of a bag of 80 image patches, randomly extracted. We also used a variant in which the body is subdivided into nine parts using a pictorial structure [1]: the torso and the upper and lower part of every limb. We obtained a MCD descriptor from each of the above *multiple part-multiple component* descriptors, as explained in Sect. 18.2. We will denote them respectively by MCD_2, MCD_1 and MCD_3.

To construct a MCD descriptor, prototypes have to be extracted first, from a given image gallery of individuals. This can be made in two different ways, depending on the application scenario. In a scenario like the off-line, forensic analysis of a fixed dataset of images (or videos), all the images are available beforehand. In this case, prototypes can be conveniently extracted from all such images (we remind the reader that this step is totally unsupervised, and thus does not require any manual labelling). In other scenarios where the images/videos are not available beforehand, prototypes must be extracted off-line from a design gallery, and can then be used to compute the dissimilarity representations of the images of individuals at operation phase. In this case, the retrieval performance of the proposed method may be affected by the representativeness of the design gallery with respect to images processed at operation phase. We chose to carry out our experiments under the latter scenario, which is the most challenging one. Nevertheless, the experimental evidences reported in [16] suggest that, if a design gallery containing a wide range of different clothing characteristics is used, prototypes can be representative of different images. Since we subdivided the considered dataset into a training set, to construct attribute detectors, and a testing set, to evaluate the performance of detectors (see below), prototypes

Fig. 18.3 Examples of images taken from the VIPER dataset. **a**–**d** positive examples for attributes related to **a** upper body clothing colours (*from left to right*: *red*, *blue*, *pink*, *white*, *black*, *green*, *grey* and *brown* upper garment); **b** lower body clothing colours (*blue*, *white*, *black*, *grey*, *brown* trousers/skirt); **c** short sleeves; **d** short trousers/skirt. **e** Examples of ambiguous images discarded from the dataset because of occlusions, shadows or low quality

were extracted from training images. We used to this aim the two stage clustering procedure of [16]. Note that in MCD_2, two different sets of prototypes were created, one for each kind of local features. We used the K-th Hausdorff distance for computing dissimilarities, with $K = 10$.

Dataset We used the VIPER dataset [8], which is one of the benchmarks for re-identification tasks. It is made up of 1264 images of 632 pedestrians, exhibiting different lighting conditions and pose variations (see the examples in Fig. 18.3). For each pedestrian, two images taken from two different cameras with non-overlapping views are present. The image size is 48×128 pixels.

Table 18.1 The 15 attributes considered in our experiments, related to upper body (*top nine rows*), and lower body (*last six rows*)

Attributes (#positive/#labelled)	MCD_1	MCD_2	MCD_3
Red upper garment (73/904)	**0.69**	0.66	0.62
Blue upper garment (84/904)	**0.48**	0.42	**0.48**
Pink upper garment (42/904)	0.50	**0.51**	0.44
White upper garment (277/904)	0.71	0.73	**0.77**
Black upper garment (298/904)	0.73	0.68	**0.74**
Green upper garment (72/904)	0.57	0.49	**0.59**
Grey upper garment (70/904)	**0.38**	0.30	0.28
Brown upper garment (71/904)	**0.46**	0.37	0.38
Short sleeves (382/1190)	0.54	0.51	**0.60**
Blue trousers/skirt (568/978)	0.87	0.85	**0.90**
White trousers/skirt (112/978)	**0.68**	0.58	0.63
Black trousers/skirt (178/978)	0.68	0.64	**0.75**
Grey trousers/skirt (52/978)	**0.30**	0.18	0.29
Brown trousers/skirt (40/978)	**0.62**	0.33	0.49
Short trousers/skirt (129/1197)	0.33	0.41	**0.58**

#positive denotes the number of images labelled as exhibiting the corresponding attribute; #labelled denotes the overall number of images labelled either as positive or negative (ambiguous images were discarded). The three right-most columns report the break-even point (BEP) of the precision-recall curves attained on testing images by the considered descriptors, averaged over the ten runs of the experiments. For each attribute, the highest BEP over the three descriptors is shown in *bold*

Attributes choice All the above descriptors enable queries related to clothing colour. In particular, MCD_1 and MCD_2 enable queries related to upper or lower body, like "individual wearing a white upper garment". MCD_3 should also enable more specific queries, like "short sleeves", since the corresponding attribute could be detected by the presence of skin-like colour in lower arms. Accordingly, we identified fifteen attributes corresponding to clothing characteristics that can be detected by the above descriptors, and are also present in at least 5 % images of the whole VIPER dataset, so that it was possible to build a training set for implementing the corresponding detector. The chosen attributes are related to the colours of the upper and lower body parts, and to the presence of short sleeves/trousers/skirts. They are reported in Table 18.1, where the corresponding number of positive samples is shown between brackets.

We then manually tagged all VIPER images for each of the considered attributes, using the following criteria. We labelled an image as positive for attributes related to a colour of upper and lower clothing, if that colour appeared approximately in at least one-third of the considered body part. In the case of short sleeves/trousers/skirt, a positive label was given if at least half of the limb was visible. Note that for some images the colour of the upper or lower body garments, or the presence of short sleeves or short trousers/skirts, was not clear, due for instance to low image quality, occlusions or shadows. We discarded these images both from training and testing samples. This is the reason why the number of labelled samples reported in Table 18.1 for each attribute is lower than the size of the VIPER dataset. We point out that this is a correct procedure for training samples, while in a real application such kind of

Fig. 18.4 Examples of prototypes obtained using the MCD_1 descriptor. Each one can be related to one of the considered attributes. From *top to bottom*, and from *left to right*: *red*, *blue*, *pink*, *white*, *black*, *green*, *grey* and *brown* upper garment (prototypes obtained from the upper body); *blue*, *white*, *black*, *grey*, *brown* trousers/skirt (*lower body*); short sleeves (*upper body*), short trousers/skirt (*lower body*)

images may appear among testing ones. In Fig. 18.3 we show one positive sample for each attribute, and some of the discarded images.All the tagged VIPER images are available in our web site.[1]

Experimental setup For each attribute, we randomly subdivided the images of the VIPER dataset into a training set and a testing set of identical size. We used stratified sampling, so that the same number of positive samples were present in the training and testing set. We then extracted the prototypes from training images. Different numbers of prototypes were considered, ranging from 5 to 300. The results reported in the following refer to 200 prototypes for MCD_1, and 100 prototypes for MCD_2 and MCD_3. We will discuss later how the performance of our method is affected by the number of prototypes. The detector of each attribute was implemented using a two-class support vector machine (SVM) classifier with radial basis function (RBF) kernel. We used the `LIBSVM` software [3] for training SVMs. We set the C parameter of the learning algorithm to the `LIBSVM` default value, and the γ parameter of the RBF kernel to 100. Since for most attributes negative samples considerably outnumbered positive ones, we set the misclassification cost of the latter ten times higher than the former. We then evaluated the performance of each detector on testing images. We repeated this procedure ten times. In the following, we report the average results over the ten runs. Since people search is a retrieval task, we used the precision–recall (P–R) curve to evaluate the performance of each detector (i.e. the retrieval performance for each basic query). It was computed by thresholding the detectors' (SVM) outputs. We also considered some complex queries, and processed them using the fuzzy logic approach described in Sect. 18.3.

18.4.2 Experimental Results

For each of the considered attributes, in Fig. 18.4 we show one example of a clearly related prototype (obtained with MCD_1). This example supports our intuition that prototypes extracted by MCD can also encode high-level visual characteristics of clothing appearance, even though their extraction is totally unsupervised.

[1] http://prag.diee.unica.it/pra/research/reidentification/dataset/viper-tagged

The P–R curves of each basic query, for the three considered descriptors, are reported in Figs. 18.5 and 18.6. Table 18.1 also shows the average break-even point (BEP) of each curve, which is the point where precision equals recall. An example of the ten top-ranked images for five out of the fifteen basic queries is shown in Fig. 18.7. The retrieval performance depends on the attribute, and on the underlying appearance descriptor that was used to build the MCD descriptor. In general, better performances were attained for attributes with a larger number of positive examples (which is reported in Table 18.1). For instance, a BEP of about 0.5 was attained for the "blue upper garment" attribute, which has 84 positive samples, while a BEP between 0.85 and 0.90 (depending on the descriptor) was attained for "blue trousers/skirt", which has 568 positive samples. The retrieval performance for "red upper garment" was very good instead, even though the positive samples were only 79. The reason is that the red colour is well separated from the other ones in the HSV space, which is used by all the considered descriptors (note that we did not consider the "red trousers/skirt" attribute, since only a few positive samples were available in the VIPER dataset). The performance was rather low for attributes related to grey, brown and pink colours (except for brown trousers/skirt, when MCD_2 was used), not only because of the small number of positive samples, but also because these colours are not well discriminated by the HSV space.

Consider now the attributes "short sleeves" and "short trousers/skirt", whose detection relies on skin colour in arms or legs. As pointed out in Sect. 18.3, MCD_3 was likely to attain the best performance on such attributes, due to its more refined body subdivision. This is confirmed by the corresponding P-R curves (see the last two plots of Fig. 18.6). Moreover, the part detection technique used by both MCD_1 and MCD_2 (see [2]) turns out to produce a better quality mask for the upper body than for the lower body, and this is the reason why a higher performance is attained by MCD_1 and MCD_2 for "short sleeves" than for "short trousers/skirt".

We also considered some examples of complex queries. Here we report some results for "white upper garment and blue trousers/skirt", and "white upper garment and short sleeves". We processed them using the fuzzy logic approach, and thus computed for both queries the minimum of the score provided by the detectors of the two embedded basic queries. The ten images exhibiting the highest combined score are shown in Fig. 18.8. Although these results are very limited, they provide some evidence that a fuzzy logic approach can be a convenient one to process complex queries, as it does not require one to set any parameter, like the thresholds mentioned in Sect. 18.3 for the classical logic approach.

We finally evaluated how the retrieval performance is affected by the number of prototypes. We observed that the performance initially grows as the number of prototypes increases, then reaches a nearly stable value. Such a value was around 100 for MCD_1 and MCD_3, and 200 for MCD_1, with small variations depending on the basic query. This behaviour can be easily explained: once the number of prototypes is large enough that most of the distinctive visual characteristics are captured by different clusters, increasing the number of prototypes has only the effect of splitting some of the previous clusters into two or more similar ones. Consequently, no additional

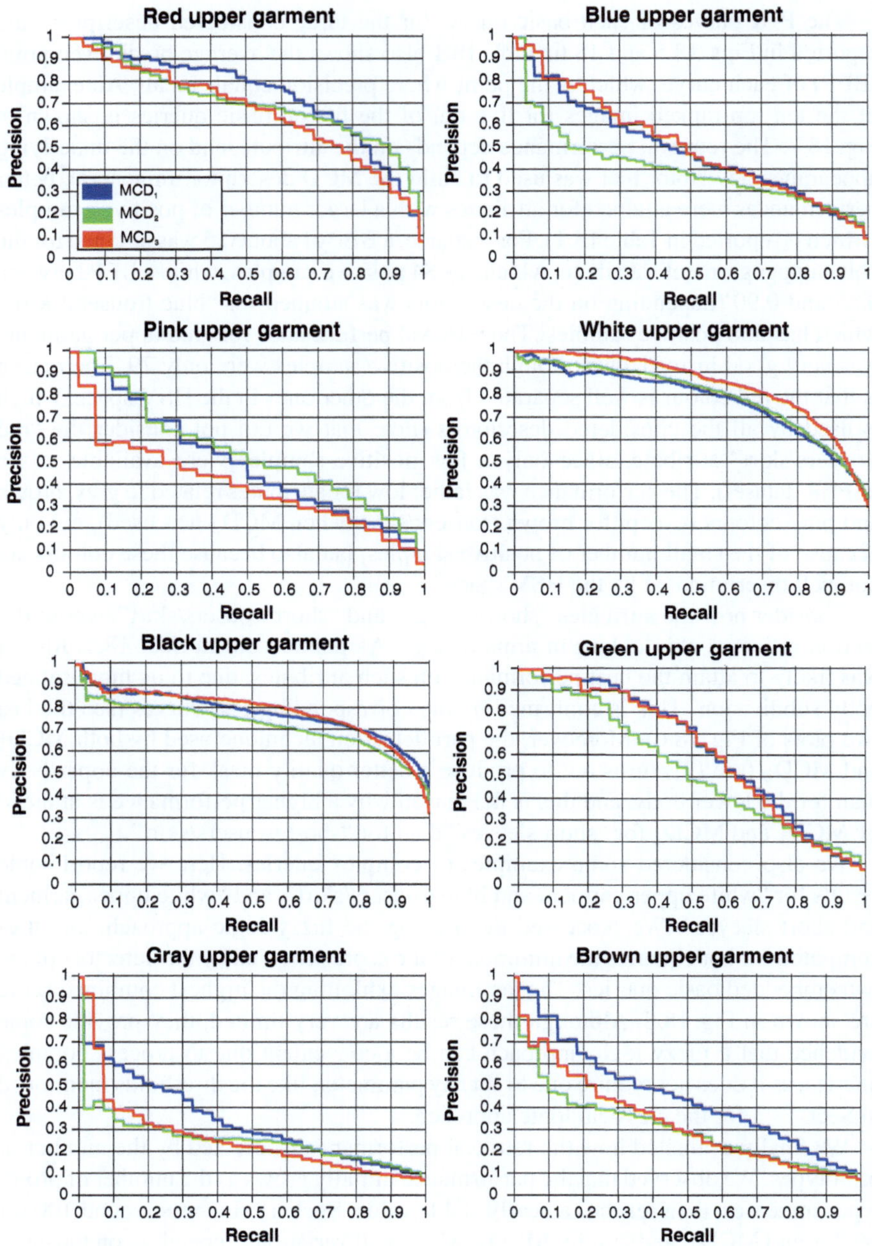

Fig. 18.5 Average P–R curves for the eight basic queries related to the clothing colours of the upper body, for the three descriptors considered.

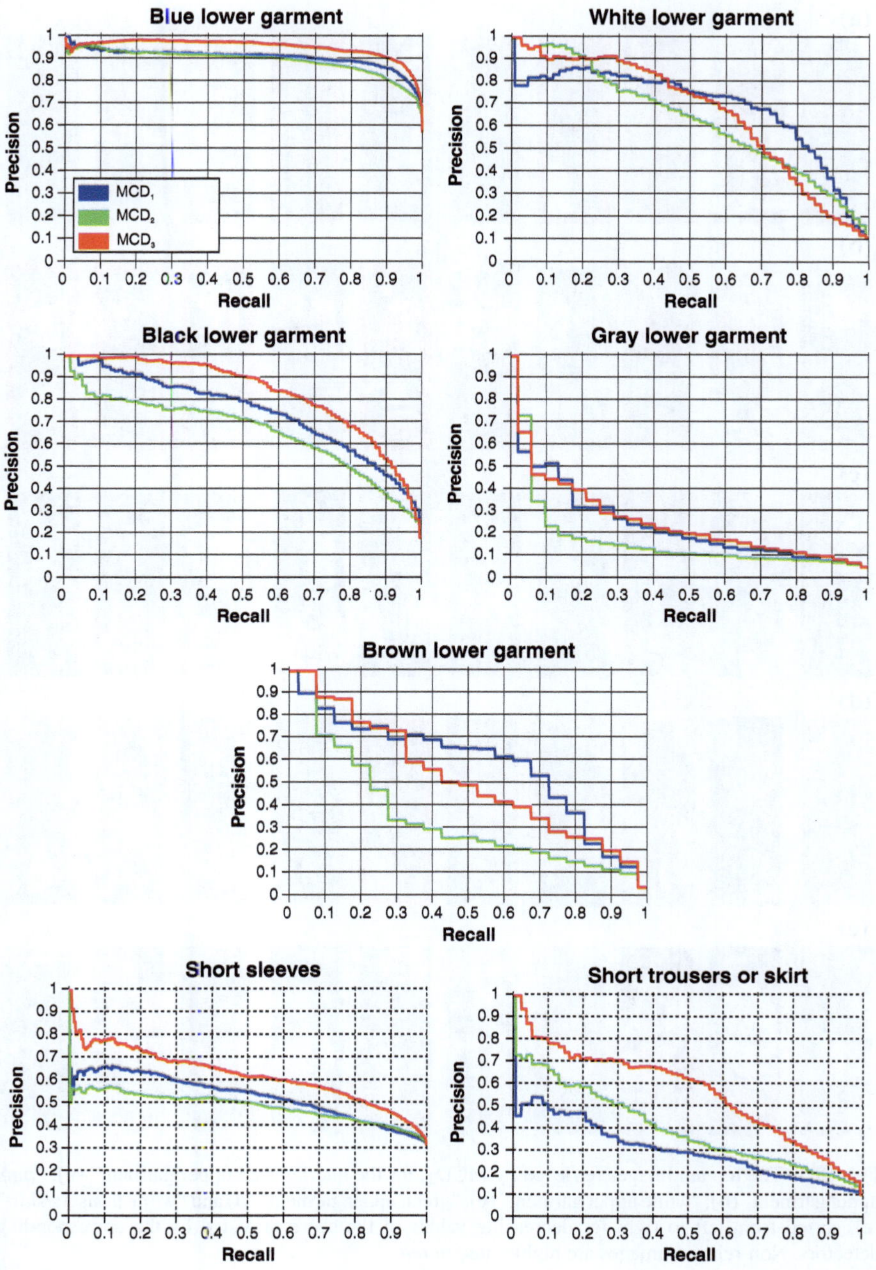

Fig. 18.6 Average P–R curves for the five basic queries related to the clothing colours of the lower body (*top five plots*), and to short sleeves and short trousers/skirts, for the three descriptors considered

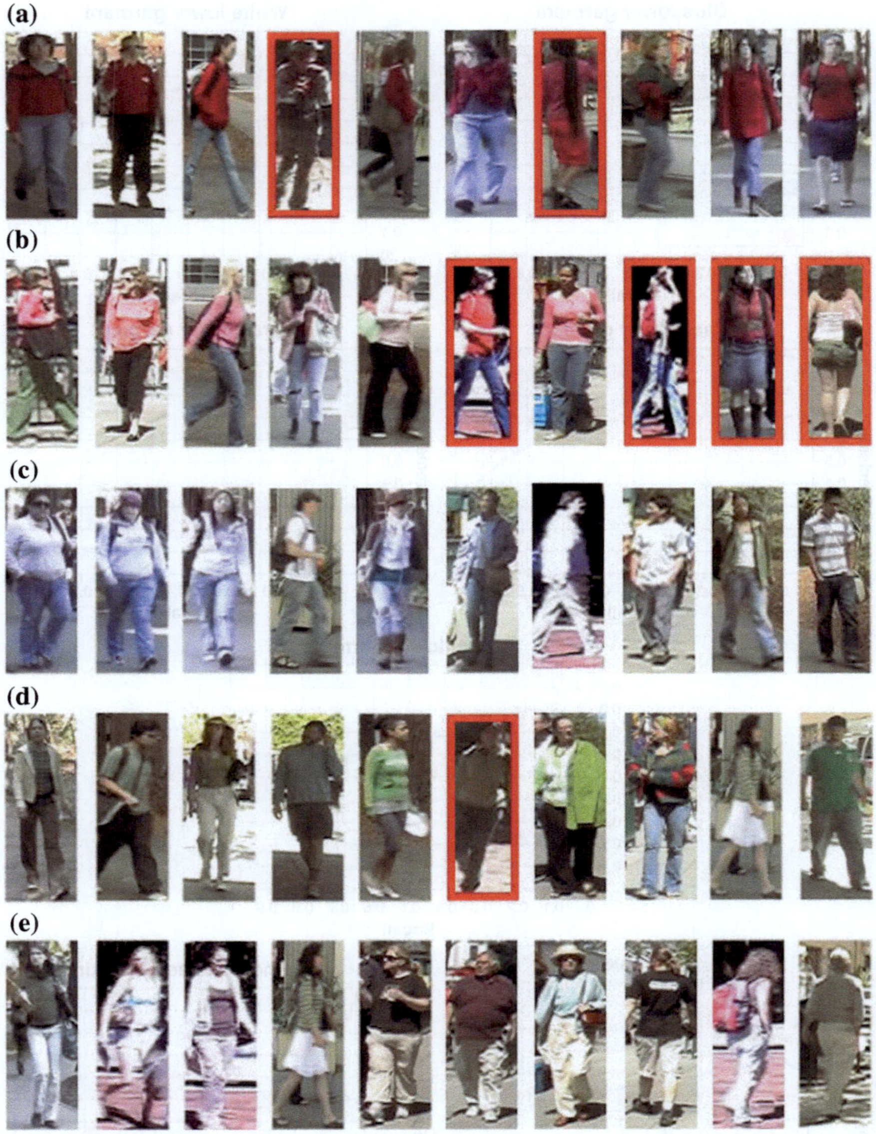

Fig. 18.7 The top ten images retrieved by MCD_2, for the queries "red upper garment" (**a**), "pink upper garment" (**b**), "white upper garment" (**c**), "green upper garment" (**d**) and "white trousers/skirt" (**e**), sorted from left to right for decreasing values of the score provided by the corresponding detectors. Non-relevant images are highlighted in *red*

Fig. 18.8 The top ten images retrieved by MCD_3, for the queries "white upper garment and blue trousers/skirt" (**a**) and "white upper garment and short sleeves" (**b**). The images are sorted from *left to right*, for decreasing values of the score computed using a fuzzy logic approach, as the minimum of the scores produced by the detectors of the basic queries. Non-relevant images are highlighted in *red*

information is encoded by new prototypes. On the contrary, a too high number of prototypes may increase the risk of over-fitting, when dissimilarity vectors are used as features of classification algorithms.

18.5 Conclusions

We proposed a general method for implementing a people search functionality with textual queries related to clothing appearance attributes, using any *multiple part-multiple component* descriptor. Such a kind of descriptor is used by most existing person re-identification techniques. This allows our method to be exploited for adding to a given re-identification system a search functionality based on textual queries, without using ad hoc features for attribute detection. This is particularly advantageous when the descriptors of images of individuals are already available (e.g. in real-time re-identification systems, where they are computed online during system operation): in this case, the input features of attribute detectors need not to be computed separately as in the case of ad hoc features [10, 11, 17, 18], thus reducing computational requirements. On the other hand, the attributes that can be detected using our method are constrained by the image descriptor used by the re-identification system at hand; in particular, they depend on the kind of low-level local features that encode each image component, and on the body part subdivision.

An interesting direction for further research is extending our approach to video sequences. To this aim one could exploit the pedestrian detection and tracking functionalities, which should be deployed as part of a person re-identification system, in addition to image descriptors. In this case, a bag of dissimilarity vectors coming from different frames is available for each tracked individual, instead of a single one. To exploit the additional information (e.g. different viewpoints of the body appearance), a Multiple Instance Learning approach [4] could be used for building detectors, where each frame corresponds to a single *instance* of the clothing appearance of an individual.

Another interesting development is adding a *semantic interpretation* step at the beginning of the pipeline, for interpreting a textual description of the individual of interest given in natural language, and automatically encoding it into a Boolean combination of the available basic queries. To this aim, Natural Language Processing (NLP) techniques [9] could be exploited. These techniques perform well when the application domain and the word ambiguities are limited, as in the case of clothing appearance attributes. People search systems would benefit from this functionality, under the viewpoint of user experience and usability.

Finally, the proposed technique could be conveniently exploited also for developing detectors for attribute-based person re-identification methods like [10, 11], where attributes are used in place of, or in addition to, low-level features and descriptors of clothing appearance.

Acknowledgments Federico Pala gratefully acknowledges the Sardinia Regional Government for the financial support of his Ph.D. scholarship (P.O.R. Sardegna F.S.E. Operational Programme of the Autonomous Region of Sardinia, European Social Fund 2007–2013—Axis IV Human Resources, Objective l.3, Line of Activity l.3.1).

References

1. Andriluka, M., Roth, S., Schiele, B.: Pictorial structures revisited: people detection and articulated pose estimation. In: Conference on Computer Vision and, Pattern Recognition, pp. 1014–1021 (2009)
2. Bazzani, L., Cristani, M., Murino, V.: Symmetry-driven accumulation of local features for human characterization and re-identification. Comput. Vis. Image Underst. **117**(2), 130–144 (2013)
3. Chang, C.C., Lin, C.J.: LIBSVM: A library for support vector machines. ACM Trans. Intell. Syst. Technol. **2**, 1–27 (2011) (Software available at http://www.csie.ntu.edu.tw/cjlin/libsvm)
4. Dietterich, T.G., Lathrop, R.H., Lozano-Pérez, T.: Solving the multiple instance problem with axis-parallel rectangles. Artif. Intell. **89**(1–2), 31–71 (1997)
5. Dollár, P., Babenko, B., Belongie, S., Perona, P., Tu, Z.: Multiple component learning for object detection. In: European Conference on Computer Vision, pp. 211–224 (2008)
6. Doretto, G., Sebastian, T., Tu, P.H., Rittscher, J.: Appearance-based person reidentification in camera networks: problem overview and current approaches. J. Ambient Intell. Humanized Comput. **2**(2), 127–151 (2011)
7. Farhadi, A., Endres, I., Hoiem, D., Forsyth, D.A.: Describing objects by their attributes. In: Conference on Computer Vision and Pattern Recognition (CVPR), pp. 1778–1785 (2009)

8. Gray, D., Brennan, S., Tao, H.: Evaluating appearance models for recognition, reacquisition, and tracking. In: IEEE International Workshop on Performance Evaluation of Tracking and Surveillance (2007)
9. Indurkhya, N., Damerau, F.J. (eds.): Handbook of Natural Language Processing, 2nd edition. Chapman and Hall/CRC, Boca Raton (2010)
10. Layne, R., Hospedales, T., Gong, S.: Person re-identification by attributes. In: British Machine Vision Conference, p. 24 (2012)
11. Layne, R., Hospedales, T.M., Gong, S.: Towards person identification and re-identification with attributes. In: ECCV 2012. Workshops and Demonstrations, pp. 402–412 (2012)
12. Pekalska, E., Paclik, P., Duin, R.P.: A generalized kernel approach to dissimilarity based classification. J. Mach. Learn. Res. **2**, 175–211 (2001)
13. Pekalska, E., Duin, R.P.: The Dissimilarity Representation for Pattern Recognition: Foundations And Applications (Machine Perception and Artificial Intelligence). World Scientific Publishing Co, River Edge (2005)
14. Satta, R., Fumera, G., Roli, F., Cristani, M., Murino, V.: A multiple component matching framework for person re-identification. In: International Conference on Image Analysis and Processing, pp. 140–149 (2011)
15. Satta, R., Fumera, G., Roli, F.: A general method for appearance-based people search based on textual queries. In: ECCV 2012. Workshops and Demonstrations, pp. 402–412 (2012)
16. Satta, R., Fumera, G., Roli, F.: Fast person re-identification based on dissimilarity representations. Pattern Recogn. Lett. **33**(14), 1838–1848 (2012)
17. Thornton, J., Baran-Gale, J., Butler, D., Chan, M., Zwahlen, H.: Person attribute search for large-area video surveillance. In: IEEE International Conference on Technologies for Homeland, Security, pp. 55–61 (2011)
18. Vaquero, D., Feris, R., Tran, D., Brown, L., Hampapur, A., Turk, M.: Attribute-based people search in surveillance environments. In: IEEE Winter conference on Applications and Computer Vision, pp. 1–8 (2009)
19. Yu, F.X., Ji, R., Tsai, M.H., Ye, G., Chang, S.F.: Weak attributes for large-scale image retrieval. In: Conference on Computer Vision and Pattern Recognition (CVPR), pp. 2949–2956 (2012)

Chapter 19
Large-Scale Camera Topology Mapping: Application to Re-identification

Anthony Dick, Anton van den Hengel and Henry Detmold

Abstract In this chapter we describe the problem of camera network topology mapping, which is a critical precursor to person re-identification in large camera networks. After surveying previous approaches to this problem we describe "exclusion", a practical, robust method for deriving a topology estimate that scales to thousands of cameras. We then consider re-identification within such networks by modelling and matching target appearance. By combining a simple appearance model with the topology estimate generated by exclusion, person re-identification can be accomplished within far larger scale networks than would be possible using appearance matching alone.

19.1 Introduction

Manual inspection is an inefficient and unreliable way to monitor large surveillance networks, particularly when coordination across observations from multiple cameras is required. In response to this, several systems have been developed to automate inspection tasks that span multiple cameras, such as target re-identification, following a moving target, or grouping together related cameras.

Although there is a growing literature on these tasks, these generally require some prior knowledge about the layout of the cameras. This may include whether

A. Dick (✉)
School of Computer Science, University of Adelaide, North Terrace, Adelaide 5005, SA, Australia
e-mail: anthony.dick@adelaide.edu.au

A. van den Hengel
University of Adelaide, Adelaide, SA, Australia
e-mail: anton.vandenhengel@adelaide

H. Detmold
Snap Network Surveillance, Adelaide, SA, Australia
e-mail: henry@snapcontrol.com

S. Gong et al. (eds.), *Person Re-Identification*,
Advances in Computer Vision and Pattern Recognition,
DOI: 10.1007/978-1-4471-6296-4_19,

the cameras overlap, and if so geometric information about their overlap region, or their coverage on a ground plane. In early surveillance systems, this information was manually specified or derived from camera calibration, but recent systems at least partly automate the process by analysing videos from the cameras. These systems are demonstrated on networks consisting of the orders of 10 cameras, but it is required that they cannot be applied to networks that are orders of magnitude larger. In this chapter, we investigate fully automatic ways of obtaining camera layout information that scale to a large number—potentially hundreds—of cameras. This is a requirement for tracking and re-identification in networks of this size [23, 31], as the difficulty of re-identification increases with the square of the number of observed targets if layout information is not used to constrain the search space. We evaluate the measures typically used to detect camera overlap and the assumptions underlying them, from the point of view of scalability and usefulness for re-identification and other applications that depend on this information. Our analysis shows that mutual information is well suited to overlap detection, in combination with an "exclusion" cost function, and we show how these can be computed for large camera networks. We then demonstrate that this overlap estimate enables accurate target re-identification in a real-world camera network using a simple appearance model, without target motion estimation.

19.1.1 Related Work

Despite the quantity of papers on cross-camera target identification, there is relatively little work on automatically obtaining the requisite prior information on camera layout. Calculating this information was originally approached as a problem of learning the probability of transitions between fields of view (FOVs) from corresponding tracks [14]. However, correspondence between tracks in different images must be supplied a priori as training data, which limits the scale of the network for which it is useful. Dick et al. [10] suggest an alternate approach whereby learn activity topology is represented by a Markov model. This does not require correspondences, but does need a training phase and does not scale well. Ellis et al. [11] do not require correspondences or a training phase, instead observing motion over a long period of time and accumulating appearance and disappearance information in a histogram. This approach has been extended by Stauffer [28], Tieu et al. [30] and Chen [6] to include a more rigorous definition of a transition based on statistical significance, and by Gilbert et al. [12] to incorporate a coarse to fine topology estimation. More recently, Chen et al. [8] use this approach to feed back into the cross-camera matching measure, while Loy et al. [20] develop a correlation measure that more robustly captures activity at each location. A topology graph is inferred from observations sampled over long-time periods in [22]. These methods show promise for larger scale applications, but have only been demonstrated on networks of approximately 10 cameras or less.

Rahimi et al. [25, 26] perform experiments on configurations of several cameras involving non-overlapping FOVs. One experiment [25] involves calculating the 2D position and 1D orientation of a set of overhead cameras viewing a common ground plane by manually recording the paths followed by people in each camera's FOV. It is shown that a simple smoothness prior is enough to locate the cameras and reconstruct a path where it was not visible to any camera. In another experiment [26], pre-calibrated cameras are mounted on the walls of a room so that they face horizontally. In this case, the 2D trajectory of a person is recovered as they move around in the room, even when the person is not visible to any camera. An EM algorithm is used to join partial tracks in [34], a particle filter in [19], kernel density estimation in [15], an incremental method in [7], and a graph optimisation in [33], although all of these papers assume some knowledge of the environment (for example, the ground plane). Anjum et al. [1] instead build environmental knowledge incrementally from camera pairs, but this restricts the scale of the camera network. In [27] the problem is reformulated, using cross-camera matching to control PTZ cameras dynamically to keep a target in view.

On a larger scale, Brand et al. [3] consider the problem of locating hundreds of cameras distributed about an urban landscape. Their method relies on having internal camera calibration and accurate orientation information for each camera, and enough cameras viewing common features to constrain the solution. Given this information the method can localise the cameras accurately due to the constraints imposed by viewing common scene points from different viewpoints.

An important step towards recovering spatial camera layout is to determine where cameras overlap. The approach taken in [18] is to estimate motion trajectories for people walking on a plane, and then match trajectories between cameras. However, this assumes planar motion, and accurate tracking over long periods of time. It also does not scale well, since track matching complexity increases as $O(k^2)$ with the number of cameras k. In [17] evidence for overlap is accumulated by estimating the boundary of each camera's field of view in all other cameras. Again, this does not scale well to large numbers of cameras, and assumes that all cameras overlap. In [16], a moving camera is used to scan the area under surveillance and thereby link together static, non-overlapping views. This works across a small area but again does not scale to large areas that may contain hundreds of cameras and cannot be viewed by a single moving camera. In [5], the reverse problem is addressed: devising a camera layout so that it is possible to hand over tracks between cameras for any target trajectory within a defined area.

Most methods for determining spatial layout start with an assumption of camera independence, and gradually accumulate evidence for connections. This relies on accurately detecting and/or tracking objects over a long time period. They also require comparisons to be made between every pair of cameras in a network. The number of pairs of cameras grows with the square of the number of cameras in the network, rendering exhaustive comparisons infeasible. An exception is [13], which proposes a measure that estimates camera connections by gradually eliminating camera pairs for which there is strong evidence to the contrary.

This chapter introduces and compares a number of alternative measures for determining camera overlap in large networks. Issues considered include scalability, robustness to poor quality data and suitability for higher level surveillance tasks such as cross-camera tracking. Several overlap measures are considered, and each is evaluated in a variety of scenarios, using both synthetic and real footage from up to 1,400 cameras. For each measure, we consider three loss functions that can be applied to the measure to decide whether camera pairs overlap. We find that a loss function we call *exclusion* is in fact more suitable in many cases than the usual loss functions used in previous work.

The method we propose is computationally fast, and does not rely on accurate tracking of objects within each camera view. In contrast to most existing methods, it does not attempt to build up evidence for camera overlap over time. Instead, it starts by assuming all cameras are connected and uses observed activity to rule out connections over time. This is an easier decision to make, especially when a limited amount of data are available. It is also based on the observation that it is impossible to prove a positive connection between cameras—any correlation of events could be coincidence—whereas it is possible to prove a negative connection by observing an object in one camera while not observing it at all in another.

19.2 Detecting Camera Overlap

Consider a set of k cameras that generates k images at time t. By applying foreground detection [29] to all images we obtain a set of foreground blobs, each of which can be summarised by an image position and camera index. Each image is partitioned into cells, and each cell can be labelled "*occupied*" or "*unoccupied*" depending on whether it contains a foreground object. A cell is thus the 'unit' of object location and therefore of overlap; it can be defined as a single pixel, an image region, or an entire image, but it should be based on the accuracy with which the detector is typically able to locate objects. In our implementation we divide each image into a regular grid of 9×12 square cells, although we have also experimented with cells that are equivalent to a camera's entire field of view.

Let the set of cells in all images be denoted $\mathcal{C} = \{c_1 \dots c_m\}$. Corresponding to each cell c_i is an occupancy vector $\mathbf{o}_i = (o_{i1}, \dots, o_{iT})'$ with o_{it} set to 1 if cell c_i is occupied at time t, and 0 if not. For conciseness, we drop the t when describing events that occur at a single time.

We define a binary random variable O_i which is 1 if cell c_i is occupied and 0 otherwise. In this section, we consider a range of measures operating on pairs of these variables for different cells: O_i, O_j, and consider their usefulness for estimating cell overlap. These measures are based entirely on occupancy as represented by instances of the random variable O: in other words, appearance information is not used beyond determining occupancy in each cell. This differs from typical matching-based approaches that are used to relate small numbers of cameras. For larger camera networks, appearance matching is often ineffectual, due to the large number of candidate matches and the presence of images with poor quality or strongly varying environmental conditions.

Cell overlap can range between 0 (no overlap) and 1 (complete overlap), and can assume a value between 0 and 1 indicating partial overlap. The value in this case is related to the area of spatial overlap between the cells, but actually measures the probability of observing a target in one cell given its appearance in the other. The measured overlap therefore depends also on the paths that people take through the environment: for example, two cells may share only 10 % of their area, but if that shared area is more frequently occupied than the remainder of the cells (e.g. a corridor with heavy traffic), their degree of overlap will be greater than 10 %. Note also that overlap is not symmetrical, as will be examined in Sect. 19.2.3.

Coupled with a decision threshold, each of the measures described in this section constitutes a discriminant function [2]. That is, each cell pair can be classified as overlapping or not by comparing the value of the overlap measure to the threshold.

The value of the decision threshold is therefore critical but also application dependent. When evaluating each measure, we consider three different loss functions and the corresponding thresholds that apply. First is a loss function to simply minimise the probability of misclassification. The others are denoted *inclusion*, in which the penalty for concluding that two cells overlap when they do not is higher than the reverse; and *exclusion*, where the penalty for deciding that cells do *not* overlap in the case that they actually do overlap is higher.

Previous work on this topic (e.g. [11]) implicitly adopts the inclusion loss function—they begin by assuming no cameras overlap and conservatively accumulate evidence for pairwise overlap. This produces estimates that approach the true solution from below, only forming connections between cells for which there is sufficient evidence. Although it is typically the case in a large camera network that overlapping cell pairs are in the minority, we find that for many applications it is better to use the exclusion loss function. For example, when tracking across cameras, it is more useful to have an estimate of camera connections that is a superset of the truth than one which is a subset, because missing links means that object tracks will be broken, whereas extra links merely increase the cost and difficulty of correctly tracking the object. We therefore consider both approaches when evaluating overlap measures, and evaluate both for our datasets.

We now propose four measures of cell pair overlap based on occupancy vectors: mutual information, lift, conditional entropy and conditional probability.

19.2.1 Mutual Information

If two cells c_i and c_j are at least partially overlapping, then O_i and O_j are not independent. Conversely, if cells c_i and c_j are not overlapping or close to overlapping, we assume that O_i and O_j at each timestep are independent. We can test for independence by calculating the mutual information between these variables:

$$MI(O_i, O_j) = \sum_{o_i \in O_i, o_j \in O_j} \Pr(o_i, o_j) \log \frac{\Pr(o_i, o_j)}{\Pr(o_i)\Pr(o_j)} \tag{19.1}$$

MI ranges between 0, indicating independence, and the entropy $H(O_i)$, indicating that O_j is completely determined by O_i.

19.2.2 Lift

In some scenarios, occupancy data can be very unbalanced; for example in low traffic areas, $O = 1$ is far less probable than $O = 0$. This means that the joint observation (0, 0) is not in fact strong evidence that this cell pair is related. This can make decisions based on calculation of the mutual information difficult, as the entropy of each variable is already low, and MI ranges between 0 (independence) and $\max(H(O_i), H(O_j))$ (complete dependence).

One solution is to measure the independence of the *events* $O_i = 1, O_j = 1$ rather than the independence of the *variables* O_i, O_j. This leads to a measure known as *lift* [4]:

$$\text{lift}(O_i, O_j) = \frac{\Pr(O_i = 1, O_j = 1)}{\Pr(O_i = 1)\Pr(O_j = 1)} \tag{19.2}$$

Lift ranges between 1, indicating independence (non-overlap) and $1/\Pr(O_j = 1)$, indicating that cells i and j are identical.

19.2.3 Conditional Entropy

Dependency between cell pairs is not necessarily symmetric. For example, if a wide angle camera and a zoomed camera are viewing the same area, then a cell c_i in the zoomed camera may occupy a fraction of the area covered by a cell c_j in the wide angle camera. In this case, $O_i = 1$ implies $O_j = 1$, but the converse is not true. Similarly, $O_j = 0$ implies $O_i = 0$, but again the converse is not true. To this end we can measure the conditional entropy of each variable:

$$CE(O_i, O_j) = -\sum_{o_i \in O_i, o_j \in O_j} \Pr(o_i, o_j) \log \Pr(o_i | o_j) \tag{19.3}$$

$$CE(O_j, O_i) = -\sum_{o_i \in O_i, o_j \in O_j} \Pr(o_i, o_j) \log \Pr(o_j | o_i) \tag{19.4}$$

CE ranges between $H(O_i)$, indicating independence, and 0, indicating that O_j is completely dependent on O_i.

Table 19.1 Occupancy counters used to calculate overlap measures. In a large network, it is common for some cameras to be unavailable at any given time due to maintenance or other factors.

Counter	Definition
T	the number of frames for which the network has been operating
O_i	the number of frames at which cell c_i is occupied: an n element vector, where n is the number of cells
U_i	the number of frames at which cell c_i is unoccupied: an n element vector
X_i	the number of frames at which the camera of c_i is unavailable: a k element vector, where k is the number of cameras
OX_{ij}	the number of frames at which cell c_i is occupied and the camera of c_j is unavailable: an $n \times k$ matrix
OU_{ij}	the number of frames at which cell c_i is occupied and cell c_j is occupied: a $n \times n$ matrix
XX_{ij}	the number of frames at which the camera of c_i is unavailable and the camera of c_j is unavailable: a $k \times k$ matrix

19.2.4 Conditional Probability

Combining the ideas of non-symmetry between cell pairs (CE measure) with that of measuring dependence only of events rather than variables (lift), we arrive at a conditional probability measure for cell overlap:

$$CP(O_i, O_j) = \Pr(O_j = 1 | O_i = 1) \quad \text{or} \quad \frac{\Pr(O_i = 1, O_j = 1)}{\Pr(O_i = 1)} \tag{19.5}$$

which can be seen to be a non-symmetric version of lift, and also analogous to conditional entropy (based on the same quantities). CP ranges between $\Pr(O_j = 1)$, indicating independence, and 1, indicating complete dependence.

19.3 Implementation

19.3.1 Calculating Cell Occupancy Probability

As we are developing a system to operate over large, continuously operating camera networks, it is important to be able to calculate cell occupancy probabilities efficiently and scalably. All the required quantities can be calculated by maintaining the counters in Table 19.1.

For example, the mutual information measure requires the calculation of each of the joint probabilities for cell pair, which can be carried out as follows:

$$\begin{aligned}
\Pr(O_i = 1, O_j = 1) &= \left(O_j - OU_{ji} - OX_{ji}\right)/A_{ij} \\
\Pr(O_i = 1, O_j = 0) &= \left(OU_{ij}\right)/A_{ij} \\
\Pr(O_i = 0, O_j = 1) &= \left(OU_{ji}\right)/A_{ij} \\
\Pr(O_i = 0, O_j = 0) &= (U_j - OU_{ij} \\
&\quad - (X_i - XX_{ij} - OX_{ji}))/A_{ij}
\end{aligned} \tag{19.6}$$

where A_{ij} is the number of frames at which both cells c_i and c_j are available, which can be calculated by summing over observed values:

$$\begin{aligned}
A_{ij} = O_j - OU_{ji} - OX_{ji} + OU_{ij} + OU_{ji} + U_j - OU_{ij} \\
- (X_i - XX_{ij} - OX_{ji})
\end{aligned} \tag{19.7}$$

The other probability terms can be calculated similarly. Although other combinations of counters are possible, those in Table 19.1 are chosen to minimise storage requirements and expected update frequency for the matrices in large networks. In particular, if all cameras are available and all cells unoccupied, no matrix elements are updated. Clearly, OU_{ij} requires by far the largest amount of storage (space complexity is $O(n^2)$), though OX_{ij} and XX_{ij} are also $O(k^2)$ in space. Thus, each measure is simple to compute, but storing the required variables in memory becomes a significant bottleneck as the number of cameras increases. Our solution to this is to partition the memory usage across multiple machines, as described in Sect. 19.4.3.

19.3.2 Camera Synchronisation

In existing camera networks, it is rare for cameras to be synchronised to a global clock. Thus, we need to allow some tolerance when detecting overlapping cells in different cameras based on "simultaneous" occupancies. We take a straightforward padding approach to this problem: two observations in different cameras that occur within a small-time interval (typically 20 frames) of each other are considered candidates for correspondence.

Although it would be possible to implement a pre-processing synchronisation procedure in some cases, for large and operational networks this is difficult. In practice, we find that a small amount of padding produces a significant increase in overlap detection without markedly diminishing from precision. Adding temporal padding also gives us the ability to detect cameras that are nearby but do not overlap. We have not analysed this quantitatively, but in general, the greater the degree of padding, the greater the gap in coverage that can be tolerated.

19.3.3 Object Summarisation

A detected object may occupy many adjacent cells within a camera, but to maintain consistency we wish to summarise its location as one cell only. We do this by calculating the lowest visible extent (LVE) of the object outline: a bounding box is estimated for the object, and the cell in which the midpoint of its lower boundary lies is used as the location for that object.

Use of the LVE rather than the centroid of the object is motivated by the desire to estimate all object locations on a single ground plane, which is consistent across all cameras regardless of the obliqueness of their field of view. Of course this is not always possible, due to occlusions or detector errors, and we find that this is an occasional source of error in cluttered environments. However, such errors do not affect the overlap estimate which is based on measurements accumulated over a significant period of time.

19.4 Results

We evaluate our algorithm and its implementation based on three criteria: (i) the accuracy of the overlap estimate; (ii) the applicability of the overlap estimate for target re-identification; and (iii) its scalability to networks containing over 1,000 cameras.

19.4.1 Ground Truth Comparison

We evaluate the accuracy of the method in terms of *precision-recall* [24] (PR) for each measure when compared with ground truth labelled overlap. Precision and recall are measured for varying decision threshold applied to each of the overlap measures in Sect. 19.2.

Five Camera Network

Initially, we test the measures on a 5-camera network for which binary values for cell-to-cell overlap have been manually measured. The measured overlap for the five cameras is shown in Fig. 19.1. This camera layout is tested with two kinds of traffic. First, a remote control car is driven around the area monitored by the cameras for approximately half an hour. Second, several people walk around in the area. This second dataset is more challenging because we are tracking the motion of people's feet, which are more difficult to localise to a particular location on the ground plane. The difference in the PR curves (shown in Fig. 19.2a, b respectively) for these two

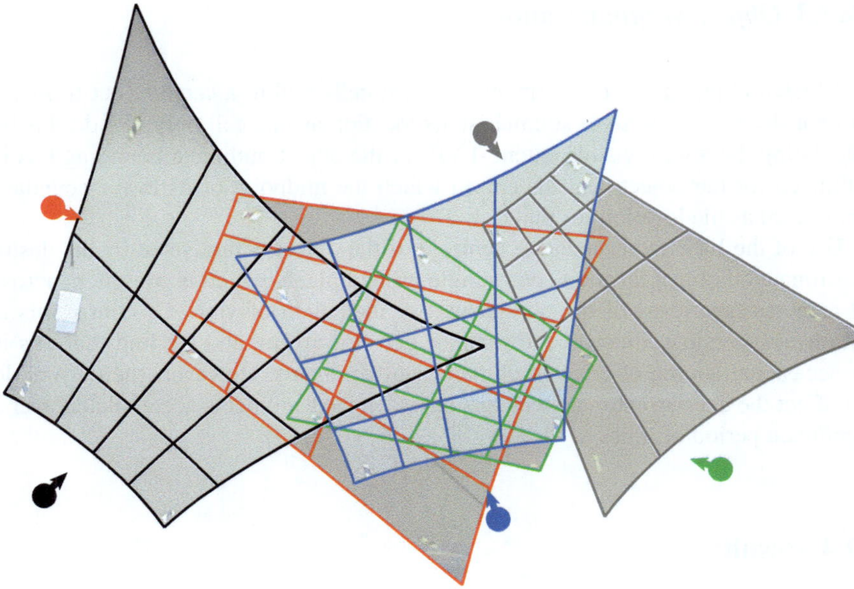

Fig. 19.1 Ground truth overlap for 5-camera network. A 3 × 4 grid of cells is superimposed on each camera

cases is due to the presence of more than one target in the scene, which introduces occlusions and matching ambiguity, and the difficulty of locating the LVE of a person with the same accuracy as a car.

In both cases, the conditional entropy (CE) measure is significantly worse than the others. This is due to its dependence on the entropy observed in each cell, which in this case is low due to the sparse nature of the observed activity. The CE measure can therefore not be compared to a fixed threshold to detect cell overlap.

We evaluate an optimal point along each precision-recall (PR) curve based on each of the min-error, inclusion and exclusion loss functions described earlier. We do this by maximising versions of the F-measure

$$F_f = \frac{(1+f^2) \times P \times R}{f^2 \times P + R} \tag{19.8}$$

to decide the optimal point on each PR curve for each loss function. Here, f denotes the relative importance of precision and recall. We set $f = 1$ for the min-error loss function, $f = 1/3$ for the inclusion loss function and $f = 3$ for the exclusion loss function.

The optimal precision and recall values for each loss function are shown in Table 19.2. In each case it can be seen that the exclusion loss function favours precision over recall: falsely removing overlap is penalised more strongly than falsely including it. Conversely, the inclusion loss function favours recall over precision.

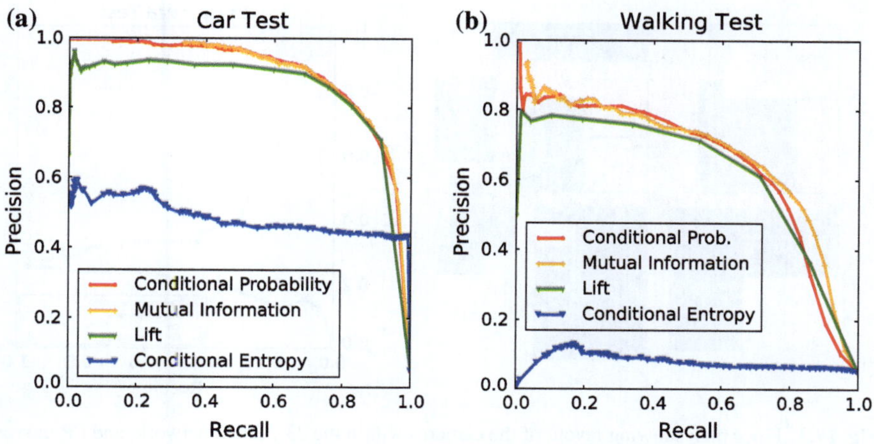

Fig. 19.2 Comparisons of overlap measures with varying threshold for the 5 camera ground truth datasets

Table 19.2 Comparison to ground truth for mutual information, using each loss function, for car dataset

	Min-error	Inclusion	Exclusion
MI	0.88, 0.82	0.77, 0.92	0.92, 0.76
CP	0.83, 0.85	0.76, 0.89	0.92, 0.75

Each entry is the corresponding precision, recall

23 Camera Office Dataset

The 23 camera dataset was obtained from a network of surveillance cameras installed in offices and corridors at a university campus. They cover a number of corridors inside the building, with a floor plan shown in Fig. 19.3 (left). The cameras recorded many people moving around the area and interacting over a four-and-a-half hour period. This dataset has a significantly higher degree of difficulty and activity than the previous two datasets. Due to the size of the data, the ground truth was determined by linking cells to other cameras rather than individual cells. The estimated topologies were evaluated with cell links being considered correct when they connected to the appropriately overlapping camera view.

Figure 19.3 (right) shows the PR curve for varying thresholds this dataset. It is comparable, and in some cases exceeds, the results for the 5-camera dataset. This is partly due to the cameras being positioned further from the targets, which makes them easier to track, but it still shows the scalability of the approach.

To illustrate the type of conditions in which we are operating, Fig. 19.4 shows a recovered topology for the 23 camera office network. Connections between views are correctly determined despite the extreme difference in viewpoints and scale between views. The four overlapping views shown in Fig. 19.5 provide an example of this

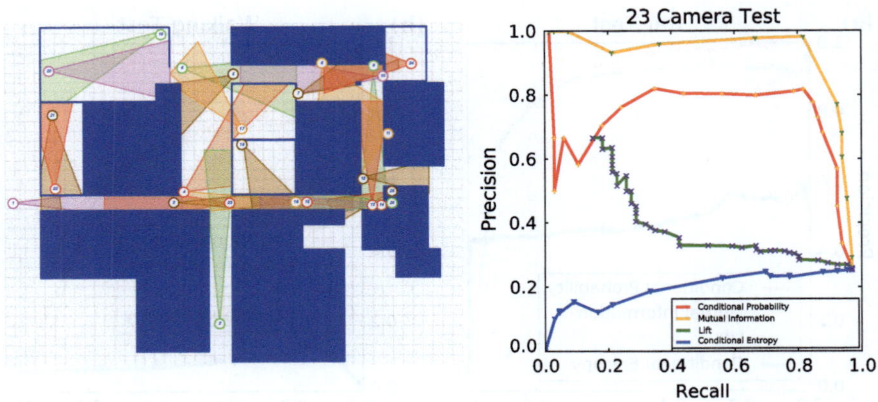

Fig. 19.3 Floor plan showing layout of the cameras within the 23 camera network, and PR curves resulting from overlap detection on this network

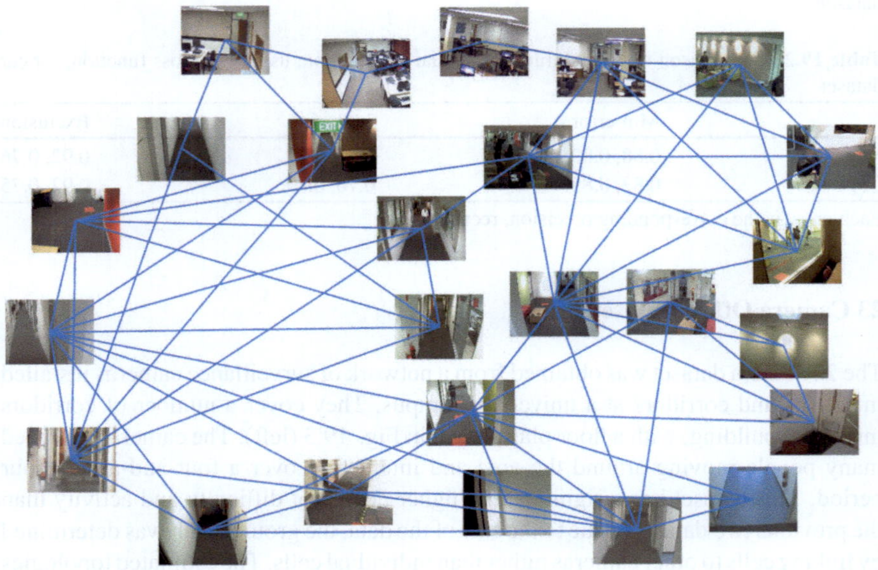

Fig. 19.4 Topology for the 23-camera network recovered by an exclusion loss function, after viewing an hour of footage. Each camera is a single cell

effect. The connection between the bottom left and top right cameras is correctly identified despite a significant difference in scale; the entire overlap area occupies one cell at the end of a darkened corridor in the bottom left camera, whereas it occupies 10 cells in the top right camera. Similarly, the green furniture in the top left and top right images are matched despite being viewed from cameras with opposing view angles.

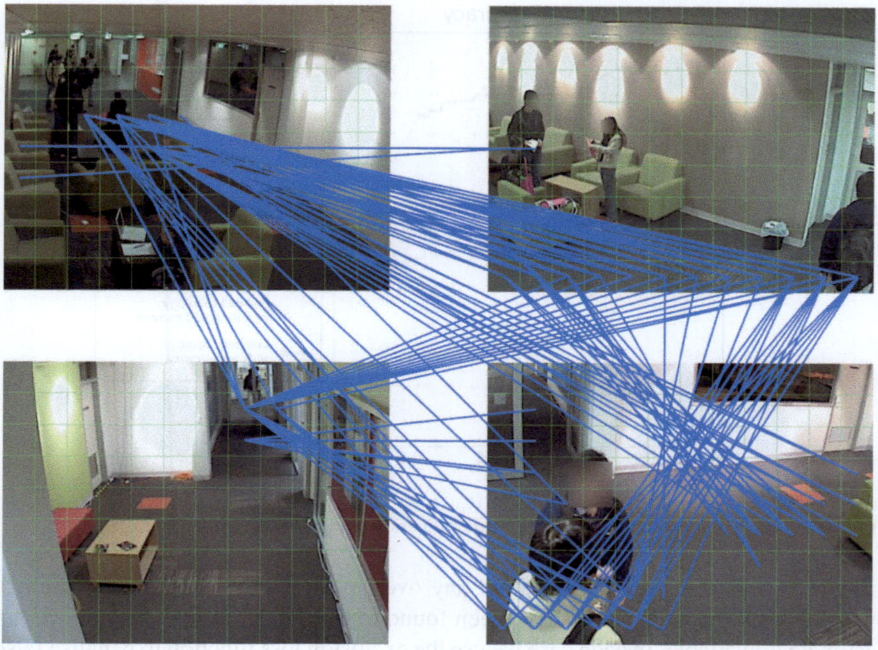

Fig. 19.5 A 4-camera subset of the topology for the 23-camera network recovered by an exclusion loss function, after viewing an hour of footage, where each camera is a 9 × 12 cell grid

19.4.2 Application to Re-identification

Target tracking across large camera networks is impeded by the need to re-identify targets that were previously seen in other cameras when they reappear. Without knowledge of camera layout, this must be done based only on target appearance, which becomes increasingly unreliable as the size of the network grows.

By using overlap information we can reduce the space within which we must search when a target enters or leaves a camera's field of view. Importantly, the size of the search space is independent of the number of cameras in the network, instead depending only on their connectivity.

In our experiments we tracked people as they moved through the office space monitored by our 23-camera network. We attempted to re-identify targets on the basis of appearance alone, and also on the basis of the overlap estimated by the mutual information (MI) method and the exclusion loss function. For comparison, we used 9 × 12 cells per camera, and also 1 cell per camera.

To measure the accuracy of re-identification, we manually labelled 500-target transitions chosen at random, and then tested how many of these were found by each method. Each method represents target appearance as a RGB colour histogram and attempts to match targets using the Bhattacharyya distance; the only difference is

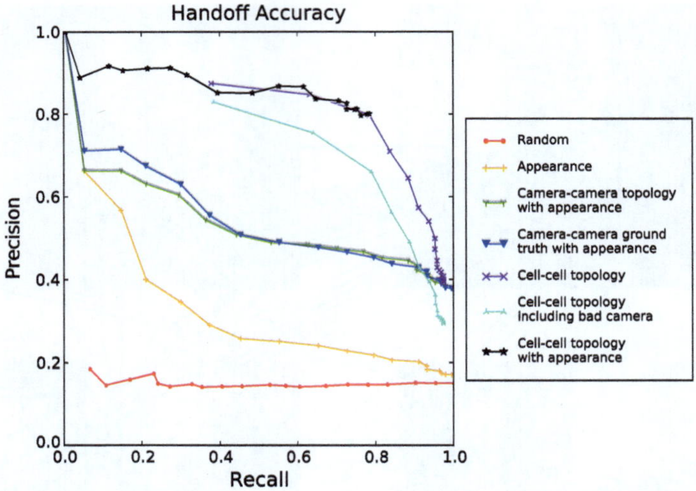

Fig. 19.6 Precision and recall of target re-identification using overlap estimation

in which cells are searched. Without any overlap estimate, all cells are searched, whereas only the cells which have been found to overlap are used when an overlap estimate is available. In each case we use the exclusion loss function to penalise false negative overlap more than false positive; this likely results in an overestimate of the overlap in the network. Note that matching at each frame is performed independently, so no frame-to-frame tracking is required. The RGB histogram is computed over all adjacent occupied cells. Results are summarised in Fig. 19.6.

This reveals three distinct levels of performance. Worst is using appearance information only, as might be expected given that all targets in the network are candidate matches. In fact, appearance on its own, without topology information, only marginally beats a randomly generated matching of available targets. Best is to use an overlap estimate based on a 9×12 grid of cells in each camera. Note that using appearance to match targets within the cell–cell topology does not significantly increase precision for a given level of recall. This suggests that even when the search space is highly restricted, the appearance of targets in different views is so different that colour histogram is unable to effectively match them. In other words, the topology estimation is the dominant factor behind matching performance. Note that one camera malfunctioned severely during the experiment, producing only intermittent and noisy images; this is included only in the 'bad camera' result. It is consistently lower than the result using only reliable cameras, but not catastrophic, indicating the robustness of the method.

Between having no topology and cell–cell topology, the overlap estimate based on entire camera fields of view constrains the search space but not as tightly as the 9×12 cell grid. Interestingly, the ground truth camera–camera overlap performs only slightly better than our automatically obtained overlap estimate, indicating that our overlap estimate is accurate.

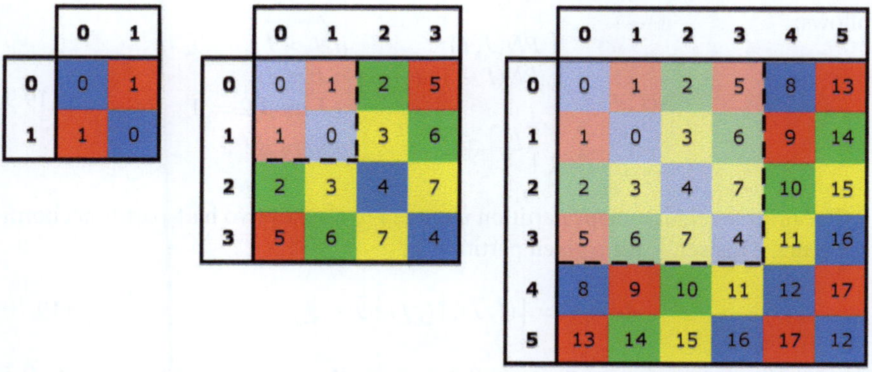

Fig. 19.7 Partitioning the *OU* matrix. This partitioning scheme allows cameras to be added and removed from the network dynamically, and to use memory distributed across a network. Each partition is numbered and shown in a different colour

19.4.3 Scalability to Large Networks

Partitioned Occupancy Matrix

With 4 byte counts and 108 cells per camera (in a 9×12 grid), the *OU* matrix required to estimate overlap for a 1,000-camera network requires over 43GB of RAM! In order to scale to networks of multiple thousands of cameras, we partition the $n \times n$ matrix *OU* across N partitions distributed across a cluster of servers. The partitioning scheme has several desirable properties, for a network of k cameras:

1. Processing for the k^2 camera pairs is distributed evenly across N partitions.
2. Symmetry: if camera pair (i, j) is in partition P, then (j, i) is also in partition P.
3. Containment: the number of partitions can be increased, and hence more cameras supported, without affecting existing partitions in any way, so that the capacity of the partitioned can be expanded whilst it remains online, an important property for surveillance systems requiring 24×7 operation.
4. The network bandwidth and memory requirements are understood in terms of simple functions of k and N [9]; these functions are verified empirically in "Analysis and Results".

Figure 19.7 illustrates the containment property for a system growing from 2 partitions to 8 partitions and then to 18 partitions. Note that to achieve the symmetry property each partition consists of two square regions (termed *half partitions*) in the matrix.

There is a requirement for an efficient bidirectional mapping between coordinates in the partitioned matrix (i.e. the logical location of the data) and integral partition numbers (which are used to locate partitions on the network, and hence physically locate the data). Specifically, the mapping must not involve any network communication. The mapping from partition coordinates, (I, J), to partition numbers is as follows:

$$PN(I,J) = \begin{cases} PN(J,I) & \text{if } J > I \\ PN(I-1,J-1) & \text{if } I = J \wedge \\ & I \bmod 2 = 1 \\ \left\lceil \frac{I^2}{2} \right\rceil + J & \text{otherwise} \end{cases} \tag{19.9}$$

The inverse relation maps partition numbers to a set of two half-partition coordinate pairs. This set, $\mathbb{P}$ for a given partition is:

$$\mathbb{P} = \{(\overline{I}, \overline{J}), (\underline{I}, \underline{J}) : \overline{J} < \underline{J}\} \tag{19.10}$$

The coordinate pair $(\overline{I}, \overline{J})$ is termed the upper half-partition, and the pair $(\underline{I}, \underline{J})$ is termed the lower half-partition; they are distinguished based on the y axis coordinate, as shown.

The x coordinate of the upper half-partition, $\overline{I}$, is a function of the partition number, P:

$$\overline{I} = \left\lfloor \sqrt{2P} \right\rfloor \tag{19.11}$$

and the y coordinate of the upper half-partition, $\overline{J}$, is a function of the partition number and the x coordinate:

$$\overline{J} = P - \left\lceil \frac{\overline{I}^2}{2} \right\rceil \tag{19.12}$$

Combining Eqs. 19.11 and 19.12 yields the following, *upper half-partition address function*:

$$UHPA(P) = \left(\left\lfloor \sqrt{2P} \right\rfloor, P - \left\lceil \left\lfloor \sqrt{2P} \right\rfloor^2 /2 \right\rceil \right) \tag{19.13}$$

Next, the lower half-partition coordinate pair, $(\underline{I}, \underline{J})$, is a function of the upper half-partition coordinate pair, $(\overline{I}, \overline{J})$, as follows:

$$(\underline{I}, \underline{J}) = \begin{cases} (\overline{I}+1, \overline{J}+1) & \text{if } \overline{I} = \overline{J} \\ (\overline{J}, \overline{I}) & \text{otherwise} \end{cases} \tag{19.14}$$

Combining Eqs. 19.11, 19.12 and 19.14 yields the following, *lower half-partition address function*:

$$LHPA(P) = \begin{cases} \left(\left\lfloor\sqrt{2P}\right\rfloor + 1, \left(P - \left\lceil \left\lfloor\sqrt{2P}\right\rfloor^2/2 \right\rceil\right) + 1\right) \\ \quad \mathit{if} \left\lfloor\sqrt{2P}\right\rfloor = P - \left\lceil \left\lfloor\sqrt{2P}\right\rfloor^2/2 \right\rceil \\ \left(P - \left\lceil \left\lfloor\sqrt{2P}\right\rfloor^2/2 \right\rceil, \left\lfloor\sqrt{2P}\right\rfloor\right) \\ \quad \mathit{otherwise} \end{cases} \tag{19.15}$$

This partitioning scheme has been deployed across a cluster of 16 commodity servers, each with 4 GB of RAM and hosting 2 partitions (i.e. 32 partitions with 2 GB per partition), giving a total of 64 GB available for storage of OU. Results for a partitioned deployment are presented in "Analysis and Results".

Analysis and Results

We next analyse the performance of the overlap estimation algorithm when distributed across multiple partitions, which are stored on separate machines.

Results are reported for running distributed exclusion for surveillance networks of between 100 and 1,400 cameras, using 32 partitions. The occupancy data is derived by running foreground detection on 2 h footage from a real network of 132 cameras and then duplicating occupancy data as necessary to synthesise 1,400 input files, each with 2 h of occupancy data. These files are then used as input for the estimation partitions, enabling us to repeat experiments. The use of synthesis to generate a sufficiently large number of inputs for the larger tests results in input that contains an artificially high incidence of overlap, since there is complete overlap within each set of input replicas. The consequence of this is that the time performance of estimation computations is slightly worse than it would be in a real network.

Our experimental platform is a cluster of 16 servers, each with two 2.0 Ghz dual-core Opteron CPUs and each server having 4 GB of memory. We instantiate up to 32 estimation partitions (of size up to 2 GB) on this platform.

Results are verified against a previous, non-partitioned implementation of the exclusion approach. As in Sect. 19.4.2, this previous implementation exhibits sufficient precision and recall of the ground truth overlap to support tracking. The partitioned implementation achieves very similar results for the same data. Minor differences arise because the distributed implementation uses a less CPU intensive approach to dealing with clock skew.

Performance Results

Figure 19.8 shows the arithmetic mean of measurements of memory usage and CPU time, for 32 8 GB partitions spread across the 16 servers. The actual measurements

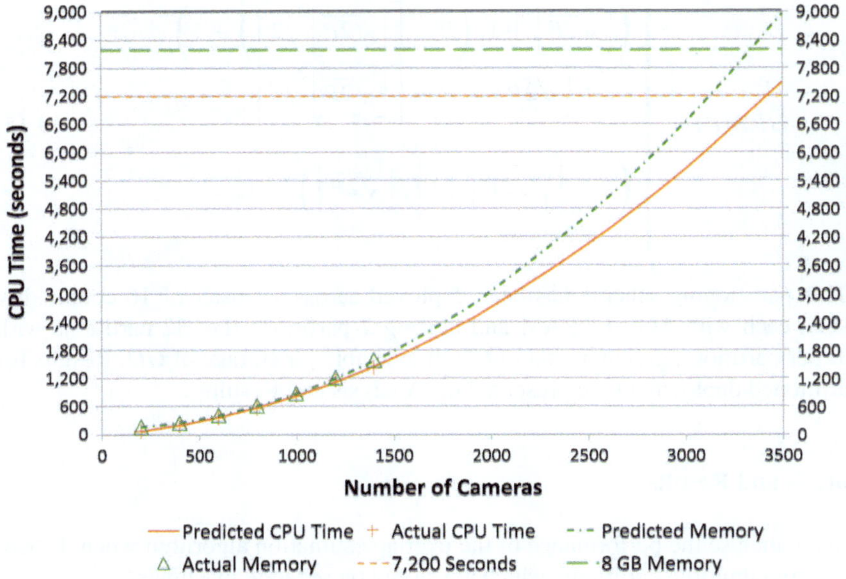

Fig. 19.8 Scale limit for 32 partition system, extrapolated past 1,400 cameras

are shown as points; a quadratic curve is fitted to each set of data points to extrapolate predicted values for larger camera networks.

Measurements of memory usage are close to uniform, with a standard deviation that is at most 1.0×10^{-3} times the magnitude of the mean. The total memory required for a given number of cameras is almost independent of the number of partitions, and the cost of this additional memory is more than outweighed by the ability for the memory to be distributed across multiple machines, thus avoiding any requirement for expensive machines capable of supporting unusually large amounts of memory.

Measurements of CPU time have a standard deviation of at most 7.4×10^{-2} times the mean, for the 32 partition/200 camera case. Partitions in this case require a mean of 77 s CPU time for 7,200 s real-time, with the consequence that CPU time sampling effects contribute much of the variance. In contrast, the 32 partition/1,400 camera case requires a mean of 1,435 s CPU time for 7,200 s real-time, and has standard deviation of 1.6×10^{-2} of the mean. Note that this is still significantly faster than real-time.

The extrapolated quadratic curves indicate that the memory curve crosses the 8 GB requirement at about 3,200 cameras, with the CPU time curve crossing 7,200 s at about 3,400 cameras, leading to the conclusion that a 16 server system can support over 3,000 cameras. The network bandwidth requirement increases linearly and is in practice never the bottleneck.

19.4.4 Lessons Learned from Practical Deployment

The system described in this chapter has been implemented and deployed in real surveillance scenarios, including installation at an international airport where it was used on a trial basis by professional security staff. This experience led us to the following conclusions:

- Cameras are often deployed in an ad hoc manner, and any image analysis must be robust to extreme changes in appearance and poor quality footage.
- Cameras are routinely added, removed or moved within the network, or simply break down. Therefore, the robustness and adaptability supported by our distributed architecture are essential for continuous operation on large-scale networks. The importance of incremental addition and removal of cameras is also noted in [21].
- Surveillance operators are more interested in methods that empower human monitors, rather than complete automation. For example, the average speed with which operators were able to detect and trace events in a test scenario quadrupled with the use of an estimated topology. This is more valuable to operators than automated event detection or re-identification, whose reliability cannot be guaranteed.
- Similarly, surveillance operators view the role of automation to be cost reduction and improvement in the working conditions of their monitoring staff. They want simple, reliable tools that allow their staff to spend more time analysing events of interest rather than searching for them.

19.5 Conclusions

In this chapter, we have described a scalable approach to automatically derive overlap topology for camera networks and evaluated its use for target re-identification. Its behaviour can be set to converge to the true present topology either from above or below, which is useful for downstream applications such as cross-camera tracking and re-identification. We demonstrate that topology has significant benefits for re-identification in camera networks and for the deployed 23-camera network it turns a difficult association problem into one for which simple colour histogram matching is mostly adequate. In future, we plan to apply our reservoir sampling approach [32] to tracking across much larger networks of cameras for which topology has been estimated in order to improve the re-identification accuracy.

References

1. Anjum, N., Cavallaro, A.: Automated localization of a camera network. IEEE Intell. Syst. **27**(5) (2012)
2. Bishop, C.: Pattern Recognition and Machine Learning. Springer, New York (2006)
3. Brand, M., Antone, M., Teller, S.: Spectral solution of large-scale extrinsic camera calibration as a graph embedding problem. In: Proceedings of 8th European Conference on Computer Vision, pp. 262–273 (2004)
4. Buehler, C.: Computerized method and apparatus for determining field-of-view relationships among multiple image sensors. United States Patent 7286157 (2007)
5. Chen, C.H., Yao, Y., Page, D., Abidi, B., Koschan, A., Abidi, M.: Camera handoff and placement for automated tracking systems with multiple omnidirectional cameras. Comput. Vis. Image Underst. **114**, 179–197 (2010)
6. Chen, K.W., Lai, C.C., Lee, P.J., Chen, C.S., Hung, Y.P.: Adaptive learning for target tracking and true linking discovering across multiple non-overlapping cameras. IEEE Trans. Multimed. **13**(4), 625–638 (2011)
7. Chen, K.Y., Huang, C.L., Hsu, S.C., Chang, I.C.: Multiple objects tracking across multiple non-overlapped views. In: Ho, Y.S. (ed.) Advances in Image and Video Technology. Lecture Notes in Computer Science, vol. 7088, pp. 128–140 (2012)
8. Chen, X., Huang, K., Tan, T.: Learning the three factors of a non-overlapping multi-camera network topology. In: Pattern Recognition, Communications in Computer and Information Science, vol. 321, pp. 104–112 (2012)
9. Detmold, H., van den Hengel, A., Dick, A., Cichowski, A., Hill, R., Kocadag, E., Yarom, Y., Falkner, K., Munro, D.: Estimating camera overlap in large and growing networks. In: Proceedings of ACM/IEEE International Conference on Distributed Smart Cameras (2008)
10. Dick, A., Brooks, M.J.: A stochastic approach to tracking objects across multiple cameras. In: Proceedings of Australian Joint Conference on, Artificial Intelligence (AI'04), pp. 160–170 (2004)
11. Ellis, T.J., Makris, D., Black, J.: Learning a multi-camera topology. In: Joint IEEE Workshop on Visual Surveillance and Performance Evaluation of Tracking and Surveillance (VS-PETS), pp. 165–171 (2003)
12. Gilbert, A., Bowden, R.: Tracking objects across cameras by incrementally learning inter-camera colour calibraion and patterns of activity. In: European Conference on Computer Vision, vol. 2, pp. 125–136 (2006)
13. van den Hengel, A., Dick, A., Detmold, H., Cichowski, A., Hill, R.: Finding camera overlap in large surveillance networks. In: Yagi, Y., Kang, S., Kweon, I., Zha, H. (eds.) Asian Conference on Computer Vision. Lecture Notes in Computer Science, vol. 4843, pp. 375–384 (2007)
14. Javed, O., Rasheed, Z., Shafique, K., Shah, M.: Tracking across multiple cameras with disjoint views. In: Proceedings of IEEE International Conference on Computer Vision, pp. 952–957 (2003)
15. Javed, O., Shafique, K., Rasheed, Z., Shah, M.: Modeling inter-camera spaceâĂŞtime and appearance relationships for tracking across non-overlapping views. Comput. Vis. Image Underst. **109**(2), 146–162 (2008)
16. Kang, J., Cohen, I., Medioni, G.: Persistent objects tracking across multiple non overlapping cameras. In: IEEE Workshop on Motion and Video, Computing, vol. 2, pp. 112–119 (2005)
17. Khan, S., Javed, O., Rasheed, Z., Shah, M.: Human tracking in multiple cameras. In: Proceedings of IEEE International Conference on Computer Vision, pp. I:331–336 (2001)
18. Lee, L., Romano, R., Stein, G.: Monitoring activities from multiple video streams: establishing a common coordinate frame. IEEE Trans. Pattern Anal. Mach. Intell. **22**(8), 758–767 (2000)
19. Lim, F., Leoputra, W., Tan, T.: Non-overlapping distributed tracking system utilizing particle filter. J. VLSI Signal Process. **49**, 343–362 (2007)
20. Loy, C.C., Xiang, T., Gong, S.: Multi-camera activity correlation analysis. In: IEEE Conference on Computer Vision and, Pattern Recognition, pp. 1988–1995 (2009)

21. Loy, C.C., Xiang, T., Gong, S.: Incremental activity modelling in multiple disjoint cameras. IEEE Trans. Pattern Anal. Mach. Intell. **34**(9), 1799–1813 (2012)
22. Marinakis, D., Dudek, G.: Self-calibration of a vision-based sensor network. Image Vis. Comput. **27**, 116–130 (2009)
23. Mavrinac, A., Chen, X.: Modeling coverage in camera networks: a survey. Int. J. Comput. Vis. **101**(1), 205–226 (2013)
24. Raghavan, V., Bollmann, P., Jung, G.S.: A critical investigation of recall and precision as measures of retrieval system performance. ACM Trans. Inform. Syst. **7**(3), 205–229 (1989)
25. Rahimi, A., Dunagan, B., Darrell, T.: Simultaneous calibration and tracking with a network of non-overlapping sensors. In: IEEE Computer Society Conference on Computer Vision and Pattern Recognition (CVPR), pp. 187–194 (2004)
26. Rahimi, A., Dunagan, B., Darrell, T.: Tracking people with a sparse network of bearing sensors. In: European Conference on Computer Vision, pp. 507–518 (2004)
27. Soto, C., Song, B., Roy-Chowdhury, A.: Distributed multi-target tracking in a self-configuring camera network. In: IEEE Conference on Computer Vision and, Pattern Recognition, pp. 1486–1493 (2009)
28. Stauffer, C.: Learning to track objects through unobserved regions. In: IEEE Computer Society Workshop on Motion and Video Computing, pp. II: 96–102 (2005)
29. Stauffer, C., Grimson, W.E.L.: Learning patterns of activity using real-time tracking. IEEE Trans. Pattern Anal. Mach. Intell. **22**(8), 747–757 (2000)
30. Tieu, K., Dalley, G., Grimson, W.: Inference of non-overlapping camera network topology by measuring statistical dependence. In: Proceedings of IEEE International Conference on Computer Vision, pp. II: 1842–1849 (2005)
31. Wang, X.: Intelligent multi-camera video surveillance: a review. Pattern Recogn. Lett. **34**(1), 3–19 (2013)
32. Yao, R., Shi, Q., Shen, C., Zhang, Y., van den Hengel, A.: Robust tracking with weighted online structured learning. In: European Conference on Computer Vision (2012)
33. Zhang, L., Li, Y., Nevatia, R.: Global data association for multi-object tracking using network flows. In: IEEE Conference on Computer Vision and, Pattern Recognition, pp. 1–8 (2008)
34. Zhao, T., Aggarwal, M., Kumar, R., Sawhney, H.: Real-time wide area multi-camera stereo tracking. In: IEEE Conference Computer Vision and, Pattern Recognition, vol. 1, pp. 976–983 (2005)

Chapter 20
Scalable Multi-camera Tracking in a Metropolis

Yogesh Raja and Shaogang Gong

This work is dedicated to Colin Lewis, in memory of his lifelong passion for pushing the boundaries in making academic research relevant to meeting real-world challenges, and for his unequivocal support in making this work possible.

Abstract The majority of work in person re-identification is focused primarily on the matching process at an algorithmic level, from identifying reliable features to formulating effective classifiers and distance metrics in order to improve matching scores on established 'closed-world' benchmark datasets of limited scope and size. Very little work has explored the pragmatic and ultimately challenging question of how to engineer working systems that best leverage the strengths and tolerate the weaknesses of the current state of the art in re-identification techniques, and which are capable of scaling to 'open-world' operational requirements in a large urban environment. In this work, we present the design rationale, implementational considerations and quantitative evaluation of a retrospective forensic tool known as Multi-Camera Tracking (MCT). The MCT system was developed for re-identifying and back-tracking individuals within huge quantities of open-world CCTV video data sourced from a large distributed multi-camera network encompassing different public transport hubs in a metropolis. There are three key characteristics of MCT, *associativity*, *capacity* and *accessibility*, that underpin its scalability to spatially large, temporally diverse, highly crowded and topologically complex urban environments with transport links. We discuss a multitude of functional features that in combination address these characteristics. We consider computer vision techniques and machine learning algorithms, including relative feature ranking for inter-camera matching,

Y. Raja
Vision Semantics Ltd, London, UK
e-mail: yraja@visionsemantics.com

S. Gong (✉)
Queen Mary University of London, London, UK
e-mail: sgg@eecs.qmul.ac.uk

S. Gong et al. (eds.), *Person Re-Identification*,
Advances in Computer Vision and Pattern Recognition,
DOI: 10.1007/978-1-4471-6296-4_20,

global (crowd-level) and local (person-specific) space–time profiling, attribute re-ranking and machine-guided data mining using a 'man-in-the-loop' interactive paradigm. We also discuss implementational considerations designed to facilitate linear scalability to an aribitrary number of cameras by employing a distributed computing architecture. We conduct quantitative trials to illustrate the potential of the MCT system and its performance characteristics in coping with very large-scale open-world multi-camera data covering crowded transport hubs in a metropolis.

20.1 Introduction

Human investigators tasked with the forensic analysis of video from multi-camera CCTV networks face many challenges, including (1) data overload from large numbers of cameras, (2) a short attention span leading to important events and targets being missed, (3) a lack of contextual knowledge indicating what to look for and (4) a lack of or inability to utilise complementary non-visual sources of knowledge to assist the search process. Consequently, there is a distinct need for technology to alleviate the burden placed on limited human resources and augment human capabilities.

As reflected in the published literature, much research effort has been expended in developing low-level methods for the automatic visual re-identification of people and other objects appearing in different places and at different time across multiple cameras. The ultimate goal is to build 'black-box' systems capable of unilaterally solving this problem. However, this is an inherently challenging task, especially if visual appearance is the only available cue for discrimination, as shown in Fig. 20.1. Much of the focus so far has been on finding the most reliable representative features to employ in constructing templates of individuals' visual appearance (e.g. major colours [12], combinations of colour and texture [14], complex structural layouts [4]), along with distance metrics (e.g. Bhattacharyya distance [14], L1-Norm [20]) or classifiers (e.g. K-Nearest Neighbour [5], Ranking SVM [2]) for matching. Such work is generally conditioned towards maximising ranking performance on small, carefully constructed *closed-world* benchmark datasets largely unrepresentative of the scale and complexity of *open-world* scenarios where the number of cameras, spatial size of the environment and numbers of people are all at a significantly larger scale, with a search space of unknown size and a potentially unlimited number of candidate matches for a target. Re-identification of targets in such open environments can potentially scale to arbitrary levels, covering huge spatial areas spanning not just different buildings but different cities, countries or even continents, leading to an overwhelming quantity of 'big data'.

To date, very little work has focused on addressing the practical question of how to best leverage the current state of the art in re-identification techniques, while tolerating their limitations in *engineering* practical systems that are *scalable* to typical real-world operational scenarios. In this work, we describe the design rationale and implementational considerations of building a prototype system known as *Multi-Camera Tracking* (*MCT*), a tool which human operators may employ for generating

Fig. 20.1 An illustration of the difficulties of visual matching across different camera views. Individuals may undergo significant variability in appearance due to changes in lighting, scale and viewpoint. Other difficulties are caused by partial or complete occlusion which results in a lack of complete visual information, and the tendency for variability between people (inter-variation) to be less than the variability for a single person at different times and camera views (intra-variation). All of these problems are compounded by spatially large environments with significant numbers of cameras and levels of crowding

a *global target trail* by retrospectively searching, 'back-tracking' and reconstructing the movements of targets of interest across multiple disjoint camera views in a large public space spanning a city. The system takes the basic approach of searching within multiple camera views for a specified target from a watchlist and producing *ranked lists* of candidate matches. Rather than attempting to solve the challenge of a fully automatic black-box winner-take-all solution for Rank-1 re-identification, the system takes the more practical approach of implementing mechanisms that: (a) quickly and effectively narrow the search space of candidates for human operators to perform target verification; and (b) incrementally increase the ranks of likely correct matches without making hard decisions that may inadvertently discard them at too early a stage during the re-identification process.

The overall design of the MCT system takes into account three key characteristics in order to systematically address the challenge of scalability: (1) *Associativity*, concerning the ability of the system to help users accurately extract targets of interest from an extremely large search space; (2) *Capacity*, relating to computational resources and the ability to process large numbers of camera inputs simultaneously; and (3) *Accessibility*, the speed with which users can conduct searches of targets and reconstruct their movements. In order to scale to arbitrarily large, busy and visually complex spaces, the MCT system requires various augmentations to satisfy these three requirements, in addition to the implementation of standard computer vision and machine learning techniques. This is addressed through a highly modular, flexible network-centric implementation able to incrementally leverage multiple hardware components in order to process arbitrary numbers of cameras, and a carefully designed user interface combining several mechanisms that support a coherent iterative *piecewise search strategy* for efficiently retracing multiple target movements through large complex multi-camera environments.

In Sect. 20.2, we discuss six key mechanisms that enable the MCT system to address the requirement of *associativity*. In Sect. 20.3, we detail implementational considerations that permit the system to address the requirements of *capacity* and *accessibility*. In Sect. 20.4, we describe a highly challenging open-world dataset encompassing two transport hubs in a metropolis. This is used to conduct quantitative trials of the MCT system, results of which are provided in Sect. 20.5. Finally, in Sect. 20.6 we conclude with lessons learned and open questions for future work.

20.2 Key Mechanisms

The *associativity* of a scalable multi-camera tracking tool is related to the efficiency and reliability with which it can aid users in locating targets of interest amongst very large numbers of individuals. Consequently, the fundamental objectives for the system are to reduce user workload by: (a) appropriately narrowing the search space and producing a minimal set of candidates containing the target; and (b) ranking the target highly within the candidate set. There are six key mechanisms of the MCT system that combine to address these objectives.

20.2.1 Relative Feature Ranking

The MCT system employs a comprehensive set of 29 types of visual features encompassing the colour and texture appearance of individuals for matching across camera views. More specifically, the colour features incorporate different colour spaces including RGB, Hue-Saturation and YCrCb, with texture features derived from Gabor wavelet responses at eight different scales and orientations, as well as thirteen differently parameterised Schmid filters [18]. Details can be found in [15]. Image patches within bounding boxes corresponding to people automatically detected by a parts-based person detector [3] are resampled to 300 pixels wide for consistency of scale. They are then split into six equal horizontal segments, with separate normalised histograms generated for each segment before concatenation into a single *feature vector*. Given 16 bins for the histogram corresponding to each of the 29 feature types for each of the 6 horizontal strips, we thus have a 2784-dimensional feature vector per bounding box, which is used as an *appearance descriptor*.

Rather than considering each feature type equally in terms of relevance, we dynamically learn the importance of each of these feature types to more strongly weight those features most relevant for matching across different cameras [15, 21, 22]. Such a model is trained from a dataset of pairs of feature vectors derived from single detections of the same person taken from different cameras [17]. More precisely, given a training set of m samples $X = (\mathbf{x}_i, y_i)_{i=1}^m$ where $\mathbf{x}_i \in R^d$ is a feature vector for a specific individual and y_i a corresponding label, and given the feature vectors $\mathbf{x}_j^+$ from the same training set X corresponding to the same person from another view (called *relevant* feature vectors) along with $\mathbf{x}_j^-$ corresponding to different people (called *irrelevant* feature vectors), we learn a ranking function δ to rank vector pair similarity such that $\delta(\mathbf{x}_i, \mathbf{x}_j^+) > \delta(\mathbf{x}_i, \mathbf{x}_j^-)$. This takes the form of a support vector machine (SVM) known as RankSVM [2, 6]. The RankSVM model is characterised by a linear function for ranking matches between two feature vectors as $\delta(\mathbf{x}_i, \mathbf{x}_j) = \mathbf{w}^\top |\mathbf{x}_i - \mathbf{x}_j|$. Given a feature vector $\mathbf{x}_i$, the required relationship between relevant and irrelevant feature vectors is $\mathbf{w}^\top(|\mathbf{x}_i - \mathbf{x}_j^+| - |\mathbf{x}_i - \mathbf{x}_j^-|) > 0$, i.e. the ranks for all correct matches are higher than the ranks for incorrect matches. Accordingly, given $\hat{\mathbf{x}}_s^+ = |\mathbf{x}_i - \mathbf{x}_j^+|$ and $\hat{\mathbf{x}}_s^- = |\mathbf{x}_i - \mathbf{x}_j^-|$ and the set $P = \{(\hat{\mathbf{x}}_s^+, \hat{\mathbf{x}}_s^-)\}$ of all pairwise relevant difference vectors required to satisfy the above relationship, a corresponding RankSVM model can be derived by minimising the objective function:

$$\frac{1}{2}||\mathbf{w}||^2 + C\sum_{s=1}^{|P|} \xi_s \tag{20.1}$$

with the constraints:

$$\mathbf{w}^\top(\hat{\mathbf{x}}_s^+ - \hat{\mathbf{x}}_s^-) \geq 1 - \xi_s \tag{20.2}$$

for each $s = 1, \ldots, |P|$ and restricting all $\xi_s \geq 0$. C is a parameter for trading margin size against training error.

20.2.2 Matching by Tracklets

For comparing individuals, the Munkres Assignment algorithm, also known as the Hungarian algorithm [7, 13], is employed as part of a multi-target tracking scheme to increase the number of samples for each individual by locally grouping detections in different frames as likely belonging to the same person. This process yields *tracklets* encompassing individual detections over multiple frames, representing short intra-camera trajectories. An individual D is accordingly represented as a tracklet $T_D = \{\alpha_{D,1}, \ldots, \alpha_{D,J}\}$ comprising a set of J individual detections with appearance descriptors $\alpha_{D,j}$. Two individuals are then matched by computing the median match score between each combination of detection pairs, one each from their respective tracklets. This approach mitigates the difficulties that might be faced by object tracking techniques in highly crowded environments, where irregular movement and regular occlusion causes tracking failure. Computing the median as a tracklet match score permits a degree of robustness against erroneous assignments, where tracklets may inadvertently comprise samples from multiple individuals.

More precisely, tracklets are built up incrementally over time, with an incomplete set updated after each frame by assigning individual detections from that frame to a tracklet according to their appearance similarity and spatial proximity. That is, given: (1) a set $S = \{\alpha_{1,f}, \ldots, \alpha_{M,f}\}$ of M appearance descriptors for detections in frame f with corresponding pixel locations $\{\beta_{1,f}, \ldots, \beta_{M,f}\}$; (2) a current set of N incomplete tracklets $R = \{\hat{T}_1, \ldots, \hat{T}_N\}$ with their most recently added appearance descriptors $\{\hat{\alpha}_{n,f_n}\}$; and (3) corresponding *predicted* pixel locations $\{\hat{\beta}_{n,f}\}$, an $M \times N$ cost matrix $\mathbf{C}$ is generated where each entry $\mathbf{C}_{m,n}$ is computed as:

$$\mathbf{C}_{m,n} = \omega_1 |\hat{\alpha}_{n,f_n} - \alpha_{m,f}| + \omega_2 |\hat{\beta}_{n,f} - \beta_{m,f}| \qquad (20.3)$$

In essence, this cost is computed as a weighted combination of appearance descriptor dissimilarity and physical pixel distance. Predicted pixel locations $\hat{\beta}_{n,f}$ for frame f are estimated by assuming constant linear velocity from the last known location and velocity. The Munkres Assignment algorithm maps rows to columns in $\mathbf{C}$ so as to minimise the cost, with each detection added accordingly to their mapped incomplete tracklets. Surplus detections are used to initiate new tracklets. In practice, an upper bound is placed on cost, with assignments exceeding the upper bound being retracted, and the detection concerned treated as surplus. Additionally, tracklets which have not been updated for a length of time are treated as complete.

For re-identification, completed tracklets are taken as a representation for an individual, though individuals may comprise several tracklets. When matching two individuals D_1 and D_2 with corresponding tracklets T_{D_1} and T_{D_2}, the score S_j for each pairing of appearance descriptors $\{(x, y) : x \in T_{D_1}, y \in T_{D_2}\}, j = 1, \ldots, J_1 J_2$ where $J_1 = |T_{D_1}|$ and $J_2 = |T_{D_2}|$ is computed using the RankSVM model as:

$$S = \mathbf{w}^\top (|x - y|) \qquad (20.4)$$

where $\mathbf{w}$ is obtained by minimising Eq. (20.1). The match score S_{D_1,D_2} for the two tracklets as a whole is computed as the median of these scores over all pairs of their appearance descriptors:

$$S_{D_1,D_2} = \text{median}\left(\{S_1, S_2, \ldots, S_{J_1 J_2}\}\right); \qquad (20.5)$$

A set of candidate matches is ranked by sorting their corresponding tracklet scores in descending order.

20.2.3 Global Space–Time Profiling

Given the inherent difficulties in visual matching when visual appearance lacks discriminability, not least in real-world scenarios where there are a very large number of possible candidates for matching, it becomes critical that higher level prior information is exploited to provide space–time context and significantly narrow the search space [10, 11, 17]. Our approach is to dynamically learn the typical movement patterns of individuals throughout the environment to yield a probabilistic model of when and where people detected in one view are likely to appear in other views. This top-down knowledge is imposed during the query process to drastically reduce the search space and dramatically increase the chances of finding correct matches, having a profound effect on the efficacy of the system.

More specifically, we employ the method proposed in [10, 11]. Each camera view is decomposed automatically into regions, across which different spatio-temporal activity patterns are observed. Let $\mathbf{x}_i(t)$ and $\mathbf{x}_j(t)$ denote the two regional activity time series observed in the ith and jth regions, respectively. These time series comprise the 2,784-dimensional appearance descriptors of detected individuals (Sect. 20.2.1). Cross Canonical Correlation Analysis (xCCA) is employed to measure the correlation of two regional activities as a function of an unknown time lag τ applied to one of the two regional activity time series. Denoting $\mathbf{x}_j(t) = \mathbf{x}_i(t+\tau)$, we drop the parameters t and τ for brevity to denote $\mathbf{x}_j = \mathbf{x_i}$. Then, for each time delay index τ, xCCA finds two sets of optimal basis vectors $\mathbf{w}_{\mathbf{x}_i}$ and $\mathbf{w}_{\mathbf{x}_j}$ such that the projections of $\mathbf{x}_i$ and $\mathbf{x}_j$ onto these basis vectors are mutually maximally correlated.

That is, given $\mathbf{x}_i = \mathbf{w}_{\mathbf{x}_i}^T \mathbf{x}_i$ and $\mathbf{x}_j = \mathbf{w}_{\mathbf{x}_j}^T \mathbf{x}_j$, the canonical correlation $\rho_{\mathbf{x}_i,\mathbf{x}_j}(\tau)$ is computed as:

$$\rho_{\mathbf{x}_i,\mathbf{x}_j}(\tau) = \frac{E[\mathbf{w}_{\mathbf{x}_i}^\top \mathbf{C}_{\mathbf{x}_i\mathbf{x}_j} \mathbf{w}_{\mathbf{x}_j}]}{\sqrt{E[\mathbf{w}_{\mathbf{x}_i}^\top \mathbf{C}_{\mathbf{x}_i\mathbf{x}_i} \mathbf{w}_{\mathbf{x}_i}]}\sqrt{E[\mathbf{w}_{\mathbf{x}_j}^\top \mathbf{C}_{\mathbf{x}_j\mathbf{x}_j} \mathbf{w}_{\mathbf{x}_j}]}} \tag{20.6}$$

where $\mathbf{C}_{\mathbf{x}_i\mathbf{x}_i}$ and $\mathbf{C}_{\mathbf{x}_j\mathbf{x}_j}$ are the within set covariance matrices of $\mathbf{x}_i$ and $\mathbf{x}_j$ respectively, and $\mathbf{C}_{\mathbf{x}_i\mathbf{x}_j}$ is the between-set covariance matrix.

The time delay that maximises the canonical correlation between $\mathbf{x}_i(t)$ and $\mathbf{x}_j(t)$ is then computed as:

$$\hat{\tau}_{\mathbf{x}_i,\mathbf{x}_j} = \text{argmax}_\tau \frac{\sum^{\Gamma} \rho_{\mathbf{x}_i,\mathbf{x}_j}(\tau)}{\Gamma} \tag{20.7}$$

where $\Gamma = \min\left(\text{rank}\,(\mathbf{x}_i), \text{rank}\,(\mathbf{x}_j)\right)$.

Given a target nominated in camera view j for searching in camera view k, the search space is narrowed by considering only tracklets from k with a corresponding time delay less than $\alpha\hat{\tau}_{\mathbf{x}_j,\mathbf{x}_k}$ (with α a constant factor) for matching. This candidate set is then ranked accordingly.

20.2.4 *'Man-in-the-Loop' Machine-Guided Data Mining*

The MCT system is an interactive 'man-in-the-loop' tool designed to enable human operators to retrospectively re-trace the movements of targets of interest through a spatially large, complex multi-camera environment by performing *queries* on generated metadata. A common-sense approach to doing so in such an environment is to employ an iterative *piecewise search strategy*, conducting multiple progressive

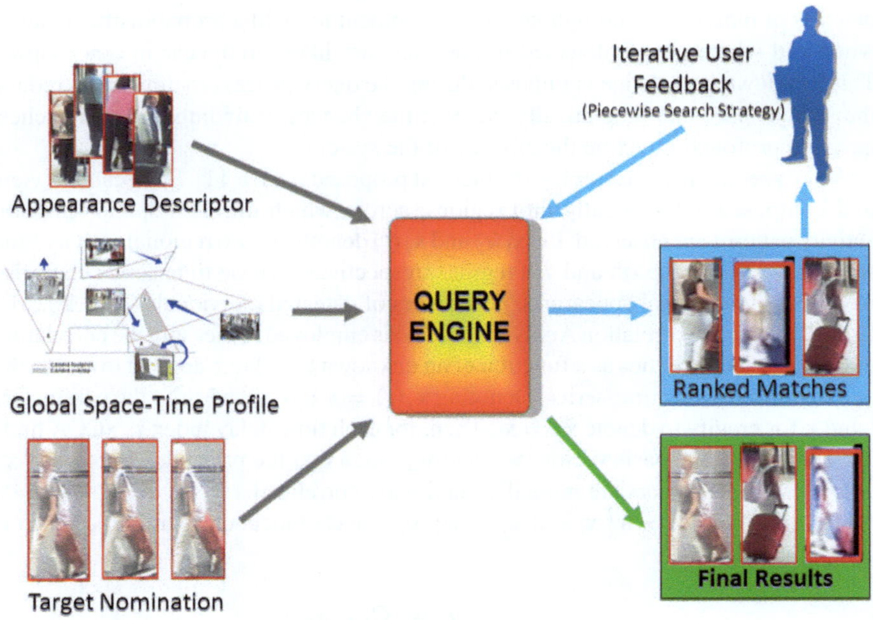

Fig. 20.2 Usage of the MCT system. Given automatically extracted appearance descriptors from across the multi-camera network along with a global space–time profile (Sect. 20.2.3), users nominate a target and then iteratively search through the network in a piecewise fashion, marking observed locations and times of the target in the process. The procedure stops when the target has been re-identified in a sufficient number of views for an automatically generated reconstruction

searches over several iterations to gradually build a picture of target movements, or *global target trail.*

More precisely, given the initial position of a nominated target, the first search is conducted in the place most likely to correspond to their next appearance, such as the adjacent camera view depending on direction of movement. Further detections of the target provide constraints upon the next most likely location, within which the next search iteration is conducted. The search thus proceeds in a manner gradually spanning out from the initial detected position, marking further detections along the way and building a picture of target movements, until the number of locations has been exhausted or the picture is sufficiently detailed for an automatically generated reconstruction of the target's movement through the environment. This approach ensures that the problem is tackled piecemeal, with the overall search task simplified and the workload on users minimised. Figure 20.2 illustrates the top-level paradigm for system usage.

Additionally, in the process of conducting a query, unexpected associations such as previously unknown accomplices may be discovered. These are not only highly relevant to the investigation at large, but may be exploited as part of the search process itself. Such associates may naturally and seamlessly be incorporated into the query, forming a parallel branch of enquiry which proceeds in the same way. This allows: (a) accomplices to aid in the detection of the target of interest, for example if the latter is not visible for the system to detect but inferable by way of their proximity to the detectable accomplice; and (b) accomplices to be tracked independently at the same time as the original target should their trajectories through the multi-camera network diverge.

The basic MCT query procedure is as follows: A user initiates a query Q_0 which comprises a nominated target with tracklets $T_0 = \{t_{0,j_0}\}$, $j_0 = 1, \ldots, J_0$ from camera view ξ_0. The first search iteration is conducted in camera view ξ_1, resulting in a set of J_1 candidate matches $T_1 = \{t_{1,j_1}\}$, $j_1 = 1, \ldots, J_1$. Any number of these can be tagged by the user, whether they correspond to the initial target or a relevant association, yielding a set R_1 of K_1 indices for 'relevant' flags $R_1 = \{r_{k_1}\}$, $k_1 = 1, \ldots, K_1$. The set $C_1 = \{t_{1,r_{k_1}}\}$ is then used to initiate the next iteration of the query Q_1 in camera view ξ_2, yielding J_2 new candidate matches $T_2 = \{t_{2,j_2}\}$, $j_2 = 1, \ldots, J_2$. These are again marked accordingly by the user, yielding a set R_2 of K_2 indices for 'relevant' flags $R_1 = \{r_{k_2}\}$, $k_2 = 1, \ldots, K_2$. The new set $C_2 = \{t_{2,r_{k_2}}\}$ is combined with the set from the previous iteration C_1 as well as the initial nomination to produce an aggregate set $\hat{C}_2 = \{T_0 \cup C_1 \cup C_2\}$. The search proceeds for as many iterations as required, finding relevant matches in each camera view. After n iterations, we have the aggregate pool of matches tagged as relevant by the user over all previous search iterations, plus the original nomination:

$$\hat{C}_n = \{T_0 \cup C_1 \cup C_2 \cup \cdots \cup C_n\} = \{t_{0,j_0}\} \cup \{t_{1,r_{k_1}}\} \cup \cdots \cup \{t_{n,r_{k_n}}\} \tag{20.8}$$

This set constitutes the final associated evidence from which a video reconstruction of target movements is automatically created by the system and instantly viewable. Note that the search process is not generally linear. The interface provides the flexibility to search in multiple cameras at once and then analyse the results from each camera one-by-one. A user may also select matches from previous iterations to conduct searches in a future iteration. This enables multiple targets to be tracked as part of a single query as well as tracking movements both backwards and forwards in time.

20.2.5 Attribute-Based Re-ranking

The RankSVM model (Sect. 20.2.1) [2, 6] employs appearance descriptors comprising a multitude of low-level feature types which are weighted by the RankSVM model. However, such a representation is not always sufficiently invariant to changes in viewing conditions, leading to blunted discriminability. Furthermore, to a human observer, such feature descriptors are not amenable to descriptive interpretation. For example, depending on the tracking scenario, human operators may focus on unambiguous characteristics of a target, such as attire, colours or patterns. Consequently, we incorporate mid-level *semantic attributes* [8, 9] as an intuitive complementary method of ranking candidate matches. Users may select multiple attributes descrip-

Fig. 20.3 Examples of images associated with semantic mid-level attributes. Some images can be associated with multiple attributes simultaneously—for example, the second example labelled 'Femal' can be also be labelled 'Skirt' (i.e. she is also wearing a skirt), and the third example labelled as 'Bald' can also be labelled 'Backpack'

tive of the target to re-rank candidates and encourage correct matches to rise, reducing the time taken for localisation.

We identify 19 semantic attributes, including but not limited to *bald, suit, female, backpack, headwear, blue* and *stripy*. Figure 20.3 shows some example images associated with these attributes. We then create a training set of 3,910 sample images of 45 different individuals across multiple camera views and for each sample j generate an appearance descriptor α_j of the form used for the RankSVM model (Sect. 20.2.1). These are manually annotated according to the 19 attributes. Given this data, a set of *attribute detectors* a_i, $i = 1, \dots, 19$ in the form of support vector machines using intersection kernels are learned [8, 9] using the LIBSVM library [1]. Cross-validation is employed to select SVM slack parameter values.

The outputs of the detectors are in the form of posterior probabilities $p(a_i|\alpha)$, denoting the probability of attribute i given an appearance descriptor α. Given I user-selected attributes $\{a_1, \dots, a_I\}$ and a set of K candidate matches $\{t_1, \dots, t_K\}$ where candidate $t_k = \{\alpha_{k,1}, \dots, \alpha_{k,J}\}$ is a set of J appearance descriptors, the score $S_{i,k}$ for each attribute a_i is computed for each candidate t_k as an average of the posterior probabilities for each of the J appearance descriptors:

$$S_{i,k} = \frac{1}{J} \sum_{j=1}^{J} p(a_i|\alpha_j) \tag{20.9}$$

Accordingly, each candidate t_k has an associated vector of scores $[S_{1,k}, \dots, S_{I,k}]^\top$. The set of candidates is then ranked separately for each attribute, averaging the ranks for each candidate and finally sorting by the average rank.

20.2.6 Local Space–Time Profiling

Global space–time profiles (Sect. 2.1.2) significantly narrow the search space of match candidates by imposing constraints learned from the observed movements of *crowds in-between camera views*. To complement this, local space–time profiles further reduce the set of candidate matches by imposing constraints implied by observed movements of *specific individuals within each camera view*. Ultimately, this may incorporate knowledge of scene structure and likely trajectories of individuals within the view, for example depending on which exit they are likely to take in a multi-exit scene.

For the MCT system, we employed a simple method of filtering known as *Convergent Querying*. For each camera view i, $i = 1, \ldots, 6$, we selected a small set of example individuals (e.g. 20) at random and manually measured the *length of time* they were visible in that view, i.e. from the frame of their appearance to the frame of their disappearance. Temporal windows τ_i were then estimated for each camera view as:

$$\tau_i = E[X_i] + 3\sqrt{(\mathrm{Var}(X_i))} \qquad (20.10)$$

where X_i denotes the random variable for the observed transition times in frames from Camera i.

Given a set $T = \{t_1, \ldots, t_J\}$ of J candidate matches (tracklets) from Camera i, the user may tag one of the matches t_j for local space–time profiling, resulting in the pruned set:

$$\hat{T} = \{\hat{t} \in T : |\phi(\hat{t}) - \phi(t_j)| \leq \tau_i\} \qquad (20.11)$$

where $\phi(t)$ is a function returning the average of the first and last frame indices of the individual detections of tracklet t. Consequently, the filter removes all tracklets lying outside the temporal window, narrowing the results to those corresponding to the tighter time period within which the specific target is expected to appear.

20.3 Implementation Considerations

The *capacity* of a multi-camera tracking system relates to the ability to process, generate and store metadata for very large numbers of cameras simultaneously. A related characteristic is *accessibility*, the ability to query the generated metadata in a speedy fashion. Accordingly, the ability of the system to scale to typical open-world scenarios where the quantity of data can arbitrarily increase depends upon careful design and implementation of the processing architecture, the user interface and in particular, the metadata storage scheme.

In order to enable on-the-fly analysis of videos streams which may be pre-recorded and finite or live and perpetual, the general top-level approach we take towards implementing the MCT prototype is to produce two independently functioning subsystems.

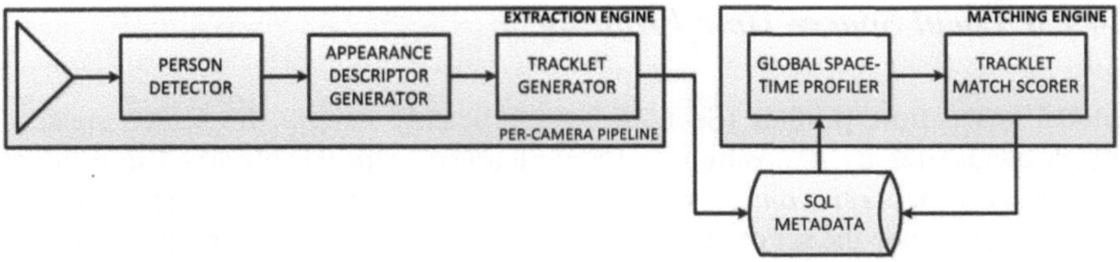

Fig. 20.4 MCT Core Engine, depicting the asynchronous Extraction and Matching engines. The Extraction Engine takes the form of a multi-threaded processing pipeline, enabling efficient processing of multiple inputs simultaneouslyon multi-core CPU platforms

First, the *Generator Subsystem* is responsible for processing video streams and generating *metadata*. This metadata includes *tracklets* of detected people in each camera view and the storage of this metadata in a backend database. Targets which may be nominated by individuals are restricted to those that can be *automatically detected* by the system rather then permitting users to arbitrarily select image regions that may correspond to objects of interest but which may not be visually detectable automatically. Second, the *Interrogator Subsystem* provides a platform for users to query the generated metadata through a secure, encrypted online browser-based interface. These two subsystems operate asynchronously, enabling users to query metadata via the Interrogator Subsystem as and when they become available by way of the Generator Subsystem functioning in parallel. The MCT system is designed to be flexible and for its components to inter-operate either locally or remotely across a network, in order to permit the incremental utilisation of off-the-shelf hardware. For example, the entire system may operate on a single server, or with each component on separate servers connected via the Internet.

Metadata is stored in an *SQL Metadata* backend database component. A *Video Streamer* provides video data from recorded or live input to an *MCT Core Engine* and multiple *User Interface (UI) Clients* that encapsulate the essential functionalities of the MCT system.

The MCT Core Engine comprises two asynchronous sub-components known as the *Extraction* and *Matching* Engines, which form the primary processing pipeline for generating metadata for the Generator Subsystem (Fig. 20.4). This MCT pipeline employs a multi-threaded approach, efficiently utilising multi-core CPUs to process multiple camera inputs simultaneously. This implementation enables additional processing resources to be added as available in a flexible manner. For example, multiple camera inputs may be processed on a single machine, or allocated as desired across several machines. Such flexibility also applies to the Extraction and Matching Engines, which can be allocated separately for each camera. This facilitates potentially unlimited incremental additions to hardware resources with ever-increasing numbers of cameras.

The User Interface (UI) Clients are Java web-based applets which interface remotely with a *Query Engine Server* to enable the search of metadata stored in the SQL Metadata component. Usage of the system only requires access to a basic

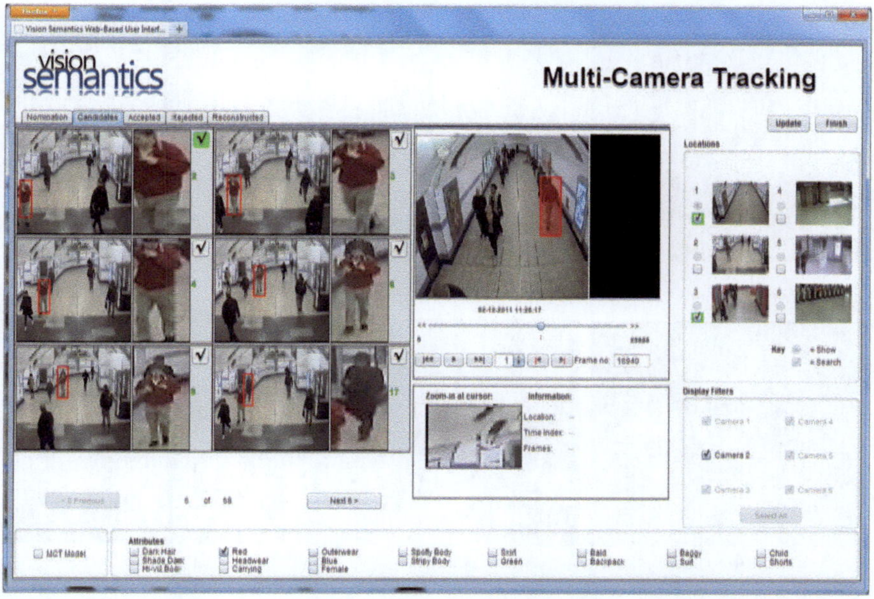

Fig. 20.5 MCT User Interface Client example screenshot. Here, users examine a paginated set of match candidates from a search iteration, locating and tagging those relevant including target associates. The local space–time profile Convergent Querying filter is employed here, using tagged candidates to immediately narrow the set being displayed to the appropriate temporal window. Attributes appropriate to the target may also be selected, which instantly re-rank the candidate list accordingly

terminal equipped with a standard web browser with a Java plugin. Security features include password protected user logins, per-user usage logging, automatic time-outs and fully encrypted video and metadata transfer to and from the Query Engine Server. The interface includes functions to support the piecewise search strategy (Sect. 20.2.4), as well as for viewing dynamically generated chronologically-ordered video reconstructions of target movements.

Figure 20.5 depicts an example screenshot of the MCT User Interface Client. This screen lists all candidate matches returned from a search iteration in paginated form. Here, users may browse through candidate matches from a search iteration, locating and tagging those which are relevant to the query. Two key features available are: (1) the *Convergent Querying filter* which is applied when tagging a candidate, instantly imposing local space–time profiling on the currently displayed set (Sect. 20.2.6); and (2) *Attribute selection* checkboxes for instant re-ranking of candidates by user-selected semantic attributes (Sect. 20.2.5).

Fig. 20.6 MCT trial dataset example video images. **a** Cameras 1, 2 and 3 (from Station A). *Left* Corridor leading from entrance to Station A; *Centre* Escalator to train platforms; *Right* Entrances to platforms. **b** Cameras 4, 5 and 6 (from Station B). *Left* Platforms for trains to Station A; *Centre* Platforms for those arriving from Station A; *Right* Ticket barriers at entrance to Station B

20.4 MCT Trial Dataset

As defined in Sect. 20.1, there are three key characteristics that influence scalability: *associativity*, *capacity* and *accessibility*. The scale of the environment concerned profoundly impinges upon all of these factors since it is correlated with the quantity of data to process, as well as the number of individuals to search through and for whom metadata must be generated and stored. We conducted an in-depth evaluation of the MCT system in order to determine its scalability in terms of these three factors.

The MCT system has previously been tested [17] using the i-LIDS multi-camera dataset [19]. The i-LIDS data comprised five cameras located at various key points in an open environment at an airport. A key limitation of this dataset is that the five cameras covered a relatively small area within a single building, where passengers moved on foot in a single direction with transition across the entire network taking at most 3 min. As such, the scale of the i-LIDS environment is limited for testing typical open-world operational scenarios. Trialling the MCT system requires an open-world test environment unlike all existing closed-world benchmark datasets.

To address this problem, we captured a new trial dataset during a number of sessions in an operational public transport environment [16]. This dataset comprises six cameras selected from existing camera infrastructure covering two different transport hubs at different locations on an urban train network, reflecting an open-world

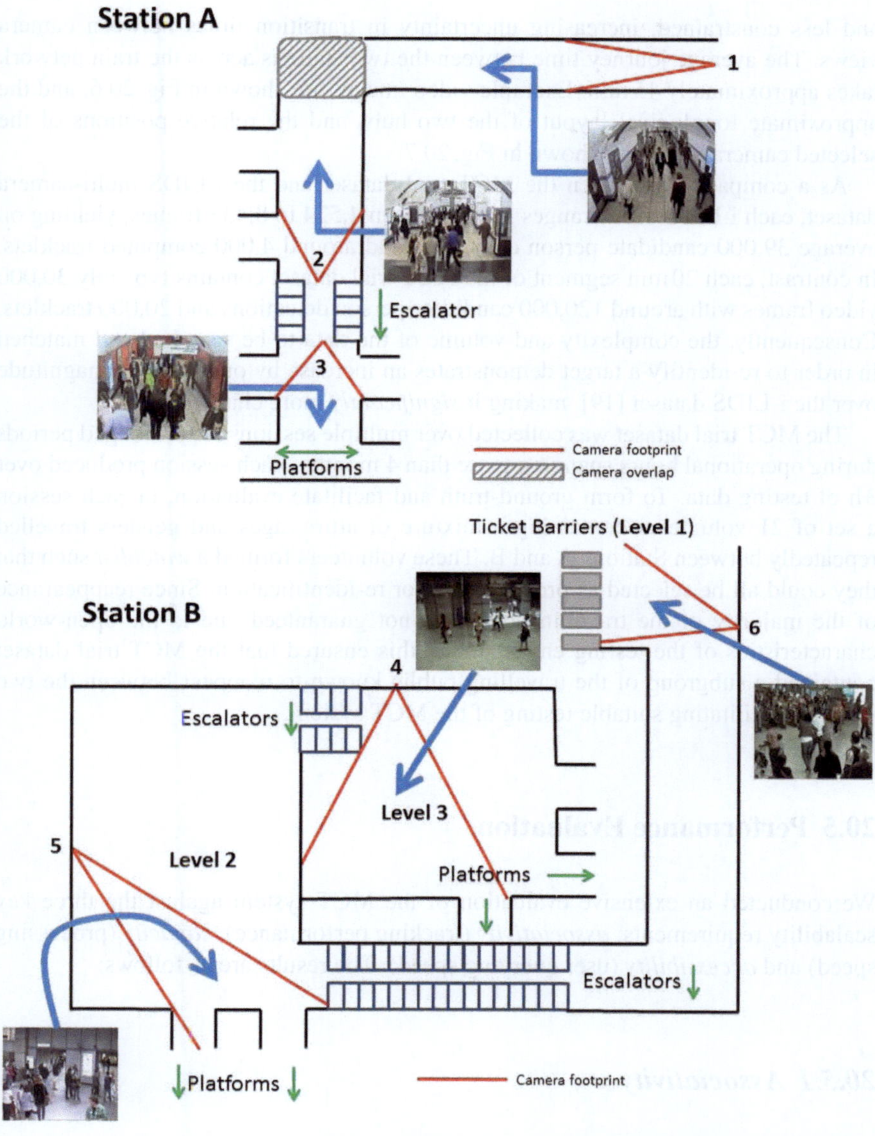

Fig. 20.7 Topological layout of Stations A and B. Three cameras (sample frames shown) were selected from each and used for data collection and MCT system testing

operational scenario. Camera locations are connected by walkways within each hub and a transport link connecting the two hubs. Lighting changes and viewpoints exhibit greater variability, placing more stress on the matching model employed by a re-identification system. Furthermore, passenger movements are multi-directional and less constrained, increasing uncertainty in transition times between camera views. The average journey time between the two stations across the train network takes approximately 15 min. Example video images are shown in Fig. 20.6, and the approximate topological layout of the two hubs and the relative positions of the selected camera views are shown in Fig. 20.7.

As a comparison between the MCT trial dataset and the i-LIDS multi-camera dataset, each i-LIDS video ranges from between 4,524 to 8,433 frames, yielding on average 39,000 candidate person detections and around 4,000 computed tracklets. In contrast, each 20 min segment of the MCT trial dataset contains typically 30,000 video frames with around 120,000 candidate person detections and 20,000 tracklets. Consequently, the complexity and volume of the data to be searched and matched in order to re-identify a target demonstrates an increase by one order of magnitude over the i-LIDS dataset [19], making it *significantly* more challenging.

The MCT trial dataset was collected over multiple sessions for prolonged periods during operational hours spanning more than 4 months. Each session produced over 3 h of testing data. To form ground-truth and facilitate evaluation, in each session a set of 21 volunteers containing a mixture of attire, ages and genders travelled repeatedly between Stations A and B. These volunteers formed a *watchlist* such that they could all be selected as probe targets for re-identification. Since reappearance of the majority of the travelling public is not guaranteed due to the open-world characteristics of the testing environment, this ensured that the MCT trial dataset contained a subgroup of the travelling public known to reappear between the two stations, facilitating suitable testing of the MCT system.

20.5 Performance Evaluation

We conducted an extensive evaluation of the MCT system against the three key scalability requirements: *associativity* (tracking performance), *capacity* (processing speed) and *accessibility* (user querying speed). The results are as follows:

20.5.1 Associativity

The performance of the MCT system in aiding cross-camera tracking, i.e. re-identification, was evaluated by conducting queries for each of the 21 volunteers on our watchlist making the test journey between Stations A and B. The total number of search iterations (see Sect. 20.2.4) conducted over all 21 examples was 95. We were primarily interested in measuring the effectiveness of the three key ranking mechanisms: *relative feature ranking* [15]; *attribute-based re-ranking* [8, 9]; and *local space–time profiling*, in increasing the ranks of correct matches, as well as gauging the more holistic effectiveness of all six mechanisms (the above in

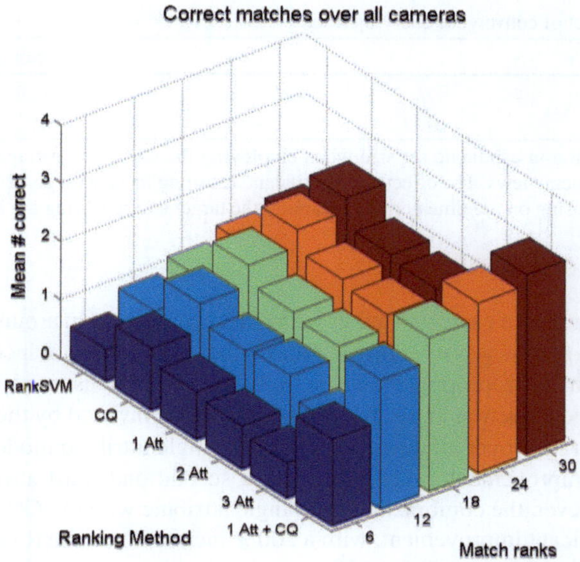

Fig. 20.8 The cumulative number of correct matches appearing in the top 6, 12, 18, 24 and 30 ranks, averaged over all search iterations and all camera views. The Convergent Querying (CQ) filter doubled the average number of correct matches in the first 6 ranks over the RankSVM model alone, from around 0.5 to 1. Selecting a single attribute was more beneficial than two or three, improving on the RankSVM model. Overall, a single attribute combined with CQ demonstrated the greatest improvement of around 200 % over the RankSVM model

addition to *matching by tracklets*; *global space–time profiling*; and *machine-guided data mining*) in tracking the targets across the multi-camera network.

We measured two criteria: (1) the *number of correct matches* in the first 6, 12, 18, 24 and 30 ranks after any given search iteration, averaged over all 95 search iterations, indicating how quickly a user will likely find the target amongst the candidates; and (2) *overall re-identification rates* in terms of the average percentage of cameras targets were successfully re-identified in, indicating tracking success through the environment overall. The exact querying procedure adhered to the iterative piecewise search strategy described in Sect. 20.2.4.

Number of Correct Matches

Figure 20.8 shows the cumulative number of correct matches that appeared in the top 6, 12, 18, 24 and 30 ranks viewed by a user, averaged overall search iterations and all camera views.

Using the RankSVM model alone [2, 6], the average number of correct matches in the first six ranks was around 0.5. Using the Convergent Querying (CQ) filter significantly improved upon the RankSVM model at all ranks, and approximately

Table 20.1 Effect of convergent querying filter

Stage of query	Mean candidate set size
Before CQ	392.9
After CQ	79.6

The mean reduction in candidate set size when employing the CQ filter, averaged over all query iterations and camera views. The effect was significant, resulting in an average 72.1 % reduction by more acutely focusing on the time period containing the target and removing the bulk of irrelevant candidates

doubled the number of correct matches in the first six ranks from around 0.5 to 1. The primary reason for this was its ability to remove the vast majority of incorrect matches by focusing on the appropriate time period. This is demonstrated by Table 20.1, showing that the reduction in the number of candidates invoked by the CQ filter was over 72 %, averaged over all query iterations. A single attribute model also showed around 50 % improvement, whereas adding a second and third attribute was less effective. However, the combination of a single attribute with the CQ filter provided the most significant improvement, with a 200 % increase over the RankSVM model. Consequently, local space–time profiling was critical for narrowing the search space more acutely and finding the right target more quickly amongst very large numbers of distractors. Combining this with a single attribute model provided an extra 50 % performance boost on average by providing an additional context for narrowing the search further.

Overall Re-identification Rates

Table 20.2 shows the percentage of watchlist targets that were explicitly detected by the system in each camera view. Apart from Cameras 4 and 5, detection rates were above 80 %. For Camera 5, the slightly larger distances to individuals resulted in slightly lower performance for the MCT person detector [3]. The profile views common in Camera 4 were responsible for lower person detection performance.

It is important to note that detection failure does *not* imply tracking failure, due to the facility for tagging visible associates of targets (refer to the piecewise search strategy in Sect. 20.2.4). Consequently, targets may still be tracked through camera views in which they may not be detected.

Table 20.3 shows the percentage of all six cameras that watchlist targets were tracked within on average; more specifically, the percentage of cameras within which users could tag matches that contained the target and which could be incorporated into a reconstruction, regardless of whether that target was *explicitly detected* by the system.

It can be seen that tracking coverage, i.e. re-identification, was very high, approaching 90 % over the entire network on average for both directions of movement. The result for the Station B to Station A journey was lessened due to the rele-

Table 20.2 Detection rates per camera

Camera	Target detection rate (%)
1	85.7
2	85.7
3	81
4	72.7
5	76.2
6	85.7

The percentage of watchlist targets explicitly detected in each camera view. Apart from Cameras 4 and 5, detection rates were above 80 %. For Camera 5, the slightly larger distances to individuals resulted in slightly lower performance for the MCT person detector [3]. The profile views common in Camera 4 were responsible for lower detection performance

Table 20.3 Overall tracking coverage

Direction of journey	Mean tracking coverage (%)
Station A to B	88
Station B to A	84.6

The average percentage of all six cameras within which a watchlist target could be found and incorporated into a reconstruction, whether or not explicitly detected by the system. Often targets were found for all cameras, with the few failures occurring due to: (a) unpredictable train times operating outside the global temporal profile, resulting in a loss of the target between stations; and (b) target occlusion due to crowding or moving outside the video frame

vance of Camera 4 for this journey (Sect. 20.4) and its corresponding lower detection reliability.

Failures were due to two main reasons. First, abnormal train waiting or transition times resulted in two watchlist targets being lost in between stations. These times fell outside the range of the learned global space–time profile, resulting in a faulty narrowing of the search space. In very large-scale multi-camera networks such as those spanning cities where different parts of the environment are connected by transportation links, this danger can be compounded by multiple unpredictable delays. This suggests that integrating live non-visual information, such as real-time train updates, should be exploited to override or dynamically update global space–time profiles in order to ensure correct focusing of the search as circumstances change. Second, lone targets could occasionally become occluded and thus remain undetected or untrackable by association, due to excessive crowding or moving outside the view of the camera. This highlights the value of careful camera placement and orientation. Nevertheless, occasional detection failure in some camera views was not a barrier to successful tracking since searches could be iteratively widened when required and the target successfully reacquired further along their trajectory.

Table 20.4 Module processing time per frame

Module	Processing time (%)
Person detector	39.8
Appearance descriptor generator	57.7
Other	2.5

The relative computational expense of key processing modules of the Extraction Engine

20.5.2 Capacity

A critical area of system scalability is the *speed* of the system in processing video data depending on the size of the multi-camera environment. Consequently, a major area of focus is the effective use of acceleration technologies such as GPU acceleration and multi-threading. Table 20.4 shows the relative time taken by two key processing modules of the Extraction Engine in the Generator Subsystem (Sect. 20.3) to process a single video frame. The most computationally expensive processing module, namely the Appearance Descriptor Generator, was re-implemented to employ GPU acceleration in order to conduct an initial exploration of this area. Additionally, multi-threading was employed to specifically exploit the computational capacity of multi-core processors.

In exploring the characteristics of processing capacity, four GPU and multi-threading configurations were evaluated in order to highlight the importance and effectiveness of applying acceleration technologies in working to achieve acceptable processing speeds: (1) single thread, no GPU acceleration; (2) single thread, GPU acceleration of the Appearance Descriptor Generator; (3) multi-threading of pipelines to parallelise the processing of individual camera inputs, no GPU acceleration; and (4) both GPU acceleration of the Appearance Descriptor Generator and multi-threading together. The hardware platform employed contained an Intel Core-i7 quad-core processor operating at 3.5 GHz, running Microsoft Windows 7 Professional with 16 GB of RAM and two Nvidia GTX-580 GPU devices.

Figure 20.9 shows the average time in seconds taken for each of the four acceleration configurations to process a frame for 2, 3, 4, 5 and 6 cameras simultaneously. It can be clearly seen that GPU accelerating the Appearance Descriptor Generator alone (requiring more than 50 % of the computational resources when unaccelerated) resulted in halving the processing time for a video frame. This amounted to the vast majority of the processing time for that component being eliminated. Significantly, it can also be seen that the use of multi-threading enabled six cameras to be processed on the same machine with negligible overhead, demonstrating that multi-threading, in addition to the distributed architecture design, facilitates scalability of the system to arbitrary numbers of cameras (i.e. the ability to process multiple video frames from different cameras simultaneously) by exploiting the multi-core architecture of off-the-shelf CPUs. A quad-core processor with hyper-threading technology is capable of processing eight cameras simultaneously with little slow-down; more cameras

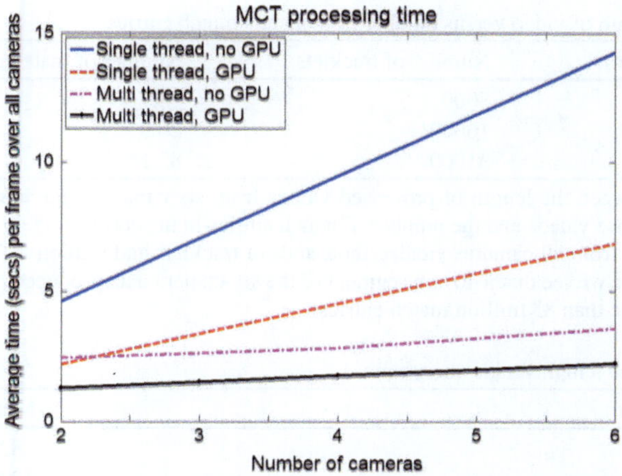

Fig. 20.9 Time taken in seconds for the MCT system to process a single frame across all camera streams for 2, 3, 4, 5 and 6 cameras in 4 different acceleration configurations. This demonstrated both the efficacy of employing multi-threading and GPU acceleration as well as the scalability of the system to arbitrary numbers of cameras. It can be seen that employing GPU acceleration dramatically improved the *time to process a single video frame*, and multi-threading facilitated the ability to *process frames from multiple cameras simultaneously*, demonstrating *linear scalability* of the system to larger camera networks

may be processed by simply adding another quad-core machine to provide another eight camera capability. Future processors with greater core numbers promise to efficiently increase scalability yet further.

20.5.3 Accessibility

The quantity of metadata generated by the system is strongly correlated with the size of the multi-camera environment, influencing the speed and responsiveness of the user interface in the course of a query being conducted. As such, this is a critical factor where scalability to typical real-world scenarios is concerned. Here, we investigate two key areas determining accessibility: (1) *query time versus database size*, relating system usability with the quantity of data processed; and (2) *local versus remote access*, comparing the speed of querying when running the User Interface Client locally and remotely in three different network configurations.

Query Time Versus Database Size

Open-world scenarios will typically present arbitrarily large numbers of individuals forming the search space of candidates during a query. The key factor in querying

Table 20.5 Length of video versus number of metadata match entries

Video length (min)	Number of tracklets	Number of match entries (millions)
10	8000	8.7
20	16000	26.9
40	31000	83.1

Relationship between the length of processed videos from six cameras, the number of extracted tracklets from those videos and the number of match entries in the corresponding metadata. A few minutes of video from all cameras yielded thousands of tracklets and millions of match entries in the database. Here we see that a 40 min segment of the six-camera data produced more than 30,000 tracklets and more than 83 million match entries

Table 20.6 Video length versus query time

Video length (min)	Mean query time (s)
10	82
20	124
40	284

The average time for the same query conducted three times for databases generated from 10, 20 and 40 min segments of the six-camera MCT trial dataset. While the 10 and 20 min segments resulted in acceptable times of around 1.5–2 min, the 40 min segment more than doubled the query time for the 20 min segment. The significant increase in query time with the quantity of video data processed highlighted a key bottleneck of the current system

time is the number of tracklets which have been generated for those individuals, and the size of a corresponding *match table* in the metadata which contains the matching results for appropriate global space–time filtered sets of tracklets between camera views. Table 20.5 shows the relationship between the number of tracklets and the corresponding number of match entries in the metadata match table for three different processed video segment lengths. It can be seen that 20 min of processed video from six cameras produced on the order of tens of thousands of tracklets and tens of millions of match entries in the database.

The question arises as to what effect this increase in the size of the database has on querying times. Table 20.6 shows the average time for the same query conducted three times over the same LAN connection for each of these three database sizes. While the 10 and 20 min segments resulted in acceptable times of around 1.5–2 min, the 40 min segment more than doubled the time over the 20 min segment. The significant increase in query time with the quantity of video data processed highlights a key bottleneck of the current system and a major focus on improving the scheme for metadata storage and access in working towards a deployable system.

Local Versus Remote Access

Table 20.7 shows the difference in query time for the same query conducted on the same metadata database accessed: (a) locally on the same machine as the Query Engine Server and SQL Metadata database; (b) remotely on a 1 Gbps local area

Table 20.7 Network access environment versus query time.

Query environment	Mean query time (s)
Local	97
LAN	125
Internet	181

Comparison of a typical query involving two feedback iterations for local, remote LAN (1 Gbps) and remote Internet (1 Mbps upload) access to the web server. Using the system over the internet with a very modest upload bandwidth resulted in almost doubling the query time over local access. A dedicated server with sufficient bandwidth would alleviate this drawback

network connected to the machine hosting the Query Engine Server and SQL Metadata database; and (c) remotely from the Internet with a server-side upload speed of approximately 1 Mbps. The query was conducted on metadata generated from a 20 min segment, and involved two search iterations examining and tagging appropriate candidates.

It can be seen that the same query took nearly twice as long over the Internet as compared to locally. The main slow-downs occurred in two places: (a) when retrieving either initial or updated candidate match lists, requiring the transmission of image data and bounding box metadata; and (b) when browsing the candidate tabs, again requiring the transmission of both image thumbnails and bounding boxes. This is a function of server-side upload bandwidth which in this case was very modest; a dedicated server offering higher bandwidth would result in lower delays and faster response times, important for open-world scenarios where highly crowded environments will typically result in larger numbers of candidates being returned after each query iteration.

20.6 Findings and Analysis

In this work, we presented a case study on how to engineer a multi-camera tracking system capable of coping with re-identification requirements in large scale, diverse open-world environments. In such environments where the number of cameras and the level of crowding are large, a key objective is to achieve *scalability* in terms of *associativity*, *capacity* and *accessibility*. Accordingly, we presented a prototype Multi-Camera Tracking (MCT) system comprising six key features: (1) *relative feature ranking* [2, 6, 15], which learns the best visual features for cross-camera matching; (2) *matching by tracklets*, for grouping individual detections of individuals into short intra-camera trajectories; (3) *global space–time profiling* [10], which models camera topologies and the physical motion constraints of crowds to significantly narrow the search space of candidates across camera views; (4) *machine-guided data mining*, for utilising human feedback as part of a *piecewise search strategy*; (5) *attribute-based re-ranking* [8, 9], for modelling high-level visual attributes such as colours and attire; and (6) *local space–time profiling*, to model the physical motion constraints of individuals to narrow the search space of candidates within each camera view.

Our extensive evaluation shows that the MCT system is able to effectively aid users in quickly locating targets of interest, scaling well despite the highly crowded nature of the testing environment. It required 3 min on average to track each target through all cameras using the remote Web-based interface and exploiting the key features as part of a piecewise search strategy. This is in contrast to the significantly greater time it would require human observers to manually analyse video recordings.

It was observed that attribute-based re-ranking was on average effective in increasing the ranks of correct matches over the RankSVM model alone. However, employing more than one attribute at a time was generally not beneficial and often detrimental. Local space–time profiling was *extremely* effective under all circumstances and combining it with a single attribute always led to a significant increase in the ranks of relevant targets, with a tripling of the average number of correct matches in the first six ranks alone. These features are critical in enabling the MCT system to cope with the large search space induced by the data by focusing on the right subset of candidates, at the right place and at the right time.

Overall, out of the 21 watchlist individuals, all but two were trackable across both stations in the MCT trial dataset. The two exceptions were lost on a single train journey. This was due to the train time falling outside the learned global space–time profile. This emphasises the utility of employing non-visual external information sources such as real-time train updates to modify global space–time profiles on-the-fly. This would permit such profiles to be tighter and more relevant over time, making them more consistently effective in narrowing the search space. This can be a critical factor in very large open-world scenarios where different parts of the multi-camera network may be connected by unpredictable and highly variable transport links.

Our testing of system speed shows that employing GPU acceleration for the most computationally intensive component resulted in a 50 % reduction in computation time per frame. Furthermore, employing multi-threading on a quad-core CPU with multi-threading enabled all six cameras of the MCT trial dataset to be processed simultaneously with negligible slow-down. This suggests that, in conjunction with the modular distributed nature of the system architecture design, the processing capacity of the system is *linearly scalable* to an arbitrary number of cameras by adding more CPUs to the system architecture (e.g. another machine on the network). Furthermore, by focusing effort on optimising each processing component of the Extraction Engine, a real-time frame rate per camera is likely achievable.

The most significant bottleneck of the entire MCT system was found to be metadata storage. Using an off-the-shelf SQL installation and basic tables, stored metadata was found to become prohibitively large over time. Querying metadata from video data longer than 20 min would result in long waiting times for the Query Engine Server to return the relevant results. Given the typical number of cameras in a highly crowded open-world scenario, this highlights the criticality of designing an appropriate storage scheme to store data more efficiently, reduce waiting times during a query and improve accessibility to metadata covering longer periods of time.

It is clear that there is great promise for the realisation of a scalable, highly effective and deployable computer vision-based multi-camera tracking and re-identification tool for assisting human operators in analysing large quantities of multi-camera video data from arbitrarily large camera networks spanning large spaces across cities. In building the MCT system, we have identified three areas worthy of further investigation. Firstl, integration with non-visual intelligence such as real-time transportation timetables (e.g. flights, trains and buses) is critical for dynamically managing global space–time profiles and ensuring that the search space is always narrowed in a contextually appropriate manner. Second, careful optimisation of individual processing components is required, which also involves a proper mediation between multi-threading and GPU resources to best harness availability in each machine comprising the distributed MCT system network. Finally, an optimised method for metadata storage is required for quick and easy accessibility regardless of the quantity being produced.

Acknowledgments We thank Lukasz Zalewski, Tao Xiang, Robert Koger, Tim Hospedales, Ryan Layne, Chen Change Loy and Richard Howarth of Vision Semantics and Queen Mary University of London who contributed to this work; Colin Lewis, Gari Owen and Andrew Powell of the UK MOD SA(SD) who made this work possible; Zsolt Husz, Antony Waldock, Edward Campbell and Paul Zanelli of BAE Systems who collaborated on this work; and Toby Nortcliffe of the UK Home Office CAST who assisted in setting up the trial environment and data capture.

References

1. Chang, C., Lin, C.: LIBSVM: a library for support vector machines. ACM Trans. Intell. Syst. Technol. **2**, 27:1–27:27 (2011)
2. Chapelle, O., Keerthi, S.: Efficient algorithms for ranking with SVMs. Inf. Retrieval **13**(3): 201–215 (2010)
3. Felzenszwalb, P., Girshick, R., McAllester, D., Ramanan, D.: Object detection with discriminatively trained part based models. IEEE Trans. Pattern Anal. Mach. Intell. **32**(9), 1627–1645 (2010)
4. Gheissari, N., Sebastian, T., Hartley, R.: Person reidentification using spatiotemporal appearance. In: IEEE Conference on Computer Vision and Pattern Recognition, vol. 2, pp. 1528–1535 (2006)
5. Hahnel, M., Klunder, D., Kraiss, K.F.: Color and texture features for person recognition. In: IEEE International Joint Conference on Neural Networks, vol. 1, pp. 647–652 (2004)
6. Joachims, T.: Optimizing search engines using clickthrough data. In: Knowledge Discovery and Data Mining, pp. 133–142 (2010)
7. Kuhn, H.: The hungarian method for the assignment problem. Naval Res. Logist. Quarterly **2**, 83–97 (1955)
8. Layne, R., Hospedales, T., Gong, S.: Person re-identification by attributes. In: British Machine Vision Conference, Guildford, UK (2012)
9. Layne, R., Hospedales, T., Gong, S.: Towards person identification and re-identification with attributes. In: European Conference on Computer Vision, First International Workshop on Re-Identification. Firenze, Italy (2012)
10. Loy, C.C., Xiang, T., Gong, S.: Multi-camera activity correlation analysis. In: IEEE Conference on Computer Vision and Pattern Recognition, pp. 1988–1995 (2009)

11. Loy, C.C., Xiang, T., Gong, S.: Time-delayed correlation analysis for multi-camera activity understanding. Int. J. Comput. Vis. **90**(1), 106–129 (2010)
12. Madden, C., Cheng, E., Piccardi, M.: Tracking people across disjoint camera views by an illumination-tolerant appearance representation. Mach. Vis. Appl. **18**(3), 233–247 (2007)
13. Munkres, J.: Algorithms for the assignment and transportation problems. J. Soc. Ind. Appl. Math. **5**(1), 32–38 (1957)
14. Prosser, B., Gong, S., Xiang, T.: Multi-camera matching under illumination change over time. In: European Conference on Computer Vision, Workshop on Multi-camera and Multi-model Sensor Fusion (2008)
15. Prosser, B., Zheng, W., Gong, S., Xiang, T.: Person re-identification by support vector ranking. In: British Machine Vision Conference, Aberystwyth, UK (2010)
16. Raja, Y., Gong, S.: Scaling up multi-camera tracking for real-world deployment. In: Proceedings of the SPIE Conference on Optics and Photonics for Counterterrorism, Crime Fighting and Defence, Edinburgh, UK (2012)
17. Raja, Y., Gong, S., Xiang, T.: Multi-source data inference for object association. In: IMA Conference on Mathematics in Defence, Shrivenham, UK (2011)
18. Schmid, C.: Constructing models for content-based image retrieval. In: IEEE Conference on Computer Vision and Pattern Recognition, pp. 30–45 (2001)
19. UK Home Office: i-LIDS dataset: Multiple camera tracking scenario. http://scienceandresearch.homeoffice.gov.uk/hosdb/cctv-imaging-technology/i-lids/ (2010)
20. Wang, H., Suter, D., Schindler, K.: Effective appearance model and similarity measure for particle filtering and visual tracking. In: European Conference on Computer Vision, pp. 606–618, Graz, Austria (2006)
21. Zheng, W., Gong, S., Xiang, T.: Person re-identification by probabilistic relative distance comparison. In: IEEE Conference on Computer Vision and Pattern Recognition, pp. 649–656, Colorado Springs, USA (2011)
22. Zheng, W., Gong, S., Xiang, T.: Re-identification by relative distance comparison. IEEE Trans. Pattern Anal. Mach. Intell. **35**(3), 653–668 (2013)

Index

S. Gong et al. (eds.), *Person Re-Identification*,
Advances in Computer Vision and Pattern Recognition,
DOI: 10.1007/978-1-4471-6296-4, © Springer-Verlag London 2014

P

Q

R

MIX
Papier aus verantwortungsvollen Quellen
Paper from responsible sources
FSC® C105338

If you have any concerns about our products,
you can contact us on
ProductSafety@springernature.com

In case Publisher is established outside the EU,
the EU authorized representative is:
Springer Nature Customer Service Center GmbH
Europaplatz 3, 69115 Heidelberg, Germany

Printed by Libri Plureos GmbH
in Hamburg, Germany